A Colour Handbook

Pests of Fruit Crops

Second Edition

David V Alford

BSc, PhD

CRC Press
Taylor & Francis Group
Boca Raton London New York

CRC Press is an imprint of the
Taylor & Francis Group, an **informa** business

CRC Press
Taylor & Francis Group
6000 Broken Sound Parkway NW, Suite 300
Boca Raton, FL 33487-2742

© 2015 by Taylor & Francis Group, LLC
CRC Press is an imprint of Taylor & Francis Group, an Informa business

No claim to original U.S. Government works

Printed on acid-free paper
Version Date: 20140520

International Standard Book Number-13: 978-1-4822-5420-4 (Hardback)

Library of Congress Cataloging-in-Publication Data

Alford, D. V.
  Pests of fruit crops : a colour handbook / David V. Alford. -- Second edition.
      pages cm
  Includes bibliographical references and index.
  ISBN 978-1-4822-5420-4 (hardcover : alk. paper) 1. Fruit--Diseases and pests. I. Title.

SB608.F8A44 2014
634'.0497--dc23
                                                    2014018311

Visit the Taylor & Francis Web site at
http://www.taylorandfrancis.com

and the CRC Press Web site at
http://www.crcpress.com

Contents

Preface   7

Chapter 1 Introduction   9

Chapter 2 Smaller insect orders   23
Order Saltatoria (crickets, grasshoppers and locusts)   23
    Family Tettigoniidae (bush crickets)   23
    Family Gryllotalpidae (mole crickets)   24
    Family Acrididae (grasshoppeeers and locusts)   25
Order Dermaptera (earwigs)   26
    Family Forficulidae   26
Order Isoptera (termites)   27
    Family Kalotermitidae   27
Order Thysanoptera (thrips)   27
    Family Thripidae (thrips)   27
    Family Phlaeothripidae   31

Chapter 3 True bugs   32
    Family Acanthosomatidae (shield bugs)   32
    Family Pentatomidae (shield bugs)   32
    Family Lygaeidae (ground bugs)   34
    Family Tingidae (lace bugs)   35
    Family Miridae (capsids or mirids)   37
    Family Cercopidae (froghoppers)   43
    Family Flatidae (planthoppers)   45
    Family Membracidae   46
    Family Cixiidae (planthoppers)   47
    Family Cicadellidae (leafhoppers)   47
    Family Psyllidae   57
    Family Carsidaridae   62
    Family Aleyrodidae (whiteflies)   63
    Family Aphididae (aphids)   66
    Family Phylloxeridae (phylloxeras)   97
    Family Diaspididae (armoured scales)   99
    Family Asterolecaniidae (pits scales)   109
    Family Coccidae (soft scales, wax scales)   109
    Family Pseudococcidae (mealybugs)   120
    Family Margarodidae (giant scales)   122

Chapter 4 Beetles 124
   Family Carabidae (ground beetles) 124
   Family Scarabaeidae (chafers) 127
   Family Buprestidae (jewel beetles) 130
   Family Elateridae (click beetles) 133
   Family Cantharidae 134
   Family Bostrychidae 135
   Family Nitidulidae 135
   Family Byturidae 136
   Family Tenebrionidae 138
   Family Cerambycidae (longhorn beetles) 138
   Family Chrysomelidae (leaf beetles) 140
   Family Attelabidae (weevils) 145
   Family Rhynchitidae (weevils) 146
   Family Apionidae (weevils) 151
   Family Curculionidae (true weevils) 152

Chapter 5 True flies 175
   Family Tipulidae (crane flies) 175
   Family Bibionidae (St. Mark's flies) 177
   Family Cecidomyiidae (gall midges) 177
   Family Tephritidae (large fruit flies) 187
   Family Drosophilidae 190
   Family Agromyzidae 191
   Family Muscidae 193
   Family Anthomyiidae 194

Chapter 6 Butterflies and moths 195
   Family Hepialidae (swift moths) 195
   Family Nepticulidae 197
   Family Tischeriidae 200
   Family Incurvariidae 202
   Family Heliozelidae 203
   Family Cossidae 204
   Family Zygaenidae 206
   Family Psychidae 208
   Family Lyonetiidae 208
   Family Gracillariidae 211
   Family Phyllocnistidae 219
   Family Sesiidae (clearwing moths) 222
   Family Choreutidae 224
   Family Yponomeutidae 227
   Family Schreckensteiniidae 238
   Family Coleophoridae (casebearer moths) 239
   Family Oecophoridae 244
   Family Gelechiidae 247
   Family Blastobasidae 250
   Family Momphidae 251
   Family Cochylidae 252
   Family Tortricidae (tortrix moths) 253
   Family Pyralidae 295
   Family Papilionidae 301
   Family Pieridae 302
   Family Lycaenidae 304

Family Nymphalidae 307
Family Lasiocampidae 309
Family Saturniidae 314
Family Thyatiridae 315
Family Geometridae (geometer moths) 315
Family Sphingidae (hawk moths) 335
Family Notodontidae 343
Family Dilobidae 344
Family Lymantriidae 345
Family Arctiidae (e.g. ermine moths and tiger moths) 351
Family Nolidae 354
Family Noctuidae 354

Chapter 7 Sawflies, ants and wasps 379
Family Pamphiliidae 379
Family Cephidae (stem sawflies) 380
Family Cimbicidae 381
Family Tenthredinidae 383
Family Cynipidae (gall wasps) 400
Family Eurytomidae (seed wasps) 401
Family Torymidae (e.g. seed wasps) 402
Family Formicidae (ants) 402
Family Vespidae (true wasps) 403

Chapter 8 Mites 405
Order Prostigmata 405
Family Phytoptidae (gall mites) 405
Family Eriophyidae (gall mites and rust mites) 406
Family Tarsonemidae (tarsonemid mites) 419
Family Tetranychidae (spider mites) 421
Family Tenuipalpidae (false spider mites) 427
Order Cryptostigmata 428
Family Mycobatidae 428

Wild or ornamental host plants cited in the text 429

Selected bibliography 433

Host plant index 435

General index 441

To Inge, Ingaret, Kerstin,
Michael, Iona and Xakiera

Preface

A Colour Atlas of Fruit Pests, published in 1984, reviewed the pests occurring on fruit crops in the British Isles and aimed to present a detailed and up-to-date account of their recognition, biology and control. Since 1984, accounts of fruit pests in Belgium (*De geïntegreerde bestrijding in de fruitteelt* by Guido Sterk), Finland (*Hedelmä- ja marjakasvien tuhoeläimet* by Tuomo Tuovinen) and the Netherlands (*Schadelijke en nuttige insekten en mijten in fruitgewassen* by the late A. van Frankenhuyzen) (all of which include colour photographs) have also appeared. To date, however, no single colour-illustrated text (apart from *Schädliche und nützliche Insekten und Milben an Kern- und Steinobst in Mitteleuropa* by the late A. van Frankenhuyzen & H. Stigter) has attempted to cover the subject on a broader international scale.

In this fully revised and renamed version of 'Fruit Pests' (itself now an updated, second edition), the scope of the original publication has been expanded to encompass pests of fruit crops throughout Europe. Accordingly, greater emphasis than formerly has been given to pests of crops such as almond, apricot, peach and grapevine, and information on pests of subtropical crops (particularly citrus, fig and olive) has been added. The main aims are to provide an illustrated account of the various pests to be found in this part of the world and to serve as a source of reference that, hopefully, will be of interest and value both within Europe and elsewhere. Indeed, many of the pest species described in this book, or their close relatives, are present in non-European countries and several pose problems on fruit crops worldwide.

For simplicity, the term 'fruit' is used to embrace not only conventional horticultural fruits, but also hops and nuts (see Chapter 1). Minor temperate fruit crops such as bilberry, cranberry and strawberry-tree, the fruits of which are largely collected from wild plants (albeit, at least in the case of bilberry and strawberry-tree, sometimes commercially), and Cucurbitace (e.g. melon and water melon, which are traditionally considered in tandem with other cucurbit crops such as cucumber, marrow and pumpkin rather than with fruit crops) are excluded, as are tropical fruit crops (such as dwarf banana and mango) which are grown commercially in Europe on only a very limited scale and then in only the hottest of regions.

Pests are no respecters of international boundaries and, over the years, the European fauna has been 'enriched' by a range of newcomers from many other parts of the world. Conversely, European pests have often found a foothold in other continents (notably in North America), and no doubt others will do so in the future. In spite of modern plant quarantine measures, movement of pests as a result of international trade or travel (and their subsequent establishment in new areas) is still an ongoing process. The establishment in Italy of the American grapevine pest *Phyllocnistis vitegenella* is a recent example; the appearance of this insect in Europe, however, can in no way be likened to the accidental introduction into Europe in the mid-1800s (again from America) of the infamous grape phylloxera, which then subsequently devastated the European viticultural industry. The natural drift of species from country to country is also a continuing phenomenon, at least in some instances probably encouraged and enhanced by modern-day climate change. Similarly, details of pest life cycles (including, perhaps, the number of generations in a season) are liable to alter in the wake of temperature changes and acccording to local conditions. Therefore, where cited, dates (months) for the appearance of the various stages of pests are intended merely as a general guide. They also apply to Europe and may well be different elsewhere, as in the Southern Hemisphere.

The range of fruit pests included in this new book cannot claim to be fully comprehensive, especially in the case of minor pests and those of only local or sporadic occurrence. Also, as indicated above, the situation is far from static and 'new' pests (i.e. new regional or national records), invasive or otherwise, are being discovered on a regular basis. Conversely, some fruit pests formerly of considerable significance are today of lesser importance: apple capsid (*Lygocoris rugicollis*) is an example. Some species that might be regarded as pests in some parts of Europe may be rare (perhaps even protected) species elsewhere. The inclusion of a particular species, therefore, does not necessarily imply that it should always be regarded as harmful.

In view of the wider international coverage, unlike its predecessor, this new book excludes information on pest control. The pesticides available and the strategies adopted to combat particular pests can differ significantly from country to country, if not from region to region; they are also under constant review. Further, particularly in a commercial situation, sustainable pest management as opposed to simplistic 'control' of individual pest species is also becoming more widely practised. Readers seeking information on pest control or pest management on fruit crops, therefore, should refer to information relevant to their regional or local circumstances. A relatively recent review of pest management on fruit crops in the UK is included in the *Pest and Disease Management Handbook*, published in 2000 by Blackwell Science Limited on behalf of BCPC (the British Crop Production Council). Other information is available in regularly updated booklets and leaflets produced by agrochemical companies and national or regional extension services. Locally relevant on-line electronic services providing information on pest control may also be available.

The inclusion of many more pest species and information from a wider geographical area has meant that there could be no chapter on natural enemies. The new text is also restricted to insects and mites. Within the various chapters, families of pests are arranged according to generally accepted systematic systems. However, for ease of reference, genera within families (and species within genera) are arranged alphabetically, divisions at subfamily level (for example) being ignored; hopefully,

specialist readers will regard this as no more than a minor irritation. Alternative names for genera and species (many of which are not strictly synonyms) are excluded from the text, but are cross-referenced in the pest index; such names are restricted mainly to those commonly found in the applied literature and to those often still in common usage in crop-protection circles. Where appropriate, vernacular names of species are included in the pest entries (several here applied for the first time); again, as in the case of scientific names, frequently used alternatives are cross-referenced in the pest index.

Wherever possible, in compiling accounts of the various pests, live specimens have been examined and immature stages collected in the field reared through to adulthood. The help of my family, especially my wife, in obtaining material during field trips to various parts of Europe, or in caring for such material during my absence, has been invaluable. My thanks are also due to Alain Fraval (INRA, Paris) for helpful suggestions, to Stuart C. Gordon, Dr Michael Maixner, Dr Michael G. Solomon and Tuomo Tuovinen for providing live examples of certain pests, and to Dr Chris Malumphy (Central Science Laboratory, York) for dealing with numerous queries relating to scale insects.

David V Alford
Cambridge

Chapter 1

Introduction

Most fruit pests are insects, a major group of invertebrate animals belonging to the phylum Arthropoda. Members of this diverse and exceedingly successful phylum occur in all kinds of habitat, on land and in water, and are characterized by their often hard exoskeleton or body shell, their segmented body and their jointed limbs. In addition to insects (class Insecta), arachnids (class Arachnida: harvestmen, mites, pseudoscorpions, spiders and ticks), crustaceans (class Crustacea: crabs, crayfish, lobsters, shrimps and woodlice) and myriapods (class Myriapoda: centipedes and millepedes) are familiar examples. Insects differ from other arthropods in possessing just three pairs of legs, usually one or two pairs of wings (all winged invertebrates are insects), and by having the body divided into three distinct regions: head, thorax and abdomen.

External features of insects

The outer skin or integument of an insect is known as the cuticle. This forms a non-cellular, waterproof layer over the body and is composed of chitin and protein, the precise chemical composition and thickness determining its hardness and rigidity. The cuticle has three layers (epicuticle, exocuticle and endocuticle) and is secreted by an inner lining of cells that form the hypodermis or basement membrane. When first produced, the cuticle is elastic and flexible, but soon after deposition it usually undergoes a period of hardening or sclerotization and becomes more or less darkened by the addition of a chemical called melanin. The adult cuticle is not replaceable, except in certain primitive insects. However, at intervals during the growth of the immature stages (larvae and nymphs), the 'old' hardened cuticle becomes too tight and is replaced by a new, initially expandable, one secreted from below.

The insect cuticle is often thrown into ridges and depressions, is frequently sculptured or distinctly coloured and may bear a variety of spines and hairs. In larvae, body hairs often arise from hardened plates or wart-like pinacula, tubercles and verrucae. In some groups, as in beetles (order Coleoptera), features of the adult cuticle (such as colour, sculpturing or texture) are of considerable value in distinguishing between species.

The basic body segment of an insect is divided into four sectors (a dorsal tergum, a ventral sternum and two lateral pleurons) which often form horny, chitinized plates called sclerites. These may give the body an armour-like appearance and are either fused rigidly together or are joined by soft, flexible, chitinized membranes to allow for body movement. Appendages, such as legs, are developed as outgrowths from the pleurons.

The **head** of an insect is composed of six fused body segments and carries a pair of sensory antennae, eyes and mouthparts. The form of an insect antenna varies considerably, the number of antennal segments ranging from one to more than a hundred. The basal segment is called the scape; the second segment is the pedicel and from this arises the many-segmented flagellum (each segment of which is termed a flagellomere). In a geniculate (elbowed) antenna, the pedicel acts as the articulating joint between the often greatly elongated scape and the flagellum; such antennae are characteristic of certain weevils, bees and wasps. Many insects possess two compound eyes, each composed of several thousand facets, and three simple eyes called ocelli, the latter usually forming a triangle at the top of the head. Compound eyes are large and especially well developed in insects, such as predators, where good vision is important. The compound eye provides a mosaic (rather than a clear) picture, but is well able to detect movement. The ocelli are optically simple and lack a focusing mechanism; they are concerned mainly with registering light intensity, enabling the insect to distinguish between light and shade. Insect mouthparts are derived from several modified, paired appendages; they range from simple biting jaws (mandibles) to complex structures for piercing, sucking or lapping. Amongst plant-feeding (phytophagous) insects, biting mouthparts are found, for example, in adult and immature grasshoppers, locusts, earwigs and beetles, but (as in butterflies and moths) may be restricted to the larval stages. Stylet-like, suctorial mouthparts are characteristic of thrips and bugs (including aphids); such insects may introduce toxic saliva into plants and cause distortion or galling of the tissue. Some biting or sucking insects transmit plant viruses or other pathogens to host plants.

The **thorax** has three segments (the prothorax, the mesothorax and the metathorax), whose relative sizes vary from one insect group to the next. In beetles, for example, the prothorax is the largest section and is covered on its upper surface by an expanded dorsal sclerite called the pronotum; in true flies, the mesothorax is greatly enlarged and the prothoracic and metathoracic segments are much reduced. Typically, each thoracic segment bears a pair of jointed legs. Their form varies considerably, but all have the same basic structure: the main components being the coxa (basally), the femur, the tibia and the tarsus (the latter subdivided into several tarsomeres). Wings, when present, arise from the mesothorax and the metathorax as a pair of forewings and hindwings respectively. Basically, each wing is an expanded membrane-like structure supported by a series of hardened veins, but considerable modification has taken place in the various insect groups. In earwigs and beetles, for instance, the forewings have become hardened and thickened protective flaps, called elytra, and only the hindwings are used for flying; in true flies, the forewings retain their propulsive function, but the hindwings have become greatly reduced in size and are modified into balancing organs known as halteres. Details of wing structure and venation are of importance in the classification and identification of insects, and the names of many insect orders are based upon them.

The **abdomen** is normally formed from 10 or 11 segments, but fusion and apparent reduction of the most anterior or posterior components is common. Although present in many larvae, abdominal appendages are wanting on most segments of adults, their ambulatory function, as found in various other arthropods, having been lost. However, appendages on abdominal segments 8 and 9 remain to form the genitalia, including the male claspers and the female ovipositor. In many groups, a pair of cerci are formed from appendages on the last body segment. These are particularly long and noticeable in primitive insects (as in crickets, stoneflies, mayflies and earwigs), but are absent in the most advanced form (at least in the adult stage). Abdominal sclerites are limited to a series of dorsal tergites and a ventral set of sternites; these give the abdomen an obviously segmented appearance.

Internal features of insects

The **body cavity** of an insect extends into the appendages and is filled with a more or less colourless, blood-like fluid called haemolymph. This bathes all the internal organs and tissues, and is circulated by muscular action of the body and by a primitive, tube-like heart that extends mid-dorsally from the head to near the tip of the abdomen.

The **brain** is the main co-ordinating centre of the body. It fills much of the head and is intimately linked with the antennae, the compound eyes and the mouthparts. The brain gives rise to a central nerve cord that extends back mid-ventrally through the various thoracic and abdominal segments. The nerve cord is swollen at intervals into a series of ganglia, from which arise numerous lateral nerves. These ganglia control many nervous functions independent of the brain, such as movement of the body appendages.

The **gut** or alimentary tract is a long, much modified tube stretching from the mouth to the anus. It is subdivided into three sections: a foregut, with a long oesophagus and a bulbous crop; a mid-gut, where digestion of food and absorption of nutriment occurs; and a hindgut, concerned with water absorption, excretion, and the temporary storage of waste prior to its disposal. A large number of blind-ending, much convoluted Malpighian tubules arise from the junction between the mid-gut and hindgut. These tubules collect waste products from the body and pass them into the gut.

The **respiratory system** comprises a complex series of branching tubes (tracheae) and microscopic tubules (tracheoles) that ramify throughout the body in contact with the internal organs and tissues. This tracheal system opens to the outside through segmentally arranged valve-like breathing holes or spiracles, present along either side of the body. Air is forced through the spiracles by contraction and relaxation of the abdominal body muscles.

The male **reproductive system** includes a pair of testes and associated ducts that lead to a seminal vesicle in which sperm is stored prior to mating. The male genitalia may include chitinized structures, such as the claspers that help to grasp the female during copulation. Female insects possess a pair of ovaries, subdivided into several egg-forming filaments called ovarioles. The ovaries enter a median oviduct and this opens to the outside on the ninth abdominal segment. Many insects have a protrusible egg-laying tube, called an ovipositor. Examination of the male or female genitalia is often essential for distinguishing between closely related species.

Reproduction and growth of insects

Sexual reproduction is commonplace in insects, but in certain groups fertilized eggs produce only female offspring and males are reared from unfertilized ones only. In other cases, male production may be wanting or extremely rare and parthenogenesis (reproduction without a sexual phase) is the rule.

Although some insects are viviparous (giving birth to active young), most lay eggs. A few, such as aphids, reproduce viviparously by parthenogenesis in spring and summer, but may produce eggs in the autumn (after a sexual phase). Insect eggs have a waterproof shell (the chorion) and many are capable of surviving severe winter conditions in exposed situations on tree bark or shoots.

Insects normally grow only during the period of pre-adult development as nymphs or larvae, their outer cuticular skin being moulted and replaced between each successive growth stage or instar. The most primitive insects (subclass Apterygota) have wingless adults, their eggs hatching into nymphs that are essentially similar to adults, but smaller and not sexually mature. The more advanced, winged or secondarily wingless, insects (subclass Pterygota) develop in one of two ways. In the division Exopterygota (also known as the Hemimetabola), there is a succession of nymphal stages in which wings usually develop externally as buds that become fully formed and functional once the adult stage is reached. In such insects, nymphs and adults are frequently of similar appearance (apart from the presence of wing buds or wings), and often share the same feeding habits. This type of development is termed incomplete metamorphosis. The most advanced insects (division Endopterygota or Holometabola) show complete metamorphosis, development including several larval instars of quite different structure and habit from the adults. Here, the transformation from larval to adult form occurs during a quiescent pupal stage. Unlike nymphs, insect larvae lack compound eyes, but they often possess one or more pairs of simple eyes (stemmata), arranged on either side of the head. Insect larvae are of various kinds. Eruciform larvae are caterpillar-like; they have three pairs of jointed thoracic legs (true legs) and a number of pairs of fleshy, false legs (prolegs) on the abdomen. Many sawfly, butterfly and moth larvae are of this type; unlike those of sawflies, the abdominal prolegs of butterfly and moth larvae are usually provided with small chitinous hooks, known as crochets. These crochets may be of one, two or more sizes (e.g. uniordinal, biordinal, multiordinal), and often have a characteristic circular, linear or elliptical arrangement. Campodeiform larvae are elongate and dorsoventrally flattened, with well-developed antennae and thoracic limbs, but without abdominal prolegs. Scarabaeiform larvae also lack abdominal prolegs, but have a thick, fleshy, often C-shaped body, with a well-developed head and thoracic legs. Campodeiform and scarabaeiform larvae are typified by many beetle grubs. Other insect larvae are apodous (without legs, legless): wasp and fly larvae are examples. Insect pupae are typically exarate (with some or all appendages – antennae, legs, mouthparts, wing buds – 'free', i.e. not fused to the body) or obtect (with the appendages fused to the body).

Classification of insects

Class **INSECTA**
Subclass APTERYGOTA

Order Thysanura	bristle-tails, silverfish
Order Diplura	diplurans
Order Protura	proturans
Order Collembola	springtails

Subclass PTERYGOTA
Division EXOPTERYGOTA (= HEMIMETABOLA)

Order Ephemeroptera	mayflies
Order Odonata	dragonflies
Order Plecoptera	stoneflies
Order Grylloblattodea	grylloblattodeans

Order Saltatoria	crickets, grasshoppers, locusts
Order Phasmida	stick-insects, leaf-insects
Order Dermaptera	earwigs
Order Embioptera	web-spinners
Order Discyoptera	cockroaches, mantids
Order Isoptera	termites
Order Zoraptera	zorapterans
Order Psocoptera	psocids
Order Mallophaga	biting lice
Order Anoplura	sucking lice
Order Hemiptera	true bugs

Order Thysanoptera	thrips

Division ENDOPTERYGOTA (= HOLOMETABOLA)

Order Neuroptera	e.g. alder flies, lacewings
Order Coleoptera	beetles
Order Strepsiptera	stylopids
Order Mecoptera	scorpion flies
Order Siphonaptera	fleas
Order Diptera	true flies
Order Lepidoptera	butterflies, moths
Order Trichoptera	caddis flies
Order Hymenoptera	e.g. ants, bees, sawflies, wasps

The main features of the groups within which pests of European fruit crops occur are summarized below.

Saltatoria: medium-sized to large, stout-bodied insects, with a large head and chewing mouthparts; compound eyes large; pronotum large and saddle-like; adults usually with two pairs of wings, but either or both pairs may be reduced or absent; forewings typically thickened (leathery) and called tegmina; tarsi usually 3- or 4-segmented; hind legs usually much enlarged and adapted for jumping; anal cerci 1-segmented. Metamorphosis incomplete; development includes egg and nymphal stages: *family Tettigoniidae* (p. 23); *family Gryllotalpidae* (p. 24); *family Acrididae* (p. 25).

Dermaptera: medium-sized, elongate, omnivorous insects, with mouthparts adapted for biting and chewing; forewings modified into very short, leathery elytra; hindwings semicircular (fan-like) and membranous, with a radial venation; legs short, tarsi 3-segmented; anal cerci usually modified into a pair of forceps-like pincers; ovipositor reduced or absent. Metamorphosis incomplete; development includes egg and nymphal stages, the latter being similar in appearance to adults, but smaller and less strongly sclerotized: *family Forficulidae* (p. 26).

Isoptera: small to medium, primarily tropical, soft-bodied, pale-coloured insects, with biting mouthparts; either with or without wings; wings, when present, long and narrow, with thickened anterior veins; forewings and hindwings of similar appearance. Termites (often called 'white ants') are social insects and inhabit colonies that contain various castes; only two species are native to Europe: *family Kalotermitidae* (p. 27).

Hemiptera: minute to large insects, characterized by piercing, suctorial mouthparts, known as a beak; forewings frequently partly or entirely hardened. Development hemimetabolous, including egg and several nymphal stages (the egg stage often omitted); nymphs often similar in appearance and habits to the adult.

Suborder Heteroptera: usually with two pairs of wings, the forewings (termed 'hemelytra') with a leathery basal area and a membranous tip; hemelytra with or without a cuneus (a triangular area between the membrane and the rest of the wing); hindwings membranous; wings held flat over the abdomen when in repose; the beak-like mouthparts (rostrum) arise from the front of the head and are flexibly attached; prothorax large; some species are phytophagous, but many are predacious: *family Acanthosomatidae* (p. 32); *family Pentatomidae* (p. 32); *family Lygaeidae* (p. 34); *family Tingidae* (p. 35); *family Miridae* (p. 37).

Suborder Auchenorrhyncha: wings (when present) typically held over the body in a sloping, roof-like posture; fore wings (termed elytra) uniform throughout and horny; hind wings membranous; mouthparts arising from the base of the head and the point of attachment rigid; entirely phytophagous. *Superfamily Cercopoidea* – tegulae absent; hind legs modified for jumping, with long tibiae bearing one or two long spines: *family Cercopidae* (p. 43). *Superfamily Fulgoroidea* – elytra with anal vein Y-shaped; antennae 3-segmented: *family Flatidae* (p. 45); also includes small phytophagous bugs with subterranean, root-feeding nymphs: *family Cixiidae* (p. 47). *Superfamily Membracoidea* – treehoppers with thorn-mimicking adults: *family Membracidae* (p. 46). *Superfamily Cicadelloidea* – leafhoppers; tegulae absent; hind legs modified for jumping, with long tibiae bearing longitudinal rows of short spines: *family Cicadellidae* (p. 47).

Suborder Sternorrhyncha: antennae long and thread-like, without a distinct arista; rostrum appearing to arise from between the forelegs; tarsi 1- or 2-segmented; wings (when present) typically held over the body in a sloping, roof-like posture; forewings and hindwings membranous and uniform throughout; mouthparts arising from a rearward position relative to the head and the point of attachment rigid; entirely phytophagous. *Superfamily Psylloidea* – antennae usually 10-segmented; tarsi 2-segmented and with a pair of claws: *family Psyllidae* (p. 57); *family Carsidaridae* (p. 62). *Superfamily Aleyrodoidea* – antennae 7-segmented; wings opaque and whitish: *family Aleyrodidae* (p. 63). *Superfamily Aphidoidea* – females winged or wingless; wings, when present, usually large and transparent with few veins; abdomen usually with a pair of tube-like siphunculi; tarsi 2-segmented and with a pair of claws: *family Aphididae* (p. 68); *family Phylloxeridae* (p. 97). *Superfamily Coccoidea* – tarsi, if present, 1- segmented and with a single claw; females always wingless, often scale-like or cushion-shaped, and usually sedentary and apodous; males usually rare or absent, with a single pair of wings (or wingless) and vestigial mouthparts, and developing through a pupal stage: *family Diaspididae* (p. 99); *family Asterolecaniidae* (p. 109); *family Coccidae* (p. 109); *family Pseudococcidae* (p. 120); *family Margarodidae* (p. 122).

Thysanoptera: minute or small, slender-bodied insects, with a distinct head, a well-developed prothorax and a long, narrow, 11-segmented abdomen; cerci absent; wings, when present, very narrow, membranous and strap-like, with few or no veins, and with hair-like fringes; antennae short, 6- to 10-segmented; tarsi 1- or 2-segmented, each with a protrusible terminal bladder-like vesicle (the arolium); mouthparts asymmetrical and adapted for piercing. Metamorphosis gradual; development intermediate between that of hemimetabolous and holometabolous insects, and includes an egg, two nymphal and two or three inactive stages (propupae and pupae); nymphs are similar in appearance to adults, but wingless, less strongly sclerotized and with fewer antennal segments; the non-feeding propupae and pupae have conspicuous wing pads and lack the tarsal vesicles (arolia) found in nymphs and adults; in pupae, the antennae are folded back over the thorax.

Suborder Terebrantia: ovipositor saw-like; tip of abdomen conical in female, bluntly rounded in male; forewings with at least one longitudinal vein extending to the apex. Development includes an egg, two nymphal, and single propupal and pupal stages: *family Thripidae* (p. 27).

Suborder Tubulifera: forewings without longitudinal veins and female without an ovipositor; tip of abdomen tubular in both sexes. Development includes an egg, two nymphal, one propupal and two pupal stages: *family Phlaeothripidae* (p. 31).

Coleoptera: minute to large insects, with biting mouthparts adapted for chewing; body usually covered with a hard, often brightly coloured exoskeleton; forewings modified into horny or leathery elytra, which usually meet in a straight line along the back; hindwings membranous and folded beneath the elytra when in repose, but often reduced or absent; pronotum normally large, shield-like and mobile. Metamorphosis complete. Eggs usually spherical, oval, egg-shaped or sausage-shaped. Larvae usually with a distinct head and three pairs of thoracic legs, but sometimes apodous; often campodeiform or eruciform, but occasionally scarabaeiform. Pupae normally exarate. The largest order of insects, with more than a quarter of a million species worldwide.

Superfamily Caraboidea – a large group of mainly predacious beetles, with the hind coxae fused rigidly to the metasternum and extending posteriorly to the hind margin of the first visible abdominal sternite; antennae 11-segmented and usually filiform or moniliform. Larvae usually with claw-bearing tarsi: *family Carabidae* (p. 124). *Superfamily Scarabaeoidea* – a large group of often very large, brightly coloured insects, some of which possess enlarged horns on the head and thorax: *family Scarabaeidae* (p. 127). *Superfamily Buprestoidea* – usually small or very small, metallic-looking beetles, with very large eyes and short, toothed antennae, and the head retracted deeply into the thorax. Larvae mainly wood-borers: *family Buprestidae* (p. 130). *Superfamily Elateroidea* – elongate beetles, with a hard exoskeleton, the head sunk into the thorax, and toothed or comb-like antennae; hind angles of the pronotum sharply pointed and often extended: *family Elateridae* (p. 133). *Superfamily Cantharoidea* – an ill-defined group of narrow, elongate, soft-bodied beetles; elytra also soft and often clothed in a short, velvet-like pubescence: *family Cantharidae* (p. 134). *Superfamily Bostrichoidea* – adults with the pronotum extended forward to form a hood over the head; most species are wood-borers, with soft-bodied, scarabaeiform larvae: *family Bostrychidae* (p. 135). *Superfamily Cucujoidea* – beetles usually with five visible abdominal segments and, often, clubbed antennae: *family Nitidulidae* (p. 135); *family Byturidae* (p. 136); *family Tenebrionidae* (p. 138). *Superfamily Chrysomeloidea* – mostly plant feeders, adults with 4-segmented tarsi (the fourth segment very small). Larvae usually with well-developed thoracic legs: *family Cerambycidae* (p. 138); *family Chrysomelidae* (p. 140).

Superfamily Curculionoidea – a very large superfamily, including weevils and bark beetles; weevils often have a very elongated snout (rostrum), which bears the mouthparts and antennae; in most weevils the antennae are geniculate, with an elongated basal segment (scape), but in some families the antennae are filiform and all segments are of a similar length. Larvae usually apodous, often with an enlarged thoracic region: *family Attelabidae* (p. 145); *family Rhynchitidae* (p. 146); *family Apionidae* (p. 151); *family Curculionidae* (p. 152).

Diptera: minute to large insects, with a single pair of membranous wings; hindwings reduced to knobbed (often drumstick-like) stalks, called halteres, that function as balancing organs during flight; mouthparts suctorial, but sometimes adapted for piercing. Metamorphosis complete. Eggs usually oval or cigar-shaped. Larvae apodous and usually maggot-like, with a reduced, inconspicuous head.

Suborder Nematocera: antennae of adults with a scape, pedicel and flagellum, the latter comprising numerous, similar-looking segments (flagellomeres), each bearing a whorl of hairs. Larvae usually (not in the family Cecidomyiidae) with a well-defined head and horizontally opposed mandibles: *family Tipulidae* (p. 175); *family Bibionidae* (p. 177); *family Cecidomyiidae* (p. 177).

Suborder Cyclorrapha: antennae of adults with a scape, pedicel and flagellum, the latter usually forming an enlarged, compound segment tipped by a short, bristle-like arista. Larvae maggot-like, often tapering anteriorly, with distinctive rasping 'mouth-hooks' (forming part of the cephalopharyngeal skeleton), but head small and inconspicuous. Pupation occurs within the last larval skin, which then forms a protective barrel-like puparium from which the adult eventually escapes by forcing off a circular cap (the operculum): *family Tephritidae* (p. 187); *family Drosophilidae* (p. 190); *family Agromyzidae* (p. 191); *family Muscidae* (p. 193); *family Anthomyiidae* (p. 194).

Lepidoptera: minute to large insects, with two pairs of membranous wings that have few cross-veins; wings and appendages usually scale-covered; adult mouthparts suctorial, often forming a long, coiled proboscis, with mandibles vestigial or absent. Metamorphosis complete. Eggs extremely variable in form and colour; often with a sculptured surface. Larvae eruciform, most often with three pairs of thoracic legs and five pairs of abdominal prolegs, the latter usually armed with crochets – prolegs are usually present on abdominal segments 3–6 and 10, the last-mentioned often being termed 'anal claspers' (descriptions of the arrangement of crochets on the abdominal prolegs usually do not apply to the anal claspers); head usually strongly chitinized, with several ocelli, and with a pair of silk glands (modified salivary glands); body often with setae or longer hairs arising from distinctive plates, pinacula or verrucae, and sometimes marked with more or less complete longitudinal stripes or lines; larvae typically phytophagous, often leaf-mining, rarely carnivorous. Pupae normally obtect, the tip of the abdomen often with a characteristic cluster of hooks, spines or bristles frequently borne on a distinctive outgrowth termed the cremaster.

Suborder Monotrysia: very small to very large moths; mandibles absent; labial palps 2- or 3-segmented; venation of forewings and hindwings usually similar.

Superfamily Hepialoidea – adults with non-functional, vestigial mouthparts and short antennae: *family Hepialidae* (p. 195). **Superfamily Nepticuloidea** – adults with wing venation reduced; ovipositor soft: *family Nepticulidae* (p. 197); *family Tischeriidae* (p. 200). **Superfamily Incurvarioidea** – small, day-flying moths; antennae of males often very long: *family Incurvariidae* (p. 202); *family Heliozelidae* (p. 203).

Suborder Ditrysia: very small to very large insects; maxillary palps 1- to 5-segmented or vestigial; labial palps 3-segmented, rarely 2-segmented; venation of hindwings reduced, but rarely reduced in forewings.

Superfamily Cossoidea – heavy-bodied moths, with a primitive wing venation: *family Cossidae* (p. 204). **Superfamily Zygaenoidea** – mainly colourful, metallic-looking moths; antennae usually clubbed or bipectinate: *family Zygaenidae* (p. 206). **Superfamily Tineoidea** – primitive moths, with narrow or very narrow wings: *family Psychidae* (p. 208); *family Lyonetiidae* (p. 208); *family Gracillariidae* (p. 211); *family Phyllocnistidae* (p. 219). **Superfamily Yponomeutoidea** – an indistinct and rather diverse group: *family Sesiidae* (p. 222); *family Choreutidae* (p. 224); *family Yponomeutidae* (p. 227); *family Schreckensteiniidae* (p. 238). **Superfamily Gelechioidea** – a large group of relatively small moths, often with long, narrow wings: *family Coleophoridae* (p. 239); *family Oecophoridae* (p. 244); *family Gelechiidae* (p. 247); *family Blastobasidae* (p. 250); *family Momphidae* (p. 251). **Superfamily Tortricoidea** – a major group of relatively small moths, with mainly rectangular forewings. Larvae mainly leaf-folding or leaf-rolling: *family Cochylidae* (p. 252); *family Tortricidae* (p. 253). **Superfamily Pyraloidea** – a very large group of mainly slender-bodied, long-legged moths, often with long, narrow forewings; *family Pyralidae* (p. 295). **Superfamily Papilionoidea** – day-flying adults (butterflies), with clubbed antennae; antennal tips unhooked: *family Papilionidae* (p. 301); *family Pieridae* (p. 302); *family Lycaenidae* (p. 304); *family Nymphalidae* (p. 307). **Superfamily Bombycoidea** – often large to very large moths, with non-functional mouthparts; male antennae strongly bipectinate: *family Lasiocampidae* (p. 309); *family Saturniidae* (p. 314). **Superfamily Geometroidea** – mainly slender-bodied moths, with broad wings; larvae usually with a reduced number of functional abdominal prolegs; *family Thyatiridae* (p. 315); *family Geometridae* (p. 315). **Superfamily Sphingoidea** – large-bodied, strong-flying moths, often with a large proboscis; larvae usually with a prominent dorsal horn: *family Sphingidae* (p. 335). **Superfamily Notodontoidea** – a small group of moths, sometimes included within the Noctuoidea: *family Notodontidae* (p. 343); *family Dilobidae* (p. 344). **Superfamily Noctuoidea** – the largest group of lepidopterous insects, with a wide variety of, mainly stout-bodied, forms: *family Lymantriidae* (p. 345); *family Arctiidae* (p. 351); *family Nolidae* (p. 354); *family Noctuidae* (p. 354).

Hymenoptera: minute to large insects, usually with two pairs of transparent, membranous wings that have relatively few veins; hindwings the smaller pair and, as in bees and wasps, often interlocked with the forewings by small hooks (hamuli) or, as in certain sawflies, by small tubercles (cenchri); mouthparts adapted for biting, but often also for lapping and sucking; females always with an ovipositor, modified for sawing, piercing or stinging. Metamorphosis complete. Eggs usually sausage-shaped. Larvae usually apodous or eruciform, and usually with a well-developed head. Pupae typically exarate (rarely obtect) and usually formed in a silken cocoon. One of the largest insect orders.

Suborder Symphyta: includes sawflies, insects with a well-developed ovipositor and the abdomen and thorax joined without a conspicuous constriction or 'waist'. Larvae mainly plant-feeding and eruciform; most possess abdominal prolegs, but (unlike those of Lepidoptera) these never bear crochets.

Superfamily Megalodontoidea – a small group of primitive sawflies, with a flattened abdomen; fore tibiae with two apical spurs; cenchri present: *family Pamphiliidae* (p. 379). **Superfamily Cephoidea** – fore tibiae with one apical spur; cenchri absent; abdomen narrow, and constricted between the first and second segments: *family Cephidae* (p. 380). **Superfamily Tenthredinoidea** – the main group of sawflies; fore tibiae with two apical spurs; cenchri present; ovipositor saw-like: *family Cimbicidae* (p. 381); *family Tenthredinidae* (p. 383).

Suborder Apocrita: the main group of hymenopterous insects, the first abdominal segment being fused to the thorax and separated from the rest of the abdomen (termed the gaster) by a narrow, waist-like constriction. Larvae apodous, translucent to whitish and usually with a well-developed head. The suborder is composed of two groups: the Parasitica (e.g. gall wasps, seed wasps and a very large number of parasitoids), in which the ovipositor is adapted for piercing animal or plant hosts; and the Aculeata (ants, bees and social wasps), in which the ovipositor is modified into a sting and has lost its egg-laying function.

Superfamily Cynipoidea – minute or very small, mainly black-bodied insects, with the gaster compressed laterally; antennae with a short scape; includes gall wasps: *family Cynipidae* (p. 400). **Superfamily Chalcidoidea** – minute or very small insects, with the gaster not laterally compressed; antennae usually geniculate, and with a long scape; wings, when present, with a much reduced venation; includes seed wasps: *family Eurytomidae* (p. 401); *family Torymidae* (p. 402). **Superfamily Formicoidea** – ants, usually readily identified by their characteristic appearance and habits; antennae geniculate, and usually 4- to 13-segmented; first or first and second gastral segments small, but often with distinctive outgrowths: *family Formicidae* (p. 402). **Superfamily Vespoidea** – wasps, with the pronotum extending back to the tegulae; eyes deeply notched (emarginate); wings folded longitudinally when at rest. Larvae carnivorous: *family Vespidae* (p. 403).

Mites

Mites (subclass Acari) form part of the Arachnida, a major class of arthropods. Unlike insects, mites have no antennae, wings or compound eyes, and are usually 8-legged. The body is composed of just two main sections: a relatively small, head-like gnathosoma; and a large, sac-like idiosoma. The gnathosoma bears a pair of sensory pedipalps and paired chelicerae, the latter being adapted for biting or piercing and functioning as mouthparts. The idiosoma is typically subdivided by a subjugal furrow into the propodosoma and the hysterosoma, each of which bears two pairs of legs. The body and limbs of a mite possess various setae, the arrangement and characteristics of which are of considerable value in the identification and classification of species. Determination of species, however, is usually a specialist task that often requires the detailed examination of microscopical features.

The respiratory system in the Acari usually includes a pair of breathing pores known as stigmata. Their position on the body, or their absence, can form a basic character for naming the various acarine orders. However, selected names of orders and other subdivisions (see below) differ widely from authority to authority.

Mite development, from egg to adult, usually includes a 6-legged larva and two or three 8-legged nymphal stages known as protonymphs, deutonymphs and tritonymphs. These immature stages are generally similar in appearance and habits to the adult, but are smaller and sexually immature. Many phytophagous mites are free living, but others (notably, in the superfamily Eriophyoidea) inhabit distinctive plant galls, formed in response to toxic saliva injected into the host during feeding (see Chapter 8).

Classification of mites

The classification of mites is subject to considerable disagreement, and is also frequently subject to modification. The following is an example in which mites are subdivided into seven orders.

Subclass **ACARI**
Superorder ANACTINOTRICHIDA

Order Notostigmata	notostigmatid mites
Order Holothyrida	holothyridid mites
Order Mesostigmata	mesostigmatid mites
Order Ixodida	ticks

Superorder ACTINOTRICHIDA

Order Prostigmata	prostigmatid mites
Order Astigmata	astigmatid mites
Order Cryptostigmata	beetle mites

The main features of the groups within which pests of fruit crops occur are summarized below.

Prostigmata: mites with stigmata located between the chelicerae, and with one or two pairs of sensory hairs (trichobothria) on the propodosoma. Most phytophagous mites are included in this order.

Superfamily Eriophyoidea – minute, sausage-shaped or pear-shaped mites, with just two pairs of legs, located anteriorly, each terminating in a branched feather-claw; body with a distinct prodorsal shield; hysterosoma more or less annulated, with a dorsal series of tergites and a ventral series of sternites: *family Phytoptidae* (p. 405); *family Eriophyidae* (p. 406). ***Superfamily Tarsonemoidea*** – minute, often barrel-shaped mites, with short, needle-like mouthparts; hind legs of females without claws; hind legs of males modified into claspers: *family Tarsonemidae* (p. 419). ***Superfamily Tetranychoidea*** – small, spider-like mites, with long, needle-like chelicerae: *family Tetranychidae* (p. 421); *family Tenuipalpidae* (p. 427).

Cryptostigmata: small, dark-bodied, hard-shelled, more or less spherical mites; chelicerae forceps-like; idiosoma with a pair of ridge-like or wing-like expansions (pteromorphs). Most members of this group are inhabitants of top soil and leaf litter, but several are associated with plants: *family Mycobatidae* (p. 428).

Host crops

Brief details of the crops under consideration in Chapters 2 to 8 are given below.

Pome fruits

Pome fruits (family Rosaceae) consist of an enlarged, firm and fleshy receptacle. This encloses a core (consisting of several united, seed-containing, carpels) that forms the central axis of the fruit; this extends from the point of attachment on the tree (by way of a stalk: the pedicel) to the remains of the calyx, the so-called 'eye'. The skin-coated receptacle forms the edible part of the fruit; the skin may or may not be edible.

Apple (*Malus pumila*): unlike the wild crab apple (*Malus sylvestris*), which is native to Europe, cultivated apple trees are of Asian origin. They are now grown throughout the world in suitably temperate regions, including most of Europe. The vast range of cultivars in production include dessert, culinary and cider apples. The fruits, either fresh, pulped or pureed, have a wide range of uses and make excellent pie fillings. Apple juice is also a major commercial product, some of which may be distilled to produce apple brandy. Apart from cider apples, nowadays, apple trees are usually grown on dwarfing rootstock.

Medlar (*Mespilus germanica*): medlar trees, although widely grown in Europe, are of little or no commercial significance; they occur mainly as individual trees in private gardens, but small orchards can be found in southern regions. Fully ripened or partly rotted fruits can be eaten fresh, but they are most often made into jam or jelly.

Pear (*Pyrus communis*): originating in central Asia, pears are now cultivated widely in central Europe. Pears do not tolerate dry conditions and also require warmer summers than apples; much of northern and southern Europe, therefore, is unsuitable for them. The fruits are eaten fresh or are canned; some cultivars are suitable for making a cider-like perry. The juice may also be distilled to produce pear brandy.

Quince (*Cydonia oblonga*): the hard, pear-like fruits are inedible unless cooked and are most often used in jams and jellies or as additional flavouring to other pome fruit dishes. Pear trees are often grafted onto quince rootstock. Large-scale commercial production in Europe is centred on Portugal and Spain.

Stone fruits

Botanically, stone fruits (family Rosaceae) are classified as drupes, in which the edible flesh (the epicarp) overlies a hard 'stone' (the endocarp) that encloses the seed (the kernel).

Apricot (*Prunus armeniaca*): of Asian origin, now widely cultivated commercially in the warmer parts of Europe and sometimes grown further north, as garden trees. The fruits are consumed fresh or are sent for canning or dried for long-term storage. The main European areas of commercial production are located in France, Hungary, Italy and Spain.

Bullace – see under Plum

Cherry (*Prunus* spp.): the marble-like fruits are an important commercial crop in Europe. Sweet cherries (*Prunus avium*) are usually produced for the fresh-fruit market, but occasionally also for making jam and for flavouring, as in the confectionery industry. Morello (= sour) cherries (*Prunus cerasus*), which have a more acid taste, are preserved (bottled or canned) and used for jam-making or in pies; they are also used to make drinks such as cherry brandy and kirsch. The best cherry-growing regions in Europe are to be found in France, Germany and Switzerland, and in various other parts of central Europe.

Cherry-plum – see Myrobalan

Damson – see under Plum

Greengage – see under Plum

Mirabelle – see under Myrobalan

Myrobalan (*Prunus cerasifera*): a minor fruit crop, the trees producing small, round (usually red) fruits with a less sweet taste than greengages. The fruits make excellent jam, and are much favoured in central and southern Europe. However, myrobalan is rarely planted for fruit production, although fruits are often collected from trees planted as hedges and windbreaks; myrobalan is often used as a rootstock for plum trees. Mirabelle, which bears larger, sweeter fruits, is a hybrid between *Prunus cerasifera* and *P. domestica*.

Nectarine – see under Peach

Peach (*Prunus persica*): of Asian origin, now cultivated commercially in all parts of the world with a warm, temperate climate. Peaches, as sweet, juicy, white- or yellow-fleshed, downy-skinned fruits, are sold for the fresh market or sent for canning. Nectarines (var. *nectarina*) are essentially similar to peaches, but smaller and smooth skinned. Southern Europe and warmer parts of central Europe are important production areas. In more northerly regions, peaches are usually grown as individual trees in sheltered private gardens.

Plum (*Prunus domestica* ssp. *domestica*): an aggregation, probably arising from a cross between wild blackthorn (*Prunus spinosa*) and myrobalan (*P. cerasifera*); plums are grown throughout most of Europe and are an important commercial crop. Damson (ssp. *insititia*) (including bullace) and greengage (ssp. *italica*) are also cultivated widely. Plums (including damsons and greengages) are often consumed fresh and they make excellent jams and pie fillings. They are also preserved (e.g. bottled and canned), and may be used to make strong alcoholic drinks, including the well-known slivovitz of central and eastern Europe. Notably in southern France, dark-skinned fruits are dried to produce prunes.

Cane fruits

Cane fruits (family Rosaceae) include a range of species and also various so-called hybrid berries. Blackberry (bramble), for example, is an aggregate of several hundred species and subspecies, and is often hybridized. Botanically, the fruits are not true berries, but are compound fruits that consist of a number of drupelets grouped on a central 'plug' (the remains of the flower receptacle) that extends beyond the calyx. Each drupelet consists of a fleshy outer layer that encloses a hard, seed-containing, pip-like achene. For fresh consumption and for culinary purposes, the achenes are usually consumed as part of the fruit, but they may be strained off when, for example, making jellies.

Blackberry (*Rubus fruticosus* agg.): a rambling plant, usually grown commercially in rows on supporting wires. A large number of cultivars are grown in cultivation, many of which produce larger, superior fruits compared with their wild equivalents. Thornless cultivars are often grown, especially in private gardens. Blackberries are consumed raw or may be used in pies or to make jam. Unlike raspberries, the fruits do not 'plug' on picking. Commercial production, although widespread, is usually on a relatively small scale.

Boysenberry (*Rubus fruticosus*): essentially similar to blackberry, but bearing larger, raspberry-flavoured fruits, often used for canning. Commercial production in Europe is limited.

Loganberry (*Rubus loganobaccus*): a vigorous, heavy-cropping shrub, with long stems that require support. The large, red fruits are more acidic than raspberries, making them particularly suitable for culinary rather than fresh use. Unlike raspberries, the fruits do not 'plug' on picking. Commercial production in Europe is limited; however, loganberries are often grown as a garden crop, especially in the British Isles.

Raspberry (*Rubus idaeus*): the main commercial cane fruit, with a wide range of cultivars grown in Europe. Summer-fruiting and autumn-fruiting cultivars are available. European raspberries typically produce red fruits (= 'red raspberries'); less vigorous, yellow-fruited cultivars (sometimes known as 'white raspberries') are also available. The so-called black-red raspberry (which produces purple berries) is also grown; this is a cross between *Rubus idaeus* and the American black raspberry (*R. occidentalis*). When ripe fruits are picked, the central plugs remain attached to the plant; this avoids the need for them to be removed prior to consumption or processing. Raspberries are consumed fresh; they freeze well, are also canned, and are widely used for flavouring (as in ice-cream).

Tayberry (*Rubus idaeus* × *Rubus fruticosus* 'Aurora'): this hybrid berry bears elongate, red fruits that are larger and more aromatic than loganberries. They are grown mainly by amateur gardeners.

Bush fruits

Bush fruits produce 'true' berries, in which the edible skin-coated flesh surrounds two or more pips or seeds that are not themselves enclosed by a stony endocarp.

Black currant (*Ribes nigrum*) (family Grossulariaceae): a small, spineless, many-stemmed shrub, producing bunches of small, round fruits that hang in racemes. Black currants are widely cultivated, especially in central and northern Europe. The fruits ripen to black, and are used in pies, for making jam and for flavouring; they may also be eaten raw. Black currants are an important source of vitamin C and are extensively grown to produce black currant syrup and juice. Fruits of red currant (*Ribes sativum* agg.) and white currant (*Ribes sativum* agg.) also have dessert and culinary uses. Red currant and white currant bushes are usually grown on single stems and are cultivated mainly in central and northern Europe, usually on a small scale. All three are popular with amateur gardeners.

Gooseberry (*Ribes uva-crispa*) (family Grossulariaceae): a hardy, spiny, bush-like shrub that grows well and is widely cultivated in central and northern Europe. The ripe fruits are eaten fresh, but are often picked early and preserved, or used in pies or for making jam. Dessert and culinary cultivars are available.

Highbush blueberry (*Vaccinium corymbosum*) (family Ericaceae): a large bush, often 1 m or more tall, producing blue-black fruits 10 mm or more in diameter. These are used in pies and for making jelly and jam. Blueberries are of American origin and form a complex aggregation of species and cultivars. They are widely grown in the USA and, nowadays, are also grown commercially in several parts of Europe.

Jostaberry (*Ribes nigrum* × *Ribes uva-crispa*) (family Grossulariaceae): a vigorous-growing cross between black currant and gooseberry, producing bunches of purple fruits (intermediate in size between currants and gooseberries) that are especially popular for jam-making.

Red currant and **White currant** – see under Black currant.

Worcesterberry (*Ribes divaricatum*) (family Grossulariaceae): an American species, with long, arching branches that bear small, purple, gooseberry-like fruits. In Europe, it is usually grown only by amateur gardeners.

Other small fruits

Grape (*Vitis vinifera*) (family Vitaceae): grapevines are woody trailing or climbing plants that require warm or hot conditions for their fruits (grapes) to ripen satisfactorily. Grapes are true berries, with the seeds surrounded by a fleshy, juicy pulp; seedless cultivars also occur. Grapes are commonly grown for the fresh market (table grapes) and for the production of wine; in addition, the mature fruits may be dried to produce currants, raisins and sultanas. Viticulture is a major world industry, with many European countries (notably France, Italy and Spain) featuring amongst the most prolific of all wine producers. In order to avoid infestations of grape phylloxera (see Chapter 3), European vines are usually grafted onto American rootstock.

Strawberry (*Fragaria* × *ananassa*) (family Rosaceae): this American hybrid forms the bulk of the world's commercial strawberry production. Botanically, strawberries are 'false' fruits, the edible flesh being formed from the swollen receptacle which bears numerous small, brown achenes ('seeds') on the surface. Strawberries are consumed fresh, as a dessert fruit, or are used to make jam or for flavouring, as in confectionery and ice-creams. Strawberries are widely cultivated throughout Europe, either as open-field crops or under protection (under cloches or plastic tunnels, or in greenhouses). In addition, there is limited commercial production of the alpine strawberry (*Fragaria vesca*), a plant producing smaller, but less acidic, fruits with a superior flavour.

Nuts

Almond (*Prunus communis*) (family Rosaceae): almond is native to North Africa and western Asia and is grown commercially in all regions with a Mediterranean climate, especially Italy and Spain. The trees are also grown on a decreasingly smaller scale through central Europe and parts of northern Europe. The velvet-skinned, somewhat hairy, fruits (drupes – see 'stone fruits' above) are grown for their kernels. The inedible kernels of bitter almonds (var. *amara*) are crushed for oil; those of sweet almonds (var. *dulcis*) are consumed as either whole or flaked nuts, or are used as a paste. Peach × almond hybrids, which bear larger, more fleshy fruits, occur; these are often used as rootstock for peach trees.

Chestnut (*Castanea sativa*) (family Fagaceae): also known as sweet chestnut or Spanish chestnut, this nut-producing tree is a native of south-east Europe and is widely grown, especially in poorer, upland regions of central Europe. The reddish-brown nuts occur in threes, within a globose, green outer casing (involucre) that bears numerous long, fine spines. Sweet chestnuts, not to be confused with the fruits (nuts) of horse-chestnut (*Aesculus hippocastanea*), are usually roasted or pureed.

Cobnut and **Filbert** – see under Hazelnut.

Hazelnut (*Corylus avellana*) (family Corylaceae): this well-known shrub occurs naturally throughout much of Europe and in parts of Asia, and nowadays is often planted commercially in plantations. Botanically, the fruits (unlike those of, for example, almond and walnut) are true nuts, in which the hardened shell is the pericarp and the edible seed the kernel. Hazelnuts are mainly consumed as dessert nuts. Cobnut or filbert (*Corylus maxima*) and Turkish hazel (*Corylus colurna*) produce larger, more elongate fruits and are widely grown commercially; in the following chapters, the term 'hazelnut' is used in its wider sense to embrace these. Major hazelnut-producing countries include France, Italy, Spain and Turkey.

Pistachio (*Pistacia vera*) (family Anacardiaceae): a subtropical plant of Middle Eastern origin, widely grown in the Mediterranean basin; in Europe, commercial production is restricted mainly to Greece, Sicily and southern Spain. The fruits are, botanically, drupes (see 'stone fruits' above) and are grown for the edible kernels. These are consumed lightly roasted and salted, and also used for culinary purposes.

Walnut (*Juglans regia*) (family Juglandaceae): a large tree, of Eurasiatic origin and widely naturalized throughout Europe; often known as English or Persian walnut. Botanically, the fruit is a drupe (see 'stone fruits' above), the edible nut (kernel) being contained within a woody, 2-valved shell (endocarp); this is surrounded by a thick, more or less leathery outer skin (epicarp) that is eventually shed. Walnuts are grown mainly for the dessert market; they also have a range of confectionery and culinary uses. Additionally, the young fruits may be pickled and then eaten; walnut oil, obtained from the kernels, is also produced. Most commercial European walnuts are grown in France. Italy, Romania and Turkey are also major producers.

Citrus fruits

Citrus fruits (family Rutaceae) are subtropical crops that require hot conditions to ripen and flourish. The fruit itself is a berry, known as an hesperidum, with a leathery outer rind (epicarp) and the inner flesh divided by skin-like septa into several distinct segments. The epicarp contains numerous aromatic oil glands, as do the leaves. The leaf stalks (petioles) are often winged, noticeably so in the case of bergamot orange, grapefruit, pomelo and sour orange.

Clementine – see under Mandarin.

Grapefruit (*Citrus paradisi*): grapefruits are grown on spreading trees, 8–15 m tall. The fruits have a characteristic, slightly bitter taste, and are consumed mainly in the form of juice, but also as fresh fruits or as canned (often fragmented) segments. Satisfactory

ripening of the fruits requires very hot summers, and most of those sold on the European market are grown in Israel. Commercial production in Europe is confined mainly to Cyprus, Greece and southern Spain.

Lemon (*Citrus limonum*): lemon trees are distinctly thorny and bushy, 3–6 m tall. The unmistakable fruits are yellow, with a nipple-like apical swelling; the flesh is acidic. Both the peel and juice have a range of uses; in addition, the pips are a source of oil (lemon pip oil). Italy and Spain are the main European producers.

Mandarin (*Citrus reticulana*): a small, often spiny tree; the pale yellow to orange-coloured fruits (the latter often called tangerines) are characteristically flattened at both base and apex. Mandarins are usually produced for the fresh market, but are also preserved as canned segments. The crop is grown commercially in various parts of southern Europe, the crop extending further north than its other, less hardy relatives. Clementines are essentially similar to mandarins, but the fruits are smaller and have deeper-coloured skins that are difficult to peel; they are widely grown in southern Europe.

Orange (e.g. *Citrus sinensis*): orange trees are of Chinese origin and grow up to 12 m in height, bearing globular fruits that are typically orange when ripe. Fruits of sweet orange (*Citrus sinensis*) are mainly eaten fresh or are crushed to produce juice. Sweet oranges are grown throughout the world in warm or tropical countries, Italy and Spain being amongst the world's top producers. The rough, thick-skinned and highly aromatic fruits of sour orange (*Citrus aurantium*) have smaller segments than those of sweet orange and the flesh is sour tasting; they are used mainly to produce marmalade, Spain being the largest European producer. Bergamot orange (*Citrus bergamia*), often considered to be merely a cultivar of sour orange, is grown as a source of essential oils; nowadays, its commercial production in Europe is restricted to southern Italy.

Pomelo (*Citrus grandis*): although sometimes consumed as a fresh fruit, and widely grown in southern Asia, this crop is upstaged by grapefruit and is of little economic importance in Europe.

Tangarine – see under Mandarin.

Miscellaneous crops

Avocado (*Persea americana*) (family Lauraceae): an evergreen, mainly tropical, tree (in cultivation usually no more than 5 m in height), with three principal races originating in Guatemala, Mexico and the West Indies. European-grown cultivars are derived from Mexican stock and are cultivated only in the hottest of Mediterranean regions. The highly nutritious, pear-shaped fruits (single-seeded berries) are usually consumed raw, as half-fruits laced with lemon juice and salt, or in mixed salads.

Chinese persimmon (*Diospyros kaki*) (family Ebenaceae): a deciduous subtropical tree, originating in China and now grown throughout the World in the subtropics. Commercial production in Europe (e.g. in southern France, Italy and Spain) is relatively recent. The fruits are large, juicy, tomato-like berries, with a large, persistent, basal calyx. They are consumed fresh or cooked, and have a range of dessert uses; they are frequently used to make jam.

Date plum – see Chinese persimmon.

Fig (*Ficus carica*) (family Moraceae): a spreading, deciduous, subtropical shrub or small tree; although widely grown in Europe, the fruits ripen only under suitably hot conditions. Botanically, the fruit is known as a syconium; this is a hollow structure, formed from the fleshy receptacle, with an apical hole (ostiolum) leading to an inner lining of florets and developing seeds. Figs are consumed fresh or dried, and may also be cooked. Commercial production in Europe is restricted to Mediterranean regions.

Hop (*Humulus lupulus*) (family Cannabaceae): hop is a perennial climber, with the bines (often arising from earthed-up 'hills') traditionally grown on an elaborate system of very tall supports, 4–5 m in height. Hops are propagated vegetatively, using stem-base cuttings (sets). Recently, dwarf cultivars have been introduced; these so-called dwarf hops are grown on trellises (c. 2 m in height), to form a hedgerow-like system. Hops are cultivated for their papery, cone-like fruits (strobili) (often known as burrs), formed from several wing-like bracts (each containing a small, basal nutlet) arranged around a central axis (the strig). Lupulin, a substance that imparts a bitter flavour to beer, is extracted from the cones during processing; lupulin also checks bacterial growth, thereby acting as a preservative. There is also a minor, specialist market for young bines emerging from the soil in the spring, which are then cut and sold as 'hop asparagus'. European centres of hop production include various parts of central Europe, northwards to the British Isles and Scandinavia, and eastwards to Russia.

Japanese persimmon – see Chinese persimmon.

Kaki – see Chinese persimmon.

Kiwi fruit (*Actinidia chinensis*) (family Actinidiaceae): kiwi fruit, also known as Chinese gooseberry or yang tao, originated in China and commercial production elsewhere (notably in New Zealand) did not begin until the 1960s. Nowadays, the crop is grown commercially in various parts of southern Europe (including France, Italy and Spain). The plant is a woody, twining climber, with large, heart-shaped leaves, and is usually grown on wood and wire trellising. The juicy fruits (berries) are usually consumed raw, after peeling; they also have limited culinary uses.

Mulberry (*Morus* spp.) (family Moraceae): black mulberry (*Morus nigra*) and white mulberry (*Morus alba*) trees grow to 10 m; they have robust trunks and branches, and relatively large, thick leaves. The fruits (botanically known as syncarps) are formed from a collection of united, seed-containing carpels (drupelets); superficially, they are similar in appearance to blackberries and raspberries, but each drupelet is subdivided into four segments. The fruits of black mulberry are virtually stalkless and usually purple or black when ripe; those of white mulberry are white, pink or purple and are borne on long (1–2 cm) stalks. Mulberries may be consumed raw, but are more frequently used for producing jam, jellies and wine. Small-scale commercial production occurs in various parts of southern Europe. Further north, mulberry trees are usually grown for home-garden use or planted in parks and gardens as ornamentals.

Olive (*Olea europaea* var. *europaea*) (family Oleaceae): cultivated olive is an evergreen tree, with narrow, lanceolate leaves (dark green above and grey below) coated with minute, silvery scales. The fruits (drupes) ripen to black and are then harvested; some, however, may be picked whilst immature and still green, pink or red. Olives are cultivated throughout the Mediterranean basin, with no more than about two per cent produced in other parts of the world. Most are grown for their oil, some for consumption as table olives, and some for both purposes.

Oriental persimmon – see Chinese persimmon.

Pomegranate (*Punica granatum*) (family Punicaceae): a large (in Europe deciduous), subtropical or tropical shrub, bearing large, rounded fruits. The fruit is known as a balusta, an indehiscent, many-celled, many-seeded berry enclosed within a leathery, inedible skin (pericarp). The pulp that surrounds the seeds is either eaten raw or used to make jam, syrup or wine. Small-scale cultivation is practised in many Mediterranean countries.

Prickly pear (*Opuntia ficus-indica*) (family Cactaceae): this plant, also known as barbary fig, was introduced into southern Europe from the Americas in the 15th century. It rapidly became established and was often used to form impenetrable boundaries and hedges. The spiny, sweet-tasting fruits are edible and are either consumed raw or cooked. However, they are of little importance in Europe and are cultivated only in the drier parts of the Mediterranean region.

Chapter 2

Smaller insect orders

Order **SALTATORIA** (crickets, grasshoppers and locusts)

Family **TETTIGONIIDAE**
(bush crickets)

Crickets with 4-segmented tarsi, a pair of short anal cerci and the antennae longer than the body. Females possess a distinctive, broad-bladed ovipositor.

Leptophyes punctatissima (Bosc) (**1–3**)
Speckled bush cricket

A minor pest of fruit trees, usually in large gardens and weedy, unsprayed orchards. Widely distributed in Europe.

DESCRIPTION

Adult: up to 18 mm long; body mainly green, finely speckled with dark purple; head and prothorax marked with a white dorsal line and a broader (sometimes yellowish) stripe that extends backwards from above each compound eye; abdomen with a pale purplish-tinged dorsal line; wings vestigial in both sexes; female with a robust ovipositor protruding backwards from the tip of the abdomen. **Nymph:** similar to adult, but smaller.

LIFE HISTORY

Eggs are laid singly, in the autumn, in the shoots or stems of various plants, including (occasionally) apple and probably other fruit trees. The eggs hatch in the following spring or early summer. The very active nymphs feed on the foliage of various plants, and reach the adult stage about 3 months later. There is just one generation annually.

DAMAGE

There are isolated reports of large numbers of these insects causing extensive damage to the foliage of stone-fruit trees, notably nectarine and peach, but such damage is generally of little or no importance and, typically, restricted to the production of large holes in the expanded leaves.

1 Speckled bush cricket (*Leptophyes punctatissima*) – female.

2 Speckled bush cricket (*Leptophyes punctatissima*) – male.

3 Speckled bush cricket (*Leptophyes punctatissima*) – damage to peach leaf.

Phaneroptera nana Fieber (**4–5**)
Four-spot bush cricket

Nymphs and adults of this often common Mediterranean species browse on the foliage of a wide range of plants, including grapevine, and the females have been recorded laying eggs in apple leaves. However, the insect is not a significant pest and damage caused to fruit crops is of little or no importance. Adult females (body 13–15 mm long) are apple-green to greenish white, finely speckled with purplish red, with very long, fully developed wings, large hind femora and a very large, curved ovipositor. Nymphs are similarly coloured. The adults occur from July to November, and are strong fliers. This bush cricket is also present in North Africa and Asia, and has been introduced into North America.

Family **GRYLLOTALPIDAE** (mole crickets)

Crickets with the forelegs (being adapted for burrowing in soil) greatly enlarged and armed with finger-like projections (dactyles); tarsi 3-segmented; abdomen with a pair of prominent anal cerci; antennae relatively short; ovipositor vestigial.

Gryllotalpa gryllotalpa (Linnaeus) (**6**)
Mole cricket

An omnivorous, mainly subterranean species that feeds on soil invertebrates (e.g. earthworms and insect larvae) and plant roots; most numerous in light soils with a high humus content. Widely distributed in the warmer parts of Europe, but usually of little or no significance as a pest unless present in large numbers. An introduced species in the USA. In parts of its natural range (as in the British Isles) this insect is protected by law, and is not a pest.

4 Four-spot bush cricket (*Phaneroptera nana*) – female.

5 Four-spot bush cricket (*Phaneroptera nana*) – nymph.

6 Mole cricket (*Gryllotalpa gryllotalpa*).

DESCRIPTION

Adult: 35–50 mm long; body stout, greyish brown to yellowish brown, with a velvet-like coating of fine hairs; antennae relatively short, about as long as the anal cerci; forelegs each with 4 dactyles; forewings (elytra) short; hindwings large and elongate, but rolled longitudinally, and relatively inconspicuous, when in repose.

LIFE HISTORY

Mole crickets burrow through the soil, forming tunnels a few centimetres beneath the surface. At night, adults may also come to the surface (especially in warm, summer evenings), and may then fly about in swarms. In common with other crickets, adult males are capable of stridulating. After mating, the adult female forms a large subterranean chamber, in which 100–300 eggs are laid. The eggs hatch within 2–3 weeks, but the female continues to tend to her brood until the nymphs (which feed on humus and young roots) reach the second instar; the nymphs then leave the maternal 'nest' to feed and develop independently. The life cycle from egg to adult is completed in about a year in southern Europe, but extends to 18 months or more in less favourable regions.

DAMAGE

Mole crickets browse indiscriminately on roots and the basal parts of stems; seedlings may be destroyed, and injured plants in nursery beds (including young fruit trees and bushes) may subsequently wilt and die. In strawberry beds, mole crickets feed on the roots and crowns; developing fruitlets or fruits may also be attacked.

Family **ACRIDIDAE**
(grasshoppers and locusts)

Saltorians with short, stout antennae and 2-segmented front tarsi; pronotum with a prominent, longitudinal, median keel.

Locusta migratoria (Linnaeus)
Migratory locust

Populations of this notorious insect, along with those of other related species, are subject to periodic explosions, and may then invade southern Europe from their homelands in Africa or the Middle East. Migratory locusts are also now established in parts of southern Europe. Although feeding mainly on grasses and cereals, locusts (especially when numerous) will graze indiscriminately on all kinds of plants, including fruit crops, and damage in orchards, plantations or vineyards may then be extensive. Adults of *L. migratoria* (40–60 mm long) are mainly brown, sometimes tinged with green; the elytra are mainly transparent, with scattered darker markings.

Order **DERMAPTERA** (earwigs)

Family **FORFICULIDAE**

Earwigs with well-developed compound eyes, and the second tarsal segment expanded (heart shaped).

Forficula auricularia Linnaeus (**7–9**)
Common earwig
A generally abundant, but minor, pest of apple, peach, plum and other crops; sometimes particularly troublesome in black currant plantations. Also a useful predator, devouring, for example, eggs of pests such as codling moth (*Cydia pomonella*). Present throughout Europe and the Mediterranean basin; introduced into North America, Australia and New Zealand.

DESCRIPTION
Adult female: 12–14 mm long; chestnut-brown; hind wings, when folded away, projecting beyond the elytra; pincers slightly curved. **Adult male:** 13–17 mm long; similar to female, but pincers distinctly curved. **Egg:** 1.3 × 0.8 mm; pale yellow. **Nymph:** whitish to brown.

LIFE HISTORY
Adults of both sexes overwinter in sheltered situations in the soil. After mating, each female lays several eggs in an earthen cell and guards over them until they hatch in February or March. The nymphs then disperse, to feed throughout the spring in a variety of situations. There are four nymphal stages, individuals reaching maturity by the early summer. Overwintered adult females may deposit a second batch of eggs in May or June. Nymphs from these eggs develop from late June or early July to September. Earwigs are nocturnal, hiding by day under loose bark, in crumpled leaves and so on.

DAMAGE
General: earwigs often hide within fruit clusters, or between crumpled leaves and fruits, and the fruits may then become soiled with accumulations of black frass. The earwigs will also bite into the exposed flesh of fruits damaged by birds and other agents, thereby enlarging the original wounds. Sometimes, they will damage sound fruit, but primary damage is rarely important. On apple, damage tends to be more prevalent on soft-skinned cultivars and on cracked or russeted fruits. Earwigs may also cause minor damage to foliage and other tissue. **Black currant:** earwigs often roost in the bushes, but are not injurious; however, they are readily dislodged during mechanical harvesting and may then become temporary contaminants in the harvested fruit.

7 Common earwig (*Forficula auricularia*) – female and associated frass.

8 Common earwig (*Forficula auricularia*) – male.

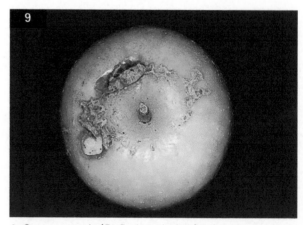

9 Common earwig (*Forficula auricularia*) – damage to mature apple.

Order **ISOPTERA** (termites)

Family **KALOTERMITIDAE** (dry-wood termites)

A primitive family characterized by the presence of compound eyes in all castes, by the absence of glands on the forehead, and by the absence of a true worker caste.

Kalotermes flavicollis Fabricius
European dry-wood termite
Although feeding mainly on dead wood, this insect is sometimes regarded as a pest in vineyards. Present in southern Europe, along the coast from Portugal eastwards to Greece, and also reaching the Middle East.

DESCRIPTION
Winged forms: body elongate, brownish black, with prothorax, legs and antennae yellow. **Soldier:** 7–9 mm long; wingless; yellowish brown, with a very large head and mandibles.

LIFE HISTORY
Colonies, which are relatively small, consist of a 'king' and a 'queen' (which together form the royal pair), and numerous 'soldiers' and so-called 'juvenile workers' (the pseudergates). Pseudergates and soldiers number no more than a few hundred individuals, the former greatly outnumbering the latter. The pseudergates are wingless and usually remain as juveniles (nymphs) throughout their lives. The main flight swarms of young kings and queens occur from September to November, primarily during the morning. The kings and queens shed their wings soon after alighting. Pairing then takes place and, eventually, new colonies are established.

DAMAGE
In vineyards, this termite will attack and destroy wooden support posts; it is also reported feeding on the wood of living vines. In addition to grapevine, various trees and shrubs may be damaged; some, however, notably olive trees, are immune from attack.

Order **THYSANOPTERA** (thrips)

Family **THRIPIDAE** (thrips)

Antennae usually 7- or 8-segmented (rarely, 6- or 9-segmented), the apical one to three segments usually forming a thin style; wings strap-like, with a pointed apex and three longitudinal veins (veins I–III); females with a saw-like, downward-curving ovipositor. Determination of species is a specialist task, that often requires the detailed examination of microscopical features. The specific descriptions provided here are given merely as a general guide.

Drepanothrips reuteri Uzel
Vine thrips
A pest of grapevine in temperate parts of Europe, but more generally associated with *Quercus robur*; also found commonly on both *Betula* and *Corylus avellana*. Widely distributed in Europe; also present in North America.

DESCRIPTION
Adult: 0.6–0.9 mm long; body yellow, shaded with brown; antennae 6-segmented (without a style), yellow, but dark tipped; wings cloudy.

LIFE HISTORY
Overwintered adults occur from early spring to early summer, with nymphs developing in July or August; new adults are active in late summer and autumn.

DAMAGE
Infested leaves are distorted and the edges may appear scorched. In addition, shoots may be stunted and rachides and stems scarred. If eggs are laid in developing grapes, then small, circular scars occur on the skins; such blemishes may lead to the downgrading of high-quality table grapes.

Frankliniella intonsa (Trybom)
Flower thrips
This generally common, polyphagous species occurs on a range of host plants and is usually of little or no importance. On strawberry, however, this species has been implicated (along with various species in the genus *Thrips*) in the development of distorted strawberry fruits, particularly on late-season cultivars. Adults are yellowish brown to reddish brown, with 8-segmented antennae and a complete row of setae along forewing veins I and II (typical of the genus).

Frankliniella occidentalis (Pergande)
Western flower thrips

In the early 1980s, this polyphagous, originally North American, thrips appeared in considerable numbers on greenhouse plants in England and various other parts of Europe, having been introduced accidentally from abroad, mainly on chrysanthemum cuttings. Eradication programmes were attempted in Europe by Plant Health authorities; however, in spite of these, the pest became established and is now widely distributed. In warmer regions it infests various outdoor crops, including apple, nectarine, peach, strawberry, grapevine and olive. The pest is now established in many parts of the world, including South Africa where it is a significant pest on fruit crops, including grapevine.

DESCRIPTION
Adult: 1–2 mm long; pale yellow to brownish yellow; antennae 8-segmented; forewings with a complete row of setae along both veins I and II (typical of genus: cf. *Thrips flavus*); individuals are most reliably distinguished from *Frankliniella intonsa* on the basis of microscopical features. **Egg:** 0.2 mm long; pearly white. **Nymph:** translucent to golden yellow; eyes reddish. **Propupa:** white; wing cases short. **Pupa:** white; wing cases long.

LIFE HISTORY
This mainly parthenogenetic species breeds continuously under suitable conditions, and the life cycle is completed in 2–3 weeks at normal greenhouse temperatures. Under suitably warm conditions, breeding and survival throughout the year are possible out of doors. There are two nymphal instars, after which individuals drop to the ground to complete their development in the soil. Although adults and nymphs are sometimes observed on the exposed surfaces of leaves and flower petals, they are of secretive habit and occur more frequently on the underside of leaves, or hidden within the shelter of flowers and beneath bud scales.

DAMAGE
Adults and nymphs cause distortion, silvering and speckling of leaves and flowers. Damage from even a relatively small number of individuals is often extensive. Deleterious effects on outdoor hosts are particularly severe in hot summers. Damage on grapevine is similar to that caused by vine thrips (*Drepanothrips reuteri*), and on olive is similar to that caused by olive thrips (*Liothrips olea*) (family Phlaeothripidae).

Pezothrips kellyanus (Bagnall)
Kelly's thrips

An Australian pest of citrus; also now established, locally, as a pest in the Mediterranean basin (including Sicily and the extreme south of the Italian mainland).

DESCRIPTION
Adult: 2–3 mm long; black and shiny.

LIFE HISTORY
This species passes through several generations annually, and development from egg to adult includes two active nymphal stages, a propupal and a pupal stage. The thrips are particularly attracted to the flowers of citrus and other fragrant plants, such as *Jasminum*, where they feed on the young tissue.

DAMAGE
On citrus, feeding by the thrips on the calyx of a developing flower often results in the presence of a circular, halo-like scar around the base of the fruit. Lemons are particularly susceptible.

Taeniothrips inconsequens (Uzel)
Pear thrips

Locally common as a minor pest of pear, but not on regularly sprayed trees; also breeds on other fruit trees, including apple, peach, nectarine and plum. Widespread in Europe; also present in North America.

DESCRIPTION
Adult female: 1.2–1.7 mm long; dark brown, with light brown to yellow tibiae and tarsi; wings greyish, paler basally; antennae 8-segmented. **Egg:** minute and white. **Nymph:** yellowish white, with dark red eyes; hind edge of abdominal segment 9 with a series of robust, wedge-shaped spines. **Propupa:** 1.3 mm long; white and translucent; wing cases short. **Pupa:** 1.4 mm long; brownish; wing cases long.

LIFE HISTORY
In late winter or early spring, as buds begin to swell, overwintered adults emerge from the soil. They then enter buds to feed on the succulent, young tissue. In May, eggs are laid in the leaf veins and blossom stalks (pedicels), and they hatch about 10 days later. Nymphs then feed during May and June on the blossoms, leaves and other tissue. When fully grown, the nymphs enter the soil and form small cells in which to pass through a propupal and then a pupal stage. Adulthood is reached in the autumn, but individuals remain within their pupal cells until the following year. Breeding is mainly parthenogenetic, males being very rare.

DAMAGE

Thrips feeding within the buds may cause weeping of sap from damaged areas. Later, brownish patches are produced at the bases of stamens and styles, and on the petals of infested blossoms; attacked leaf tissue also becomes discoloured. Damage, which is particularly severe in dry conditions, may lead to distortion of petals, russeting and distortion of fruitlets, and a reduction in fruit set.

Thrips atratus Haliday
Carnation thrips

This widely distributed species is associated mainly with flowers of Caryophyllaceae, but is also regarded, along with *Thrips fuscipennis* and *T. major*, as a causal agent of malformed strawberry fruits. Unlike the other strawberry-infesting species of thrips cited here the antennae of *T. atratus* are 8-segmented and the antennal style 2-segmented.

Thrips flavus Schrank
Yellow flower thrips

A minor pest on cultivated blackberry; also found on loganberry. Generally common throughout Europe, and locally abundant on a wide range of flowering plants.

DESCRIPTION

Adult female: 1.2–1.6 mm long; body yellow to orange yellow; legs yellow; forewings pale yellow and with the row of setae along vein I incomplete (typical of genus); hind margin of pronotum with four long setae; antennae 7-segmented. **Adult male:** smaller and paler than female. **Nymph:** white to pale yellow.

LIFE HISTORY

Adult females occur throughout the year; however, in north western Europe, they are most abundant from July to September. Males are present from July to October. There are two to three, if not more, generations in a season, nymphs feeding in open or partially open flowers from May to September. The change from nymph to adult, through a propupal and a pupal stage, takes place in the soil beneath host plants.

DAMAGE

Feeding by adults and nymphs can cause heavily infested blossoms to shrivel and darken, and developing fruitlets to wither. However, crop losses are usually small.

Thrips fuscipennis Haliday (**10**)
Rose thrips

Abundant locally, adults feeding on a wide variety of flowering plants, particularly Rosaceae. Nymphs occur on many hosts, including apple, almond, blackberry and strawberry. Widely distributed in Europe; also present in North America.

DESCRIPTION

Adult female: 1.2–1.6 mm long; yellowish brown to dark brown; legs brown; forewings dark greyish brown, paler basally; comb of setae on hind margin of abdominal tergite 8 incomplete centrally (cf. *Thrips tabaci*); antennae 7-segmented, including a 1-segmented style. **Adult male:** smaller and paler than female. **Nymph:** white to pale yellow.

LIFE HISTORY

Adult females overwinter under bark or amongst herbage, emerging in the spring. Eggs are laid from May onwards, often in young apple shoots. Nymphs occur from May to August or to September, feeding on leaves and shoots, and in flowers. Pupation occurs on the host plant or in the soil. Adult males are present from June to October; there are up to four generations in a season.

DAMAGE

Adults and nymphs cause discoloration of petals, silvering of young leaves, and also leaf bronzing and distortion. On strawberry, infestations on the blossoms can result in the development of malformed fruits.

10 Rose thrips (*Thrips fuscipennis*) – damage to almond leaf.

Thrips major Uzel
Rubus thrips

This widely distributed species is associated with the flowers of many plants, but especially Rosaceae, and is yet another thrips implicated in the malformation of strawberry fruits. Adults have 7-segmented antennae and are mainly brown, but the colour of the body and wings in this species is rather variable. Adult females are active mainly from early spring to early autumn; nymphs are present from May to September.

Thrips meridionalis Priesner
Peach thrips

This species occurs on the flowers of Rosaceae, particularly apple, pear, almond, apricot, cherry, nectarine, peach and plum. Also reported as a pest of grapevine, e.g. in France. Widely distributed in the warmer parts of mainland Europe.

DESCRIPTION
Adult female: body brownish black; antennae 8-segmented, including a 2-segmented style. **Adult male:** similar to female, but paler in colour.

LIFE HISTORY
In southern Europe there are typically three overlapping generations annually. Adults overwinter in the shelter of dead leaves and become active from mid-February onwards. They invade early-flowering hosts, such as almond. Later, the pest migrates to other hosts as these reach the flowering stage. Eggs are laid in the flowers over a period of several weeks, development from egg to adult taking about a month. Eggs of later generations are laid in leaves or fruits. Adults of the final generation mate in the autumn and the impregnated females then take up their winter quarters.

DAMAGE
Particularly on nectarine (and, to a lesser extent, on peach), attacks on blossoms can reduce fruit set (owing to damage to the stamens), or can result in the development of necrotic patches on fruitlets (following damage to the ovaries). Necrosis can also result from direct attacks on very young fruitlets; these damaged areas enlarge as the fruitlets grow and may then split open, allowing gum to exude from the cracks. Damaged fruitlets may also become distorted, or they may wither and die without reaching maturity. Further, eggs laid in the skins of fruit are unwelcome contaminants.

Thrips minutissimus Linnaeus

Although typically associated with the flowers of *Quercus*, this species has been found, occasionally, damaging the blossoms of apple trees. Widely distributed and locally common in Europe.

DESCRIPTION
Adult: body light brown to dark brown; antennae 7-segmented, with segments 1–3 distinctly lighter in colour than segments 4–7; antennal style 1-segmented.

LIFE HISTORY
Adults occur from late March to early June, and nymphs from May to July.

DAMAGE
Parts of infested petals and stamens become distorted and turn brown.

Thrips tabaci Lindeman
Onion thrips

In northern Europe, this widespread and generally abundant species is often a pest of protected crops such as cucumber and tomato, the thrips causing extensive silvering of infested tissue. Adult females also occur outdoors throughout the year, breeding from May to September or October on a wide variety of host plants. In addition to vegetable crops (including leek and onion), grapevines are sometimes infested; damage on grapevine is similar to that caused by vine thrips (*Drepanothrips reuteri*). Adult females are 1.0–1.3 mm long, pale yellow (marked with brown) to dark brown, with yellowish-brown forewings and 7-segmented antennae; the comb of setae on abdominal tergite 8 is complete centrally (cf. *Thrips fuscipennis*).

Family **PHLAEOTHRIPIDAE**

A mainly tropical family, the active stages of which feed mainly on dead wood or leaf litter; leaf-feeding species, however, do occur and these can cause noticeable galling of infested tissue; females lack an ovipositor, and both sexes have a tubular final abdominal segment; development includes an egg, two nymphal, one propupal and two pupal stages. Eggs are hard shelled and elongate-oval (cf. family Thripidae).

Gynaikothrips ficorum (Marchal)
Cuban laurel thrips
A tropical species that attacks various hosts, including citrus and fig. Found mainly in Central America and in the southern parts of the USA, but also now present in southern Europe.

DESCRIPTION
Adult: 3 mm long; black bodied; pronotum distinctly sculptured and with a single pair of very long, posteriorly directed setae.

LIFE HISTORY
Nymphs and adults occur mainly on the leaves and young shoots. Breeding is continuous so long as conditions remain favourable, with many overlapping generations annually.

DAMAGE
The thrips cause leaf rolling. Infested shoots also become discoloured and distorted, and severely damaged leaves fall off prematurely.

Liothrips oleae Costa (**11**)
Olive thrips
This pest occurs on olive throughout the Mediterranean basin.

DESCRIPTION
Adult female: 1.9–2.5 mm long; black and stout bodied; antennae 8-segmented. **Adult male:** 1.4–1.8 mm long; otherwise, similar in appearance to female. **Egg:** 0.4 × 0.2 mm long; whitish and reticulate. **Nymph:** mainly yellowish white, with tip of abdomen black and a pair of large black spots on the thorax; legs and antennae blackish.

LIFE HISTORY
Adults overwinter in cracks and crevices in the bark of olive trunks and branches. Activity is resumed in the spring and eggs are then laid in sheltered situations on the bark or on the leaves. Eggs hatch within about 2 weeks, depending on temperature. Nymphs then feed on the young leaves, new shoots, buds and developing fruits for about 2 weeks before entering the propupal and pupal stages. The very active adults emerge a further 10 to 12 days later. There are usually three generations annually. However, four generations are possible under particularly favourable conditions. Populations reach their peak in late June or early July, but at the height of summer (from July to August), when conditions are very hot, adults wander away from the foliage to aestivate on the bark.

DAMAGE
Feeding causes a yellowing of young leaf tissue, followed by desiccation and hardening. As the leaves continue to expand, they become deformed; the photosynthetic activity of infested foliage is also impaired. Damage to terminal buds can disrupt shoot development, and attacks on fruit buds and young fruits may cause them to drop off. Infested fruits also become deformed and may develop necrotic patches. They may also be smaller than normal and may drop prematurely. Heavy infestations lead to significant crop and oil losses.

11a, b Olive thrips (*Liothrips oleae*) – damage to young shoots.

Chapter 3
True bugs

Family **ACANTHOSOMATIDAE**
(shield bugs)

Medium-sized to large, shield-like insects (often called 'stink bugs'), with the lateral margins of the head concealing the base of the antennae; scutellum large and triangular; abdomen with a distinct ventral keel; antennae 5-segmented; tarsi 2-segmented (cf. family Pentatomidae).

Acanthosoma haemorrhoidale (Linnaeus) (**12**)
Hawthorn shield bug
Although associated mainly with *Crataegus monogyna*, this species is found occasionally on various other hosts, including apple and other fruit trees. Widely distributed in northern Europe.

DESCRIPTION
Adult: 13–15 mm long; green and shiny, with reddish markings and numerous black punctures; pronotum much extended laterally; hemelytra with membrane brownish; legs and antennae light green to yellowish brown.

LIFE HISTORY
Adults of this phytophagous bug overwinter and reappear in the spring. Eggs are laid from late April to early June. Nymphs feed and develop during the summer and reach the adult stage by about August. In the absence of *Crataegus* berries, upon which the bugs preferentially feed, foliage is also attacked.

DAMAGE
When on fruit trees, adults and nymphs may puncture foliage and other tissue; however, effects on the plants are usually of little or no importance.

12 Hawthorn shield bug (*Acanthosoma haemorrhoidale*).

Family **PENTATOMIDAE**
(shield bugs)

Medium-sized to large, shield-like insects (often called 'stink bugs'), with the lateral margins of the head concealing the base of the antennae; scutellum large, U-shaped or triangular; antennae usually 5-segmented; tarsi 2-segmented during the nymphal stages, but 3-segmented in the adult (cf. family Acanthosomatidae).

Dolycoris baccarum (Linnaeus)
Sloe bug
A minor pest of loganberry, raspberry and strawberry; also sometimes present on damson, pear, wild *Prunus spinosa* and a wide range of herbaceous weeds. This species is both phytophagous and predacious. Palaearctic. Widely distributed in Europe; also reported from North America.

DESCRIPTION
Adult: 11–12 mm long; reddish brown to yellowish brown and pubescent, with an overall purplish sheen; apex of scutellum white. **Egg:** 1.2 × 0.7 mm; barrel-shaped; dirty yellowish white to purplish white. **Nymph:** dark brown to reddish brown and noticeably pubescent; antennae 4-segmented.

LIFE HISTORY
Adults occur from August onwards and again, after hibernation, in the spring. Mating occurs in late June. Eggs are then laid in small batches on the upper surface of leaves of various plants. Following egg hatch, the nymphs feed for several weeks before attaining the adult stage. There are five nymphal instars.

DAMAGE
When present on loganberry, raspberry and strawberry the bugs can impart an obnoxious odour to the fruits, which may then have to be discarded. Also, direct feeding by adults and nymphs on developing fruits can cause malformation; however, this is rarely of importance.

Nezara viridula (Linnaeus) (**13–14**)
Green vegetable bug

A minor pest of citrus. In addition, various herbaceous plants (including cereals and vegetable crops) are also attacked. Widely distributed in southern Europe; currently extending its range northwards, having recently reached southern England. Also present in Africa, America, Asia and Australasia.

DESCRIPTION

Adult: 15 mm long; usually apple-green, but varying in colour to reddish brown; front margin of scutellum usually marked with three small, pale, irregular spots. **Egg:** 1.2 × 0.8 mm; barrel-shaped; white when laid, later becoming pinkish. **Nymph:** mainly green, with prominent purplish and white patches on the abdomen.

LIFE HISTORY

Adults, which form the overwintering stage, become active in spring, and eggs are then laid in batches of about 50 on the underside of leaves. Following egg hatch the first-instar nymphs remain clustered together; they move away to feed on the sap of host plants only after moulting to the next instar. Nymphs pass through a further three instars before attaining the adult stage. Development is slow and the number of generations in a season (no more than two or three in southern Europe) is dependent on temperature.

DAMAGE

On citrus, infestations lead to malformation of developing fruits and fruitlets; piercing of tissue also results in necrotic spotting of the fruit surfaces.

Palomena prasina (Linnaeus) (**15**)
Green shield bug

A minor pest of apple, pear, raspberry and hazelnut, particularly in the close vicinity of broadleaved woodlands. The bug also occurs on many other trees, shrubs and herbaceous plants. Palaearctic. Widely distributed and often common in Europe.

DESCRIPTION

Adult: 12–14 mm long; mainly green, marked with black punctures, with the lateral margins of the pronotum yellowish red; hemelytra with membrane dark brown; antennae 5-segmented, mainly green, but with segments 4 and 5 reddish. **Egg:** green and barrel-shaped, with a circle of short, fine, white setae around the operculum. **Nymph:** mainly dull green, extensively punctured with black; antennae 4-segmented; early instars often marked with black, and with black legs and antennae.

LIFE HISTORY

Adults overwinter in various sheltered situations and reappear in the spring, usually in April or early May. After mating, eggs are laid in groups of about 20 on the leaves of various plants. The eggs hatch about 3 weeks later. Nymphs pass though five instars, individuals reaching the adult stage in about 6 weeks. There is just one generation annually. Young adults are active in late summer or autumn; they then enter hibernation.

13 Green vegetable bug (*Nezara viridula*).

14 Green vegetable bug (*Nezara viridula*) – nymph.

15 Green shield bug (*Palomena prasina*).

DAMAGE

When present on raspberry, the bugs often impart an obnoxious odour and a disagreeable taste to the fruit. On apples and pears, feeding punctures (owing to the injection of toxic saliva) result in distortion and the development of hollows on the fruit surface. On hazelnut, feeding on developing fruits results in deformation of the kernels.

Pentatoma rufipes (Linnaeus) (**16**)

Forest bug

This generally common and widely distributed insect is sometimes a nuisance at harvest-time in cherry orchards and, occasionally, in raspberry plantations, especially in grassy situations. The insects produce a sticky secretion and a powerful, obnoxious smell. As a result, fruit pickers may be disturbed by its presence and the fruits themselves can become contaminated and rendered unpalatable. The mature insect is 12–15 mm long, reddish brown to bronzy brown in colour, with a more or less prominent pale yellow or orange spot on the back, at the apex of the scutellum. Apart from its role as a contaminant, the bug is basically harmless; it is sometimes beneficial to the fruit grower, as it will often feed on pests such as caterpillars.

Family **LYGAEIDAE**
(ground bugs)

Small, mainly phytophagous bugs, sometimes marked with bright red; ocelli present; rostrum 4-segmented; tarsi 3-segmented; scutellum triangular; hemelytra without a cuneus.

Oxycarenus lavaterae (Fabricius) (**17–18**)

This mainly Mediterranean species feeds occasionally on olive, but is also associated with various other trees and shrubs, including *Corylus avellana*, *Tilia* and various species in the family Malvaceae. The bugs tend to aggregate on the young shoots, and are then very conspicuous, but are not known to be damaging. Adults (6–7 mm long) are mainly jet black, with a red and black abdomen; the hemelytra are mainly red, pink and black, with a transparent membrane. This species appears to be increasing in numbers and extending its range within Europe, having been reported in recent years in, for example, Austria and Slovakia.

17 *Oxycarenus lavaterae* – adult.

16 Forest bug (*Pentatoma rufipes*).

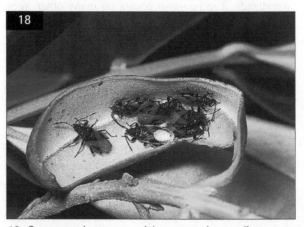

18 *Oxycarenus lavaterae* – adults aggregating on olive.

Family **TINGIDAE**
(lace bugs)

Flattened, entirely phytophagous bugs, without ocelli; pronotum with reticulate, fin-like or wing-like extensions; scutellum usually completely covered by the pronotum; hemelytra with a reticulate, lace-like pattern and the membrane indistinct; tarsi 2-segmented.

Monostira unicostata (Mulsant) (**19–20**)
Almond lace bug
A southern European pest of almond, notably in Portugal and Spain; infestations also occur on pear and *Populus*, including *P. tremula*.

DESCRIPTION
Adult: overall length 3 mm; body mainly black; hemelytra broad, transparent and reticulate, marked with black; antennae 4-segmented. **Egg:** 0.75 × 0.17 mm; elliptical and shiny white. **Nymph:** body whitish and translucent, with brown to black markings; abdomen with numerous spine-like processes (tubercles).

LIFE HISTORY
Overwintered adults become active in the spring. Eggs are then laid singly or in groups, typically inserted into the foliage close to the midrib. They hatch in about a fortnight. Nymphs develop rapidly and pass through five instars before attaining the adult stage. There are up to four generations annually, from May to September.

DAMAGE
On infested leaves, the upper surface becomes chlorotic and speckled with yellow, and the underside covered in black specks of liquid excrement; heavy infestations lead to premature leaf fall and this will have an adverse effect on fruit development. Attacks by nymphs of the summer generations are particularly serious. In addition to causing direct damage, egg-laying punctures also allow pathogenic fungi to invade the foliage.

19 Almond lace bug (*Monostira unicostata*) – infested leaf.

20 Almond lace bug (*Monostira unicostata*) – leaf damage.

Stephanitis pyri Fabricius (**21–23**)

Pear lace bug

An important pest of apple and pear in mainland Europe, but also damaging to cherry and plum; also occurs on other (mainly rosaceous) trees and shrubs, including *Crataegus monogyna*. Widely distributed in the Palaearctic region and particularly common in southern Europe, the range extending through Russia to Japan. An introduced pest in North Africa.

DESCRIPTION

Adult: overall length 3 mm; body 2 mm long and mainly black; pronotum partly yellowish and with a pair of upturned, wing-like lobes laterally, a bladder-like process extending over the head and three dorsal keel-like processes, the central 'keel' large and extending upwards as a fin – these processes mainly transparent and reticulate, partly marked with dark brown; hemelytra broad, transparent and reticulate, with dark-brown patches; legs and antennae brownish grey; antennae 5-segmented. **Nymph:** body whitish and translucent, with black markings; abdomen with numerous spine-like processes (tubercles).

LIFE HISTORY

Overwintered adults appear in early spring following bud-burst. They then congregate on the underside of the leaves and commence feeding. In southern Europe, oviposition begins at the beginning of May, each egg being inserted in the leaf tissue along either side of the midrib. Following egg hatch the nymphs feed for about 3 weeks before moulting into adults, usually in June. There are five nymphal instars. A second brood of nymphs occurs during June and July, and a third in August and September. There are typically three generations annually, adults of the third overwintering in the shelter provided by dead leaves, bark crevices and so on. However, in less favourable parts of the pest's range there are just two generations: first-generation nymphs occurring in June and July, and second-generation nymphs in August and September.

DAMAGE

Infested foliage becomes chlorotic, significantly interfering with photosynthetic activity. The underside of damaged leaves also becomes covered with extensive quantities of excrement that block the stomata and upon which sooty moulds develop. The cast-off skins of nymphs often remain attached to the leaves and these are an immediate clue to the cause of the damage in the absence of live insects.

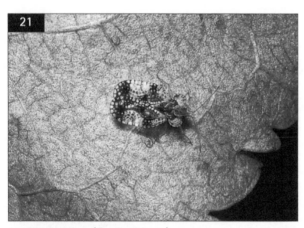

21 Pear lace bug (*Stephanitis pyri*).

22 Pear lace bug (*Stephanitis pyri*) – infested apple leaf.

23 Pear lace bug (*Stephanitis pyri*) – damage to apple leaf.

Family **MIRIDAE** (capsids or mirids)

Active, soft-bodied bugs, with usually elongate hemelytra and long, probing, needle-like mouthparts; rostrum curved and held against the underside of the body when in repose; antennae 4-segmented; tarsi 3-segmented; ocelli absent; hemelytra with a cuneus. Some members are important plant pests, but many species are useful predators; several primarily predatory species are also partly phytophagous and, under certain conditions, may cause damage to fruit crops.

Apolygus spinolae (Meyer-Dür)

A minor pest of hop; also associated with various herbaceous plants, including *Filipendula ulmaria*, *Rubus fruticosus* and *Urtica*. Eurasiatic. Widely distributed in Europe.

DESCRIPTION
Adult: 5.5–6.0 mm long; body mainly green; hemelytra green, with tip of cuneus black; tibiae with prominent black spines.

LIFE HISTORY
Eggs overwinter within the buds and stems of host plants and hatch in the spring. Nymphs then feed for several weeks, reaching the adult stage by late June or early July. There are five nymphal stages, wing buds being noticeable in the final two. Adults occur mainly in July and August, with some individuals still active in September.

DAMAGE
Similar to that caused by hop capsid (*Closterotomus fulvomaculatus*).

Atractotomus mali (Meyer-Dür) (**24**)

Black apple capsid
A partly phytophagous, mainly predatory species that feeds on the eggs and active stages of spider mites and small insects. Although usually of no more than minor pest status on apple and pear in Europe, important damage to fruitlets has occurred in North American apple orchards following the adoption of integrated pest management systems. Holarctic. Locally common in central and northern Europe.

DESCRIPTION
Adult: 3–4 mm long; body and hemelytra relatively broad; dark reddish brown to black; each antenna with the first (basal) segment somewhat triangular and the second segment distinctly swollen, both coated with black hairs. **Nymph:** mainly red, but dark brown anteriorly; body clothed with short, white hairs; antennae as in adult.

LIFE HISTORY
Eggs are laid in groups placed deeply in the young wood of, for example, apple trees and *Crataegus monogyna*. They hatch in the spring. Nymphal development occurs during May and June, and adults are present mainly in June and July. There is only one generation annually.

DAMAGE
Significant damage on fruit trees is restricted to the developing fruitlets, the bugs causing deformation and the development of surface calluses. Effects are particularly severe when feeding has occurred on ovaries during blossom time and on young fruitlets within the first 2 weeks of their development.

Campylomma verbasci (Meyer-Dür)

Mullein capsid
A mainly predatory species, but also partly phytophagous and known to cause damage in apple orchards. Holarctic. Locally distributed in Europe.

DESCRIPTION
Adult: 2.8–3.1 mm long; greyish green to orange yellow; legs with black spines. **Nymph:** stout-bodied and noticeably smaller than those of other apple-inhabiting mirid species; pale bluish green; legs with black spots and black spines; body with black and white hairs.

LIFE HISTORY
Eggs overwinter in young wood and usually hatch in May. There are two generations annually, nymphs of the first attaining the adult stage in June and those of the second becoming adults in August.

DAMAGE
Feeding on apple fruitlets results in distortion and the development of corky surface spots.

24 Black apple capsid (*Atractotomus mali*).

Capsodes sulcatus (Fieber)

A pest of grapevine in parts of south-western Europe (e.g. France and Spain), but associated mainly with herbaceous weeds. Present in the warmer parts of western Europe, particularly in Mediterranean areas.

DESCRIPTION

Adult: 6–7 mm long; mainly black, with yellowish or pale-orange markings; pronotum with a pale central line and with distinctly concave sides; pubescence short; female brachypterous (wings short), but male macropterous (wings fully developed).

LIFE HISTORY

In vineyards, eggs are often laid in crevices in wooden supporting posts. Here they overwinter, eventually hatching in late March or early April. Nymphs then feed for about 2 months before reaching the adult stage. Although usually feeding on weeds, the bugs will also attack the inflorescences of grapevines. Adults are active from late May onwards and often survive into August.

DAMAGE

On grapevines, feeding punctures in rachides and other tissue result in distortion and the appearance of brown or black lesions; flowers are often killed and fruits aborted.

Closterotomus fulvomaculatus (Degeer) (**25**)

Hop capsid

A potentially important pest of hop. Other crops, including apple, pear, peach, currant, raspberry, strawberry and grapevine, are also attacked. Formerly locally common in hop-growing areas and most abundant in 'pole-worked' as opposed to 'wire-worked' gardens. Its pest status in hop gardens, however, has been greatly reduced following the routine use of insecticide sprays against aphids. Widely distributed in Europe; also present in North America.

DESCRIPTION

Adult: 8 mm long; body olive brown, with a yellow pubescence; hemelytra yellowish, suffused with reddish; legs partly speckled with dark brown. **Egg:** 1.4 mm long; banana-shaped; creamy white in colour. **Nymph:** greenish yellow or reddish brown with a yellowish-green stripe down the back; legs more or less speckled with brown.

LIFE HISTORY

Eggs, laid during August in soft, dead wood (such as hop poles), overwinter and hatch towards the end of May. Nymphs then feed on the leaves and shoots of host plants, and become adults in late June or July. This species is very active and individuals immediately scurry away if disturbed.

DAMAGE

Hop: attacked foliage tends to split into numerous, irregular holes, particularly along the major veins; if damaged near their growing points, bines become twisted and scarred and unwanted lateral shoots may then develop. **Pear:** attacks on developing fruitlets can result in the development of stony pits (stony-pit symptom). **Grapevine:** infestations can lead to flower and fruit abortion, resulting in crop losses.

Closterotomus norvegicus (Gmelin) (**26**)

Potato capsid

A polyphagous species that attacks a wide range of herbaceous plants, including certain crops; also partly predacious. Attacks have been reported in southern Europe on peach and in Scandinavia on strawberry. Widely distributed in Europe; also present in Canada.

DESCRIPTION

Adult: 6–8 mm long; mainly green, the pronotum often marked with a pair of black spots and the hemelytra sometimes tinged with reddish brown; hemelytra with

25 Hop capsid (*Closterotomus fulvomaculatus*) – nymph.

26 Potato capsid (*Closterotomus norvegicus*).

membrane dusky. **Nymph:** mainly green to yellowish green, with black hairs.

LIFE HISTORY

Eggs are laid in late July and August in cracks in woody or semi-woody stems of various plants. They hatch in the following May or early June. The nymphs then feed on the buds, growing points, flowers and foliage of plants such as *Trifolium*, *Urtica* and various members of the family Asteraceae, including *Anthemis*, *Carduus*, *Cirsium* and *Matricaria*. Nymphs reach maturity in late June or July. There is just one generation annually.

DAMAGE

Attacks on fruit crops are most likely to occur in weedy sites or adjacent to crops such as carrot, linseed, lucerne and potato. On peach, attacked fruitlets develop cracks from which a sticky gum exudes. Attacked strawberry fruitlets become severely malformed.

Lygocoris pabulinus (Linnaeus) (27–30)
Common green capsid

A frequent pest of fruit crops, including apple, pear, cherry, plum, bush fruits, cane fruits, strawberry and grapevine. Widely distributed and often abundant in Europe; also present in North America and parts of Asia.

DESCRIPTION

Adult: 5.0–6.5 mm long; bright green and somewhat shiny, with a pale, dusky-yellow pubescence; pronotum lightly punctured and with moderate callosities anteriorly; antennae comparatively long, with the fourth, third and apex of the second segment black. **Egg:** 1.3 mm long; banana-shaped, cream, smooth and shiny. **Nymph:** light green to bright green; tips of antennae orange red.

LIFE HISTORY

The winter is passed as eggs inserted in the bark of first- or second-year shoots of currant and other woody hosts. They hatch in April or May, earlier or later according to the region and usually over a period of several weeks. The active nymphs then feed on the young foliage until about early May when, in their second, third or fourth instar, they migrate to herbaceous hosts such as potato, strawberry and various weeds. Here (except on

27 Common green capsid (*Lygocoris pabulinus*).

28 Common green capsid (*Lygocoris pabulinus*) – damage to thornless blackberry.

29 Common green capsid (*Lygocoris pabulinus*) – damage to young apple shoot.

30 Common green capsid (*Lygocoris pabulinus*) – damage to apples.

strawberry, which is a temporary host) they continue their development to the adult stage. Adults are very active and frequently fly in sunshine (cf. those of apple capsid, *Lygocoris rugicollis*). In late June and July, eggs are laid in the stems of herbaceous plants such as *Convolvulus*, *Lamium*, *Rumex*, *Senecio*, *Solanum*, *Taraxacum officinale*, *Urtica* and various other weeds. A second generation of nymphs then develops, reaching adulthood by the autumn. There is then a return migration to woody hosts, where winter eggs are eventually laid.

DAMAGE

Apple: punctured leaves and fruitlets develop small, reddish or brown markings similar to those caused by apple capsid (*Lygocoris rugicollis*); corky scars are also formed on damaged shoots and fruits. **Pear:** small, black punctures appear on the foliage, often in straight lines along former folds, which may turn into brownish-rimmed holes. Attacked fruits develop corky, pitted scars and may become malformed, often severely. **Cherry:** characteristic brown spots, which later develop into holes, appear on the leaves; corky scars may also occur on damaged new shoots. **Plum:** damage is similar to that on cherry, attacks being most frequent on sucker growth. **Currant:** brownish spots appear on the foliage and these later develop into holes. Severely damaged young leaves fail to expand properly and may die. Older leaves are attacked from either side, especially on the basal half near the midrib. Most severe damage is caused by young nymphs of the first generation. Attacks can reduce crop yields and result in blind shoots, with considerable proliferation of lateral growth. Red currant is particularly susceptible to capsid damage. **Gooseberry:** foliage and shoot damage is similar to that caused on currant. In addition, light green patches, which later split, are formed on older leaves. Heavy attacks cause foliage to shrivel and fall off. Fruits are also punctured; they become cracked and scarred, and frequently drop prematurely. **Blackberry:** leaf damage is often severe, characteristic brownish spots being formed; these later develop into holes. In some localities, winter and summer generations occur on this host without a migration to other hosts. Damage by nymphs of the second brood is particularly serious; fruits are also attacked. **Loganberry and raspberry:** leaf and shoot damage is similar to that on currant and gooseberry, although often less noticeable unless infestations are heavy. In some areas, two generations occur on raspberry and there is no migration to alternate, herbaceous hosts. **Strawberry:** damage is limited to the leaves and occurs following immigration of young nymphs from their winter hosts. Attacked foliage becomes speckled with brown and, later, black punctures appear; these eventually develop into holes and are most numerous towards the base of the leaves near the midrib. Nymphs disperse from strawberry in June, before attaining the adult stage; strawberry is, therefore, merely a transitory host. Second-generation nymphs will also feed on strawberry. **Grapevine:** foliage damage is similar to that on currant and gooseberry.

Lygocoris rugicollis (Fallén) (**31**)
Apple capsid

Formerly a major pest of apple; pear, currant and gooseberry are also attacked. Nowadays, however, usually of minor importance as a fruit pest. Widely distributed in Europe, particularly in northern regions.

DESCRIPTION

Adult: 5.5–6.8 mm long; bright green, with front of head and sides of pronotum and hemelytra often yellow; legs yellow; pronotum with transverse wrinkles and prominent callosities. **Egg:** 1.4 mm long; banana-shaped, with the apical half narrowed; cream, smooth and shiny. **Nymph:** yellowish green; tips of antennae reddish brown.

LIFE HISTORY

Eggs are laid in young apple shoots during late June and July. They are sometimes also deposited in the bark of branches and main stems. The eggs hatch in the following spring, usually in advance of those of common green capsid (*Lygocoris pabulinus*). The nymphs then begin feeding on the upper surface of the foliage of new shoots, spurs and blossom trusses. Soon after fruit set, they also puncture the skins of developing fruitlets. The nymphs, especially younger ones, are very active and move about rapidly in sunny weather. There are five nymphal stages. The adult stage is attained in June. The adults remain on the trees throughout July, but rarely fly unless disturbed.

31 Apple capsid (*Lygocoris rugicollis*) – damage to apple fruitlets.

DAMAGE

Apple: the small feeding punctures on apple foliage turn brown and then blackish; they are particularly obvious near the midrib. Badly damaged leaves become deformed and may die. Attacked shoots are stunted and development of secondary, lateral growth can occur and seriously deform young trees. Punctures on fruitlets at first turn brown. However, as fruitlets enlarge, the damaged areas develop into rough, corky patches on the skin. Most fruit damage is caused by fourth- and final-instar nymphs, and takes place before the fruitlets have grown to more than 25 mm across (at an earlier stage than is the case with common green capsid). Attacked fruitlets may remain small and extremely distorted; they may then drop prematurely, but most reach maturity. **Pear:** feeding punctures in developing fruitlets can result in the development of stony pits (stony-pit symptom). **Currant and gooseberry:** leaf damage is characterized by the development of small holes with brown margins.

Lygus rugulipennis Poppius (**32**)
European tarnished plant bug

A polyphagous pest of currant, raspberry and strawberry; infestations may also occur on quince, pear, peach and other fruit trees. Attacks on fruit crops are most likely to occur in weedy sites. Widely distributed and often abundant in Europe.

DESCRIPTION

Adult: 5.0–6.5 mm long; broad-bodied, extremely variable in colour, varying from light green through yellowish brown and reddish brown to black; upper surface pubescent; pronotum with distinct punctures; apex of hind femora ringed with black. **Nymph:** green to brownish, with a pair of black dots dorsally on each thoracic segment.

LIFE HISTORY

Adults overwinter in debris on the ground or in other suitable shelter, emerging in the spring. Eggs are then laid in the buds and stems of herbaceous weeds such as *Rumex*, *Senecio* and *Urtica*. Nymphs feed on various host plants, new adults appearing from July or August onwards. Except in more northerly districts, a larger second brood of nymphs is produced; these nymphs reach adulthood from September onwards.

DAMAGE

General: adults sometimes produce a localized discoloration of leaves and other tissue, with brown or black necrotic spots marking the position of the feeding punctures; attacks on young tissue may also result in leaves or flowers becoming puckered and distorted. Nymphs can cause noticeable damage to young shoots, which then become twisted, swollen and often blind. **Pear:** puncturing of fruitlets results in the development of stony pits (stony-pit symptom). **Peach:** feeding on developing fruitlets results in malformation and the exudation of gum (gummosis); attacks on mature fruits may cause black necrotic spots to appear on the surface. **Strawberry:** foliage damage is of little or no importance, but attacks often result in the development of malformed fruits; in regions where the bug has two generations annually, damage is most likely to occur on autumn-fruiting crops and is particularly prevalent in weedy sites. Symptoms of fruit damage are similar to those resulting from thrips (see Chapter 2) and inadequate pollination; in the case of capsid damage, however, some achenes in distorted parts of the fruits are usually well developed.

Orthotylus marginalis Reuter (**33**)
Dark green apple capsid

This mainly predatory, but also partly phytophagous, species occurs on various trees and shrubs (notably

32 European tarnished plant bug (*Lygus rugulipennis*).

33 Dark green apple capsid (*Orthotylus marginalis*).

Alnus and *Salix*), and is often also present on apple, pear and currant. Although usually of little or no significance as a pest, the bugs are known to cause damage on pear. Widely distributed in northern and central Europe.

DESCRIPTION
Adult: 6–7 mm long; bright green and slender; head and thorax usually with yellowish markings; pronotum shiny and lacking a distinct collar; hemelytra with veins of membrane usually bright green. **Nymph:** green or bluish green, with a distinct, orange abdominal gland dorsally.

LIFE HISTORY
Eggs are laid in young wood during the late summer. Most hatch in the following May. The nymphs feed for several weeks and adults appear from mid-June onwards, surviving until the end of August.

DAMAGE
Attacks on pear fruitlets result in the development of so-called stony-pits (stony-pit symptom).

Plagiognathus arbustorum (Fabricius)
A mainly predatory species, but also partly phytophagous. Often present in orchards and soft-fruit plantations, but associated mainly with *Urtica*. At least in Scandinavia, regarded as a pest of strawberries; minor damage may also occur on raspberry and other cane fruits. Palaearctic. Widely distributed in Europe.

DESCRIPTION
Adult: 4.0–4.5 mm long; light brown or brownish yellow to blackish; thorax and hemelytra with black hairs; basal two antennal segments mainly black; legs with black markings and black spines. **Nymph:** bluish green, with dark bases to the antennae; legs with black markings and black spines.

LIFE HISTORY
Eggs are laid from late summer to early autumn and begin to hatch in the following May, on strawberry typically coinciding with the start of the flowering period. Nymphs occur mainly from May to July, and adults from late June or early July to September.

DAMAGE
As for European tarnished plant bug (*Lygus rugulipennis*).

Plagiognathus chrysanthemi (Wolff)
This widely distributed and often common species causes damage to strawberries, but is far less important than *Plagiognathus arbustorum*; it is also implicated in the development of stony pits (stony-pit symptom) on pear fruits. Adults (3.5–4.0 mm long) are greenish, with a black pubescence and strong black spines on the tibiae. The bugs are particularly associated with wild Asteraceae, including *Achillea millefolium*, *Matricaria maritima* and *Senecio*.

Psallus ambiguus (Fallén) (**34**)
Red apple capsid
Although regarded mainly of importance as a predator, this species is also partly phytophagous. In orchards, fruit damage is reported on both apple (notably in Switzerland) and pear. Widely distributed in Europe.

DESCRIPTION
Adult: 4–5 mm long; dark reddish brown to black; cuneus of each hemelytron mainly reddish, especially basally; antennae black, with a pale tip.

LIFE HISTORY
Eggs are laid singly in young wood during the early summer, most often at the base of buds or at the junctions of shoots and branches. They hatch in the following April or May, earlier than most other related species. Nymphs develop rapidly, reaching adulthood within a few weeks. Adults occur mainly from May to the end of July or early August.

DAMAGE
Attacks on flowers and young fruitlets result in deformation of fruits, the skins of which often develop distinctive, corky cracks and scars.

34 Red apple capsid (*Psallus ambiguus*).

Family **CERCOPIDAE**
(froghoppers)

Small to medium-sized bugs, with the hind legs adapted for jumping; forewings hardened (= elytra); hind tibiae cylindrical and bearing just a few stout spines (cf. family Cicadellidae). Nymphs develop within a protective mass of froth (often called 'cuckoo-spit'), a secretion produced from the anus and through which air bubbles are forced from a special canal by abdominal contractions.

Cercopis vulnerata Illiger in Rossi (**35–37**)
Red & black froghopper
An infrequent and usually minor pest in orchards and hop gardens; most often present in weedy sites. Generally common and widely distributed in Europe.

DESCRIPTION
Adult: 10–11 mm long; black, with bright red patches on the elytra.

LIFE HISTORY
The adult is extremely active, and feeds on a variety of plants during May and June. Eggs are laid on *Rumex* and other herbaceous plants. Later, the pale yellow nymphs feed on the stems and roots within a protective mass of 'spittle' ('cuckoo-spit'). There is just one generation annually.

DAMAGE
In apple orchards, the adults cause characteristic damage to the upper surface of the foliage. These typically angular leaf markings are at first greenish yellow, but eventually turn brownish. There may be several hundred such marks on a single leaf, visible from both sides. This so-called 'angular leaf spot' was once thought to be caused by a fungus which often subsequently develops on the damaged tissue. Similar leaf damage is caused to other fruit crops and to hop. However, on pear, damage is confined to fruit stalks, which become twisted; this sometimes results in malformed fruits. Fruitlet damage has also been recorded on cherry.

35 Red & black froghopper (*Cercopis vulnerata*).

36 Red & black froghopper (*Cercopis vulnerata*) – damage to apple leaf, viewed from above.

37 Red & black froghopper (*Cercopis vulnerata*) – damage to apple leaf, viewed from below.

Philaenus spumarius (Linnaeus) (**38–39**)

Common froghopper

nymph = cuckoo-spit bug

Generally abundant on a wide variety of trees, shrubs and low-growing plants. Frequently a minor pest on strawberry and other fruit crops, including apple, cherry and raspberry. Holarctic. Widely distributed in Europe.

DESCRIPTION

Adult: 5–7 mm long; extremely variable in colour from yellowish, through greenish and brown to blackish, and often distinctly mottled; head bluntly wedge-shaped; eyes prominent; elytra with convex sides; each hind tibia with an apical ring of spines. **Nymph:** plump, whitish to light green and soft-bodied; wing buds yellowish green.

LIFE HISTORY

Adults occur throughout the summer and survive into the autumn. During September, eggs are laid in batches of up to 30 in the tissue of a wide range of host plants. The eggs, which form the overwintering stage, hatch in the spring. Unlike adults, the sap-feeding nymphs are sedentary and coat themselves in a frothy mass of protective 'spittle' ('cuckoo-spit'). The nymphs pass through five instars and reach the adult stage by about June.

DAMAGE

Feeding damage on host plants is usually of little or no significance. However, particularly on crops such as strawberry, the presence of masses of spittle produced by the nymphs may be a nuisance during fruit picking.

38a, b Common froghopper (*Philaenus spumarius*).

39 Common froghopper (*Philaenus spumarius*) – nymph within 'spittle' ('cuckoo-spit').

Family **FLATIDAE** (planthoppers)

Bugs with wings longer than the body and held almost vertically at the sides of the body when in repose; forewings hardened (= elytra), with a dense network of veins, including numerous cross-veins.

Metcalfa pruinosa (Say) (**40–45**)
Citrus planthopper

This North American species first appeared in Europe in the late 1970s, and is now well established in both central and northern Italy, and in southern France; in recent years the pest has also been reported elsewhere in

40 Citrus planthopper (*Metcalfa pruinosa*).

41 Citrus planthopper (*Metcalfa pruinosa*) – young nymph.

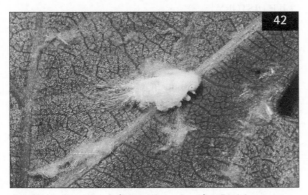

42 Citrus planthopper (*Metcalfa pruinosa*) – final-instar nymph.

43 Citrus planthopper (*Metcalfa pruinosa*) – infested strawberry leaflet.

44 Citrus planthopper (*Metcalfa pruinosa*) – infested fig shoot.

45 Citrus planthopper (*Metcalfa pruinosa*) – infested unfurling foliage of fig.

Europe, e.g. in Switzerland. Infestations occur on a wide range of woody plants, including apple, pear, apricot, peach, plum, grapevine, kiwi fruit, citrus, fig, olive and Chinese persimmon. Crops adjacent to infested wasteland sites are particularly liable to be attacked.

DESCRIPTION
Adult: 5–8 mm long (overall length from head to wing tips); body mainly grey; elytra whitish grey to dark grey, marked with black. **Nymph:** up to 3.2 mm long; whitish to light green, with distinct wing buds and conspicuous tufts of white wax at the hind end of the abdomen.

LIFE HISTORY
This pest has one generation per year and overwinters in the egg stage, the eggs being deposited in the woody parts of host plants. The eggs hatch in May; nymphs then feed gregariously on the leaves and shoot tips, amongst masses of mealy wax. The nymphs often appear somewhat sedentary, but will readily crawl or jump away if disturbed. Adults occur from July to October.

DAMAGE
Heavy infestations cause noticeable discoloration and death of foliage, as well as distortion of the tips of young shoots. In addition, leaves and developing fruits are contaminated by wax and by honeydew upon which sooty moulds may develop. On grapevine, this pest might be a vector of phytoplasmic diseases such as grapevine yellows.

Family **MEMBRACIDAE** (46)

Bugs with 3-segmented tarsi and a hood-like pronotum that extends backwards over the body; antennae very short and with a terminal arista; forewings hardened (= elytra). A mainly tropical family, with few European representatives.

Stictocephala bisonia Kopp & Yonke
Buffalo treehopper
A polyphagous North American pest of trees, shrubs and herbaceous plants; often common on fruit trees, particularly in orchards with a ground cover of leguminous plants. Accidentally introduced into Europe in the early 1900s and now established in several areas, including the Balkan States, France, Hungary, Spain and Switzerland.

DESCRIPTION
Adult: 8–10 mm long; body mainly green; pronotum very large and drawn into four (one dorsal, two lateral and one posterior) thorn-like projections; eyes prominent. **Egg:** 1.5 mm long; oval and white.

LIFE HISTORY
Eggs overwintering beneath the bark of trees and shrubs hatch in late April or early May (but much later in some areas). The nymphs then feed briefly on the trees, imbibing sap from the leaf buds. However, they soon depart to continue their development on herbaceous plants, particularly *Medicago sativa* and other members of the Fabaceae. Here they feed for 2–3 months before becoming adults, usually from July onwards. After mating, the egg-laying females invade nearby trees and shrubs on which they form small oviposition wounds (slits) in the twigs and younger branches, usually 1.5–2.0 m above ground level. Eggs are then inserted side by side beneath the bark, in arc-like groups of about 15. Most eggs are laid from August to October. There is just one generation annually.

DAMAGE
Heavy infestations weaken hosts. Also, by the spring, the slit-like oviposition wounds will have developed into open, oval-shaped lesions, and these often predispose trees to infection by fungal pathogens. The pests themselves may also accidentally introduce such pathogens.

46 A typical membracid.

Family **CIXIIDAE**

A small family of planthoppers. Antennae very short, with a terminal arista; wings large and more or less membranous, with strongly demarcated veins; first and second anal veins (veins 1A and 2A) joined distally.

Hyalesthes obsoletus Signoret (**47**)

A pest of grapevine in various parts of mainland Europe, particularly in dry, hot and sunny, weedy sites.

DESCRIPTION
Adult female: 4–5 mm long; body mainly dark brown to brownish black, with outlines of head, thorax and abdominal tergites partly creamy white; tip of abdomen with a distinct pad of white wax; wings translucent, with brown veins and a distinct stigma. **Adult male:** 3–4 mm long; similar in appearance to female, but body truncated, noticeably tapered posteriorly and lacking a waxen pad.

LIFE HISTORY
Although adults commonly occur on grapevine, they do not breed on this host. Instead, eggs are laid (and nymphs then feed) on the roots of herbaceous plants, including *Artemisia vulgaris*, *Convolvulus arvensis*, *Ranunculus bulbosus*, *Solanum nigrum* and *Urtica dioica*. The nymphs pass through five instars, overwintering in the fourth. Adults occur from late May or early June to July or August. There is just one generation annually.

DAMAGE
This pest is not directly damaging to grapevines, but the adults are of importance as vectors of grapevine bois noir (stolbur) phytoplasma.

Family **CICADELLIDAE**
(leafhoppers) (**48–49**)

Small, very active plant bugs, with the hind legs adapted for jumping and, characteristically, bearing two rows of fine spines (cf. family Cercopidae: froghoppers); forewings hardened (= elytra), but although often tinted or partly patterned, frequently more or less transparent. Leafhoppers feed on non-vascular leaf tissue, usually from the lower surface. They cause damage by puncturing the cells and withdrawing the sap; this often produces a flecking or silvering of the foliage, visible from above and from below. In addition, leaves and other surfaces may be contaminated by liquid excrement. Nymphs, often accompanied by cast-off skins of earlier instars, occur mainly on the underside of leaves and scurry away if disturbed.

Reliable identification of most species involves examination of wing venation and genitalia, which is a specialist task; some species, however, are distinguishable by characteristic markings on the body or the elytra.

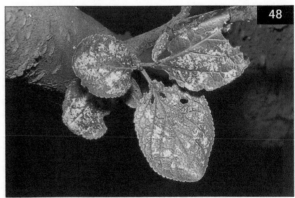

48 Leafhopper damage on plum leaves.

47 *Hyalesthes obsoletus* – female.

49 Leafhopper damage on grapevine leaf.

The following are examples of the large number of species that occur on fruit crops in Europe. Notably on grapevine, several species (not all of which necessarily breed on the crop) are vectors of phytoplasmas – disease-causing organisms, formerly known as MLOs (mycoplasma-like organisms); examples are included amongst the range of leafhoppers considered here.

Aguriahana stellulata (Burmeister) (**50**)
Cherry leafhopper
A minor pest of cherry; also associated with pear, plum and *Tilia*. Holarctic. Widely distributed in Europe.

DESCRIPTION
Adult: overall length (from head to wing tips) 3.9–4.4 mm; head distinctly narrower than pronotum; head, pronotum and scutellum whitish; elytra white, with prominent blackish-brown markings.

LIFE HISTORY
This attractive-looking species is often abundant on cherry and plum trees. Nymphs feed on the underside of the leaves during the spring and summer months, adults occur from July to October. This species also breeds on pear, but in lesser numbers.

DAMAGE
Although reported as causing severe leaf mottling on cherry, this leafhopper is usually of little or no economic importance; it does not appear to be harmful to pear or to plum.

Alnetoidia alneti (Dahlbom) (**51**)
A fruit tree leafhopper
Often abundant on apple; also breeds on cherry, plum and various other kinds of tree, including *Alnus* and *Corylus avellana*. Present in most of Europe; also present in Japan.

DESCRIPTION
Adult: overall length 3.5–4.0 mm; mainly yellow to whitish yellow.

LIFE HISTORY
This leafhopper overwinters in the egg stage. The eggs hatch from mid-May to the end of June; the nymphs then feed on the underside of the leaves during June and July, passing through five instars. Adults occur from July to mid-September or October, there being one generation annually. Eggs are laid under the bark of smaller branches and twigs from September onwards.

DAMAGE
Infested leaves are flecked with areas of pale tissue and, when the pest is numerous, foliage and fruit may become contaminated by liquid excrement.

50a, b Cherry leafhopper (*Aguriahana stellulata*).

51 A fruit tree leafhopper (*Alnetoidia alneti*) – female.

Aphrodes bicinctus (Schrank) (52)
Strawberry leafhopper

A generally common species, responsible for introducing and transmitting green petal virus disease of strawberry. Widespread in Europe. Most numerous in damp situations.

DESCRIPTION
Adult: 5–8 mm long; colour extremely variable, ranging from greyish or greenish yellow (peppered with black) to light or dark brown, reddish brown or black; body short and broad.

LIFE HISTORY
Adults occur on clover, strawberry and various other plants during the summer months. Particularly in drought conditions, they often migrate in numbers from *Trifolium* to irrigated strawberry fields, where they feed on the bases of the petioles in the sheltered crowns of the plants.

DAMAGE
Direct feeding damage is unimportant. However, this insect is regarded as the main vector of green petal virus disease on strawberry and is responsible for introducing infections to this crop. Incidence of the disease is most likely to increase under dry weather conditions following the migration of infective individuals from clover into irrigated strawberry fields.

Cicadella viridis (Linnaeus) (53–54)
A minor pest of grapevines and young fruit trees, including apple, pear, cherry, peach, plum and mulberry. Palaearctic. Widely distributed in Europe and often abundant in damp, grassy habitats.

DESCRIPTION
Adult female: overall length 5.5–9.0 mm; body robust; elytra mainly emerald green, with a white costa, the

53a, b *Cicadella viridis* – female.

52 Strawberry froghopper (*Aphrodes bicinctus*).

54 *Cicadella viridis* – male.

veins bordered narrowly with black; head brownish to yellowish, with a pair of black spots between the eyes; pronotum and scutellum yellowish green; abdomen bluish black. **Adult male:** similar to female, but smaller; elytra usually bluish black, but occasionally green.

LIFE HISTORY

In temperate parts of Europe, eggs are laid from the end of August to early November. They are deposited in groups of about 10, usually in plants such as *Agropyron repens*, *Erigeron canadensis*, *Juncus*, *Phragmites communis* and *Scirpus lacustris*, but sometimes also in the stems of bushes and other woody plants. The eggs hatch in the spring. Nymphs reach the adult stage by the end of June or in July. In warmer parts of Europe there are two, and in particularly favourable regions three, generations annually.

DAMAGE

Egg-laying in the shoots or stems of grapevines and young fruit trees results in the development of cankers on the bark. The insect is also a vector of the bacterium *Xylella fastidiosa*, the causal agent of Pierce's disease which develops in the xylem of plants and is known to be lethal to grapevines.

Edwardsiana crataegi (Douglas) (**55**)

A fruit tree leafhopper

A widespread and common pest of fruit trees, including apple, pear, cherry, damson and plum; wild hosts include *Crataegus monogyna* and *Sorbus aucuparia*. Often one of the more important leafhoppers in orchards. Present in most of Europe. Also found in North America and Australasia.

DESCRIPTION

Adult: overall length 3.5–3.9 mm; pale yellow to greenish white. **Nymph:** whitish to greenish white, and translucent.

LIFE HISTORY

This species overwinters as eggs laid in the autumn beneath the bark of twigs and smaller branches of fruit trees. Eggs hatch from mid-April to early May. Nymphs then feed until June, July or August. The new adults appear in July or August, laying eggs in or near the midrib on the underside of leaves. Second-brood nymphs occur from July or August onwards and become adults in the autumn.

DAMAGE

Feeding causes a distinct mottling of the upper surface of leaves, the importance of which depends upon the level of infestation. Damage tends to be most serious under dry conditions. This species is also known to attack, and to cause russeting on, developing plum fruitlets.

Edwardsiana prunicola (Edwards) (**56**)

A fruit tree leafhopper

A minor pest of currant (the only leafhopper known to breed on black, red and white currant); also associated with cane fruit and with fruit trees, including apple, damson, peach and plum. Present throughout most of Europe; also present in North America.

DESCRIPTION

Adult: 4 mm long; pale yellow to whitish yellow. **Nymph:** whitish.

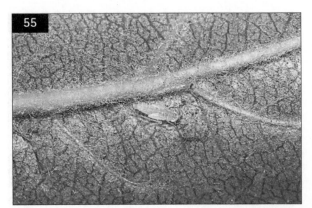

55 A fruit tree leafhopper (*Edwardsiana crataegi*) – nymph.

56 A fruit tree leafhopper (*Edwardsiana prunicola*).

LIFE HISTORY

Nymphs feed in spring and summer. Adults are present from mid-summer to late autumn, and are often abundant in the autumn on damson and *Prunus spinosa*.

DAMAGE

In common with other species, adults and nymphs cause mottling of foliage.

Edwardsiana rosae (Linnaeus) (**57–60**)

Rose leafhopper

Often abundant on a wide range of hosts, including apple, cherry, plum, peach, gooseberry (rarely a host of leafhoppers), blackberry, raspberry, strawberry, hop and hazelnut; sometimes also present as a minor pest on fig (e.g. in southern France and Spain). The primary host, however, is *Rosa*. Widely distributed in Europe; also present in North America and parts of Asia.

DESCRIPTION

Adult: overall length 3.4–4.0 mm; mainly whitish to pale yellow. **Nymph:** whitish.

LIFE HISTORY

Eggs are laid in autumn under the epidermis of young shoots on *Rosa*. These eggs hatch in the following spring and the nymphs then feed for 4–6 weeks before becoming adults. A migration of adults to summer hosts, including fruit trees, takes place in June. On apple, for example, summer eggs are then laid in leaf tissue within or near the major veins. These eggs hatch from early July onwards and the nymphs feed on the underside of the leaves. They reach adulthood in mid-August or September and eventually return to *Rosa* before leaf fall. In some situations, a primary host (i.e. *Rosa*) may not feature in the life cycle.

DAMAGE

If numerous, this species causes considerable leaf flecking and silvering. In severe cases, premature leaf fall may also result. The leafhoppers are often common on strawberry, but they cause only slight damage and do not act as virus vectors. On hazelnut, infested leaves develop small, ragged holes.

57 Rose leafhopper (*Edwardsiana rosae*).

58 Rose leafhopper (*Edwardsiana rosae*) – early-instar nymph.

59 Rose leafhopper (*Edwardsiana rosae*) – nymphs.

60 Rose leafhopper (*Edwardsiana rosae*) – damage to fig leaf.

Empoasca decipiens Paoli (**61**)

A green leafhopper

Common on a wide variety of plants, including apple, cherry, plum, peach, gooseberry (rarely a host of leafhoppers) and hazelnut. Widely distributed in central and southern Europe; also present in North Africa and western Asia.

DESCRIPTION

Adult: overall length 3.5–4.1 mm; body greenish, with whitish marks on the head and thorax; elytra greenish, with apical part of the central cell clear and colourless; eyes purplish.

LIFE HISTORY

Adults overwinter in the shelter of evergreen plants, including *Hedera helix*, reappearing in the spring. In June, eggs are laid on the underside of the leaves of host plants. They hatch in about 2 weeks and the nymphs feed for about 5 weeks, passing through five instars before reaching adulthood. The nymphs can be very active and immediately scurry over the foliage if disturbed. Unlike the adults, however, they do not jump. Eggs of a second brood are laid in August. Under favourable conditions, there may be at least a partial third generation.

DAMAGE

Attacked leaves become flecked and sometimes partly necrotic; symptoms on hosts such as apple and plum may also resemble those of silver-leaf disease. On hazelnut, attacked leaves become peppered with small, ragged holes.

Empoasca vitis (Göthe) (**62**)

A green leafhopper

A pest of various crops, including apple, cherry, plum, blackberry, grapevine, hop and hazelnut. Widespread in Europe.

DESCRIPTION

Adult: overall length 3.1–4.0 mm; body green to whitish; elytra greenish, with the whole of the central cell clear and colourless; eyes purplish.

LIFE HISTORY & DAMAGE

As for *Empoasca decipiens*.

Euscelis incisus (Kirschbaum)

This yellowish-brown, broad-bodied species, which is about 5 mm long, occurs commonly in meadows and on strawberry, but appears to do no harm. Although known to be an effective virus vector on *Trifolium* and capable of transmitting green petal virus from infected strawberry plants to other hosts, it is apparently unable to introduce the disease to strawberry. The same is true of *Conosanus obsoletus* (Kirschbaum), a slightly larger and paler (greyish-yellow) species, which also feeds on *Trifolioum* and strawberry.

Evacanthus interruptus (Linnaeus)

Hop leafhopper

Locally considered a pest of hop, particularly in gardens with chestnut supporting poles. Often common in Europe on a wide range of plants.

61 A green leafhopper (*Empoasca decipiens*).

62 A green leafhopper (*Empoasca vitis*).

DESCRIPTION
Adult: 5.0–6.5 mm long; body robust; bright yellow, with black, elongate markings on the elytra. **Egg:** 2.0 × 0.5 mm; creamy white and shiny. **Nymph:** greenish to pale yellow.

LIFE HISTORY
Egg laying takes place during the late summer and early autumn in dead wood, including stumps, posts, fences and wooden support poles in hop gardens. The eggs hatch in the following spring. Nymphs feed in May and June on the underside of leaves of hop and many other plants. They also feed near the tips of hop bines, but do not attack the developing burrs (strobili). Adults appear in July and may be found in hop gardens until September.

DAMAGE
Infested foliage turns yellow or brown, and may split and curl. Bines fail to spiral, their growth is checked and lateral shoots are produced. When infestations are heavy, crop yields are significantly reduced.

Fieberiella florii (Stål)
Associated with various plants, including apple, cherry and apricot, and a well-known vector of phytoplasmic diseases. Widely distributed in central Europe. An introduced pest in the USA.

DESCRIPTION
Adult: 6–7 mm long; mainly dark reddish brown to golden brown, with the basal abdominal segments whitish; head triangular, being noticeably narrowed anteriorly; body robust. **Nymph:** bright green to yellow, finely speckled with black; abdomen with a reddish, spine-like tip.

LIFE HISTORY
The nymphs feed on the foliage of a wide range of broadleaved trees and shrubs from about May onwards, and usually reach the adult stage by the end of June. Adults occur mainly from July to September, and then overwinter. Eggs are deposited in the spring.

DAMAGE
This pest is especially important as a vector of, for example, the apple proliferation (AP) phytoplasma (that causes witches' broom, and other symptoms, on apple) and chlorotic leaf roll disease on apricot.

Jacobiasca lybica (Bergevin & Zanon)
Burning grape leafhopper
This African species is often a common pest of grapevines in hot Mediterranean areas, e.g. southern Portugal and southern Spain.

DESCRIPTION
Adult: 2.6–2.9 mm long; greenish yellow, with distinct white spots on the head, pronotum and scutellum.

LIFE HISTORY
Adults overwinter on trees such as *Alnus*, *Malus* and *Quercus*. In May they migrate to summer hosts, upon which feeding and breeding takes place. Although regarded as polyphagous, this species often shows a preference for grapevine. Eggs are laid in the leaf veins and petioles. The eggs hatch about 2 weeks later. Nymphs then feed for about 3 weeks before attaining the adult stage, each passing through five instars. There are typically up to five overlapping generations annually.

DAMAGE
Leaves become distinctly sickly in appearance, the damage mirroring that caused by, for example, mineral deficiencies, mites and viruses. Eventually, the leaf margins become desiccated; the leaves then appear scorched (resembling symptoms of sun scorch) and often fall off prematurely. If developing grapes are then exposed to full sun they may be scalded. Infested plants are weakened, and yields of grapes are adversely affected in terms of both quality and quantity. Attacks in one year will also have a deleterious effect on the following year's crop.

Oncopsis alni (Schrank) (**63**)
Although breeding only on *Alnus*, this species is recorded as a secondary pest of grapevines. Widely distributed and generally common in Europe.

63 *Oncopsis alni* – female.

DESCRIPTION

Adult female: 5.5–6.0 mm long; head and pronotum yellowish brown, with darker markings; scutellum reddish brown, marked with black or dark brown; abdomen (dorsally) brown to blackish, tinged with red; elytra mainly translucent, suffused with brown; venation mainly brown, but with part of the inner marginal vein distinctly white; legs mainly pale, with inner and outer side of hind tibiae striped with black. **Adult male:** similar to female, but with darker markings; abdomen (dorsally) black, with hind margin of tergites reddish.

LIFE HISTORY

There is just one generation annually, with nymphs feeding in the spring and adults appearing in the summer.

DAMAGE

Although not directly harmful, in vineyards adults of this species are vectors of grapevine yellows.

Ribautiana debilis (Douglas) (**64**)

A fruit tree leafhopper

A pest of blackberry and loganberry; also reported on apple and on various other trees. Locally distributed in Europe, including the British Isles, France and Italy.

DESCRIPTION

Adult: overall length 3.2–3.5 mm; elytra with basal two-thirds mainly pale yellowish, the apical third slightly dusky, and tips of apical veins slightly darkened; body mainly whitish to yellowish, with (at least in female) two distinct black spots between the face and vertex; in female, scutellum also often dark apically and pronotum often with a median black spot on the anterior margin.

LIFE HISTORY

Eggs overwinter on *Rubus* hosts and hatch in the spring. Nymphs then feed on the underside of the leaves, reaching the adult stage in June. A summer generation occurs on a wider range of hosts. Adults may remain active on blackberry and loganberry well into the late autumn and early winter.

DAMAGE

Infested leaves are mottled extensively with silvery markings, and photosynthetic activity is reduced.

Ribautiana tenerrima (Herrich-Schaeffer) (**65**)

Loganberry leafhopper

An often common pest of *Rubus*, including cultivated blackberry and loganberry. Also associated with various trees and shrubs, including apple, plum and hop. Nearctic. Widely distributed in Europe; also present in Australasia.

DESCRIPTION

Adult: overall length 3.1–3.5 mm; elytra with basal two-thirds mainly bright yellow, the apical third transparent, with a large, dusky, triangular patch over the cross-veins and distinct black marks at the tips of three of the four apical veins; body creamy white to yellowish, the abdomen mainly blackish dorsally.

LIFE HISTORY & DAMAGE

As for *Ribautiana debilis*.

64 A fruit tree leafhopper (*Ribautiana debilis*).

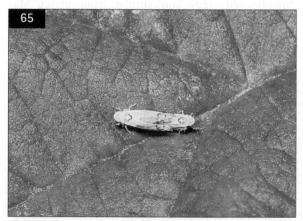

65 Loganbery leafhopper (*Ribautiana tenerrima*) – adults in copula.

Scaphoideus titanus (Ball) (**66–68**)
Vine leafhopper
An introduced North American pest of grapevine; now established in various parts of mainland Europe, including France, Italy, Spain and Switzerland.

DESCRIPTION
Adult: overall length 4.5–5.0 mm; elytra yellowish brown, with dark brown markings. **Nymph:** pale yellowish white, with dark markings and a pair of dark brown spots at the tip of the abdomen; abdomen armed with distinctive spines, and noticeably tapered posteriorly; younger nymphs whitish to pale yellowish white, with a pair of dark brown spots at the tip of the abdomen.

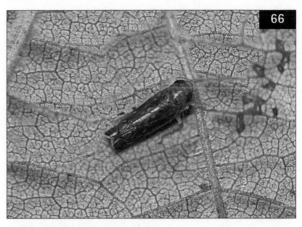

66 *Scaphoideus titanus* – adult.

LIFE HISTORY
Eggs overwinter on the bark of host plants and hatch over an extended period, from mid-May to July. Nymphs, which are very active and jump away if disturbed, feed for about 5 weeks before attaining the adult stage. There are five nymphal instars. Adults occur from late July to September; there is just one generation annually.

DAMAGE
Although not directly harmful to grapevine, this leafhopper is an important vector of grapevine flavescence dorée phytoplasma; however, it is not a vector of grapevine bois noir (stolbur) phytoplasma.

67 *Scaphoideus titanus* – final-instar nymph.

68 *Scaphoideus titanus* – young nymphs on grapevine.

Typhlocyba quercus (Fabricius) (**69**)

A fruit tree leafhopper

An often abundant species on *Quercus*; also infests a wide range of other hosts, including apple, pear, cherry, damson, peach, plum, strawberry, hop and hazelnut. Nearctic. Present in most of Europe.

DESCRIPTION

Adult: overall length 3.0–3.5 mm; elytra creamy white, with characteristic yellowish-orange or brick-red and brownish-green markings; pronotum and scutellum creamy white, more or less marked with orange or brownish-green patches.

LIFE HISTORY

This species overwinters in the egg stage and is single brooded. Nymphs occur in the spring and early summer and adults from July to September or October.

DAMAGE

On most crops the pest occurs in relatively small numbers. However, it may be sufficiently numerous on damson and plum during the summer, in company with other species of leafhopper, to cause extensive mottling of leaves and premature defoliation. In addition, excretions voided by the leafhoppers sometimes cause foliage to become sticky and may allow sooty moulds to develop.

Zygina flammigera (Geoffroy in Fourcroy) (**70**)

A fruit tree leafhopper

An often common, but minor, pest of apple, cherry and plum. Nearctic. Widely distributed throughout most of Europe.

DESCRIPTION

Adult: overall length 3.0–3.4 mm; yellow to whitish; prothorax usually marked with two red, longitudinal stripes; elytra each with a long, red, zigzag-shaped stripe.

LIFE HISTORY

Adults overwinter in the shelter of evergreens. In late spring they migrate to summer hosts, including fruit trees, where eggs are laid. Nymphs occur from June onwards, but are most numerous from early July to August. Young adults appear from mid-July onwards and may remain on the trees until leaf fall. Under favourable conditions there may be at least a partial second generation.

DAMAGE

As for *Alnetoidia alneti*.

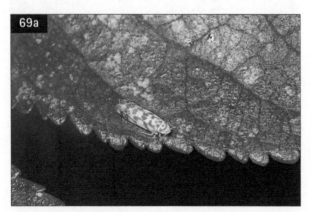

69a, b A fruit tree leafhopper (*Typhlocyba quercus*).

Zygina rhamni (Ferrari)

This widely distributed European species is a pest of grapevine, especially in northern Italy, with typically two to three generations annually. Adults (3.0–3.5 mm long) are mainly ochreous yellow, with a reddish-orange zigzag-shaped stripe along each elytron. Heavy infestations have a deleterious effect on photosynthetic activity.

Family **PSYLLIDAE**

Very active bugs (commonly called 'suckers'), with relatively large, membranous wings; wing venation is prominent and without cross-veins; hind legs are enlarged and modified for jumping. Eggs of psyllids often have a filamentous (tail-like) pedicel and a ventral (beak-like) peduncle. Nymphs are flat and scale-like, with conspicuous wing buds (pads) in the later instars. The usually slow-moving nymphs frequently occur in tight, more or less sedentary groups, and often produce copious quantities of white wax and masses of sticky honeydew.

Cacopsylla mali (Schmidberger) (**71–74**)
Apple sucker

A serious pest of apple, but confined mainly to unsprayed trees. Palaearctic. Widely distributed and often abundant in Europe; an introduced pest in North America.

71 Apple sucker (*Cacopsylla mali*).

70 A fruit tree leafhopper (*Zygina flammigera*).

72 Apple sucker (*Cacopsylla mali*) – eggs.

DESCRIPTION

Adult: 2.5–3.0 mm long; apple-green to yellowish green (in spring and summer) or brownish orange to chestnut-red (in autumn); wings transparent, with veins of autumn specimens dark. **Egg:** 0.4 mm long; elongate-oval and creamy yellow, with a very short, filamentous pedicel and a thick ventral peduncle. **Nymph:** apple-green, with red eyes; body flattened, broad and ovate, with conspicuous wing buds in later instars; youngest nymphs yellowish to pale golden brown, with brownish legs.

LIFE HISTORY

This species overwinters in the egg stage. The eggs hatch over an extended period in the spring, but usually in April or early May. Young nymphs then crawl to the opening or still unopened buds and, as soon as they can gain access, begin feeding on the sap by piercing the inner tissue with their mouthparts. The nymphs also feed in the blossom trusses and leaf buds. They are comparatively inactive and secrete white iridescent waxen threads; they also expel sticky globules of liquid honeydew from the hind end of the body, giving them a very characteristic appearance and at once betraying their presence. When fully fed, usually from late May onwards, each nymph anchors itself to a leaf with its mouthparts; the nymphal skin then splits and the adult insect emerges. Adults occur on the trees throughout the summer and autumn. They are very active and fly readily when disturbed. Mating usually takes place in August or September. Eggs are then laid on the spurs, mainly along leaf scars or amongst empty egg shells of previous generations. Less frequently, eggs are deposited at the base of leaf buds and at random on the shoots. In some seasons, egg laying continues into October or November. There is just one generation annually.

DAMAGE

Nymphs cause considerable damage during the pre-blossom period, petals on partially opened buds turning brown; flowers of seriously infested trusses may all be killed within a matter of days, often before an infestation has been noticed. Such damage is often mistaken for that caused by frost. Damage to foliage later in the spring is less serious. Adults are probably harmless.

Cacopsylla melanoneura Förster (**75**)

Hawthorn sucker

Although associated mainly with *Crataegus*, infestations of this often abundant species are also reported in parts of mainland Europe on apple and pear. Widely distributed in Europe. Also found in Japan.

DESCRIPTION

Adult: 2.5 mm long; body extremely variable in appearance, ranging from reddish brown to purplish, brown or black; thorax with whitish longitudinal lines

73 Apple sucker (*Cacopsylla mali*) – nymph.

74 Apple sucker (*Cacopsylla mali*) – damage to young apple shoot.

75 Hawthorn sucker (*Cacopsylla melanoneura*) – nymph-infested shoot.

dorsally. **Nymph:** greenish, with yellowish-orange or blackish markings and a pale longitudinal line dorsally.

LIFE HISTORY

Adults overwinter on host plants or, more frequently, on non-host trees (such as conifers). They become active early in the year, populations in orchards in the warmer parts of Europe often reaching a peak from mid-February to mid-March. Eggs are then deposited, and they hatch shortly afterwards. Nymphs, which feed on the young shoots during April or May, produce considerable quantities of honeydew, and also secrete long threads of wax which are especially noticeable in dry conditions. Nymphs complete their development within a few weeks, with new adults appearing from May onwards. There is just one generation annually. In cooler regions, activity and development is delayed, and new adults do not appear until June or July.

DAMAGE

Direct damage on fruit trees is usually of little or no significance, but the insects are known to be vectors of both apple proliferation (AP) phytoplasma and pear decline.

Cacopsylla pyri (Linnaeus) (**76**)

A widely distributed pest of pear in mainland Europe, including Scandinavia. Eurasiatic.

DESCRIPTION

Adult: 2–3 mm long; orange red to blackish, with white, longitudinal stripes on the thorax dorsally; forewings transparent (but often somewhat smoky apically), with dark veins. **Egg:** 0.3 mm long; elongate-oval and yellowish orange, without a pedicel (cf. other pear-infesting psyllids) and with a very short ventral peduncle. **Nymph:** yellowish, with

purplish-red eyes; later instars purplish to reddish brown, with blackish markings and whitish longitudinal stripes on the head and body; forewing pads each with a single marginal capitate seta.

LIFE HISTORY

This psyllid overwinters in the adult stage, sheltering on the bark of host plants. The adults become active in early spring, from the beginning of March onwards. Eggs are then deposited in crevices on the bark of twigs and spurs of host trees. The eggs hatch about 3 weeks later and, from early spring onwards, first-brood nymphs feed on the twigs and shoots. There are up to three or more generations annually, depending on temperature. Overwintering adults are usually produced in November and early December.

DAMAGE

As for pear sucker (*Cacopsylla pyricola*).

Cacopsylla pyricola Förster (**77**)
Pear sucker

An important pest of pear. Widely distributed in Europe and often common throughout the Palaearctic region, its range extending to Japan; an introduced species in North America.

DESCRIPTION

Adult: 1.5–2.0 mm long; dark brownish black, the head and thorax marked with reddish brown; head very broad; abdomen short and tapering; wings of the summer form (= *typica*) mainly transparent, but forewings of the winter form (= *simulans*) with several darkly suffused areas between the veins. **Egg:** 0.3 mm long; elongate-oval, with a long pedicel and a relatively short ventral peduncle; pale lemon-yellow, becoming orange just before hatching. **Nymph:** orange to pinkish, with darker

79 *Cacopsylla pyri* – nymphs on pear shoot.

77 Pear sucker (*Cacopsylla pyricola*) – nymph and excreted honeydew.

markings (nymphs producing the overwintering generation of adults are darkest), the youngest nymphs pale orange, with pinkish-red eyes; forewing pads each with three to five marginal capitate setae.

LIFE HISTORY

Pear suckers overwinter as adults, resting on the trunks, branches and spurs of host trees. They may also occur in hedgerows and in other suitable sheltered situations in the vicinity of pear orchards. The adults become active in sunny conditions and will often fly around, even in mid-winter. Eggs are laid on pear shoots and spurs during the early spring, attached by the ventral peduncle; they hatch from late March to about petal fall. Nymphs feed on the opening buds and in the blossom trusses, piercing the tissue with their mouthparts and imbibing the sap. The adult stage is reached several weeks later, after five nymphal instars. There are several generations annually (up to six in particularly favourable districts), populations (if unchecked) becoming progressively larger as the season develops. Eggs of the summer broods are normally deposited along the midrib on the upper surface of leaves and often occur in considerable numbers. The nymphs excrete large quantities of honeydew, attracting many insects such as ants, bees, wasps and various flies.

DAMAGE

Heavy infestations lead to the development of small, misshapen fruits and premature leaf fall or fruit drop. Blossoms attacked by first-generation nymphs turn brown and die; nymphs feeding later in the season may affect the following season's pear crop by weakening or even killing the developing fruit buds. The succulent tissue at the tips of new shoots is particularly liable to be attacked, and tree growth can be severely checked. Sooty moulds grow profusely on honeydew excreted by the nymphs, often coating the shoots, foliage and developing fruitlets in an unsightly and unpleasant black layer. At harvest, contaminated fruits may have to be washed or wiped before consumption, sale or storage.

Cacopsylla pyrisuga Förster (**78–79**)
Large pear sucker
A pest of pear. Eurasiatic. Widely distributed in central and northern Europe (including Scandinavia); found only once in the British Isles, where it is not a pest.

DESCRIPTION

Adult: 3.5–4.0 mm long; body green, with head and thorax more or less yellow (spring/summer individuals), darkening to rusty red in autumn. **Egg:** unlike those of other pear-infesting psyllids, without a pedicel; ventral peduncle very short. **Nymph:** mainly green, more or less marked with black, but otherwise similar to that of *Cacopsylla pyricola*.

LIFE HISTORY

Essentially similar to that of pear sucker (*C. pyricola*), the adults overwintering, but with just one generation annually.

DAMAGE

As for pear sucker (*C. pyricola*).

78 Large pear sucker (*Cacopsylla pyrisuga*).

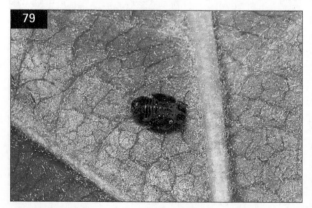

79 Large pear sucker (*Cacopsylla pyrisuga*) – nymph.

Euphyllura olivina (Costa) (**80–82**)
Olive sucker
This pest occurs only on olive. Present throughout the Mediterranean area, the distribution extending to central Asia.

DESCRIPTION
Adult: 2.0–2.5 mm long; dull greenish grey to brownish grey; forewings partly clouded and speckled with yellowish brown. **Egg:** 0.3 mm long; creamy white. **Nymph:** flat bodied; dull whitish green, with purple eyes; abdomen with two pairs of large, black, wax glands.

LIFE HISTORY
Adults overwinter in sheltered situations, e.g. in cracks and crevices, on the trunks of olive trees. In spring, they reappear and eggs, up to 100 per female, are then deposited on the developing shoots. The eggs hatch 1–2 weeks later. The nymphs, which secrete waxen threads and also excrete considerable quantities of honeydew, feed for up to 5 weeks before attaining the adult stage. There are typically three generations annually, nymphs of the first feeding from March onwards on young leaves and those of the second feeding from May onwards on flower buds and amongst the developing flower clusters. Adults of the second generation aestivate under hot, summer conditions (temperatures in excess of 27°C), usually resuming activity in September. The final generation of nymphs eventually gives rise to the winter adults.

DAMAGE
In addition to direct damage, accumulations of honeydew (upon which sooty moulds develop) and wax have an adverse effect on growth and may cause buds and flowers to abort. Most significant damage is caused by nymphs of the second generation.

Psylla costalis Flor
Summer apple sucker
A minor pest of apple in parts of mainland Europe.

DESCRIPTION
Adult: 3 mm long; initially yellowish green or light green, but eventually turning reddish brown. **Egg:** yellowish. **Nymph:** green.

LIFE HISTORY
Unlike the more widely known apple sucker (*Cacopsylla mali*), this species overwinters in the adult stage. In spring, eggs are laid on the unfurled leaves, on leaf petioles and on the underside of expanded leaves. These eggs hatch from the end of the blossom period onwards. Nymphs then feed and reach the adult stage in the summer. These adults eventually overwinter, there being just one generation annually.

DAMAGE
Infested foliage is coated in honeydew, upon which sooty moulds develop. The pest is also a vector of apple proliferation (AP) phytoplasma.

71 Olive sucker (*Euphyllura olivina*).

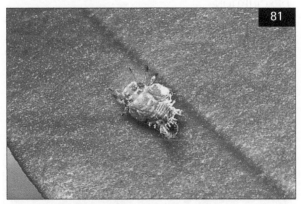

72 Olive sucker (*Euphyllura olivina*) – nymph.

73 Olive sucker (*Euphyllura olivina*) – infested shoot tip.

Family **CARSIDARIDAE**

A small family of psyllids, mostly of tropical distribution, with characteristically flattened, noticeably hairy antennae.

Homotoma ficus (Linnaeus) (**83–85**)
Fig sucker

This minor, but often common, pest occurs locally on wild and cultivated fig in various parts of southern Europe. Also introduced on ornamental *Ficus* to more northerly areas, including Austria, England, France, Jersey and Switzerland.

DESCRIPTION
Adult: forewings 3.4–4.3 mm long, transparent and often suffused with brown around the veins; veins mainly yellow; body mainly yellowish brown, with dark brown markings; antennae brown, grading to black apically; a green-bodied form also occurs. **Egg:** minute, elongate-oval; white and opaque when laid, later becoming orange. **Nymph:** flat bodied, light green.

LIFE HISTORY
This species overwinters in the egg stage. Nymphs develop from spring onwards, feeding on the underside of the leaves in association with the major veins. The adult stage is reached from May or June onwards. Adults occur throughout the summer, there being just one generation annually.

DAMAGE
Infestations cause little or no significant damage, although feeding by the nymphs may result in slight distortion of tissue.

83 Fig sucker (*Homotoma ficus*).

84 Fig sucker (*Homotoma ficus*) – nymphs.

85 Fig sucker (*Homotoma ficus*) – infested leaf.

Family **ALEYRODIDAE**
(whiteflies)

Small, moth-like insects, more or less coated with an opaque, white, waxy powder; wings soft, broadly rounded, with a reduced venation; antennae 7-segmented; tarsi 2-segmented. Nymphs are flattened, sedentary and scale-like. Development includes a quiescent, non-feeding pseudo-pupal stage. Whiteflies are not important pests of temperate fruit crops in Europe, but some species occasionally attract attention, mainly on blackberry and strawberry. In subtropical areas, however, whitefly infestations on crops such as citrus can be particularly damaging.

Aleurolobus olivinus Silvestri (**86**)
Olive whitefly

A minor pest of olive; infestations also occur on *Phillyrea latifolia*. Present in Mediterranean areas, including southern Europe and North Africa.

DESCRIPTION
Adult: 1.6–1.7 mm long; body creamy white; body and wings coated in white, powdery wax. **Egg:** 0.2 mm long; greyish yellow; more or less elliptical, with a pedicel. **Nymph:** mainly black; more or less circular, flattened and scale-like, with an outer fringe of fine setae. **Pseudo-pupa:** black and scale-like.

LIFE HISTORY
Adults of this species occur in June and July. Eggs, up to 60 per female, are then deposited singly in small groups on the upper side of the leaves of host plants, each attached to the surface by the pedicel. Following egg hatch, typically about 2 weeks later, each nymph moves away before settling on a suitable leaf to begin feeding. Development is slow, and the adult stage is not reached until the following year, there being just one generation annually.

DAMAGE
Infested foliage is contaminated by sooty moulds that develop on honeydew excreted by the nymphs.

Aleurothrixus floccosus (Maskell) (**87**)
Woolly whitefly

A polyphagous, subtropical species of South American origin. Infestations occur on various broadleaved evergreen plants, including citrus fruits (especially mandarin) and fig. Nowadays, established in southern Europe (including both France and Italy), where it was first found (in Spain) in 1966.

DESCRIPTION
Adult: 1.5 mm long; body yellowish and heavily dusted with white, powdery wax; wings relatively narrow. **Egg:** oval, with a distinct pedicel. **Nymph:** oval-bodied, scale-like, light green and more or less transparent, covered in the later stages by numerous strands of wax. **Pseudo-pupa:** 1.5 mm long; scale-like, covered by woolly strands of wax.

LIFE HISTORY
Adults are active from February or early March onwards. Eggs are then laid in circles or semicircles on the underside of leaves and hatch shortly afterwards. Nymphs develop rapidly, imbibing considerable quantities of sap and excreting masses of, characteristically, viscous honeydew. Breeding continues throughout the spring, summer and autumn,

86 Olive whitefly (*Aleurolobus olivinus*) – pseudo-pupae.

87 Woolly whitefly (*Aleurothrixus floccosus*).

with four or five generations being completed annually. The winter is passed in the egg or nymphal stages. Nymphs often occur, along with the honeydew-coated cast-off skins (exuviae) of previous generations, in dense colonies on the underside of leaves, populations becoming particularly large by the autumn. Third-instar nymphs produce long waxen threads that persist throughout the pseudo-pupal stage, colonies then appearing to be covered by masses of white wool.

DAMAGE

In addition to causing direct damage to foliage, the pest also excretes considerable quantities of sticky honeydew, upon which sooty moulds develop; this reduces photosynthetic activity and thereby weakens host plants.

Aleyrodes lonicerae Walker

Honeysuckle whitefly

Commonly breeds on cultivated blackberry; also occurs on strawberry. Wild hosts include *Chamaenerion angustifolium*, *Lonicera periclymenum*, *Rubus fruticosus* and *Symphoricarpos rivularis*. Widely distributed in Europe.

DESCRIPTION

Adult: 1 mm long; body yellow, spotted with grey and more or less coated with white waxen powder; wings white, with an indistinct greyish spot on each forewing. **Egg:** 0.27 × 0.1 mm; oblong, yellowish white and thickly dusted with wax. **Nymph:** oval or elliptical; pale yellow, surrounded by a waxy fringe. **Pseudo-pupa:** 1.0 × 0.75 mm; flat and oval; white, yellow or golden yellow, surrounded by a thick, waxy fringe.

LIFE HISTORY

Eggs are laid singly on the underside of leaves during the spring and summer months. The nymphs feed for several weeks, passing through three instars before pupating. In summer, adults emerge shortly afterwards, about 2 months after eggs were laid. There are several overlapping generations in a season and, since adults are long-lived, all stages of the insect may occur together. The winter is spent as adults or as pseudo-pupae.

DAMAGE

Adverse effects on host plants are minimal, although foliage may become contaminated by honeydew and by patches of white wax.

Aleyrodes proletella (Linnaeus) (**88–89**)

Cabbage whitefly

This notorious pest of vegetable brassicas occasionally occurs on cultivated strawberry, mainly as overwintering adults. Although eggs are laid on strawberry leaves, cabbage whitefly does not reach pest status on this crop. Unlike those of honeysuckle whitefly (*Aleyrodes lonicerae*), the eggs are deposited in characteristic, subcircular groups.

Asterobemisia carpini (Koch)

Hornbeam whitefly

A minor pest of blackberry; polyphagous on various trees and shrubs, including *Carpinus betulus* and *Corylus avellana*. Present throughout Europe.

DESCRIPTION

Adult: 1.0–1.3 mm long; yellow or light orange, with white, translucent wings. **Egg:** 0.24 × 0.10 mm; oblong; pale yellow to dark brown and slightly dusted with wax. **Nymph:** elliptical or oval; mainly yellowish green, with a narrow waxy fringe. **Pseudo-pupa:** 1.2 × 0.9 mm; elliptical; yellowish green to whitish.

88 Cabbage whitefly (*Aleyrodes proletella*).

89 Cabbage whitefly (*Aleyrodes proletella*) – egg batch on strawberry leaflet.

LIFE HISTORY

Adults occur in spring and are comparatively short-lived. Eggs are laid on the underside of leaves in May and June. Nymphs then feed during the summer months, pupating by the autumn. The winter is passed in the pseudo-pupal stage, there being just one generation annually.

DAMAGE

As for honeysuckle whitefly (*Aleyrodes lonicerae*).

Dialeurodes citri (Ashmead) (90–92)

Citrus whitefly

This polyphagous species is mainly a problem on citrus fruits, although it is also found on fig, olive, pomegranate, plum and various other hosts, including *Forsythia*, *Fraxinus excelsior*, *Ligustrum vulgare* and *Syringa vulgaris*. The pest, which is of East Asian origin, was first noted in Europe (Italy and southern France) in the 1950s. It also occurs in America.

DESCRIPTION

Adult: 1.2–1.4 mm long; creamy white, mainly coated with white wax. **Egg:** 0.2–0.3 mm long; oval and pale yellow, with a pedicel. **Nymph:** up to 1.5 mm long; pale greenish yellow, flat, oval and scale-like. **Pseudo-pupa:** 1.2–1.5 mm long; flat, oval and scale-like.

LIFE HISTORY

Adults, which first appear in April or May, are short-lived, each surviving for about 10 days. Eggs are laid in batches on the underside of young leaves and hatch 1–4 weeks later, depending on temperature. The scale-like immature stages feed on the underside of leaves for 3–4 weeks; they then enter a quiescent pseudo-pupal stage which lasts anything from 2 to 42 weeks or more. Infestations are often very heavy, with several hundred individuals clustered on a single leaf. The pest overwinters as either third- or fourth-instar nymphs. Under favourable conditions, there are four generations annually.

DAMAGE

Hosts are weakened by heavy, persistent infestations and by sooty moulds that accumulate on honeydew excreted by the pest. Contamination of leaves and fruits on infested trees is also a problem.

90 Citrus whitefly (*Dialeurodes citri*).

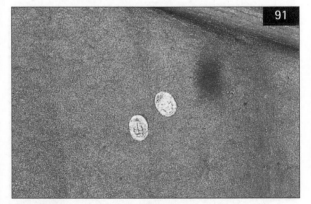

91 Citrus whitefly (*Dialeurodes citri*) – pseudo-pupae.

92 Citrus whitefly (*Dialeurodes citri*) – infested leaf.

Family **APHIDIDAE**
(aphids)

The major family of aphids. Antennae are long, usually 6-segmented, and have a long (rarely very short) terminal process; the siphunculi are typically tubular, often very long, sometimes flanged apically and sometimes swollen; the tail-like cauda is usually tongue-like or finger-like. Alternation of generations between winter and summer hosts is commonplace and life cycles of individual species are often complex.

Aphids (often called blackflies or greenflies) are well-known insect pests, occurring in both winged (alate) and wingless (apterous) forms: known as alatae and apterae, respectively. Wings, when present, are usually large and transparent, with few veins. Adults and nymphs have similar habits, often living together in dense colonies. They feed on sap by piercing plant tissue with their needle-like mouthparts, usually penetrating the phloem. Several species cause galling or other host plant reactions and some are important vectors of plant viruses. Aphid life cycles are often complex and may involve a range of forms (such as egg-laying females, known as oviparae). Many species show an alternation of generations, having a primary (winter) host upon which asexual and sexual reproduction occurs and eggs are laid, and secondary (summer) hosts where development is entirely parthenogenetic and viviparous. Migration between these primary and secondary hosts is usually achieved following the production of winged forms. In deriving English common names of several such aphids, the primary host has precedence; hence, the adoption of, for example, 'damson/hop aphid' rather than the archaic 'hop/damson aphid'.

Many aphids produce wax from glands on their bodies; aphids may also excrete an abundance of honeydew, much sought after by ants and other insects. Features of the siphunculi and the cauda are useful for distinguishing between species. Identification of species in the absence of adults and host information, however, is often difficult, if not impossible, even for the specialist. The general text descriptions of aphids given here refer mainly to wingless, viviparous, adult females (apterae).

Aphids are arranged below under the following subfamilies: Aphidinae (p. 66–91), Callaphidinae (p. 92–93), Eriosomatinae (p. 94–95) and Lachninae (p. 96).

Subfamily **APHIDINAE**

Acyrthosiphon rogersii (Theobald)
A minor pest of strawberry, of little importance except as a potential virus vector. Widely distributed in northern and western Europe.

DESCRIPTION
Adult viviparous female (aptera): 1.5–2.7 mm long; medium-sized, green and shiny; siphunculi pale, long, thin and flanged; cauda elongate and finger-like.

LIFE HISTORY
Colonies of wingless aphids occur on the young foliage, particularly along the petioles, reaching their peak of development in June. Winged aphids are produced in these colonies from May onwards and are responsible for spreading infestations to maiden, and other, strawberry plants. After June or July, aphid numbers decline rapidly. This species often occurs on strawberry in company with potato aphid (*Macrosiphum euphorbiae*).

DAMAGE
This species causes little or no direct damage to strawberry, but may spread strawberry mottle virus.

Amphorophora idaei (Börner) (**93**)
Large European raspberry aphid
Frequently present on cultivated raspberry, but never particularly numerous and important only as a virus vector. Widely distributed in Europe.

DESCRIPTION
Adult viviparous female (aptera): 2.6–4.1 mm long; medium-sized to large, pale to yellowish green, and shiny; legs and antennae long; siphunculi long, pale, slightly swollen on apical half and flanged at tip; cauda elongate.

LIFE HISTORY
Eggs hatch in early March and aphids then feed at the tips of the leaf buds. Later, they live on the underside of the leaves. However, they do not form large colonies and are typically present in only small numbers. In June and July, after two generations of wingless aphids,

93 Large European raspberry aphid (*Amphorophora idaei*).

winged forms appear. These then spread infestations elsewhere on the original host plants or migrate to new canes. Wingless aphids are then produced including, in October, the oviparae; males are winged. Eggs are deposited on the canes from October to December, most frequently within 30cm of the ground.

DAMAGE
Direct feeding by this species is of no consequence, but the aphids can act as vectors of black raspberry necrosis virus, raspberry leaf mottle virus, raspberry leaf spot virus and rubus yellow net virus. Cultivars vary considerably in their susceptibility to aphid-borne virus diseases: for example, cvs Gaia, Glen Rosa and Julia are resistant; cvs Glen Ample and Glen Shee are partially resistant; cv. Lloyd George is particularly susceptible.

Amphorophora rubi (Kaltenbach) (**94**)
Large blackberry aphid
This aphid is virtually identical to the large European raspberry aphid *Amphorophora idaei*, but occurs on blackberry not raspberry. Both aphid species are host specific (although *A. rubi* also infests *Rubus caesius*) and the host-plant relationship serves as the simplest method of distinguishing between them. The pest is widely distributed in Europe and is also now established in New Zealand.

Anuraphis farfarae (Koch) (**95–96**)
Pear/coltsfoot aphid
A common, but minor, pest of pear. Eurasiatic. Widely distributed in Europe.

DESCRIPTION
Adult viviparous female (aptera): 1.5–2.0 mm long; small, plump, dark purplish brown; antennae short; siphunculi dark, short and tapered; caudal short and broad. **Nymph:** yellowish green.

LIFE HISTORY
Eggs overwintering on pear hatch in the spring. Aphids then feed on the underside of leaves on the vegetative spurs. In April and May, small colonies develop within the folded leaves; in late May, after one wingless generation, winged forms appear and these subsequently disperse to *Tussilago farfara*, where breeding takes place on the roots. Colonies on pear then die out. Winged aphids return to pear in the autumn where, following the production of oviparae, eggs are eventually laid.

DAMAGE
Infested leaves are often twisted and folded back along the midrib so that each half meets. The leaves usually remain green whilst the tissue is still living, although they may become tinged with red (cf. the following species). Although several leaves of a cluster may show symptoms, aphids are not usually present on all of them. Also, a single colony usually completes its full development on a single leaf, without spreading further. Since colonies are small

95 Pear/coltsfoot aphid (*Anuraphis farfarae*) – old, abandoned habitation.

94 Large blackberry aphid (*Amphorophora rubi*).

96 Pear/coltsfoot aphid (*Anuraphis farfarae*) – inside of old, abandoned habitation.

and occur only on spur leaves, they are of little or no significance. Infested leaves turn brown and are dead by mid-summer, enclosing the cast-off (tell-tale) skins (exuviae) of the former inhabitants.

Anuraphis subterranea (Walker) (**97–98**)
Pear/parsnip aphid

This Eurasiatic species is very similar to pear/coltsfoot aphid (*Anuraphis farfarae*) and, on pear, causes similar leaf symptoms; the living leaf tissue, however, is turned characteristically reddish. Colonies of *A. subterranea* are also distinguished by the presence of brownish-green to brownish-black (as opposed to yellowish-green) nymphs. The summer hosts of this aphid are roots of umbelliferous plants, such as *Heracleum sphondylium* and *Pastinaca sativa*. Although widely distributed in Europe, it is not of economic importance.

Aphis craccivora Koch (**99**)
Cowpea aphid

In warmer regions a minor pest of peach and grapevine. Also associated with a wide range of other hosts, particularly *Robinia pseudoacacia* and many other Fabaceae. Present in most warmer parts of the world, especially the tropics; often common in southern Europe.

DESCRIPTION

Adult viviparous female (aptera): 1.4–2.2 mm long; small, black and shiny; siphunculi moderately long and tapered; cauda tongue-like, with a relatively pointed tip. **Nymph:** black, with a fine dusting of white wax.

LIFE HISTORY

This polyphagous species is mainly parthenogenetic. Colonies typically occur on the young shoots, and are attended by ants.

DAMAGE

Colonies on grapevines are very noticeable but cause little or no damage.

Aphis fabae Scopoli (**100**)
Black bean aphid

An often abundant and notorious cosmopolitan pest of field bean, sugar beet and various other cultivated plants. Ant-attended colonies are sometimes established on apple, pear, quince and other fruit trees during the summer months. They occur most frequently on rootstocks and young trees, causing leaf curl and, sometimes, stunting of growth or even the death of shoots. This species usually overwinters in the egg stage on its primary host, *Euonymus europaeus*. Adults (1.5–3.0 mm long) are dull black or dark olive green, often with irregular patches of whitish wax on the back; the siphunculi are rather short and slightly tapered.

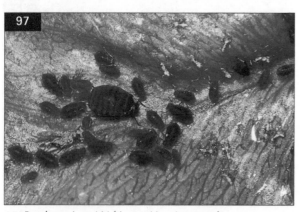

97 Pear/parsnip aphid (*Anuraphis subterranea*).

98 Pear/parsnip aphid (*Anuraphis subterranea*) – damage to young pear shoot.

99 Cowpea aphid (*Aphis craccivora*) – infested grapevine.

Aphis forbesi Weed
Strawberry root aphid

A North American pest of strawberry; well-established in France, where it was first found in the late 1920s, and now present elsewhere in mainland Europe. Also present in Japan and South America.

DESCRIPTION
Adult viviparous female (aptera): 1.2–1.6 mm long; small, dark, bluish green; eyes red; siphunculi relatively short, broad-based, distinctly tapered and paler than body; legs and antennae pale. **Nymph:** pale yellowish green. **Egg:** oval and distinctly globular; black and shiny.

LIFE HISTORY
Overwintered eggs hatch in March. Dense, ant-attended colonies then develop on the petioles and crowns of strawberry plants. Colonies also develop on the roots. Winged forms appear from May onwards and these aid spread of infestations. Sexual forms are produced in the autumn and, eventually, the winter eggs are deposited. Males are wingless.

100 Black bean aphid (*Aphis fabae*).

DAMAGE
Direct damage is of only minor significance and likely to occur only on light, sandy soils.

Aphis gossypii Glover (**101–102**)
Melon & cotton aphid

An extremely polyphagous, tropical species, occurring on many crops including (especially in hot Mediterranean areas) citrus, avocado and pomegranate. In more temperate parts of Europe, this pest can infest strawberries growing in greenhouses and under plastic tunnels; also, under favourable conditions of temperature and humidity, infestations sometimes occur on outdoor fruit crops, including pear, peach, strawberry and kiwi fruit. Cosmopolitan, surviving the winter in cooler regions only in greenhouses or under other such protection.

DESCRIPTION
Adult viviparous female (aptera): 1.5–1.8 mm long; small, yellowish green, mottled with darker green, through dark green to greenish black; siphunculi dark, stout and of moderate length; cauda tongue-like and usually pale.

LIFE HISTORY
In Europe, this aphid breeds parthenogenetically, with a very large number of generations annually. Population development is favoured by warm, damp conditions. Both winged and wingless adults are produced during the summer. However, if conditions are very hot and if colonies become overcrowded, very small (c. 1 mm long), pale yellow to whitish, wingless adults are produced. Colonies are often attended by ants.

DAMAGE
On citrus, in addition to causing direct damage to foliage (infested foliage becoming flecked with yellow, distorted

101 Melon & cotton aphid (*Aphis gossypii*).

102 Melon & cotton aphid (*Aphis gossypii*) – heavily infested citrus shoot.

and even killed), the aphids are also vectors of citrus tristeza closterovirus. Foliage on infested trees is often contaminated by honeydew, amongst which cast-off nymphal skins accumulate and upon which sooty moulds develop. This has an adverse effect on the photosynthetic activity of leaves; developing fruits are also contaminated.

Aphis grossulariae Kaltenbach
European gooseberry aphid
A pest of gooseberry and, occasionally, currant. Present throughout Europe.

DESCRIPTION
Adult viviparous female (aptera): 1.5–2.2 mm long; small, dark green to greyish green, with a mealy, waxy coating; siphunculi relatively short; cauda tongue-like.

LIFE HISTORY
Eggs hatch in March and early April. At first, wingless aphids feed on the developing fruit buds; later, infestations spread to the tips of the young shoots where dense colonies are eventually established. Breeding on gooseberry continues throughout the summer months, although there is also a migration of winged forms to smaller kinds of *Epilobium*, e.g. *E. montanum* and *E. obscurum*, the secondary (i.e. summer) hosts. A return flight to gooseberry occurs in the autumn. The winter is passed in the egg stage on the shoots.

DAMAGE
Infestations lead to leaf curl and to the development of tufts of distorted leaves at the growing points. Attacked shoots are twisted and stunted, and the internodes are shortened. The aphids are also virus vectors.

Aphis idaei van der Goot (**103–104**)
Small European raspberry aphid
Common on raspberry and important as a vector of virus disease; loganberry is also a host. Present throughout Europe; an introduced pest in New Zealand.

DESCRIPTION
Adult viviparous female (aptera): 1.5–2.0 mm long; small, light green or yellowish green, covered with waxy powder; siphunculi dark, moderately long and slender, and curved slightly outwards.

LIFE HISTORY
Overwintered eggs hatch in March and dense colonies then develop on the blossom trusses and lateral shoots. Early attacks, however, rarely occur on the new canes. After two generations of wingless aphids, winged forms are produced. From early June to late July, these spread infestations to the new canes or elsewhere on the original host plants. Unlike the spring generations, the progeny of the winged forms are pale cream in colour; they are also smaller (0.8–1.0 mm long) and occur singly, usually hidden in the junction of two veins on the underside of leaves. Wingless sexual forms are produced from October onwards. Oviparae are then produced and (after mating) these lay eggs in the axils at the base of buds, usually on the upper half of the canes; the oviparae persist in diminishing numbers until mid-December. Colonies of this species are frequently attended by ants.

DAMAGE
Spring infestations cause leaf curl. However, the aphids are important only as potential vectors of raspberry vein chlorosis virus; direct feeding damage is of little or no consequence.

103 Small European raspberry aphid (*Aphis idaei*) – infested loganberry shoot.

104 Small European raspberry aphid (*Aphis idaei*) – infested raspberry bud.

Aphis nerii Boyer de Fonscolombe (**105**)
Oleander aphid

A pest of citrus, although associated mainly with Asclepiadaceae and Apocynaceae (e.g *Nerium oleander*). Widely distributed in subtropical and tropical regions, including southern Europe.

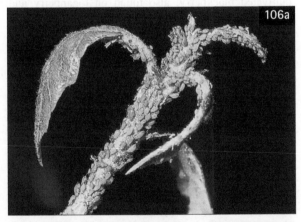

105 Oleander aphid (*Aphis nerii*).

106a, b Green apple aphid (*Aphis pomi*) – infested apple shoots.

DESCRIPTION
Adult viviparous female (aptera): 1.5–2.6 mm long; medium sized, bright lemon-yellow (occasionally tinged with green), with mainly dark legs and antennae; siphunculi black, stout and moderately long; cauda black and tongue-like.

LIFE HISTORY
This species breeds entirely parthenogenetically. The aphids occur mainly on the young, succulent growth, where very large, dense, ant-attended colonies develop on the shoots and along the midribs of expanded leaves. Winged aphids are produced during the summer and these aid spread of infestations.

DAMAGE
Direct damage is of little significance. However, the aphids produce considerable quantities of honeydew upon which sooty moulds develop, and this can reduce photosynthetic activity. The aphids are also vectors of citrus tristeza closterovirus.

Aphis pomi Degeer (**106–107**)
Green apple aphid

A common pest of apple; also attacks pear, quince and various other woody rosaceous plants. Eurasiatic. Widely distributed in Europe; also present in North America.

DESCRIPTION
Adult viviparous female (aptera): 1.3–2.2 mm long; small, bright green or yellowish green; antennae short; siphunculi black or dark brown and moderately long; cauda tongue-like.

107 Green apple aphid (*Aphis pomi*) – leaf curl on apple.

LIFE HISTORY

Overwintered eggs hatch in April. Aphid colonies then develop beneath the leaves, particularly at the tips of young shoots. Winged aphids begin to appear in early June and are produced in greater or lesser numbers throughout the summer months. It is the winged migrants from other apple trees that are largely responsible for initiating the heavy infestations that often build up rapidly in June or July. Breeding continues until the autumn when wingless sexual forms occur. Oviparae are then produced and (after mating) these lay eggs on the bark of young shoots. The life cycle of this species does not include a migration to an alternative summer host.

DAMAGE

Spring attacks are usually unimportant. Summer infestations, however, cause considerable leaf curling; shoot tips may be stunted or even killed. Damage is rarely important on mature trees, but is often serious on nursery stock and on young trees.

108 Pomegranate aphid (*Aphis punicae*).

109 Permanent blackberry aphid (*Aphis ruborum*) – infested blackberry shoot.

Aphis punicae Passerini (**108**)
Pomegranate aphid

A minor pest of pomegranate. Present throughout the Mediterranean basin and also further to the east.

DESCRIPTION

Adult viviparous female (aptera): 1.2–3.0 mm long; yellowish green to bluish green, with dark green markings on the thorax and abdomen; siphunculi usually mainly light green, but often black apically and sometimes entirely darkened; antennae, legs and cauda more or less colourless.

LIFE HISTORY

Colonies occur on the leaves from April to September; the aphids may also spread onto the flowers.

DAMAGE

Infestations, although often obvious, have little or no adverse effect on host plants.

Aphis ruborum (Börner) (**109**)
Permanent blackberry aphid

Generally uncommon, attacking both wild and cultivated blackberry; also found on loganberry and, rarely, on strawberry. Widely distributed in Europe; also present in parts of Asia and North Africa.

DESCRIPTION

Adult viviparous female (aptera): 1.5–2.5 mm long; small to medium-sized, pale yellowish to dark green; siphunculi usually pale, but sometimes dusky at base or apex, and relatively stout.

LIFE HISTORY

Colonies occur mainly on the underside of leaves and at the shoot tips throughout the spring and summer; when colonies are large, they may also spread down the shoots. As with small raspberry aphid (*Aphis idaei*), very small (0.8–1.0 mm long), whitish, wingless aphids occur singly on the underside of the leaves during the summer months; they also occur on the flowers and developing fruits. The winter is spent in the egg stage at the base of buds.

DAMAGE

Infestations may cause slight leaf curling, but are otherwise harmless.

Aphis schneideri (Börner) (**110**)
European permanent currant aphid
A local European pest of currant; most frequently found on black currant and red currant.

DESCRIPTION
Adult viviparous female (aptera): 1.2–2.2 mm long; small, dark blue green, with a bluish-white, waxy covering; siphunculi pale and of moderate length; cauda tapered.

LIFE HISTORY
Eggs hatch in the spring, and wingless aphids at first invade the flower trusses. Later, dense colonies develop at the tips of young shoots until, in June, winged forms are produced. These then spread to various kinds of currant, both fruiting and ornamental, and are also responsible for initiating infestations on young cuttings. Breeding of wingless offspring from the migrants continues into the autumn when, following production of sexual forms, eggs are eventually deposited on the shoots.

DAMAGE
Infested shoots become severely distorted and stunted, with the internodes shortened and twisted. Tight bunches of distorted leaves develop at the shoot tips, each leaf being characteristically bent downwards at its point of attachment to the petiole. Attacks on nursery stock, cuttings and young bushes are particularly devastating.

Aphis spiraecola Patch (**111–112**)
Green citrus aphid
An important pest of citrus; also associated with many other host plants, including apple and other rosaceous fruit trees. Probably of Far Eastern origin; now present in the Mediterranean region, where it was first noted in the late 1930s. Also accidentally introduced during the 1900s into Africa, North America, South America, Australia and New Zealand.

DESCRIPTION
Adult viviparous female (aptera): 1.2–2.2 mm long; small, bright yellowish green to green, with a brown head; legs and antennae pale; siphunculi brownish black or black and of moderate length; cauda brownish black or black and elongate, with a basal constriction.

LIFE HISTORY
In some parts of the world, this species includes a sexual phase in the life cycle and overwinters in the egg stage on its primary host (*Spiraea*). In southern Europe, however, this species breeds entirely parthenogenetically, adult females overwintering on the shoots of citrus and other secondary host plants. Colonies on citrus develop from spring to autumn, with a very large number of overlapping generations annually; there is no period of summer diapause.

110 European permanent currant aphid (*Aphis schneideri*).

111 Green citrus aphid (*Aphis spiraecola*).

112 Green citrus aphid (*Aphis spiraecola*) – damage to foliage.

DAMAGE
Spring infestations are particularly damaging, leaves becoming severely distorted, and infested flowers falling off before setting fruit. Foliage also becomes contaminated by honeydew, upon which sooty moulds develop. The aphids are vectors of citrus tristeza closterovirus.

Aphis triglochinis Theobald
Red currant/arrow-grass aphid
This uncommon, small, brownish-green species is occasionally noted on red currant and, less frequently, black currant, where it causes the leaves to curl downwards. However, it is not an important pest. In summer, the aphids migrate to various plants of wet habitats, including *Triglochin palustris*, with migrant aphids returning to currant in the autumn. This species eventually overwinters on currant in the egg stage. Infestations are found mainly in northern Europe.

Aulacorthum solani (Kaltenbach) (**113–114**)
Glasshouse & potato aphid
Small breeding colonies of this common, and now virtually worldwide, polyphagous species occur occasionally on strawberry, particularly under protection. Although the aphids may be found on strawberry throughout the year, they are not considered harmful. Adults (2–3 mm long) are yellowish green, with darker patches at the base of the long, thin, slightly tapered, dark-tipped and distinctly flanged siphunculi; the antennae are long and the front of the head emarginate. The nymphs are shiny green with a rather bright green or yellow patch at the base of each siphunculus.

Brachycaudus cardui (Linnaeus)
Thistle aphid
A usually minor pest of apricot, damson, myrobalan and plum. Widely distributed in Europe; also present in parts of North Africa, North America and Asia.

DESCRIPTION
Adult viviparous female (aptera): 1.8–2.5 mm long; colour varying from green through yellow to reddish, with a dark, shiny dorsal patch on the abdomen (cf. *Brachycaudus helichrysi*); siphunculi short and flanged; cauda broadly rounded.

LIFE HISTORY
This species overwinters in the egg stage on primary hosts. In spring, colonies develop on the shoots, winged forms eventually dispersing to summer hosts. These are either Asteraceae (e.g. *Carduus*, *Chrysanthemum*, *Cirsium*, *Matricaria* and *Senecio*) or Boraginaceae (e.g. *Cynoglossum*, *Myosotis* and *Symphytum*). A return migration to primary hosts occurs in the autumn.

DAMAGE
Spring infestations on apricot and plum cause leaf rolling. The aphids are also vectors of plum pox ('Sharka'), an important viral disease of damson, plum, apricot, peach and certain other plants.

Brachycaudus helichrysi (Kaltenbach) (**115–117**)
Leaf-curling plum aphid
A common and often a serious pest of damson, myrobalan and plum. Of Palaearctic origin and now distributed throughout the world.

113 Glasshouse & potato aphid (*Aulacorthum solani*) – apterous females.

114 Glasshouse & potato aphid (*Aulacorthum solani*) – infested strawberry leaflet.

DESCRIPTION

Adult viviparous female (aptera): 1–2 mm long; small, rounded, yellowish green ('early spring' forms brownish) and shiny; abdomen without a dark patch dorsally (cf. *Brachycaudus cardui*); antennae short; siphunculi pale, short and flanged; cauda broadly rounded.

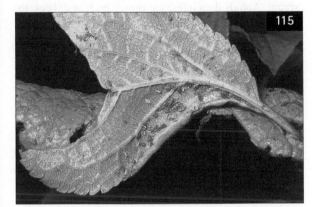

115 Leaf-curling plum aphid (*Brachycaudus helichrysi*) – infested leaf.

116 Leaf-curling plum aphid (*Brachycaudus helichrysi*) – severely damaged shoot.

117 Leaf-curling plum aphid (*Brachycaudus helichrysi*) – damage to branch.

LIFE HISTORY

Overwintered eggs hatch very early in the spring, before bud burst. In many areas, eggs may hatch as early as November and December. At first, the young nymphs feed at the base of fruit buds. However, as fruit and leaf buds eventually open the young tissue of both is attacked. Later, infestations also develop on the foliage of young shoots. Winged forms are produced in May, and these migrate to various summer hosts such as *Aster*, *Chrysanthemum* and *Trifolium*. Breeding on the primary host will also continue whilst young growth is still available, but such colonies do eventually die out. In autumn, winged aphids fly back to plum from the summer hosts and, following a sexual phase, eggs are finally laid on the shoots and spurs.

DAMAGE

This species causes severe leaf curl and distortion, considerably stunting growth; this is particularly serious on young trees and nursery stock, and the effects can be devastating. The aphid is also a vector of plum pox ('Sharka').

Brachycaudus persicae (Passerini) (**118**)

Black peach aphid

A generally common species on apricot, nectarine, peach and plum, but not a serious pest. Widely distributed in Europe. Also found in parts of Africa, North America, South America, Australia and New Zealand.

DESCRIPTION

Adult viviparous female (aptera): 1.5–2.2 mm long; small, brownish black to black, and shiny; siphunculi black, moderately short and flanged; cauda short and broad. **Nymph:** yellowish brown to dark brown.

118 Black peach aphid (*Brachycaudus persicae*) – infested leaves.

LIFE HISTORY

This species occurs on the roots of host trees, overwintering as wingless aphids. In spring and early summer, large, dense, ant-attended colonies develop on the suckers and young shoots; the aphids also occur on the underside of leaves, but typically only in small numbers. Unlike other aphids on peach and plum, this species usually does not pass through a sexual phase and oviparae (and, hence, eggs) are rarely produced.

DAMAGE

This pest is usually of importance only on nursery stock; also, infested foliage is distorted only slightly (cf. peach aphid, *Brachycaudus schwartzi*); see also peach/potato aphid (*Myzus persicae*) and peach leaf-roll aphid (*Myzus varians*).

Brachycaudus schwartzi (Börner) (**119**)
Peach aphid

An often common pest of peach. Widely distributed in Europe. Also found in North America, South America, India and Iran.

119a, b Peach aphid (*Brachycaudus schwartzi*) – damage to young shoots.

DESCRIPTION

Adult viviparous female (aptera): 1.4–2.1 mm long; small, blackish brown; siphunculi dark, very short and flanged; cauda dark, short and broad. **Nymph:** green to reddish brown.

LIFE HISTORY

Eggs are laid in autumn on the shoots and they hatch in the spring. Aphids then infest the flower buds and, later, the leaves at the tips of young shoots. Winged aphids appear in June. These migrate to young shoots on the same or other peach trees. Here, breeding then occurs until, in the autumn, sexual forms arise and winter eggs are eventually laid. In warmer regions, viviparous females may overwinter in bark crevices on the branches of host trees.

DAMAGE

Unlike black peach aphid (*Brachycaudus persicae*), this aphid causes severe leaf curl. Growth of terminal shoots is also stunted.

Chaetosiphon fragaefolii (Cockerell)
Strawberry aphid

An important pest of strawberry and potentially very serious as a virus vector. Of American origin, but now cosmopolitan. Widely distributed in Europe.

DESCRIPTION

Adult viviparous female (aptera): 0.9–1.8 mm long; small, pale whitish, with red eyes and conspicuous capitate body hairs; siphunculi moderately long.

LIFE HISTORY

Wingless aphids breed on strawberry throughout the year, except during extremely cold winter weather. Populations reach their height in early summer, but then decline rapidly. On first-year plants, however, greatest numbers are usually present in September. The aphids occur mainly on the underside of young leaves, infested foliage soon becoming sticky with honeydew. Winged aphids, which spread infestations to new plantations, occur in May and June; a smaller, but less significant, generation of winged forms is also present from October to December. At least in the field, there is no sexual stage in the life cycle and eggs are not laid.

DAMAGE

The aphids, although often numerous, do not distort the foliage of infested plants. However, they do make it sticky with honeydew. This species is most important as a potential vector of strawberry crinkle virus, strawberry yellow-edge virus and various other strawberry viruses.

Corylobium avellanae (Schrank) (**120**)
Large hazel aphid

A common species on wild *Corylus avellana* and a minor pest on cultivated hazelnut. Widely distributed in Europe. An introduced pest in Canada.

DESCRIPTION
Adult viviparous female (aptera): 1.7–2.7 mm long; medium-sized, spindle-shaped, green, suffused with reddish, to very light green; body with numerous, capitate hairs arising from distinct tubercles; eyes red; siphunculi long and thin (cf. *Myzocallis coryli*, subfamily Callaphidinae). **Nymph:** light green.

120 Large hazel aphid (*Corylobium avellanae*).

LIFE HISTORY & DAMAGE
Extensive colonies often appear on the suckers, shoots and leaf petioles during the summer months, but infestations in hazelnut plantations are rarely of importance. Winged viviparous females occur mainly from June to late August.

Cryptomyzus galeopsidis (Kaltenbach) (**121–122**)
European black currant aphid

A pest of black currant; also, occasionally, associated with gooseberry and red currant. Widely distributed in Europe. An introduced pest in North America.

DESCRIPTION

121 European black currant aphid (*Cryptomyzus galeopsidis*).

Adult viviparous female (aptera): 1.3–2.6 mm long; small to medium-sized, delicate, creamy white or pale yellowish green to green, with an indistinct green dorsal stripe; body with capitate hairs; siphunculi thin and frequently less than twice as long as the cauda.

LIFE HISTORY
This species may occur on various kinds of *Ribes* throughout the year, usually on the foliage at the tips of shoots. More frequently, however, the life cycle includes a summer migration to secondary labiate hosts such as *Galeopsis tetrahit* and *Lamium purpureum*.

DAMAGE
Although, unlike red currant blister aphid (*Cryptomyzus ribis*), this species does not blister leaves, infested foliage becomes crinkled and yellowish; later, leaves at the tips of infested shoots may turn brown and die. By June, populations on unsprayed bushes are sometimes sufficiently large to contaminate fruit and foliage with honeydew.

122 European black currant aphid (*Cryptomyzus galeopsidis*) – damage to young shoot.

123 Red currant blister aphid (*Cryptomyzus ribis*).

124 Red currant blister aphid (*Cryptomyzus ribis*) – galled leaves of red currant.

125 Red currant blister aphid (*Cryptomyzus ribis*) – galled leaves of jostaberry.

Cryptomyzus ribis (Linnaeus) (**123–125**)
Red currant blister aphid

A common pest on red currant; jostaberry, white currant and, less frequently, black currant are also attacked. Widely distributed in Europe; also present in North America, China and Japan.

DESCRIPTION
Adult viviparous female (aptera): 1.2–2.5 mm long; small to medium-sized, rather plump, delicate and shiny, and creamy white to pale yellowish green; body with capitate hairs; siphunculi comparatively long and thin, more than three times as long as the cauda.

LIFE HISTORY
Eggs are laid during the autumn on the shoots of currant. They hatch in the spring and aphid colonies then become established on the underside of the leaves. In summer, winged aphids are produced and these disperse to *Stachys sylvatica* and related hosts; eventually, a return migration of winged forms to currant takes place later in the year.

DAMAGE
This aphid causes characteristic reddish or purplish blisters on the leaves of red currant and white currant. Heavy infestations lead to considerable discoloration and distortion of foliage. On black currant, the blisters are yellowish green. In addition, the fruit and foliage of infested bushes may become sticky with honeydew upon which sooty moulds develop.

Dysaphis anthrisci Börner (**126–127**)
Apple/anthriscus aphid

This widely distributed, but local, species occurs on apple in several parts of Europe (e.g. England, Germany, the Netherlands and Switzerland), inhabiting similar galls to those initiated in the spring by rosy leaf-curling aphid (*Dysaphis devecta*). Unlike the latter species, however, *D. anthrisci* has a summer (secondary) host (namely, *Anthriscus sylvestris*); also, unlike rosy leaf-curling aphid, most (if not all) second-generation individuals are winged forms (alatae), and colonies on apple die out soon after these are produced. On apple, inhabited galls of apple/anthriscus aphid and those of rosy leaf-curling aphid are usually readily distinguished by the presence in the latter of wingless females (apterae) and characteristic 'alatiform apterae'.

Dysaphis chaerophylli (Börner)
Apple/chervil aphid

At least in Germany and, to a lesser extent in the Netherlands, this aphid is considered a minor pest of apple, forming leaf galls in the spring that are essentially similar to those caused by apple/anthriscus aphid (*Dysaphis anthrisci*) and rosy leaf-curling aphid (*D. devecta*). The summer host is *Chaerophyllum* (e.g. *C. bulbosum* and *C. hirsutum*).

Dysaphis devecta (Walker) (**128–130**)
Rosy leaf-curling aphid

A very local, but persistent, pest of apple, and usually present on only a few trees in an orchard. Known only from eastern Europe (e.g. Romania) and western Europe (e.g. England & Wales, France, Germany and the Netherlands).

DESCRIPTION
Adult viviparous female (aptera): 1.8–2.4 mm long; medium-sized, grey to dark bluish grey, dusted with white, waxy powder; siphunculi black, short, tapering, and flanged apically; cauda black, short and triangular. **Nymph:** pinkish to bluish grey.

LIFE HISTORY
This species overwinters as eggs laid under loose bark or bark flakes and in deep crevices in rough bark on the trunk and main branches. The eggs hatch in early spring and, by the green-cluster stage, gall-inhabiting colonies are often already apparent on the rosette leaves. Later, infestations spread to the young shoots. In the second

128 Rosy leaf-curling aphid (*Dysaphis devecta*).

126 Apple/anthriscus aphid (*Dysaphis anthrisci*) – galled apple leaf.

129 Rosy leaf-curling aphid (*Dysaphis devecta*) – galled apple leaves.

127 Apple/anthriscus aphid (*Dysaphis anthrisci*) – underside of galled apple leaf.

130 Rosy leaf-curling aphid (*Dysaphis devecta*) – underside of galled apple leaf.

and third generations, both winged (alate) and wingless (apterous) females are produced (and also so-called 'alatiform apterae' – wingless females with a partly sclerotized, alate-like body form). The 'normal' apterae then produce winged males, whereas the alatae and 'alatiform' apterae produce wingless oviparae. The oviparae eventually mate, whilst still within the maternal galls; they then emerge to deposit eggs, usually from mid-June to mid-July. Colonies die out early in the season, most live aphids disappearing by the end of July. In spite of the production of winged female forms, there is no apparent migration; spread of infestations from tree to tree is slow and usually achieved by walking aphids.

DAMAGE

Infested leaves are curled downwards and become characteristically bright red. Attacks are largely confined to older trees with rough bark and tend to occur on the same trees year after year. The vigour of heavily infested hosts is reduced.

Dysaphis plantaginea (Passerini) (**131–134**)
Rosy apple aphid

A serious pest of apple, but usually uncommon in well-sprayed orchards. Widely distributed in Europe; also present in Africa, North America, South America and Asia.

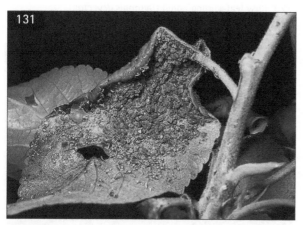

131 Rosy apple aphid (*Dysaphis plantaginea*).

132 Rosy apple aphid (*Dysaphis plantaginea*) – damage to young shoot.

133 Rosy apple aphid (*Dysaphis plantaginea*) – damage to branch.

134 Rosy apple aphid (*Dysaphis plantaginea*) – damage to fruits.

DESCRIPTION

Adult viviparous female (aptera): 2.1–2.6 mm long; medium-sized, pink to dark bluish grey, with a mealy powdering of white wax; siphunculi black, of moderate length, tapered, with flanged tips; cauda dark, short and triangular.

LIFE HISTORY

Eggs, which are laid in the autumn in bark crevices on spurs and smaller branches and at the base of buds, hatch in the spring. At first the developing aphids attack the buds, spur leaves and rosette leaves; later, infestations spread to the young shoots. By late May or June, numerous large colonies may be present on infested trees. In June or July, winged aphids are produced and these disperse to *Plantago*, particularly *P. lanceolata*. Breeding on apple will continue into August, even after the production of winged forms, if new growth is still available on host trees. A return flight from *Plantago* to apple occurs in the early autumn.

DAMAGE

Infested leaves are severely curled downwards and distorted. They sometimes turn yellowish or brown, but never red (cf. damage caused by rosy leaf-curling aphid, *Dysaphis devecta*). Attacked shoots are stunted and twisted. Heavy attacks may also lead to premature leaf fall and death of shoots. Fruits from infested trusses are small and malformed, with uneven surfaces; they also 'ripen' prematurely. Aphids feeding directly on fruits cause reddish blotches to develop on the skin.

Dysaphis pyri (Boyer de Fonscolombe) (135–136)

Pear/bedstraw aphid

The most important aphid on pear. Widespread and often common in Europe; also present in North Africa and parts of Asia.

DESCRIPTION

Adult viviparous female (aptera): 1.3–2.5 mm long; medium-sized, plump, pinkish and covered with mealy wax; antennae much shorter than body; siphunculi of moderate length and flanged apically.

LIFE HISTORY

This species overwinters in the egg stage on the spurs and branches of pear. Eggs usually hatch by the white-bud stage and the aphids at first colonize the rosette leaves. However, as soon as young shoots are produced, these too are invaded. By the end of May, numerous large colonies may be present and these persist throughout June and often longer. Where attacks are severe, infestations may even spread to adjacent trees. Winged aphids are produced from June onwards and these fly away to *Galium*, particularly *G. aparine* and *G. mollugo*, where breeding continues. A return migration to pear occurs in the early autumn.

DAMAGE

Infested leaves are severely distorted, the twisted foliage turning yellowish. Heavy infestations considerably check plant growth, the aphids producing vast quantities of honeydew.

135 Pear/bedstraw aphid (*Dysaphis pyri*).

136 Pear/bedstraw aphid (*Dysaphis pyri*) – infested pear shoot.

Hyalopterus amygdali (Blanchard) (**137–138**)
Mealy peach aphid

This Eurasiatic species is essentially similar to mealy plum aphid (*Hyalopterus pruni*) and has a similar life cycle, but the primary hosts are apricot and peach; it is also reported on almond. Infestations occur widely on peach trees in the Mediterranean basin. It is readily distinguished from other leaf-infesting aphids on peach by the very short siphunculi and by the characteristic mealy-coated appearance of colonies. The aphids typically form dense colonies on the underside of leaves, without causing leaf curl.

Hyalopterus pruni (Geoffroy) (**139–141**)
Mealy plum aphid

A pest of damson and plum; other primary hosts include apricot and *Prunus spinosa*. Cosmopolitan. Widely distributed in Europe.

DESCRIPTION

Adult female (aptera): 1.5–2.6 mm long; small to medium-sized, light green, marked with dark green and with a bluish-grey tinge, being coated profusely with white, mealy wax; siphunculi, legs and antennae with dusky tips; siphunculi short, rounded at apex and unflanged; cauda finger-like.

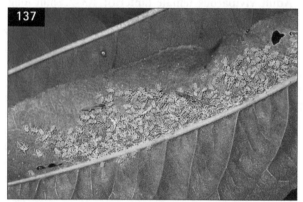

137 Mealy peach aphid (*Hyalopterus amygdali*).

138 Mealy peach aphid (*Hyalopterus amygdali*) – damage to leaf.

140 Mealy plum aphid (*Hyalopterus pruni*) – colony and associated wax.

139 Mealy plum aphid (*Hyalopterus pruni*).

141 Mealy plum aphid (*Hyalopterus pruni*) – infested, wax-coated shoots.

LIFE HISTORY

Eggs are laid during the autumn at the base of buds on the young shoots. They hatch in the spring, usually by the white-bud stage. Colonies of wingless aphids then develop on the underside of leaves. During the spring, aphid numbers are generally low. By June or July, however, the underside of leaves and the stems of infested shoots are often completely covered with the aphids. The foliage of an infested tree quickly becomes sticky with honeydew, upon which sooty moulds develop. Although production of wingless forms may continue into August, winged aphids begin to appear from late June or early July onwards. These then migrate to *Phragmites communis* and waterside grasses. A return migration to damson and plum takes place in the autumn.

DAMAGE

This species does not cause leaf curl (cf. leaf-curling plum aphid, *Brachycaudus helichrysi*); however, when attacks are severe, infested leaves may turn yellow and fall prematurely. In addition to direct damage, the development of sooty moulds on the aphid honeydew may reduce the photosynthetic activity of leaves. The aphid is a weak vector of plum pox ('Sharka').

Hyperomyzus lactucae (Linnaeus) (142–143)
Currant/sowthistle aphid

The most abundant and frequently noted aphid pest of black currant; also found on red currant and white currant. Originally Palaearctic. Also now present in North America, South America, Australia and New Zealand. Widespread and common in Europe.

DESCRIPTION

Adult viviparous female (aptera): 2.0–2.7 mm long; medium-sized, spindle-shaped and green, with pale legs and siphunculi; antennae grey apically, and much shorter than the body; siphunculi of moderate length, distinctly swollen in the middle, with dark tips.

LIFE HISTORY

Eggs hatch in March or early April and the aphids immediately infest the nearest leaf buds. Soon afterwards, they enter the unfurling clusters of flower buds, feeding until the late grape stage under a protective cover of bud scales. Later, the aphids invade the shoot tips where, by May, large colonies often develop on unsprayed bushes. Adults of the third generation are winged and these migrate to *Sonchus* where, during the summer months, they breed on the flower heads. Winged females fly back to currant bushes in the autumn and then produce a generation of oviparae. These subsequently mate with winged males which return from *Sonchus*. The overwintering eggs are then laid in the axils of buds.

DAMAGE

Spring infestations on currant cause a down-curling and a yellow mottling of the leaves; in addition, shoot growth may be stunted.

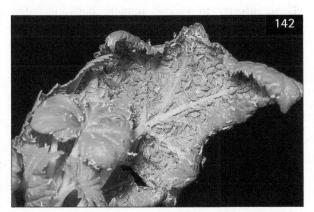

142 Currant/sowthistle aphid (*Hyperomyzus lactucae*).

143 Currant/sowthistle aphid (*Hyperomyzus lactucae*) – damage to black currant shoot.

Hyperomyzus pallidus Hille Ris Lambers (**144**)
Gooseberry/sowthistle aphid
A pest of gooseberry. Widely distributed in Europe. An introduced pest in the USA.

DESCRIPTION
Adult viviparous female (aptera): 2.1–3.0 mm long; medium-sized, light green, elongate, with pale antennae, legs and siphunculi; siphunculi of moderate length and distinctly swollen.

LIFE HISTORY
Eggs overwinter on the shoots of gooseberry bushes and hatch in the spring. Colonies of aphids then develop on the underside of leaves, particularly at the tips of young shoots. In late spring or early summer, winged aphids migrate to *Sonchus arvensis* where breeding continues on the lower leaves until the return migration to gooseberry in the autumn.

DAMAGE
Spring infestations on young shoots cause stunting and leaf curl.

Hyperomyzus rhinanthi (Schouteden)
Currant/yellow-rattle aphid
A generally uncommon, if not rare, pest of red currant. Present throughout Europe, but occurring mainly in cooler, damper regions; a well-established pest in Iceland.

DESCRIPTION
Adult viviparous female (aptera): 2.4–2.9 mm long; greenish or yellowish, with a large, shiny black patch on the abdomen dorsally; siphunculi black and distinctly swollen.

144 Gooseberry/sowthistle aphid (*Hyperomyzus pallidus*) – damage to young gooseberry shoot.

LIFE HISTORY
In Iceland, this species occurs on red currant throughout the year. Elsewhere, the life cycle includes a summer migration to secondary hosts such as *Euphrasia officinalis*, *Rhinanthus crista-galli* and *R. major*. On these secondary hosts it infests the flowers but remains concealed inside the inflated calyxes. Winged aphids return to currants in early autumn and, following a sexual phase, overwintering eggs are eventually laid on the shoots.

DAMAGE
Spring infestations on young shoots and suckers can cause leaf curl, but this species is rarely sufficiently numerous to be regarded as an economic pest.

Macrosiphum euphorbiae (Thomas)
Potato aphid
A polyphagous, minor pest of strawberry; also found, occasionally, on pear, loganberry, raspberry and hop, but more frequently associated with herbaceous plants such as lettuce, potato and a wide range of ornamentals (including those grown in greenhouses). Of North American origin, first introduced into Europe in the early 1900s, and now cosmopolitan. Present throughout Europe.

DESCRIPTION
Adult viviparous female (aptera): 1.7–3.4 mm long; medium-sized to large, pyriform or spindle-shaped, often shiny, yellowish green or pinkish; antennae long; siphunculi very long, cylindrical and sometimes dark tipped; cauda thin and elongate.

LIFE HISTORY
Under suitable conditions, this species may breed on strawberry throughout the year. In spring, populations develop rapidly on the young foliage and by late April, when numbers are greatest, winged aphids are reared. These disperse to various summer hosts, including potato crops, while small numbers of wingless forms may remain on strawberry throughout the summer. Mixed colonies of this species and *Acyrthosiphon rogersii* often occur on strawberry.

DAMAGE
This aphid usually causes little direct damage to fruit crops. On hop, it may transmit hop mosaic virus.

Macrosiphum funestum (Macchiati)
Scarce blackberry aphid
Associated mainly with wild *Rubus fruticosus*, but also a minor pest of cultivated blackberry. Widely distributed in Europe, but local and generally uncommon.

DESCRIPTION
Adult viviparous female (aptera): 2–4 mm long; medium-sized to large, spindle-shaped, dull green or reddish, with very long legs and antennae; siphunculi dark, very long, slightly curved outwards and flanged.

LIFE HISTORY
This species overwinters in the egg stage. In spring, colonies develop at the tips of the young canes; these become most obvious in May and June. Small numbers of wingless, sexual forms occur in the autumn and eggs are eventually laid on the canes.

DAMAGE
Infestations on cultivated blackberry have little or no effect on growth or cropping.

Melanaphis pyraria (Passerini) (**145–146**)
Pear/grass aphid
A minor pest of pear. Although widely distributed in pear-growing areas of mainland Europe, the pest is generally uncommon in the British Isles.

145 Pear/grass aphid (*Melanaphis pyraria*).

DESCRIPTION
Adult viviparous female (aptera): 1.3–2.1 mm long; small, dark brown to blackish; siphunculi dark, cylindrical and slightly shorter than the relatively short, black cauda; antennae of moderate length, and pale, with the base and tips brown; legs with the femora and tarsi mainly dark and the tibiae mainly pale. **Nymph:** pale purplish to brownish black.

LIFE HISTORY
This species overwinters on pear as eggs. These eggs mostly hatch by the white-bud stage. Ant-attended colonies then develop during the spring and early summer on the underside of leaves and on the young shoots. Breeding often continues on pear until August, with winged forms migrating to grasses, particularly *Brachypodium sylvaticum* and *Poa annua*, from June onwards. A return migration to pear occurs in the autumn.

DAMAGE
Spring infestations cause the leaves at the tips of shoots to curl; such damage is particularly evident in May and June. Attacks, however, are relatively unimportant and are usually confined to a few scattered trees.

Myzus ascalonicus Doncaster (**147**)
Shallot aphid
In some seasons an important pest of strawberry, but better known on agricultural crops such as potato and sugar beet. Cosmopolitan. Widely distributed in Europe.

DESCRIPTION
Adult viviparous female (aptera): 1.1–2.2 mm long; small to medium-sized, shiny, light brown, yellowish brown or greenish brown; front of head emarginate, the

146 Pear/grass aphid (*Melanaphis pyraria*) – leaf curl on pear.

147 Shallot aphid (*Myzus ascalonicus*) – damage on strawberry plant.

antennal tubercles only slightly convergent; siphunculi distinctly swollen towards the tip; cauda bluntly triangular, barely visible from above.

LIFE HISTORY
Winged individuals often migrate to strawberry in the autumn, where colonies of wingless aphids develop. The aphids breed throughout the winter and spring; in May or early June, winged forms are produced and these soon depart for various summer hosts, including potato, shallot and sugar beet. Colonies on strawberry then decline rapidly and finally die out. Breeding is entirely parthenogenetic.

DAMAGE
Unlike other aphids on strawberry, feeding by this species invokes a noticeable host-plant reaction. Infested plants become severely stunted when growth commences in the spring, the petioles being shortened and the leaves curled and twisted. Blossom trusses on infested plants are similarly affected. Cropping is much reduced and fruits remain small, are of poor quality and often quite unmarketable. The presence of stunted plants becomes increasingly obvious from late April onwards. The symptoms tend to appear first on individual plants or in small patches within the plantation. However, if infestations are heavy, damage soon spreads until large areas, or even complete beds, are affected. Survival of this species on strawberry during the winter period is dependent upon favourable weather conditions; below-average temperatures, particularly in January and February, greatly reduce the likelihood of damage occurring.

Myzus ornatus Laing (**148–149**)
Violet aphid
A highly polyphagous pest. Sometimes found on strawberry, but mainly a pest of ornamentals. Of worldwide distribution, it is widespread in Europe.

DESCRIPTION
Adult viviparous female (aptera): 1.0–1.7 mm long; small or very small; oval, slightly dorsoventrally flattened, pale yellow or green; abdomen with dark green or brownish markings; siphunculi pale, moderately long, stout and cylindrical.

LIFE HISTORY
This species breeds parthenogenetically throughout the year, in cooler regions surviving the winter in greenhouses. The aphids live singly on the leaves; although sometimes numerous, they do not form dense colonies.

DAMAGE
Infested plants are often contaminated by honeydew, amongst which cast-off nymphal skins accumulate and upon which, sooty moulds develop. The aphid is a vector of strawberry crinkle virus.

Myzus persicae (Sulzer) (**150–151**)
Peach/potato aphid
A polyphagous pest of almond, nectarine and peach; although a vector of plum pox ('Sharka'), this aphid rarely occurs on plum. In southern Europe (e.g. Italy), infestations are also reported on strawberry. Of worldwide occurrence.

DESCRIPTION
Adult viviparous female (aptera): 1.2–2.1 mm long; small to medium-sized, whitish green, through green, reddish green to pink or reddish black; front of head emarginate; siphunculi moderately long, usually pale, and sometimes with darkened tips (cf. *Myzus varians*). **Nymph:** green, yellowish, pinkish or reddish green.

148 Violet aphid (*Myzus ornatus*).

149 Violet aphid (*Myzus ornatus*) – infested strawberry leaf.

LIFE HISTORY

Eggs on peach hatch in February or March and colonies of typically green, wingless aphids develop on the spur leaves. Later, infestations spread to the young shoots. Winged forms are produced in May and June, such individuals developing from reddish-green nymphs. These winged aphids migrate to various summer hosts, including many weeds and crops such as potato and sugar beet. A return migration to peach occurs in the autumn where, following a sexual phase, eggs are laid. This species also continues to breed viviparously throughout the winter on various brassicaceous plants (including vegetable brassicas) and on other secondary hosts, such as potato and sugar beet, upon which it is a notorious virus vector.

DAMAGE

Infested peach leaves are severely curled and shoot growth is stunted or even halted. Attacks are often sufficiently serious to affect fruit quality and quantity; they may also have a detrimental effect on tree vigour and reduce the following year's crop.

Myzus pruniavium Börner (152–154)

Sweet-cherry aphid

An often common pest of cherry, both fruiting and ornamental. Present throughout Europe. Also found in parts of Asia, Australia, New Zealand and North America. For many years, infestations on cherry (both sweet and sour) were attributed to *Myzus cerasi*

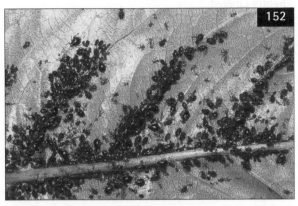

152 Sweet-cherry aphid (*Myzus pruniavium*).

150 Peach/potato aphid (*Myzus persicae*).

153 Sweet-cherry aphid (*Myzus pruniavium*) – infested fruitlet.

151 Peach/potato aphid (*Myzus persicae*) – damage on peach.

154 Sweet-cherry aphid (*Myzus pruniavium*) – damage to young shoot.

(Fabricius). However, the main species (that infests sweet cherry – namely, *Prunus avium*), is now known to be *M. pruniavium*; *M. cerasi* (here cited as the sour-cherry aphid) is less harmful and infests sour cherry – namely, *Prunus cerasus*.

DESCRIPTION
Adult viviparous female (aptera): 1.5–2.5 mm long; medium-sized, shiny, dark brown to black; front of head emarginate, siphunculi black, moderately long and slightly tapered.

LIFE HISTORY
Eggs are laid in the autumn at the base of buds and in bud axils on the spurs and young shoots. They hatch in March or early April, before the white-bud stage, and ant-attended colonies of wingless aphids quickly develop on the underside of the young leaves. From mid-May onwards, by which time colonies may be very large, winged forms are produced and these migrate to summer hosts such as *Euphrasia officinalis*, *Galium* and *Veronica*. Breeding colonies of wingless individuals may persist on cherry throughout July, and even into August, but eventually die out. In autumn, there is a return migration from summer hosts to cherry where, following a sexual phase, oviparae eventually deposit the winter eggs. The life cycle of *M. cerasi* is similar to that of *M. pruniavium*, but winged aphids usually migrate to the summer hosts slightly later (from the end of May or early June onwards).

DAMAGE
Sweet cherry: leaves infested by sweet-cherry aphid become severely curled, and damage to nursery stock and to young shoots on older trees can be severe. With heavy infestations, shoots become stunted and their tips may die. Developing fruits and leaves are also contaminated by honeydew and sooty moulds. Persistent infestations during the summer months are particularly harmful in nurseries. **Sour cherry:** dense colonies of sour-cherry aphid develop on the underside of leaves; infested leaves may become slightly cupped downwards, but they do not become curled.

Myzus varians Davidson (155–156)
Peach leaf-roll aphid
A pest of peach in various parts of Europe, including the British Isles. Of eastern Asian origin; the first European records were from Switzerland in 1947. Also present in North America, although not on peach.

DESCRIPTION
Adult viviparous female (aptera): 1.7–2.3 mm long; medium-sized; light green to green; front of head emarginate; antennae dark tipped; basal half of siphunculi green, otherwise black (cf. *Myzus persicae*).

LIFE HISTORY
This species overwinters in the egg stage on peach, the primary host. The eggs hatch in the early spring. Following the development of colonies of wingless aphids, winged forms are produced, typically in late May or early June. These aphids migrate to *Clematis*, the secondary host; aphid colonies may also persist on peach throughout the summer. A return migration to peach occurs in the late autumn (October or November) where, following a sexual phase, oviparae eventually deposit the winter eggs.

DAMAGE
On peach, aphids (which inhabit the underside of the leaves) cause tight, cigar-like rolling of the foliage; infested tissue later turns red. The aphid is also a vector of plum pox ('Sharka').

155 Peach leaf-roll aphid (*Myzus varians*).

156 Peach leaf-roll aphid (*Myzus varians*) – damage.

Nasonovia ribisnigri (Mosley)
Currant/lettuce aphid

Widespread and common on gooseberry; other primary hosts include black currant and red currant bushes. A well-known pest of cultivated lettuce, one of its summer hosts. Eurasiatic. Present throughout Europe. An introduced pest in North and South America.

DESCRIPTION
Adult viviparous female (aptera): 1.3–2.7 mm long; medium-sized, spindle-shaped, light green to dark green, and shiny; siphunculi moderately long, dark-tipped, and apically flanged; cauda elongate; antennae and legs with blackish tips.

LIFE HISTORY
This species usually overwinters as eggs laid at the base of buds on gooseberry bushes. In spring, following egg hatch, colonies develop at the tips of young shoots and become noticeable in late April or early May. In late May or June, winged aphids disperse to cultivated lettuce and to a wide variety of wild summer hosts, including *Crepis*, *Hieracium* and *Veronica*. Aphids return to gooseberry in the autumn, although some may remain on greenhouse-grown lettuce throughout the winter.

DAMAGE
On gooseberry, the aphids cause slight leaf curl at the shoot tips, but no mottling of the foliage. Heavy infestations may check the growth of bushes and shorten the distance between shoot nodes. The aphids are also vectors of gooseberry vein-banding virus.

Phorodon humuli (Schrank) (**157–158**)
Damson/hop aphid

A major pest of hop and often regarded as one of the main limiting factors to crop production. Primary (winter) hosts include damson, peach and plum, but on these the aphid is usually not an important pest. Present throughout Europe. Also present in North Africa and parts of Asia, and an introduced pest in North America and New Zealand.

DESCRIPTION
Adult viviparous female (aptera): 1.1–1.8 mm long (on hop, in summer); 2.0–2.6 mm long (on plum, in spring), small to medium-sized; whitish to pale yellowish green, relatively shiny, and marked with three, dark green longitudinal stripes; head with a characteristic pair of elongate projections (antennal tubercles) arising from the antennal bases; siphunculi moderately long.

157 Damson/hop aphid (*Phorodon humuli*) – infested plum leaf.

158 Damson/hop aphid (*Phorodon humuli*) – infested hop leaf.

LIFE HISTORY
This species overwinters as eggs on spurs and shoots of damson, plum and other kinds of *Prunus*, including *P. spinosa*. In spring, the eggs hatch and aphid colonies subsequently develop on the underside of the leaves (which they do not distort). The aphids are often found on damson or plum leaves distorted by plum leaf-curling aphid (*Brachycaudus helichrysi*) and this can lead to misidentification of the causal agent. From mid-May onwards, winged forms are produced and these migrate to wild and cultivated hop. Usually, after feeding on several hop plants, individuals settle down at the tips of the bines or lateral shoots where colonies of wingless aphids soon develop. Migration to hop may continue into August, but is usually at its height in the second half of June. In late summer, aphids occur on the older leaves and may also attack the developing cones. In September, there is a return migration of winged forms to winter hosts where, following a sexual phase, eggs are eventually laid.

DAMAGE

Damson and plum: apart from slight leaf curl in spring, little or no damage is caused, but the aphid is implicated in the transmission of plum pox ('Sharka'). **Hop:** infestations seriously disrupt growth. Also, as colonies rapidly develop, honeydew contaminates the foliage, allowing sooty moulds to become established. Where aphids attack the hop cones, considerable damage can result and this may not be noticed until harvest-time. The aphids are also virus vectors, including hop mosaic virus.

Rhopalosiphoninus ribesinus (van der Goot)
Currant stem aphid

A minor pest of currant bushes growing in shaded situations. Widely distributed in northern Europe, including the British Isles, Germany, the Netherlands, Poland and Scandinavia.

DESCRIPTION

Adult viviparous female (aptera): 2.0–2.5 mm long; medium-sized, plump, dull reddish brown to brownish black; antennal tubercles well developed; antennae long and thin; siphunculi strongly swollen between their constricted basal and strongly flanged apical sections; cauda short and triangular.

LIFE HISTORY

This species occurs only on *Ribes*, overwintering in the egg stage. The aphids typically infest the old wood, but colonies may also spread onto the young shoots and underside of the leaves. There are several generations annually. Males are wingless.

DAMAGE

Although overwintering eggs are sometimes present on the shoots in considerable numbers (and may then cause concern), this species is not of economic importance.

159 Apple/grass aphid (*Rhopalosiphum insertum*) – infested apple bud.

Rhopalosiphum insertum (Walker) (**159**)
Apple/grass aphid

Often abundant on apple, particularly fruiting trees, but rarely a serious pest. Medlar, pear and quince are also attacked, as are other rosaceous plants such as *Crataegus*, *Cotoneaster* and *Sorbus*. Palaearctic. Widely distributed in Europe. Also now established in the Azores and in Australia.

DESCRIPTION

Adult viviparous female (aptera): 2.0–2.5 mm long; medium-sized, oval and shiny; bright green to yellowish green, with a darker green longitudinal dorsal stripe; siphunculi short, light green, and flanged apically; antennae short and dark tipped. **Nymph:** greenish yellow.

LIFE HISTORY

Overwintered eggs hatch in spring by the green-cluster stage; the nymphs then feed on the underside of the young rosette leaves or amongst the flower buds. At the pink-bud stage, the aphids may sometimes invade the petals. Although the first generation consists of wingless aphids, most of the second are winged. From early to mid-May, at the height of the blossom period, the first winged aphids migrate to grasses, particularly *Poa annua*; with the production of these summer forms, populations on apple rapidly decline. In autumn, winged females return to fruit trees and their wingless progeny, the oviparae, mate with winged males that also return from grasses. Eggs are eventually deposited on spurs and branches, just before leaf fall.

DAMAGE

There is little or no risk of damage from this species after mid-blossom. However, earlier in the spring the aphids may cause slight leaf curl. Damage is rarely important unless populations are very large. Heavy infestations tend to occur only following summers with sufficient rainfall to promote and maintain good growth of grasses.

Rhopalosiphum nymphaeae (Linnaeus)
Water-lily aphid

This aphid is mainly of significance as a pest of aquatic plants, such as *Nuphar lutea* and *Nymphae alba* which (along with plants such as *Alisma plantago-aquatica*, *Sagittaria sagittifolia* and *Typha*) are its summer hosts. The primary hosts (upon which winter eggs are laid) are *Prunus spinosa* and certain other species of *Prunus* (including, at least in parts of mainland Europe, cultivated apricot, peach and plum). On fruit trees, dense colonies of the dark greenish-brown aphids (partly flecked with white wax), may develop in the spring on the young shoots, causing noticeable distortion of the foliage. Such colonies usually die out by the end of June. A return migration from summer hosts to *Prunus* occurs in the autumn, although under favourable conditions colonies may survive on secondary hosts throughout the year.

Sitobion fragariae (Walker)
Blackberry/cereal aphid

A common European species, attacking wild and cultivated blackberry. Eurasiatic. An introduced pest in South Africa, North America and South America.

DESCRIPTION
Adult viviparous female (aptera): 2–3 mm long; moderately large, spindle-shaped, green and shiny; legs and antennae of moderate length; siphunculi black, long, and flanged apically.

LIFE HISTORY
Eggs, having overwintered on blackberry canes, hatch from early February to March. Two generations of wingless aphids then occur. By May, large, dense colonies develop on the leaves of fruiting canes, such infestations often spreading down the shoots. Winged aphids are reared in May and June. These then depart for grasses and, occasionally, *Juncus*, where breeding continues until a return migration to blackberry takes place in the autumn. Eggs are eventually deposited on the canes in November and December. Small numbers of returning migrants may fly to loganberry, raspberry and strawberry, usually in October. Although oviparae appear on these plants, and viable eggs are then laid, infestations fail to develop in the spring, apparently because the aphids that emerge from eggs deposited on such hosts are unable to mature.

DAMAGE
Heavy infestations in spring will produce severe leaf curl and substantially reduce yields of blackberries.

Toxoptera aurantii (Boyer de Fonscolombe)
Black citrus aphid

A worldwide pest of citrus. Well established in southern Europe. In tropical regions also found on many other hosts.

DESCRIPTION
Adult viviparous female (aptera): 1.1–2.0 mm long; oval-bodied, reddish brown, through brownish black to black; oval-bodied; antennae distinctly banded black and white; siphunculi black; cauda black; peg-like hairs on the hind tibiae and ridges located on the underside of the abdomen enable the aphid to stridulate. **Adult (alata):** median vein of forewing usually with just one branch (cf. *Toxoptera citricida*). **Nymph:** brownish.

LIFE HISTORY
The aphids overwinter in small groups on the buds of host trees. From spring onwards, dense, ant-attended colonies develop on the underside of leaves and may also spread to the young shoots and flower buds. When disturbed, a distinct scraping sound may be emitted from large colonies. Breeding continues throughout the spring, summer and autumn, with many overlapping generations annually, and is especially favoured by ambient temperatures of 20–25°C; in hotter regions there may be a period of summer diapause. Alates occur commonly, and readily spread infestations. Breeding is entirely parthenogenetic, there being no sexual forms.

DAMAGE
The aphid is of particular importance in the spring, infested flower buds failing to develop and dropping off. Symptoms of lesser importance include slight distortion of leaves, resulting from damage to the midrib. The aphids, which excrete considerable quantities of honeydew upon which sooty moulds develop, are also virus vectors.

Toxoptera citricida (Kirkaldy)
Brown citrus aphid

A tropical pest of citrus and an important virus vector. Widely distributed in Africa, South America, Asia and Australasia. Also reported, occasionally, in citrus-growing regions of Europe (where it is treated as a quarantine pest), but supposed specimens usually turn out to be black citrus aphid (*Toxoptera aurantii*).

DESCRIPTION
Adult viviparous female (aptera): 2.0–2.8 mm long; dark reddish brown to black. **Adult (alata):** median vein of forewing with two branches (cf. *T. aurantii*).

LIFE HISTORY
In common with black citrus aphid (*T. aurantii*), this species is parthenogenetic, but the aphids do not stridulate.

DAMAGE
Infested flowers usually fail to open, young shoots are distorted, and leaves become brittle, wrinkled and curled downwards. The aphids are also important vectors of citrus tristeza closterovirus.

Subfamily CALLAPHIDINAE

A small subfamily of usually delicate-looking aphids, with short, stumpy siphunculi and a short, lobed or constricted cauda; antennae usually 6-segmented. All species are monoecious, lacking an alternation of generations between primary and secondary hosts (see under family Aphididae).

Callaphis juglandis (Goeze) (**160–161**)
Large walnut aphid
A minor pest of walnut. Eurasiatic. Locally common in Europe. An introduced pest in North America.

DESCRIPTION
Adult viviparous female (alata): 3.5–4.3 mm long; large, bright greenish yellow, with dark orange and brown markings on the abdomen; body with long, thin hairs; veins of forewings dusky bordered.

LIFE HISTORY
This species overwinters in the egg stage. Eggs hatch in the spring and aphids then occur on the upper surface of leaves, usually feeding in double rows along the midrib. Following a sexual phase, eggs are eventually laid in October at the base of buds.

DAMAGE
Although infested leaves may turn partially yellow and perhaps fall prematurely, mature trees are not harmed. However, the vigour of young plants may be reduced by heavy infestations.

Chromaphis juglandicola (Kaltenbach)
Small walnut aphid
A generally uncommon pest of walnut; although sometimes numerous, of importance only on young trees and nursery stock. Eurasiatic. Widely distributed in Europe. An introduced pest in North America.

DESCRIPTION
Adult viviparous female (alata): 1.5–2.5 mm long; small, pale yellow, with red eyes; siphunculi short, conical and flanged; veins of forewings not conspicuously dark bordered (cf. *Callaphis juglandis*). **Nymph:** whitish, with capitate body hairs.

LIFE HISTORY
Overwintered eggs hatch in April and aphids then feed on the underside of walnut leaves, excreting liberal quantities of honeydew. This honeydew appears to have an adverse effect on colonies of large walnut aphid (*Callaphis juglandis*) and probably explains why infestations of the two species rarely occur together. In summer, the aphids may also infest the developing nutlets. In late autumn, following the production of sexual forms, eggs are eventually laid at the base of buds, or amongst hairs or close to leaf scars on the young shoots.

DAMAGE
Infested hosts are contaminated by honeydew and heavy attacks may reduce the vigour of young trees; significant infestations, however, are rare. On mature trees, infestations which spread to the nutlets may affect nut development. However, they are not of economic importance.

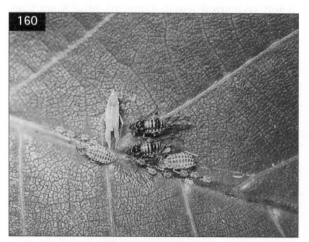

160 Large walnut aphid (*Callaphis juglandis*).

161 Large walnut aphid (*Callaphis juglandis*) – ant-attended nymphs.

Myzocallis coryli (Goeze) (**162**)
Hazel aphid

An abundant species on wild *Corylus avellana*, and a minor pest of cultivated hazelnut. Cosmopolitan. Widely distributed in Europe; also present in North and South America.

DESCRIPTION
Adult viviparous female (aptera): 1.3–2.2 mm long; small, yellowish white and shiny, with large red eyes; antennae with black tips; siphunculi short, stout and cone-like (cf. *Corylobium avellanae*, subfamily Aphidinae). **Nymph:** whitish, with long, capitate body hairs.

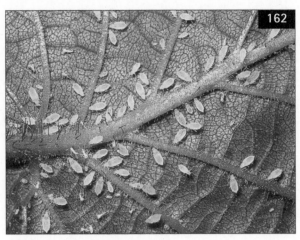

162 Hazel aphid (*Myzocallis coryli*).

LIFE HISTORY
Overwintered eggs hatch in the spring. Aphids then feed and breed on the underside of the leaves of host plants, forming open colonies of scattered individuals. The aphids excrete sticky honeydew, which drops down and contaminates the underlying foliage. A sexual phase occurs in late autumn, following which the small, wingless oviparae deposit eggs on the shoots.

DAMAGE
Foliage on infested trees is not distorted. However, leaves may become contaminated by honeydew and blackened by sooty moulds that reduce photosynthetic activity. Otherwise, infestations are of little or no importance.

Subfamily **ERIOSOMATINAE**

Aphids often covered with flocculent masses of wax secreted from gland plates; terminal process of antennae short; siphunculi usually indistinct or absent; compound eyes reduced to three facets; tip of cauda broadly rounded; antennae of winged forms often with annulated segments. Species are often gall-forming, the galled structures on some hosts being very elaborate.

Eriosoma grossulariae (Schüle)
Gooseberry root aphid

A minor pest of gooseberry, the aphids infesting the roots during the summer months. Infestations also occur on the roots of *Ribes sanguineum*. The primary (winter) host is *Ulmus*. Eurasiatic. Present throughout Europe. An introduced pest in North America.

DESCRIPTION
Adult viviparous female (aptera) (on gooseberry): 1.0–1.5 mm long; small, oval, pink to yellowish white, surrounded by white, woolly masses of wax; siphunculi pore-like; cauda small.

LIFE HISTORY
This species overwinters as eggs on *Ulmus*. In spring, colonies develop amongst curled leaves, protected by masses of flocculent wax. Winged forms are produced in June and July, and these migrate to gooseberry where nymphs are deposited on or in the soil. The nymphs then attack the underground parts of the stems and roots. Colonies of wingless aphids develop during the summer, coated by masses of waxy 'wool'. In autumn, winged aphids are reared and these depart for *Ulmus* where, following a sexual phase, oviparae eventually deposit the winter eggs.

DAMAGE
Infestations on gooseberry are rarely noticed and do little harm to fruiting bushes. However, nursery stock and young plants may be weakened, stunted or even killed.

Eriosoma lanigerum (Hausmann) **(163–164)**
Woolly aphid

An important and notorious pest of apple; other hosts include quince, pear, *Chaenomeles*, *Cotoneaster* and *Sorbus*. Of worldwide distribution and generally common in Europe.

DESCRIPTION
Adult viviparous female (aptera): 1.2–2.6 mm long; small to medium-sized, purplish brown, covered with masses of white, mealy wax; body with numerous wax plates; antennae short; siphunculi pore-like; cauda small.

LIFE HISTORY
This aphid overwinters on apple trees as naked (i.e. wax-less) nymphs, sheltering in cracks in the bark or under loose bark. In late March or April, the nymphs become active and by the end of May breeding colonies are established beneath masses of conspicuous, sticky, white 'wool'. These colonies occur mainly on the spurs and branches, particularly where the bark is broken. A series of generations occurs during the summer months, with infestations also developing on the new growth; watershoots arising from the trunk or main branches are most liable to be attacked. A few winged aphids are produced in July. These may infest other apple trees. However, most natural spread is by young, wingless nymphs which often crawl, or are blown, from tree to tree. Populations show a decline in August, but rise again in September with the production of wingless, egg-laying females. Although these aphids may each deposit a single egg, there is no further development following a sexual phase. Survival of the species from year to year, therefore, is entirely dependent upon the

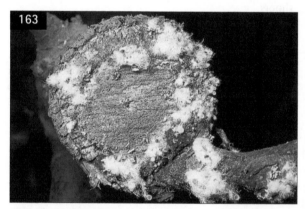

163 Woolly aphid (*Eriosoma lanigerum*) – colony on old apple pruning scar.

164 Woolly aphid (*Eriosoma lanigerum*) – old galls on apple trunk.

wingless, viviparous, parthenogenetic forms. In Europe, unlike the situation in South Africa, America and Australasia, the aphid does not infest apple roots (except, occasionally, those accidentally appearing above ground) and only rarely does it occur on the subterranean basal part of the trunk.

DAMAGE

The aphid does little direct harm to mature trees, but galls are often produced on infested wood; on young trees and nursery stock, galling may seriously disfigure the plants. Should the galls split, they may allow entry of pathogenic fungi such as gloeosporium rots (*Gloeosporium album* and *G. perennans*); apple canker (*Nectria galligena*) may also develop. The sticky masses of 'wool' produced by this species may contaminate foliage and developing fruits and are often a nuisance at harvest-time or when summer pruning is done.

Eriosoma lanuginosum (Hartig)

Elm balloon-gall aphid

This uncommon European aphid attacks pear, feeding on the roots or basal section of the trunk, but is not important as a pest. The primary host is *Ulmus*, the aphids migrating to pear in early summer and returning in the autumn. Adult viviparous females (apterae) on pear roots are 2.0–2.7 mm long, elongate-oval, pinkish or yellowish to reddish, and surrounded by masses of white, wax 'wool'. Colonies are established mainly on the smaller fibrous roots, but can also occur on the subterranean parts of watershoots.

Eriosoma ulmi (Linnaeus) (**165–166**)

Currant root aphid

A minor pest of currant. Eurasiatic. Widely distributed in Europe.

DESCRIPTION

Adult viviparous female (aptera) (on currant): 1.0–1.5 mm long; small, oval, brownish red; body with numerous wax plates; siphunculi pore-like; cauda small. **Alata:** body bluish grey to black; antennae annulated.

LIFE HISTORY

The primary host is *Ulmus*, upon which in spring the aphids inhabit tightly curled and galled leaves. In early summer, winged migrants depart for *Ribes*. Here, colonies of wingless aphids develop on the roots, often amongst masses of white or bluish wax. With the exception of the summer host, the life cycle is essentially similar to that of gooseberry root aphid (*Eriosoma grossulariae*).

DAMAGE

Infestations on currant roots are rarely noticed and do little harm to fruiting bushes. However, nursery stock and young plants may be weakened.

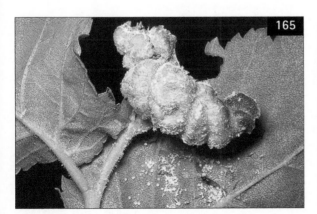

165 Currant root aphid (*Eriosoma ulmi*) – leaf gall on *Ulmus*.

166 Currant root aphid (*Eriosoma ulmi*) – infested currant roots.

Subfamily **LACHNINAE**

Aphids with terminal process of antennae very short; siphunculi usually short, very hairy cones; cauda broadly rounded. See also general comments on aphids given under the family Aphididae.

Pterochloroides persicae (Cholodkovsky) (**167a, b**)
Brown peach aphid
A pest of almond, apricot and peach; other hosts include plum and *Prunus spinosa*. Originating in central Asia and currently extending its range westwards into Europe, including Mediterranean areas; in 1994, the pest had reached south-eastern Spain, where it is now well established.

DESCRIPTION
Adult viviparous female (aptera): 2.7–4.2 mm long; large, oval-bodied, shiny, dark brown to black; abdomen with two longitudinal rows of large tubercles dorsally; antennae short; siphunculi cone-like.

LIFE HISTORY
Ant-attended colonies occur on the bark of host trees, particularly on the underside of branches. Colonies may also develop on the trunks of host trees. Although both winged and wingless aphids are produced, in Europe there are no sexual stages in the life cycle.

DAMAGE
Colonies are particularly damaging to peach and heavy infestations may cause death of branches.

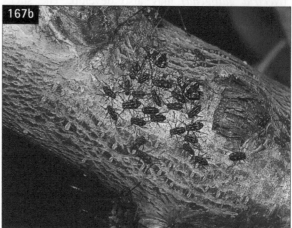

167a, b Brown peach aphid (*Pterochloroides persicae*) – on almond.

Family **PHYLLOXERIDAE** (phylloxeras)

Small to minute, aphid-like insects, without siphunculi; antennae 3-segmented; wings, when present, are held flat (i.e. horizontally) over the body when in repose.

Aphanostigma piri (Cholodkovsky)
Pear phylloxera

An important pest of pear in warmer parts of Europe, including France, Italy, Portugal and Switzerland. Also reported from other parts of the world, including Israel and Taiwan.

DESCRIPTION

Adult viviparous female (aptera): 0.8–1.0 mm long; body very small, yellow and pear-shaped, being broadest anteriorly and tapering posteriorly; spiracles absent on abdominal segments 2–5. **Sexual forms:** 0.5 mm long; ovoid and whitish to yellowish; devoid of mouthparts. **Egg (summer):** minute, yellowish green.

LIFE HISTORY

Where the life cycle is complete (i.e. where sexual forms occur), winter eggs hatch in the spring, after which there are several generations of parthenogenetic females (virginiparae). In September, sexuparae are produced and these deposit 'summer' eggs; these give rise to the non-feeding sexual forms. After mating, the females deposit the 'winter' eggs in bark crevices. In regions or years where sexual forms are not produced, breeding is entirely parthenogenetic and viviparous females form the overwintering stage.

DAMAGE

In spring, infested buds may be destroyed, such damage being particularly severe on late-flowering cultivars. Later in the season, on infested fruits, black necrotic areas form around the calyx and elsewhere. Such damage occurs mainly during fruit maturation and also during the post-harvest storage period. In the case of heavy infestations, more than half of the crop may be unsaleable.

Viteus vitifoliae (Fitch) (**168–171**)
Grape phylloxera

A notorious and potentially important pest of grapevine. Of North American origin, but now present in many grape-growing regions of the world. Widely distributed in mainland Europe, having been introduced accidentally from America in the mid-19th century. Absent from the British Isles, although sometimes accidentally introduced.

168 Grape phylloxera (*Viteus vitifoliae*) – infested leaf viewed from above.

169a, b, c Grape phylloxera (*Viteus vitifoliae*) – leaf galls.

DESCRIPTION
Adult female (gallicicola): 1.6–1.8 mm long; translucent and broadly pear-shaped, tapering posteriorly. **Egg (in leaf gall):** minute, oval and translucent. **Adult female (radicicola):** 1.0–1.4 mm long; yellow to yellowish green and pear-shaped, broadest anteriorly, with numerous tubercles dorsally on the head, thorax and abdomen. **Leaf gall:** a green to reddish, somewhat hairy, warty swelling (5–7 mm across), open to the underside (cf. galls formed by *Janetiella oenophila*, Chapter 5).

LIFE HISTORY
Grape phylloxera has a highly complex and flexible life cycle that varies considerably according to climatic conditions and host plant origin. In early summer, under conditions suitable for a sexual phase in the life cycle (rare in Europe), winged sexuparae are produced. These fly a short distance to neighbouring vines or may be carried by the wind to invade host plants some distance away. Eventually, the sexuparae deposit eggs and these develop into apterous (wingless) sexual forms (males and fmales). Each mated female lays a single egg in a crack or cranny on the wood of a two- or three-year-old vine. The egg hatches in the spring, at about bud burst, into a nymph that moves onto the underside of an expanding leaf, where it initiates a rounded gall on the lamina. The nymph soon develops within the gall into an apterous, parthenogenetic female (fundatrix). This fundatrix continues to inhabit the gall, in which she deposits a very large number of eggs (up to 1,200 in her lifetime). These eggs hatch, 8–10 days after being laid, into apterous aphids (gallicicolae) that, when mature, initiate further leaf galls and deposit their own eggs, there being several generations of gallicicolae in a season. As time progresses, instead of invading the foliage, more and more of the progeny of the gallicicolae drop to the ground and enter the soil. Here they attack the roots and cause nodule-like galls to develop. When mature, these aphids (as radicicolae) deposit eggs and thereby initiate further generations of root-inhabiting radicicolae. In Europe, where the production of sexuparae is very rare, the pest usually breeds continuously as radicicolae, with the occasional production of leaf-inhabiting gallicicolae without the intervention of a sexual phase.

DAMAGE
Infestations on the roots result in the development of large, irregular swellings and a blackening of the rootlets; infested plants may be killed. Leaf galling can result in considerable distortion and puckering of leaves and young shoots, but is of lesser significance. European vine (*Vitis vinifera*) (see Chapter 1) is very prone to attack, whereas American species are more or less resistant. Nowadays, therefore, grapevines in Europe (i.e. scions of *V. vinifera*) are largely grafted onto American rootstock and the pest is rarely seen in established vineyards. However, infestations often occur on adventitious vines arising in abandoned sites.

170 Grape phylloxera (*Viteus vitifoliae*) – lateral view of leaf gall.

171 Grape phylloxera (*Viteus vitifoliae*) – eggs.

Family **DIASPIDIDAE**
(armoured scales)

Small, flattened insects, with sedentary, wingless and legless adult females that remain attached permanently to the host plant. Individuals are protected by a hard, scale-like covering (commonly called a 'scale', but more correctly termed a 'test'), formed from cast-off nymphal skins (exuviae) and secreted wax (cf. family Coccidae).

First-instar nymphs of scale insects, often termed 'crawlers', are typically very mobile; following egg hatch, they usually swarm over host plants before becoming sedentary and moulting into the next developmental stage. Males, when present in the life cycle, develop through a pupa-like stage, sheltered beneath a scale-like covering (here termed the 'test'); they are minute insects, and very short-lived as adults; they possess only one pair of wings or are wingless.

Specific identification of scale insects is based primarily on features of adult females. However, appreciation of structural differences between closely related species requires microscopical examination and is a specialist task. Here, descriptions of species refer mainly to gross features, such as overall appearance, and are intended merely as a general guide.

Aonidiella aurantii (Maskell) (**172–173**)
California red scale

A pest of citrus in various parts of the Mediterranean basin. Also polyphagous on many other plants. Present in the warmer parts of Europe (e.g. France, Greece, Italy and Spain) on hosts such as pear, almond, plum and grapevine; infestations also occur in cooler regions on various greenhouse plants. Widely distributed in tropical and subtropical parts of the world.

DESCRIPTION
Female test: 1.5–2.0 mm across; circular, waxy and transparent. **Adult female:** broadly reniform, with the abdominal lobes extending back on either side of the pygidium; brownish red and clearly visible through the covering test. **Male test:** 1.2 mm long; oval and dark reddish brown. **Adult male:** 1 mm long; yellowish; winged. **First-instar nymph ('crawler'):** elongate-oval, reddish brown.

LIFE HISTORY
This species is favoured by stable, hot and dry conditions. The adult females are viviparous, giving birth to up to 150 nymphs over a period of about 2 weeks. The first-instar nymphs invade the stems, but will also settle on the foliage and developing fruits. Development to adulthood involves three nymphal stages and takes from 1 to 2 months. There are several overlapping generations annually, all stages often occurring together.

DAMAGE
Attacked trees are weakened; infested branches develop distinct lesions and become desiccated, the leaves turning yellow and falling off; developing fruits may also drop prematurely. If the pest settles on leaves or fruits, these may become noticeably speckled; invaded fruits are often unmarketable. Colonies of *A. aurantii* and citrus mussel scale (*Lepidosaphes beckii*) often occur together.

Aonidiella citrina (Coquillett)
Yellow scale

A subtropical and tropical pest of citrus, especially mandarin and orange; other hosts include olive and various rosaceous fruit trees. Of Far Eastern origin, and

172 California red scale (*Aonidiella aurantii*) – infested lemon.

173 California red scale (*Aonidiella aurantii*) and citrus mussel scale (*Lepidosaphes beckii*) – infestation on lemon.

widely distributed in Asia and many other parts of the world, including Africa, North America, South America and Australia. Now established in Italy and also present in parts of southern France.

DESCIPTION
Similar in appearance to *Aonidiella aurantii*, but adult female distinctly yellow.

LIFE HISTORY
Similar to that of *Aonidiella aurantii*, but ovoviviparous (producing eggs, with a definite shell, which hatch whilst still within the maternal body).

DAMAGE
Infested leaves turn yellow and may shrivel, resulting in die-back of heavily infested shoots. The pest is also a vector of virus diseases.

Aspidiotus nerii Bouché (**174–176**)
Oleander scale

A polyphagous pest of citrus and olive; also found on pomegranate and on various temperate host plants, including certain fruit crops (e.g. mulberry and plum). Widely distributed in southern Europe; also present in Africa, America and Australasia. Often present in cooler regions, including northern Europe, on protected (e.g. greenhouse-grown) plants, particularly evergreen ornamentals such as *Nerium oleander*.

DESCRIPTION
Female test: 1.8–2.2 mm across; almost circular, whitish to greyish brown, with a slightly eccentric apex. **Male test:** similar to that of female, but smaller and slightly elongate. **Adult male:** 1 mm long; yellowish, with darker legs and antennae; winged. **Egg:** minute, oval and yellow. **First-instar nymph ('crawler'):** pale yellow.

LIFE HISTORY
In southern Europe there are typically three generations annually, with young adult females (and sometimes also nymphs) overwintering. This species breeds both parthenogenetically and sexually, activity typically commencing in March and April, with a second brood being initiated in summer and a third in the autumn; generations may, however, overlap. In spring and summer, development from egg to adult takes about 2 months. The first-instar nymphs often swarm in large numbers before settling down to feed. This species excretes considerable quantities of honeydew upon which sooty moulds develop.

DAMAGE
Infested hosts are weakened, shoots may wilt and leaves may wither and fall off. Plants also become contaminated by honeydew and sooty moulds. Particularly on olive, attacks can also lead to deformation of young leaves and fruits, and subsequent yield reductions.

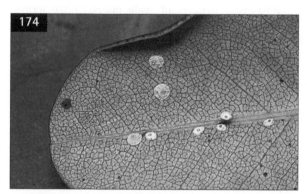

174 Oleander scale (*Aspidiotus nerii*).

175 Oleander scale (*Aspidiotus nerii*) – infested young shoot of olive.

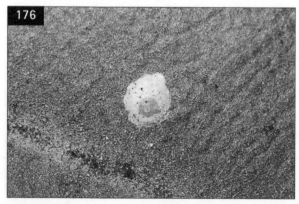

176 Oleander scale (*Aspidiotus nerii*) – female test.

Aulacaspis rosae (Bouché) (**177**)

Rose scale

A local pest of blackberry, but most frequently found on *Rosa*. Of virtually worldwide distribution and present throughout Europe.

DESCRIPTION
Female test: 1.5–2.5 mm across; oyster-shell-shaped and mainly greyish white. **Adult female:** 1.0–1.1 mm long; pear-shaped, brownish red. **Male test:** 0.8–1.0 mm long; flat, white, with distinct longitudinal ribs. **Adult male:** 0.7 mm long; reddish orange, with a distinct caudal spine; winged. **Egg:** 0.2 mm long; oval and reddish. **First-instar nymph ('crawler'):** oval, flat, and brownish orange to reddish.

LIFE HISTORY
This pest inhabits the canes of host plants, forming encrustations on the bark. Adults appear from May onwards. Mating occurs, and eggs are then laid under the female scales, although usually not until July or early August. Nymphs hatch shortly afterwards and then begin feeding. Male nymphs moult and settle down on the canes before winter; however, females usually wander about throughout the winter and do not moult into a sedentary form until the spring. At least in some regions, eggs are also stated to be an overwintering stage. Under European conditions there is typically one generation annually.

DAMAGE
Heavily infested canes become encrusted with scales, growth is checked and crop yields reduced. In the case of persistent attacks, death of plants is possible.

Chionaspis salicis (Linnaeus) (**178–179**)

Willow scale

A minor pest of currant bushes; apple and pear trees are also attacked. Associated mainly with a wide range of broadleaved forest trees and shrubs, including *Salix*. Widely distributed in Europe, particularly in the north and in montane districts; also present in North Africa and parts of Asia.

DESCRIPTION
Female test: 2–3 mm long; broadly mussel-shaped, convex, white to greyish white, with a yellowish apex. **Adult female:** 1.3–1.7 mm long; oval to elongate-oval; orange yellow to red. **Male test:** 1.0 mm long; white, elongate and very narrow, with one central and two lateral longitudinal ribs. **Adult male:** 1 mm long; orange to bright reddish orange; legs and antennae bright yellow; caudal spine long; winged or wingless. **Egg:** 0.3 mm long; oval and red. **First-instar nymph ('crawler'):** 0.4 mm long; oval and dark red.

178 Willow scale (*Chionaspis salicis*) – infested bark.

177 Rose scale (*Aulacaspis rosae*).

179 Willow scale (*Chionaspis salicis*) – swarming first-instar nymphs.

LIFE HISTORY

Overwintering eggs, sheltered beneath the remains of the maternal test, hatch in the spring, usually in late April or early May. Nymphs then feed, either beneath the maternal test or elsewhere (including, sometimes, on the leaves), and eventually moult into the adult stage. Adult males emerge from mid-June to mid-July. Eggs are laid by the young females from late August onwards. Fecundity is low, each female depositing about 10–12 eggs. Breeding is both parthenogenetic and sexual, with a single generation annually.

DAMAGE

Heavy infestations weaken host plants, but attacks on fruit crops are usually of little or no importance.

Chrysomphalus aonidum (Linnaeus) (**180**)

Florida red scale

This tropical and subtropical pest infests citrus and a wide range of other plants, particularly palms and other non-deciduous ornamentals. Although the pest occurs outdoors on citrus crops growing in parts of the southern and south eastern Mediterranean basin (e.g. North Africa and the Middle East), in Europe it is normally reported only in greenhouses, hot houses and so on. The female tests are 2–4 mm in diameter, and dull purplish red with a pale apex; male tests are of similar appearance, but smaller and more elongate. Eggs are deposited by mated females over an extended period and hatch within a few hours into active, yellow nymphs. There are several overlapping generations annually, all stages in the life cycle often occurring together.

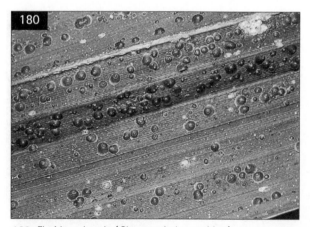

180 Florida red scale (*Chrysomphalus aonidum*).

Chrysomphalus dictyospermi (Morgan)

Palm scale

A polyphagous, southerly distributed pest of citrus, fig and olive. Present throughout the Mediterranean basin; also present in many other parts of the world, including America, Asia, Australia and New Zealand.

DESCRIPTION

Female test: 1.8–2.2 mm across; more or less circular; pale brownish yellow, marked with brown. **Adult female:** 1.5–2.0 mm long; broadly pear-shaped, yellow. **Male test:** 1.2–1.3 mm long; similar to that of female, but elongate. **Adult male:** 1 mm long; body yellowish, with blackish-brown legs and antennae; winged. **Egg:** minute, oval and golden yellow.

LIFE HISTORY

The winter is passed as either young adult females or as second-instar nymphs. Breeding commences in March, each female depositing approximately 150 eggs over a period of about 6 weeks. Eggs hatch almost immediately and the active first-instar nymphs wander away and soon settle on the old leaves to commence feeding. Nymphs develop for about 3 weeks before becoming adults, each passing through three instars. There is then a delay of several weeks before, eventually, mating of the adult males and females takes place and eggs are laid. There are usually three generations annually. However, under particularly favourable conditions, there may be a partial fourth. Peaks of activity by the young crawlers typically occur in late March or early April, in June and in late August. Members of the second and third generations usually settle on the young leaves and developing fruits.

DAMAGE

Feeding by the young scales on leaves and fruits causes the surrounding tissue to turn yellow. If infestations are heavy, branches begin to wilt, and the foliage eventually dries out and drops to the ground. Infested fruits also become deformed. In addition to direct feeding damage, foliage and fruits are contaminated by honeydew excreted by the nymphs and upon which sooty moulds develop.

Epidiaspis leperii (Signoret)

Italian pear scale

A pest of pear and plum in the warmer parts of central, southern and western Europe; also associated with apple, currant, mulberry, peach, walnut and several other hosts. Outside Europe, it is present in the Middle East, North Africa, North America and South America.

DESCRIPTION
Female test: 1.2–1.8 mm across; circular and slightly convex; white to yellowish white or greyish white, with a yellow or dark red, central or subcentral, apex. **Adult female:** 0.7–1.0 mm long; broadly pear-shaped, pinkish red, red or orange yellow. **Male test:** 0.8 mm long; elongate and mainly white. **Adult male:** wingless.

LIFE HISTORY
Infestations occur on the bark of the twigs and branches of host trees. Eggs, up to 40 per female, are laid from late May onwards, over a period of 1–2 months or more. They hatch within a week. Nymphs then settle down to feed, usually very close to, and often beneath, the old maternal scales. The adult stage is reached in late July or early August. There is one generation annually, with mated females overwintering.

DAMAGE
Particularly on pear and plum, infested shoots and branches are deformed; they may also split and readily snap off at the points of attack. Growth of infested shoots is also interrupted and fruit production reduced. On peach, and to a lesser extent on plum, gum is exuded from damaged tissue. Infestations on apple are relatively unimportant.

Lepidosaphes beckii (Newman) (**173, 181–182**)
Citrus mussel scale

A common pest of citrus in southern Europe; also widely distributed in tropical and subtropical parts of Africa, America and Asia.

DESCRIPTION
Female test: 2.5–4.5 mm long; elongate and broadly mussel-shaped; purplish brown. **Adult female:** 1.6 mm long; whitish. **Male test:** 1 mm long; similar to female test, but more elongate. **Egg:** minute, oval and white. **First-instar nymph ('crawler'):** 1.3 mm long; elongate-oval and flattened; brownish white.

LIFE HISTORY
After mating, each female deposits a batch of 50–100 eggs beneath the shelter of her test and then dies. The first-instar nymphs move onto the leaves, shoots and other parts of host plants, where they soon settle down to feed and mature; they also occur on the fruits. The life cycle from egg to adult is completed in 2–4 months; there are several generations annually.

DAMAGE
Heavy infestations cause premature leaf fall and death of young shoots. Infested fruits may be unmarketable. The insects often occur amongst colonies of California red scale (*Aonidiella aurantii*).

Lepidosaphes gloverii (Packard)
Citrus scale

A pest of citrus throughout the world. Widely distributed in southern Europe.

DESCRIPTION
Female test: 3–4 mm long; mussel-shaped, but relatively long and narrow; brown. **Adult female:** 1.5 mm long; spindle-shaped, pinkish. **Egg:** minute, oval and pinkish.

LIFE HISTORY & DAMAGE
Essentially, as for citrus mussel scale (*Lepidosaphes beckii*).

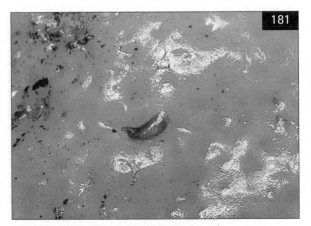

181 Citrus mussel scale (*Lepidosaphes beckii*).

182 Citrus mussel scale (*Lepidosaphes beckii*) – infested leaf.

Lepidosaphes minima (Newstead) (**183–184**)

Fig mussel scale

A common pest of fig in southern Europe, including France, Greece, Italy, Spain and Turkey. Also present in North Africa, North America, South America and Asia.

DESCRIPTION

Female test: 1.8–2.8 mm long; yellowish brown; relatively narrow, elongate and mussel-shaped, straight or slightly curved, and somewhat expanded posteriorly.

LIFE HISTORY

Dense colonies occur on the underside of leaves and on the twigs of host plants. Mated females overwinter and deposit eggs in the early spring. First-instar nymphs appear in late March or early April, and eventually reach adulthood in June. A second brood of nymphs develops during the summer. The next generation of adults appears in the autumn, when mating takes place.

DAMAGE

The pest causes noticeable, and extensive, pale mottling of the foliage.

Lepidosaphes ulmi (Linnaeus) (**185–186**)

Mussel scale

A highly polyphagous pest of apple, pear, cherry, plum, quince, walnut and many other trees and shrubs; currant, gooseberry and hazelnut are attacked occasionally. Also found throughout the Mediterranean basin on olive. Of almost worldwide distribution and generally abundant throughout Europe.

DESCRIPTION

Female test: 2.0–3.5 mm long; flat and mussel-shaped; grey to yellowish brown. **Egg:** 0.3 mm long; oval and white. **First-instar nymph ('crawler'):** 0.4 mm long; oval, pale yellowish brown.

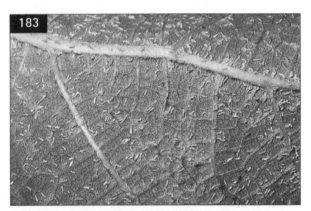

183 Fig mussel scale (*Lepidosaphes minima*).

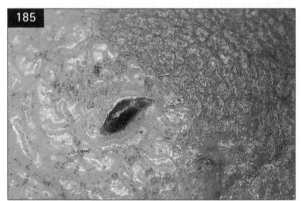

185 Mussel scale (*Lepidosaphes ulmi*).

184 Fig mussel scale (*Lepidosaphes minima*) – damage to leaf.

186 Mussel scale (*Lepidosaphes ulmi*) – infested apple.

LIFE HISTORY
Overwintered eggs hatch in late May or early June and the nymphs then wander over the host plant until finally settling down on the bark. Each then moults to a second-instar and then to a third-instar nymph. The adult stage is reached in late July. In late August or September, each female deposits up to 80 eggs beneath her test and then dies. The test, however, remains attached to the bark and protects the eggs throughout the winter. Although male forms appear in some races of this species, only females occur on fruit crops and reproduction is entirely parthenogenetic.

DAMAGE
This pest is rarely of importance and heavy infestations are usually found only on old, unsprayed fruit trees. Occasionally, nymphs will settle on fruits, eventually causing pale, mussel-shaped blotches that may downgrade the crop. On olive, scales may occur on the developing fruits, causing marks to develop on the surface. Infested olives may also become deformed. Such fruits are unsuitable for dessert use; infestations on the fruits also result in lower yields of oil.

Parlatoria oleae (Colvée) (**187–189**)
Olive parlatoria scale
A pest of citrus and olive in Mediterranean areas; also occurs on pomegranate and, in central and southern Europe (including Bulgaria, Romania and the former Yugoslavia) on hosts such as apple, pear and peach. Infestations also occur on grapevine, walnut and many other hosts. Widely distributed in the warmer parts of Europe; also found in South Africa, North America and Asia.

DESCRIPTION
Female test: up to 2.5 × 1.6 mm; subcircular and convex, white or brownish grey, with the brownish to brownish-black nymphal exuviae forming an eccentric apex. **Adult female:** 0.9–1.0 mm long; broadly elongate, purplish or mauve. **Male test:** 1.5 × 0.5 mm long; elongate, with an eccentric, circular apex. **First-instar nymph ('crawler'):** 0.2 mm long; purplish.

LIFE HISTORY
Infestations occur on the leaves, trunks and branches of host trees, and will also spread onto the fruits. Adult females overwinter and eventually deposit eggs in the spring. There are two or more generations annually, depending on temperature. Adult males are short-lived and do not overwinter.

DAMAGE
Heavy infestations weaken host plants. Scales on developing fruits cause discoloration. The feeding sites of nymphs on developing olives, for example, soon become surrounded by a broad circle of purple speckles. At harvest, affected crops may be down-graded or unmarketable.

187 Olive parlatoria scale (*Parlatoria oleae*) – damage to peach.

188 Olive parlatoria scale (*Parlatoria oleae*).

189 Olive parlatoria scale (*Parlatoria oleae*) – damage to fruitlets.

Parlatoria pergandii (Comstock) (**190**)
Chaff scale
A pest of citrus in various parts of the world, including southern Europe. Also associated with a wide range of other host plants, including *Camellia* and *Euonymus*.

DESCRIPTION
Female test: 1.5–2.0 mm across; subcircular, greyish brown or blackish, with a mainly posterior whitish fringe. **Male test:** 1 mm long; greyish brown. **Egg:** minute, oval and purplish.

LIFE HISTORY
Colonies develop on the upper surface of leaves. Although often widely scattered, the scales tend to cluster mainly along the midrib. Development from egg to adult takes about 2 months and, depending on conditions, there are two or more generations annually.

DAMAGE
Heavily infested hosts are weakened and growth is affected adversely.

Parlatoria ziziphi (Lucas)
Black citrus scale
A pest of citrus. Widely distributed in the Mediterranean region. Also found in various other parts of the world, and probably of Far Eastern origin.

DESCRIPTION
Female test: 2.0–2.5 mm long; elongate-oval and more or less parallel-sided; mainly black, but with a whitish or brownish-white fringe, especially posteriorly. **Adult female:** 0.9–1.0 mm long; broadly elongate. **Male test:** 1 mm long; elongate and mainly black to brownish white. **Egg:** minute, oval and violet.

LIFE HISTORY
Each female deposits about 20 eggs. These remain sheltered beneath the female test for an extended period before eventually hatching. Large numbers of the seed-like scales may develop during the summer on the leaves and fruits. In southern Europe, there is probably just one generation annually.

DAMAGE
Heavily infested hosts are weakened and infested fruits may be unmarketable.

Pseudaulacaspis pentagona (Targioni-Tozzetti) (**191**)
White peach scale
A polyphagous pest of mulberry, peach and plum in warmer parts of the world, including southern Europe; infestations are also reported on other crops, including almond, apricot, currant, pear, kiwi fruit and walnut. Of Oriental origin; its range in Europe currently extends northwards into parts of central Europe, including Bulgaria, Germany, Hungary, Switzerland and the former Yugoslavia.

DESCRIPTION
Female test: 2.2–2.8 mm across; circular or subcircular, white to creamy white, with an eccentric, reddish-brown apex. **Adult female:** 1 mm long; broadly pear-shaped or broadly oval; orange yellow to bright yellow. **Male test:** 1.5–2.0 mm long; elongate and mainly white.

LIFE HISTORY
This species develops on the woody parts of host plants, often forming dense colonies on the twigs and branches amongst accumulations of whitish wax. Egg-laying commences in April, each female depositing up to 100 eggs. First-instar nymphs appear from late May

190 Chaff scale (*Parlatoria pergandii*) – infested leaf of mandarin.

191 White peach scale (*Pseudaulacaspis pentagona*) – infested mulberry shoot.

onwards. In Mediterranean areas there are usually two generations annually, but sometimes a partial third; individuals overwinter either as final-instar nymphs or as unmated adult females.

DAMAGE
Infestations cause considerable weakening of hosts, restricting growth and reducing cropping potential.

Quadraspidiotus marani Zahradnik
Zahradnik's pear scale

A pest of apple, pear and plum in various parts of mainland Europe, from France to the former Yugoslavia; also found in Turkey. Other hosts include *Crataegus monogyna*, *Fraxinus excelsior*, *Prunus* and *Sorbus*.

DESCRIPTION
Female test: 1.8–2.1 mm across; circular, convex and dark grey. **Adult female:** 1.5 mm long; broadly pear-shaped, dark yellow. **Male test:** 1 mm long; elongate and grey. **Adult male:** 0.4 mm long; mainly orange yellow, with a prominent, shiny-black crossband (scutellum) and a short caudal spine; winged. **Egg:** minute, oval and yellowish. **First-instar nymph ('crawler'):** flat, oval and yellowish.

LIFE HISTORY
Infestations occur on the trunks and branches. Eggs (c. 100 per female) are laid in June and July. The eggs hatch within a few days and nymphs eventually reach adulthood in the autumn. Mating then takes place and the mated females overwinter (cf. *Quadraspidiotus ostreaeformis*, *Q. perniciosus* and *Q. pyri*). There is just one generation annually.

DAMAGE
Heavy infestations weaken host trees.

Quadraspidiotus ostreaeformis (Curtis)
Yellow plum scale

A polyphagous pest of apple, cherry, peach, pear, plum and currant, and many other trees and shrubs. *Betula* is believed to be the original wild host. Of virtually worldwide distribution and often very common in Europe.

DESCRIPTION
Female test: 1–2 mm across; roughly circular and convex; blackish to yellowish grey. **Adult female:** 1.2–1.5 mm long; broadly pear-shaped, lemon-yellow. **Male test:** similar to female scale, but smaller and more elongate. **Adult male:** similar to that of the previous species, but with a relatively short scutellum.

LIFE HISTORY
In central Europe, adults appear from late April onwards. After mating, up to 200 eggs are laid by each female, usually from late June or early July to the end of July. Males are short-lived, surviving for no more than a few days. Eggs hatch within about 3 days and the first-instar nymphs usually settle near or even immediately beneath the mother scale. Consequently, spread of colonies is slow and old infested wood tends to be heavily encrusted with distorted scales, live individuals sometimes feeding beneath several layers of dead scales from previous generations. A few nymphs, however, do successfully spread infestations to new growth. A protective, brownish test (formed from secreted wax) gradually develops over the soft body of each nymph and during growth this is later reinforced by cast-off nymphal skins. The young scales become noticeable on the wood during September. The winter is passed as second-instar nymphs beneath their protective scales. Feeding and growth are resumed in the early spring. A thin layer of wax is present beneath each insect and this remains as a prominent white patch on the bark if an individual is dislodged. There is one generation annually.

DAMAGE
Heavy infestations, which are most often reported on plum, weaken hosts trees. The pest rarely occurs in sprayed orchards.

Quadraspidiotus perniciosus (Comstock)
San José scale

A notorious subtropical pest of apple, pear, peach, plum and many other plants, especially Rosaceae; currant (particularly red currant) and gooseberry are also attacked. Of Far Eastern origin, but now cosmopolitan in warmer regions; widely distributed in parts of central and southern Europe (where it was first accidentally introduced from America in the 1920s), but it has been successfully eliminated from some regions.

DESCRIPTION
Female test: 1.5–2.2 mm across; more or less circular and slightly convex; light grey to dark grey, with a yellowish or whitish, central or subcentral, apex. **Adult female:** 1.0–1.2 mm long; broadly pear-shaped to circular, yellow to reddish yellow. **Male test:** 1.0 × 0.5 mm; similar to that of female, but smaller and more elongate. **Adult male:** similar to that of the previous two species, but with a shorter scutellum.

LIFE HISTORY
Unlike other species of *Quadraspidiotus* considered here, this species is viviparous and overwinters as first-instar

nymphs, each hidden beneath a small, black test. With the onset of winter, older nymphs and adults die. Overwintered nymphs continue their development in the spring and moult twice before reaching the adult stage, usually in May. Each adult female produces about ten nymphs over a period of about 6 weeks. These nymphs are very active, but soon settle down to complete their development. In central Europe there are usually two or more generations annually, depending on temperature, but up to five generations are completed annually in warmer areas.

DAMAGE

Attacked plants are weakened and heavily infested twigs and branches, if not complete hosts, may be killed. The skins of infested fruits often develop red blotches.

Quadraspidiotus pyri (Lichtenstein) (**192**)
Yellow pear scale

A pest of apple, peach and pear, but often confused with the previous species. Various other broadleaved trees are also hosts. Widely distributed in Europe; also found in Australia, the Middle East and North Africa.

DESCRIPTION

Female test: 1.8–2.1 mm across; roughly circular and flattened; blackish brown to greyish brown, with a prominent, orange-yellow, eccentric apex. **Adult female:** 1.5 mm long; broadly pear-shaped, yellow. **Male test:** similar to that of female, but smaller and more elongate. **Adult male:** similar to that of the previous three species, but the scutellum is distinctly longer.

LIFE HISTORY

The biology of this species is similar to that of yellow plum scale (*Quadraspidiotus ostreaeformis*), but females lay fewer eggs, usually from 40 to 70 (although sometimes considerably more), and the period of egg-laying is longer, extending from May into September. In consequence, there is frequent overlap of the developmental stages and hibernating individuals are not always second-instar nymphs. Although developing mainly on the trunk and branches of host trees, scales sometimes occur on the fruits. There is one generation annually. Reproduction is either sexual or parthenogenetic.

DAMAGE

Scales settling on developing fruits cause reddish spots on the skin, spoiling the appearance of the crop. Heavy infestations on the bark may weaken host plants, but usually occur only on neglected trees.

Unaspis yanonensis (Kuwana)
Japanese citrus fruit scale

An important pest of citrus in south-east Asia, and established in south-east France since the 1960s; in Europe, also reported from Italy.

DESCRIPTION

Female test: 2.5–3.5 mm long; oyster-shell-shaped, blackish brown, with a paler margin and a brownish-yellow apex. **Male test:** 1.5 mm long; white, elongate, with two or three longitudinal ridges.

LIFE HISTORY

In southern France, where infestations occur mainly on the northern aspect of trees, there are two generations annually, with adult females forming the overwintering stage.

DAMAGE

Infestations on leaves and branches result in leaf withering and leaf fall; they also result in die-back and death of branches, which may affect the whole tree (possibly as a result of the injection of toxins by the pests). Fruits with scales developing on the skin may be unmarketable.

192 Yellow pear scale (*Quadraspidiotus pyri*).

Family **ASTEROLECANIIDAE**
(pit scales)

Scales with the sedentary adult female enclosed in a hard, semitransparent or transparent test, fringed with waxen threads.

Pollinia pollini Costa
Olive pit scale
A pest of olive in the Mediterranean basin, particularly in areas subject to periodically cool conditions. Also present in North and South America.

DESCRIPTION
Female test: 1.5 mm across; globular and transparent. **Adult female:** orange red to reddish brown and clearly visible through the test.

LIFE HISTORY
Adult females occur in cracks on the shoots, often utilizing parts of host plants damaged by hailstones. Male tests, however, occur mainly in association with buds and leaves. In late spring, after mating, each female deposits eggs (typically from 30 to 70) beneath the test, the eggs hatching shortly afterwards. First-instar nymphs wander over the host plants and eventually settle down to feed and develop. They reach adulthood early in the following year, there being just one generation annually.

DAMAGE
Growth and development of the shoots and terminal buds of infested plants is impeded, and infested foliage becomes deformed and withered. The pest also has a detrimental effect on fruiting. Further, pit-like lesions occur at the location of the scales and these can result in subsequent deformation of shoots.

Family **COCCIDAE**
(soft scales, wax scales)

Small insects, with sedentary, wingless and legless adult females that remain permanently attached to the host plant; the female body often becomes sclerotized to form a smooth, bare or wax-coated, often tortoiseshell-shaped, 'scale'. Eggs are often deposited in a large, cottonwool-like ovisac. In some species the ovisac is formed beneath the egg-laying female; in others it may completely enclose the insect. See also, general comments on scale insects given under the family Diaspididae.

Ceroplastes ceriferus (Fabricius)
Japanese wax scale
In 2001, this tropical and subtropical pest was found in northern Italy on apple and various ornamental plants. The pest is extremely polyphagous, and widely distributed in subtropical and tropical parts of America and Asia; it is also reported from Africa and New Zealand. Outside Europe, in addition to various tropical fruits, non-ornamental hosts include pear, plum, quince, citrus, fig and avocado.

DESCRIPTION
Adult female (scale): 12 × 10 mm; hemispherical, coated with a cap-like layer of white wax, that typically includes an anterior, horn-like projection. **Nymph:** oval and dark red, with several discrete, projecting patches of white wax.

LIFE HISTORY
This species is parthenogenetic. Large numbers of eggs are laid beneath the mature female scale and they hatch 2–3 weeks later. The nymphs then disperse, before eventually settling down to feed. Development includes three nymphal stages. The life cycle varies from region to region. In the USA, for example, adults overwinter and deposit eggs from April to June; in Asia, first-instar nymphs form the overwintering stage, and adults occur in July and August.

DAMAGE
Heavy infestations weaken host plants and may also result in significant discoloration of foliage, followed by premature leaf fall. Infested hosts are also contaminated by honeydew, upon which sooty moulds develop.

Ceroplastes rusci (Linnaeus) (**193–195**)
Fig wax scale
An important pest of fig. Other hosts of this polyphagous pest include pistachio and many wild and

ornamental plants; the natural host is *Myrtus communis*. Widely distributed in the Mediterranean basin, including southern Europe.

DESCRIPTION
Adult female (scale): up to 5 mm long, 4 mm wide and 3 mm high; barnacle-like, formed from eight fused plates; whitish-marble, variably marked with brown. **Male test:** 2.0–2.2 mm long; reddish. **Egg:** 0.32 × 0.23 mm; oval; pale brownish yellow when laid, later becoming reddish brown. **First-instar nymph ('crawler'):** 0.3 mm long; oval-bodied and rusty red in colour.

LIFE HISTORY
After mating, each female deposits many hundreds of eggs over a period of several weeks, typically from May onwards. These eggs remain under her body and hatch 3–4 weeks later. Nymphs pass through three instars before attaining the adult stage, differences between the sexes becoming apparent in the final instar. The scales tend to occur along the major leaf veins or in clusters along the shoots. The pest excretes considerable quantities of honeydew and colonies are often attended by ants. In particularly favourable (e.g. coastal) districts there may be two generations annually; elsewhere, there is just one. The winter is usually passed as second-instar nymphs.

DAMAGE
Infested fig-trees are weakened, and heavily infested leaves and shoots may be killed. In addition, sooty moulds developing on honeydew excreted by the pest can reduce photosynthetic activity of the leaves.

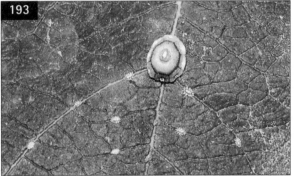

193 Fig wax scale (*Ceroplastes rusci*) – female and second-instar nymphs.

194 Fig wax scale (*Ceroplastes rusci*) – infested shoot.

196 Citrus wax scale (*Ceroplastes sinensis*).

195 Fig wax scale (*Ceroplastes rusci*) – damage to leaf.

197 Citrus wax scale (*Ceroplastes sinensis*) – infested shoot.

Ceroplastes sinensis del Guercio (**196–197**)
Citrus wax scale

This minor citrus pest, of Chinese origin, occurs in various Mediterranean areas, including Italy and Spain, where it is associated with mandarin and orange. More frequently, however, it occurs on non-citrus hosts such as *Ilex crenata*, *Rhus*, *Schinus molle* and ornamental *Solanum*.

DESCRIPTION
Adult female (scale): 5–6 mm long; barnacle-like, rectangular in outline and strongly convex; pinkish white when young, but later becoming reddish. **First-instar nymph ('crawler'):** purplish red and oval-bodied.

LIFE HISTORY
Adult females occur on the branches and stems of host plants, eggs (sometimes several thousand per female) being deposited in the late spring. The first-instar nymphs eventually invade the leaves, where they attach themselves to the upper surface and begin feeding. Later, as third-instar nymphs, individuals migrate back to the branches and stems. Here, they become sedentary and then overwinter, eventually attaining the adult stage.

DAMAGE
Infested shoots and branches may be weakened, but infestations are localized and generally of only minor significance.

Chloropulvinaria floccifera (Westwood) (**198**)
Cushion scale

A polyphagous pest, occurring frequently on citrus in southern Europe; nowadays also found in many other parts of the world. Widely distributed in Europe, occurring outdoors in favourable areas on a wide range of plants; in cooler regions an often common pest of greenhouse-grown ornamentals, particularly *Camellia japonica* and orchids.

DESCRIPTION
Adult female (scale): 2.5 × 2.0 mm; oval and yellowish brown to dark brown. **Egg:** minute, oval and pinkish.

LIFE HISTORY
Infestations occur on the foliage and shoots of host plants. Each female produces an elongated, white ovisac, within which large numbers of eggs are deposited. The maternal scale then dies. In suitably warm conditions, there is a succession of generations throughout the year.

DAMAGE
Heavily infested plants are weakened, and the shoots and foliage disfigured by accumulations of flocculent waxen 'wool'.

Coccus hesperidum Linnaeus (**199–201**)
Brown soft scale

A pest of citrus in various parts of the world, including the Mediterranean region. Also occurs on a wide range of other hosts, including ornamentals such as *Ficus*, *Laurus nobilis* and *Nerium oleander*; often found in more northerly districts (including northern Europe) in greenhouses and on house plants.

199 Brown soft scale (*Coccus hesperidum*).

198 Cushion scale (*Chloropulvinaria floccifera*).

200 Brown soft scale (*Coccus hesperidum*) – infested citrus leaf.

201 Brown soft scale (*Coccus hesperidum*) – sooty moulds developing on heavily infested mandarin tree.

202 Wisteria scale (*Eulecanium excrescens*) – mature female test.

DESCRIPTION
Adult female (scale): 3.5–5.0 mm long; very flat and elongate-oval, but shape varying considerably depending on the size and shape of the substratum; translucent-yellow to brown, with an often distinct median longitudinal ridge and horizontal, rib-like ridges.

LIFE HISTORY
This species is viviparous, and usually parthenogenetic, each female producing about a thousand nymphs over a period of 2–3 months. The first-instar nymphs wander over host plants for a few days before settling down to feed on the leaves, individuals then clustering along the midribs and other major veins; the scales commonly overlap one another, forming dense colonies. Breeding is continuous, so long as conditions remain favourable, the complete life cycle from birth to maturity occupying about 2 months at temperatures of around 25–30°C. The insects excrete considerable quantities of honeydew and are often attended by ants.

DAMAGE
Heavy infestations weaken host plants. Sooty moulds developing on excreted honeydew also have an adverse effect on photosynthetic activity and plant vigour.

Eulecanium excrescens (Ferris) (202–203)
Wisteria scale

An extremely polyphagous Asian species, in China attacking apple, pear, apricot, cherry, and peach, and a wide range of ornamental trees and shrubs. Unknown in Europe until infestations were found in 2001 in Central London, England, on *Wisteria* and certain other ornamentals (including *Prunus*). An introduced pest in the USA, where it is associated mainly with *Wisteria*, but also with various broadleaved trees (including rosaceous fruit trees). Although a potential threat, the

203 Wisteria scale (*Eulecanium excrescens*) – immature female tests.

pest has not yet been reported on fruit trees in the UK; nor has it been found elsewhere in Europe.

DESCRIPTION
Adult female (scale): 13 mm long; globular, strongly concave, with an irregular surface; dark brown to blackish, often dusted with a greyish waxen bloom. **Egg:** 0.5 mm long; pinkish orange. **First-instar nymph ('crawler'):** orange (older nymphs are brown, with distinct waxen patches).

LIFE HISTORY
In England, the pest has just one generation annually. It overwinters in the early nymphal stage, the nymphs eventually reaching maturity in late April or early May. The exceptionally large female scales deposit masses of eggs beneath their bodies, and then die. Following egg hatch, first-instar nymphs emerge en masse from beneath the maternal scales and swarm over the foodplant. Initially, these highly mobile nymphs feed on the foliage. Later, they move to the bark of shoots and stems, where they become sedentary and, eventually, overwinter.

DAMAGE

Infested hosts are weakened and soiled by honeydew, upon which sooty moulds develop.

Eulecanium tiliae (Linnaeus) (**204–205**)

Nut scale

A minor pest of apple, pear and, less frequently, plum; it is also associated with many other broadleaved trees and shrubs. Widely distributed in Europe; also present in the Middle East, North Africa, North America and Australia (Tasmania).

DESCRIPTION

Adult female (scale): 5–6mm long; dark chestnut-brown to greyish brown, sometimes marked with yellow; broadly oval, strongly convex and almost spherical. **Male test:** 2.0–2.5 mm long; elongate-oval, greyish and semitransparent. **Adult male:** 2 mm long; reddish crimson, with a relatively short caudal spine and a pair of very long, white, caudal filaments; winged. **Egg:** minute, pale yellowish white and shiny. **First-instar nymph ('crawler'):** pinkish to orange yellow; flat and oval, with a distinct anal cleft.

204 Nut scale (*Eulecanium tiliae*).

LIFE HISTORY

This species reproduces sexually, adult males often appearing in very large numbers. Mating occurs in the spring, and eggs (often over 1000 per female) are laid from early May onwards. The eggs hatch from mid-June onwards and, at first, the nymphs feed on the underside of leaves (cf. nymphs of plum lecanium scale, *Sphaerolecanium prunastri*). In the autumn, second-instar nymphs migrate to the twigs and shoots, where they settle down and eventually overwinter. Young overwintering male and female scales are similar in appearance (each about 1.5–2.0 mm long), but the more elongate and flattened appearance of the former becomes obvious in the spring once development has recommenced. After a further nymphal stage, individuals eventually moult into adults. There is a single generation annually. This pest excretes honeydew, and the colonies are often attended by ants.

DAMAGE

Attacks usually have little or no adverse effect on hosts, but heavy infestations may retard growth and cause death of shoots and branches.

Filippia follicularis (Targioni-Tozzetti) (**206–207**)

Olive cushion scale

A locally common pest of olive. Present in various Mediterranean countries, including France, Greece, Israel, Italy, Spain and Turkey.

DESCRIPTION

Adult female (scale): 5.0–6.5mm long; elongate-oval, yellowish brown to greyish brown. **Egg:** 0.4mm long; oval, brownish orange. **First-instar nymph ('crawler'):** 0.45 mm long; yellowish brown and translucent, with reddish speckles forming two longitudinal bands along the back; eyes black; tip of abdomen with a slight

205 Nut scale (*Eulecanium tiliae*) – infested shoot.

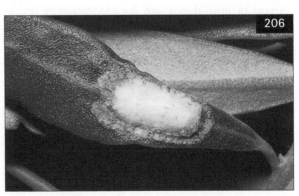

206 Olive cushion scale (*Filippia follicularis*) – ovisac.

207 Olive cushion scale (*Filippia follicularis*) – female and eggs within ovisac.

208 Viburnum cushion scale (*Lichtensia viburni*) – developing female scale.

cleft and a prominent, whitish, waxen tubule. Older nymphs typically have accumulations of white wax on their backs.

LIFE HISTORY

Adult females occur on the leaves and branches of olive during the summer months. Each produces a white, elongate-oval ovisac, 8–10 mm long, which completely encases the insect and in which large numbers of eggs are laid. The female then dies. The eggs hatch a few weeks later and the active first-instar nymphs disperse over host plants before settling down to feed. Nymphs eventually overwinter, usually in their third instar, before finally reaching the adult stage. There is typically one generation annually.

DAMAGE

The insects produce considerable quantities of honeydew, upon which sooty moulds develop. Heavily infested trees are weakened and the quality and quantity of the crop (fruit and oil) reduced. Infestations may also cause leaf distortion.

Lichtensia viburni Signoret (**208–210**)

Viburnum cushion scale

A pest of olive throughout the Mediterranean region. Also associated with other Oleaceae and with members of several other families, e.g. *Hedera helix* (Araliaceae) and *Viburnum tinus* (Caprifoliaceae). Widely distributed in Europe.

DESCRIPTION

Adult female (scale): 2.5–3.0 mm long; apart from size, very similar to that of *Filippia follicularis*. **Egg:** 0.3 mm long; oval, pale yellow. **First-instar nymph ('crawler'):** 0.4 mm long; yellowish brown and dorsoventrally flattened.

209a, b Viburnum cushion scale (*Lichtensia viburni*) – ovisacs.

210 Viburnum cushion scale (*Lichtensia viburni*) – female and eggs within ovisac.

LIFE HISTORY

Essentially similar to that of olive cushion scale (*Filippia follicularis*), but usually with two generations annually. Second-generation nymphs overwinter and reach the adult stage in the spring. In cooler regions, e.g. northern Europe, this species is single-brooded.

DAMAGE

As for olive cushion scale (*F. follicularis*).

Neopulvinaria innumerabilis (Rathvon) (**211**)

Cottony maple scale

A polyphagous North American species, now established on grapevine in Armenia, Azerbaijan, France, Georgia, Italy and Russia; in Europe, most often reported in north-eastern Italy.

DESCRIPTION

Adult female (scale): 6–7 mm long; heart-shaped to roughly circular and saddle-like; dark brown to blackish, with a relatively smooth surface. **Egg:** 0.3 mm long; creamy white. **First-instar nymph ('crawler'):** 0.4 mm long; brownish white.

LIFE HISTORY

This species occurs on the bark of host plants. Eggs, deposited in the spring, hatch in June. First-instar nymphs then wander over host plants and eventually become sedentary and begin to feed. Maturity is reached in the following spring. After mating, each female deposits a large batch of eggs that is protected by a prominent white ovisac. In Europe, there is just one generation annually.

DAMAGE

Heavy infestations cause a marked reduction in growth of the vines. They also have an adverse effect on flowering potential for the following year. The pest is also a virus vector.

Parthenolecanium corni (Bouché) (**212**)

European brown scale

A pest of many temperate fruit crops, including blackberry, currant, gooseberry, grapevine, plum, peach and raspberry. Infestations also occur on hazelnut and walnut, and on many broadleaved forest trees and ornamentals. Present in most temperate parts of the world; widely distributed in Europe and often common in unsprayed plantations and gardens.

DESCRIPTION

Adult female (scale): 4–6 mm long; more or less oval, very convex and often somewhat roughened; shiny reddish brown or chestnut-brown. **Egg:** 0.3 mm long; oval, brownish white and shiny. **First-instar nymph ('crawler'):** 0.4 mm long; oval, flat, light greenish to orange or brownish, with a distinct anal cleft.

LIFE HISTORY

Most races of this species, including those on fruit crops, are parthenogenetic; however, males do occur on some hosts in certain areas. In north-western Europe, eggs hatch from about mid-June to early or mid-July. After a few days, the first-instar nymphs wander to the youngest shoots and leaves to begin feeding. The second nymphal stage is attained in August. This is also a mobile feeding stage and, following leaf senescence and leaf fall, such nymphs migrate to the smaller branches and twigs; here, they settle down for the winter. During this time, they change colour from greenish to orange or brownish, but do not feed. Activity is resumed in March, individuals

211 Cottony maple scale (*Neopulvinaria innumerabilis*).

212 European brown scale (*Parthenolecanium corni*) – infested currant stem.

reaching the adult stage in April. Each then settles permanently on a twig or small branch and grows rapidly, while the back becomes increasingly hardened and convex to form the familiar, protective scale. In May and June, each female deposits several hundred eggs and then dies. The dead maternal scale remains in situ, protecting the eggs until they hatch, and may persist on the host plant for several years before eventually falling off. Details of the life cycle of this species vary somewhat according to conditions; two or three generations in a year may occur under glass and in warm southerly regions.

DAMAGE
Heavy infestations weaken host plants and cause premature leaf fall. Attacks tend to be heaviest in sheltered, unsprayed sites.

Parthenolecanium persicae (Fabricius) (**213**)
European peach scale
A pest of nectarine and peach; infestations also occur on grapevine and on a wide range of ornamental trees and shrubs. Virtually of worldwide distribution; present throughout much of central and southern Europe, but in more northerly regions found mainly on plants growing under protection or against sheltered, sunny walls.

DESCRIPTION
Adult female (scale): 5–10 mm long; oval to elongate-oval and convex; shiny yellowish brown to dark brown, with greyish or black markings. **Egg:** 0.3 mm long; oval, whitish and shiny. **First-instar nymph ('crawler'):** 0.4 mm long; yellowish brown, with greyish or black markings; elongate-oval, relatively flat, with a distinct anal cleft; later instars with a broad longitudinal

keel running along the back and fine, glassy threads radiating outwards from the body.

LIFE HISTORY
Nymphs infest the underside of leaves and, in their second instar, eventually migrate to the twigs and branches, where they overwinter. Development continues in the spring, the insects excreting considerable quantities of honeydew. The adult stage is reached in May and eggs (each female depositing anything from 1000 to 3000) are laid beneath the maternal scale from early June onwards. The eggs hatch in the summer and the first-instar nymphs then invade the foliage to begin feeding. In Europe, there is just one generation annually and reproduction is parthenogenetic.

DAMAGE
Heavily infested hosts are weakened, and fruit and foliage may be contaminated by accumulations of honeydew and sooty moulds.

Protopulvinaria pyriformis (Cockerell) (**214**)
Pyriform scale
This species is a pest of citrus, and is also reported on avocado. Present in southern Europe, including France, Italy, Portugal and Spain. Also found in many other parts of the world, including Africa, America and Asia.

DESCRIPTION
Adult female (scale): 5–7 mm across; mainly light brown, with a distinctly fan-shaped outline.

LIFE HISTORY
This species breeds parthenogenetically, with typically two generations annually. Dense colonies often occur on the underside of the foliage of infested plants, the scales

213 European peach scale (*Parthenolecanium persicae*).

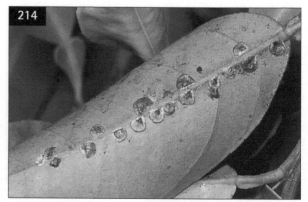

214 Pyriform scale (*Protopulvinaria pyriformis*).

tending to cluster along the midrib. The pest excretes copious quantities of honeydew.

DAMAGE

Infested host plants soon become extensively blackened by sooty moulds that develop on the excreted honeydew, and this has an adverse effect on photosynthesis and growth. Fruits are also contaminated, and may be unmarketable.

Pulvinaria hydrangeae Steinweden (**215**)
Hydrangea scale

This polyphagous species occurs on various trees and shrubs, including cherry, plum and other rosaceous fruit trees, but is most frequently noted on *Acer* and *Hydrangea*. The adult female scales (3–4 mm long) occur during the spring and summer, each eventually producing a large, white ovisac in which (usually in June) eggs are deposited. The eggs hatch in July, and the first-instar nymphs ('crawlers') then wander over the host plant before settling down to feed and develop. Nymphs overwinter in their second instar, and complete their development in the early spring. Dense colonies may develop on the shoots, leaves and developing fruits of fruit trees, and heavily infested plants may be weakened; contaminated fruits are unmarketable. The pest is now widely distributed in Europe, having relatively recently extended its range into more northerly areas (e.g. Belgium, France, Germany and the Netherlands); it also occurs in Australia, New Zealand and the USA.

Pulvinaria vitis (Linnaeus) (**216–219**)
Woolly vine scale

A pest of currant, gooseberry and grapevine; also associated with fruit trees (especially apricot and peach), and with various forest trees and shrubs. Holarctic. Widely distributed in Europe. Also recorded from Canada (on peach). *Pulvinaria vitis* now includes *P. betulae* (Linnaeus) and *P. ribesiae* Signoret, that were once regarded as distinct species.

DESCRIPTION

Adult female (scale): 5–7 mm long; heart-shaped to oval or circular, strongly convex and saddle-like; dark brown and wrinkled. **Male test:** 2 mm long; elongate-oval and brownish, with black markings and distinct paler ridges. **Adult male:** 1.5 mm long; delicate, pinkish red, with brownish legs and antennae, a short caudal spine and a pair of long caudal filaments; winged. **Egg:** 0.3 mm long; pale purplish red. **First-instar nymph ('crawler'):** 0.5 mm long; light purplish brown to orange yellow, flat, elongate-oval, with a distinct anal cleft.

LIFE HISTORY

Adults appear in September or October. The short-lived males die after mating. Females, however,

216 Woolly vine scale (*Pulvinaria vitis*).

215 Hydrangea scale (*Pulvinaria hydrangeae*).

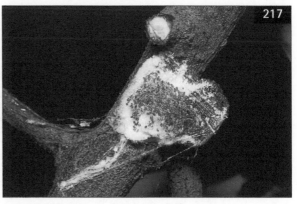

217 Woolly vine scale (*Pulvinaria vitis*) – eggs.

are longer-lived and eventually overwinter. In early spring, the surviving females begin to feed and grow, becoming distinctly convex and darker in colour, while their backs gradually harden to form a protective scale. From mid-April or May onwards, when fully mature, each female spins a white, cushion-like ovisac within which, over a period of 2–3 weeks, 1000 or more eggs are deposited. The female then dies. The bulk of the egg mass forces the scale away from the substratum, with the hind end tilted upwards. The presence of the pest then becomes particularly obvious as strands of waxen 'wool' from beneath the scales are wafted about by the wind, often covering many of the shoots and branches of infested plants. Eggs hatch from late May or early June onwards. The young, first-instar nymphs, which typically appear in swarms, then wander over the young shoots and leaves. Eventually, the nymphs disperse to the one-year-old wood where they settle down to continue their development. When feeding, this pest excretes considerable quantities of honeydew upon which sooty moulds may develop. There are three nymphal instars, and the adult stage is reached in the autumn. Although there is one generation annually, details of the life cycle vary according to conditions. Also, populations may include both sexual and asexual races.

DAMAGE
Heavy infestations, which occur on the leaves and branches, and may also extend down the main stems onto the roots, weaken host plants. In addition, the woolly secretion may envelop the fruit and be a nuisance during harvesting. Honeydew and sooty moulds also contaminate leaves, fruit and other parts of host plants, and can have a detrimental effect on photosynthesis; further, the honeydew often attracts large numbers of flies, bees and wasps to infested bushes, and this may also be troublesome to fruit pickers.

[?] *Pulvinaria* sp. (220)
Very dark, greyish to blackish, elongate-oval (6–7 mm long), *Pulvinaria*-like scales have been found in Italy on the shoots of olive. Here, the ovisac is produced from beneath the hind end of the female, forcing the body away from the substratum. This suggests that the insect is neither olive cushion scale (*Fillipia follicularis*) nor viburnum cushion scale (*Lichtensia viburni*). No further information is available.

Saissetia oleae (Bernard) (221)
Mediterranean black scale
This southerly distributed species occurs on apricot in southern Europe, but is most often reported infesting subtropical crops such as citrus, fig and olive. Other hosts include *Hedera helix*, *Nerium oleander*, *Pittosporum* and pomegranate. Originally Palaearctic, but now virtually cosmopolitan. Widely distributed in southern Europe; also present in North America, South America, Asia, Australia and New Zealand.

218 Woolly vine scale (*Pulvinaria vitis*) – developing scales.

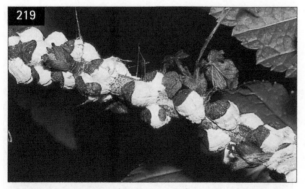

219 Woolly vine scale (*Pulvinaria vitis*) – scales and ovisacs on black currant stem.

220 [?] *Pulvinaria* sp. – female scales, one producing ovisac.

DESCRIPTION
Adult female (scale): up to 5 mm long and 3 mm wide; very convex, dark brown to black, with ridges forming a distinctive H-like pattern on the back. **Egg:** minute, oval, whitish to pale brownish yellow. **First-instar nymph ('crawler'):** pinkish, elongate-oval and dorsoventrally flattened.

LIFE HISTORY
Although males are known to occur, reproduction is mainly parthenogenetic. Infestations occur on leaves and branches, at maturity each female depositing up to 2,000 or more eggs. The eggs hatch 2–3 weeks later. The first-instar nymphs occur mainly on the young shoots and underside of leaves. Nymphal development is usually completed in 2–3 months. However, should conditions become unfavourable, the nymphs will enter a period of diapause. Under ideal conditions the complete life cycle last for 3 or 4 months and there are up to two generations annually. In the more northerly parts of its outdoor range this species has a single generation annually. Considerable quantities of honeydew are excreted by the developing nymphs, allowing sooty moulds to develop on infested hosts.

DAMAGE
Heavily infested hosts may be weakened and shoots and leaves may wither. The presence of honeydew and sooty moulds is also debilitating to host plants.

Sphaerolecanium prunastri (Fonscolombe)
Plum lecanium scale
A pest of peach and plum; other recorded hosts include almond, apricot, cherry and, less frequently, apple, pear and quince. *Prunus spinosa* is a frequent wild host. Widely distributed in central Europe; also found in parts of southern Europe (e.g. Greece and Spain), North America and Asia (e.g. Turkey).

DESCRIPTION
Adult female (scale): 2.0–2.5 mm across; almost circular and strongly convex, dark brown to brownish black.

LIFE HISTORY
Infestations occur on the branches and trunks of host plants. Adult females deposit eggs from early May onwards, and these hatch a few hours later. The first-instar nymphs do not invade the foliage. Instead, in common with the adults, they inhabit the bark of host trees (cf. nymphs of nut scale, *Eulecanium tiliae*). Moulting to the second instar, the overwintering stage, takes place in the autumn. In spring, nymphs continue their development, female nymphs passing though a further two instars; in males, there are five nymphal stages. This species excretes quantities of honeydew, and colonies are often attended by ants.

DAMAGE
Heavily infested hosts are weakened and are also contaminated by honeydew.

221 Mediterranean black scale (*Saissetia oleae*).

Family **PSEUDOCOCCIDAE**
(mealybugs)

Small to medium-sized insects. Females elongate-oval and superficially woodlouse-like, with distinct body segmentation; body more or less covered by a flocculent or mealy, waxen secretion; antennae from 5- to 9-segmented, but poorly developed; legs relatively well developed. Males rare and in many species unknown.

Heliococcus bohemicus Šulc
Bohemian mealybug
A minor pest of grapevine. Recorded from various parts of Europe, including the Czech Republic, France, Germany, Hungary, Italy, Poland and Spain.

DESCRIPTION
Adult female: 3.0–3.5 mm long; oval, pink and coated with glassy, spine-like, waxen filaments; antennae 9-segmented.

LIFE HISTORY
This species has two generations annually and overwinters as young nymphs. The females, which inhabit the woody parts of host plants, are ovoviviparous (i.e. they produce eggs with a definite shell, but these hatch whilst still within the mother's body). Adult males occur mainly from late March to early April, and also in early July. First-generation nymphs occur from late May or early June onwards, and those of the second from August or early September onwards.

DAMAGE
Infestations weaken host plants, but attacks are usually of little significance.

Planococcus citri (Risso) (**222**)
Citrus mealybug
A pest of citrus in southern Europe; other hosts include grapevine and pomegranate. Polyphagous and present in mainly tropical and subtropical regions; in temperate regions (including central and northern Europe) established only in greenhouses and in other suitably protected environments.

DESCRIPTION
Adult female: 3.5–4.5 mm long; oval and pinkish grey, coated with white, flocculent wax and with a fringe of short waxen filaments (processes), those at the hind end of the body (caudal processes) being the longest; antennae 8-segmented. **Egg:** 0.3 mm long; oval, creamy yellow and shiny.

LIFE HISTORY
Infestations occur mainly on the foliage and around the base of developing fruits. Large numbers of eggs are laid within large, white, flocculent, cottonwool-like masses of wax, that form the ovisacs. These ovisacs often totally enclose the maternal females. Nymphs are very active and also produce considerable quantities of honeydew during the course of their development. This often attracts ants and also enables sooty moulds to develop. There are several generations annually and all stages may overwinter.

DAMAGE
Heavily infested leaves may turn yellow and wilt; in addition, secreted wax and excreted honeydew cause extensive contamination of fruits and foliage. Attacks in vineyards are most serious towards the end of the season, when infested clusters of grapes are contaminated by secreted waxen 'wool' and by large quantities of excreted honeydew; subsequently, the ripening grapes also become coated with unsightly sooty moulds.

222 Citrus mealybug (*Planococcus citri*).

Pseudococcus calceolariae (Maskell) (**223**)
Citrophilus mealybug

This local, but widely distributed, species is a minor pest of currant bushes, particularly red currant. Infested shoots and fruit trusses are contaminated by honeydew and sooty moulds, but damage is rarely significant. Other hosts include *Juniperus* and ornamentals such as *Ceanothus*, *Forsythia* and *Laburnum*. Adults (3–4 mm long) are broadly oval and reddish, and coated in white, mealy wax; the caudal filaments are very short. Unlike related species, this mealybug is indigenous to northern Europe, including Scandinavia.

Pseudococcus longispinus (Targioni-Tozzetti) (**224**)
Long-tailed mealybug

A pest of citrus in southern Europe, including Greece, Italy, Portugal and Spain. Established in many tropical and subtropical parts of the world on a wide range of host plants. In temperate regions found only in greenhouses (e.g. on ornamental plants) and in other suitably protected situations.

DESCRIPTION
Adult female: 2–3 mm long; oval and pinkish grey to dull grey, with a coating of white, flocculent wax and a fringe of waxen filaments, those at the hind end of the body (caudal processes) very long; antennae 8-segmented.

LIFE HISTORY & DAMAGE
As for citrus mealybug (*Planococcus citri*).

Pseudococcus viburni (Signoret) (**225**)
Glasshouse mealybug

In southern Europe, infestations of this generally common mealybug occur out of doors on citrus trees; further north, the insect is a well-known, polyphagous pest of greenhouse-grown ornamentals and house plants. Adult females are 4 mm long and covered with white mealy wax; there is also a fringe of waxen processes, the caudal processes being about half as long as the body; antennae are 8-segmented. The white ovisacs are very large and conspicuous.

223 Citrophilus mealybug (*Pseudococcus calceolariae*).

224 Long-tailed mealybug (*Pseudococcus longispinus*).

225 Glasshouse mealybug (*Pseudococcus viburni*) – ovisacs.

Family **MARGARODIDAE**
(giant scales)

A small group of mainly tropical insects; females typically mobile, with abdominal spiracles, simple eyes and well-developed legs and antennae. Antennae 6- to 13-segmented. Males relatively large, with compound eyes. Nymphs and females are coated with white wax, but waxy plates are lacking.

Gueriniella serratulae (Fabricius) (**226–227**)
A polyphagous, but little-known species. Recorded on olive, *Pinus* and *Quercus*, and on various herbaceous plants (including *Foeniculum vulgare*, *Medicago*, *Salvia* and *Verbascum*). Locally, but widely, distributed in the Mediterranean basin, including Algeria, Corsica, France, Israel, Italy, Majorca, Rhodes, Sardinia and Spain; also recorded from Armenia and the Crimean Republic.

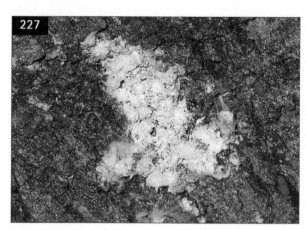

227 *Gueriniella serratulae* – infested bark of olive tree.

226a, b *Gueriniella serratulae* – adult.

DESCRIPTION
Adult: 5–8 mm long; body red, with relatively short, white waxen processes; antennae black and 11-segmented; legs black, with 1-segmented tarsi. **Egg:** 0.5 mm long; red and oval. **First-instar nymph:** 0.7 mm long; body red; legs and antennae black.

LIFE HISTORY
The adults are relatively active and occur in groups on the trunks and larger branches of olive trees. Eggs are laid in June on the bark, protected by masses of white wax, and hatch a few weeks later. There is just one generation annually.

DAMAGE
Heavy infestations can weaken hosts and are likely to be of particular significance on young plants.

Icerya purchasi Maskell (**228–231**)
Fluted scale
A destructive pest of citrus; other hosts include pomegranate. Also associated with *Acacia*, *Hedera helix*, *Mimosa*, *Robinia pseudoacacia* and many other ornamentals. Of Australian origin, but now well established in most warmer parts of the world, including southern Europe; a pest of greenhouse plants in cooler regions.

DESCRIPTION
Adult female (scale): 4–5 mm long; oval, light brown to brick-red, with dark brown to black legs and antennae; antennae 11-segmented. **Egg:** minute, oval and reddish. **Nymph:** 0.5 mm long; flat, oval and reddish, with a white, waxen coating and with long waxen tubules projecting from the hind end; hind end also with three pairs of long setae; antennae black, 6-segmented and clubbed; legs black.

LIFE HISTORY

This pest usually overwinters as third-instar nymphs. Oviposition typically commences in February, each female depositing several hundred eggs within a large, white ovisac; the ovisac, which pushes the female away from the substratum, is characterized by the presence of a series of longitudinal grooves, giving it a fluted appearance. Development from egg to adult takes at least 3 months, allowing the completion in southern Europe of no more than two or three generations annually. Adult females inhabit the twigs or branches of host plants, but the nymphs usually feed on the foliage. In their final stages of development, however, the nymphs (which excrete considerable quantities of honeydew, transported away from the anus via the waxen anal tubules) usually return to the twigs or branches of host plants. The adult females are hermaphroditic and fertilize their own eggs. Males, which arise from unfertilized eggs, are rarely found.

DAMAGE

Accumulations of honeydew, and the subsequent presence of sooty moulds, are deleterious, and may weaken hosts. Direct feeding is also important; heavy infestations can lead to distortion of shoots and branches, reduced cropping and, ultimately, death of trees.

228 Fluted scale (*Icerya purchasi*) – active female.

229 Fluted scale (*Icerya purchasi*) – young ant-attended colony.

230 Fluted scale (*Icerya purchasi*) – oviscas.

231 Fluted scale (*Icerya purchasi*) – infested citrus branches.

Chapter 4
Beetles

Family **CARABIDAE**
(ground beetles)

A large family of small to large, mainly ground-dwelling beetles; adults mainly black, but many with a metallic blue, green or coppery lustre; pronotum shield-like; elytra fused and usually each with nine longitudinal ridges (intervals), separated by distinct furrows (striae) or a series of punctures; antennae 11-segmented and filiform, and clearly inserted at the sides of the head, between the eyes and the mandibles.

Ground beetles are well-known predators of insects and other invertebrates; adults of many, however, are omnivorous whereas some are largely, if not entirely, phytophagous. A few species attack developing strawberry fruits and in some seasons they may be important pests, particularly in weedy, overgrown plantations. Such attacks are sporadic and unpredictable, but they can be serious on occasions. Adult ground beetles, which may live for up to 2 years or more, are active at night, particularly in warm weather. They are fast movers and, when disturbed, quickly run away and seek shelter under stones, clods of earth and herbage. Many species are wingless, but even when wings are present they may not be functional. In very dry weather, the beetles tend to remain inactive and do not feed. After rain, however, activity and feeding will usually reach its peak. It is under such conditions during the fruiting season that damage to strawberries is most likely to occur. Larvae of ground beetles, even if phytophagous, do not damage fruit crops.

Adults of various species cause damage in strawberry beds in Europe. The following are common examples:

Abax parallelepipedus (Piller & Mitterpacher) (**232**)
Parallel-sided ground beetle
A common species in shady habitats and an occasional pest in strawberry beds. Widely distributed in Europe.

DESCRIPTION
Adult: 18–22 mm long; mainly shiny black, but the elytra dull in female; pronotum broad, with distinct

232 Parallel-sided ground beetle (*Abax parallelepipedus*).

hind angles and well-marked, elongate basal depressions; elytra distinctly parallel-sided; head with two punctures above the compound eye (supra-orbital punctures).

LIFE HISTORY
Breeding occurs in spring, summer and autumn, the carnivorous larvae overwintering.

DAMAGE
In common with several other species of ground beetle, e.g. *Pterostichus* spp., adult beetles make holes in ripe strawberries, usually confining their attacks to the widest part of the fruit. Damage is often superficial, but holes may be up to 5 mm across and 15 mm deep, and there are frequently two or more in a single fruit. Sometimes, beetles tunnel into the fruit until only the tip of the abdomen is visible. Unlike similar holes in strawberries caused by birds and small mammals, those produced by ground beetles usually occur in parts of the fruit close to the ground.

Calathus fuscipes Goeze (**233**)
A common European species in dry, open areas, including cultivated land. Although known to attack ripe strawberries, it is not an important pest. Adults are 10–14 mm long and black, usually with reddish legs; the

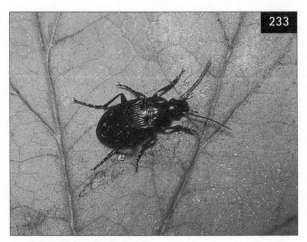

233 *Calathus fuscipes* – adult.

234 Strawberry seed beetle (*Harpalus rufipes*).

235 Strawberry seed beetle (*Harpalus rufipes*) – damage.

pronotum is relatively straight-sided and has distinct hind angles.

Harpalus affinis (Schrank)
Very common and widespread, particularly in arable habitats, but usually found in strawberry beds in only small numbers and, therefore, of little significance as a pest. Adults are 8–12 mm long, shiny metallic, and green, coppery or blue to blue black, with the legs and antennae reddish brown; females have dull elytra. The life history is similar to that of strawberry seed beetle (*Harpalus rufipes*), and both species cause similar damage. *H. affinis* is basically an autumn breeder, with the larvae overwintering; adults show some daytime activity.

Harpalus rufipes (Degeer) (**234–235**)
Strawberry seed beetle
Very common, particularly in weedy situations, and usually the most important ground beetle pest of strawberry. Palaearctic. Widely distributed in Europe.

DESCRIPTION
Adult: 11–17 mm long; dull black, the elytra with a yellowish pubescence; legs and antennae reddish brown; head with a single puncture above the compound eye (supra-orbital puncture).

LIFE HISTORY
Eggs are laid in weedy soil, either singly or in small groups, from July to late September or early October. Larvae then feed from August to the following July, passing through three instars and eventually pupating in the soil, usually at a depth of 15–45 cm. Most ground beetle larvae are voracious carnivores, but those of *H. rufipes* are seed feeders; however, unlike the adults, they do not damage strawberries. They are partial to seeds of *Chenopodium album* and *Lolium perenne*, and the presence or absence of such weeds considerably influences the distribution of the pest. Overwintering adults and larvae are usually confined to the edges and grassy surroundings of strawberry fields. Adults then move into plantations during the fruiting period, where they are most numerous from strawing until after harvest. Although possessing wings, the beetles rarely fly.

DAMAGE
Adults remove the seeds from strawberry fruits and, particularly on cultivars where these are deeply set, this usually results in damage to the surrounding flesh. The beetles usually eat the endosperm and reject the husks,

which are then left on the soil surface. Seed beetle damage is frequently confused with that caused by linnets, but the latter typically remove seeds cleanly (i.e. without harming the surrounding flesh) and attack the more exposed parts of the fruit.

Nebria brevicollis (Fabricius) (236)
Common black ground beetle

This widespread and generally common ground beetle is mainly a woodland species, but will sometimes invade strawberry beds and then damage the fruits. The insect breeds in the autumn and overwinters as larvae. Adults are 10–14 mm long, dark brown or blackish, with reddish legs and antennae; the elytra bear prominent, strongly punctured striae; the pronotum is short and has bulbous, convex sides.

Poecilus cupreus (Linnaeus) (237)
A strawberry ground beetle

A common, but local, species that usually frequents moist habitats. Adults are diurnal; they overwinter and breed in the spring. The beetles occasionally damage

strawberry fruits, but are far less important than, for example, *Pterostichus* spp. Adults are 9–13 mm long, and typically green, reddish brown or purplish (rarely black), with a metallic sheen.

Pterostichus madidus (Fabricius) (238)
A strawberry ground beetle

A pest of strawberry, but under dry conditions only. Very common, particularly in pastures and grassland, and widely distributed in Europe.

DESCRIPTION

Adult: 13–17 mm long; body black; legs black, but usually reddish at the base; pronotum noticeably rounded and narrowed towards the elytra; head with two punctures above the compound eye (supra-orbital punctures).

LIFE HISTORY

Young adults appear in the early summer and numbers in strawberry fields increase considerably at the onset of fruiting. Eggs are usually laid in grassland from August to November. The larvae, which are carnivorous, feed throughout the winter, except in the severest weather. They become fully grown in the late spring and then pupate. Most adults die before the winter; however, a few may overwinter successfully and then breed early in the spring.

DAMAGE

As for parallel-sided ground beetle (*Abax parallelepipedus*).

236 Common black ground beetle (*Nebria brevicollis*).

237 A strawberry ground beetle (*Poecilus cupreus*).

238 A strawberry ground beetle (*Pterostichus* sp.) – damaging strawberry fruit.

Pterostichus melanarius (Illiger)
A strawberry ground beetle
A frequent inhabitant of strawberry beds in some seasons and an important pest locally. Generally common in arable areas; however, unlike *Pterostichus madidus*, it avoids dry habitats. Widespread in Europe.

DESCRIPTION
Adult: 12–18 mm long; body and legs shiny black; pronotum narrowed only slightly towards the elytra, and not especially rounded; head with two punctures above the compound eye (supra-orbital punctures).

LIFE HISTORY
Similar to that of *P. madidus*, but larvae pupate slightly earlier in the season.

DAMAGE
As for parallel-sided ground beetle (*Abax parallelepipedus*).

Family **SCARABAEIDAE**
(chafers)

A very large family of relatively large, often brightly coloured beetles; tip of antennae 8- to 10-segmented and typically lamellate in form; abdomen with six visible sternites and with tergite 8 typically drawn into a more or less pointed pygidium that sometimes protrudes beyond the elytra (as in the genus *Melolontha*). The very sluggish larvae (often termed 'chafer grubs') are scarabaeiform, large and fleshy, with a distinct head, powerful mouthparts, three pairs of strong legs and the tip of abdomen distinctly swollen. They feed in the soil on decaying matter and on a wide variety of plant roots, and are particularly abundant in light, well-drained soils near heathlands or woodlands; the shape of the anal slit and the arrangement of chitinized spines on the final abdominal segment are often useful guides for distinguishing between species.

Amphimallon solstitialis (Linnaeus) (**239–240**)
Summer chafer
A minor pest of a wide range of fruit crops, particularly strawberry. Also polyphagous on many other plants. Eurasiatic. Widely distributed in Europe.

239 Summer chafer (*Amphimallon solstitialis*).

240 Summer chafer (*Amphimallon solstitialis*) – larva.

DESCRIPTION

Adult: 14–18 mm long; yellowish chestnut-brown; elytra shiny, with broadly spaced longitudinal ridges dorsally; thorax and pygidium dull and noticeably hairy. **Egg:** 2.5 × 2.0 when laid, but then enlarging; white and shiny. **Larva:** up to 30 mm long; body white and plump; head and legs yellowish brown; anal slit set in a triangular area and surmounted by two strongly divergent rows of spines.

LIFE HISTORY

The adults are active on warm evenings in June and July, but may also occur from May onwards and into August. Eggs are laid in the soil and the larvae feed on the roots of various low-growing plants from July or August onwards. Development is normally completed within 2 years. Fully fed larvae then pupate in earthen cells and adults emerge 2–4 weeks later.

DAMAGE

The larvae occasionally damage roots of strawberry plants and other fruit crops, but they are usually present in only small numbers and likely to be harmful only on small plants. Although adults may browse on leaves and other aerial parts of plants, the damage caused is unimportant.

Cetonia aurata (Linnaeus) (**241**)
Rose chafer

A locally common, but minor, pest of fruit crops, including raspberry and strawberry. Palaearctic. Widely distributed in Europe.

DESCRIPTION

Adult: 14–20 mm long; somewhat flattened, metallic golden green dorsally, purplish red ventrally; elytra with a few irregular silvery markings. **Larva:** up to 30 mm long; body whitish and rather plump, with reddish hairs in transverse rows along the body; head and legs reddish; anal segment with two close-set, more or less parallel, longitudinal rows of spines that meet anteriorly; in repose, the head is held close to the anal segment.

LIFE HISTORY

Adults appear from May onwards, and in sunny weather often visit the open flowers of Apiaceae and other plants. Eggs are laid in the soil during the early summer and hatch in 2–4 weeks, depending on temperature. Although feeding mainly on plant roots, larvae are able to survive on decaying vegetable matter. They develop over two (sometimes three) summers, hibernating during the winter months. When fully grown they pupate, each in a strong, oval, earthen cell.

DAMAGE

As for summer chafer (*Amphimallon solstitialis*).

Hoplia philanthus (Fuessly) (**242**)
Welsh chafer

Adults of this widely distributed species often swarm in sunny weather in July, and are then sometimes abundant in orchards close to grassland. Adults browse on the leaves of various plants; larvae, which feed mainly on the roots of grasses, are occasionally unearthed in fruit plantations, particularly in sites with an abundance of grass weeds. The life cycle occupies 2 years, with larvae hibernating over two winters. The adults (8–9 mm long) are mainly black or brownish black, with reddish legs and 10-segmented antennae. The larvae (up to 15 mm long) are white and translucent, with the dark gut contents clearly visible, and with numerous reddish hairs on the back and on the anal segment; the head is yellowish brown and the anal segment bears numerous scattered chitinous spines; the body is held very close to the head when in repose.

Melolontha hippocastani Fabricius
Melolontha melolontha (Linnaeus) (**243**)
Cockchafers

Polyphagous pests of fruit crops, particularly apple, raspberry and strawberry, and many other plants; the adults are often damaging to forest trees and shrubs, and the larvae are pests of grassland and pasture. Eurasiatic.

241 Rose chafer (*Cetonia aurata*).

242 Welsh chafer (*Hoplia philanthus*) – larva.

Widely distributed in temperate Europe, but the range of *M. hippocastani* extends further north and eastwards across much of Scandinavia and central Asia.

DESCRIPTION

Adult: 20–30 mm long; head and thorax black; elytra chestnut-brown, each with four, broad, hairy, longitudinal furrows, and variously covered in short whitish hairs; pygidium triangular and terminating in a blunt, downwardly directed spine; antennae with six (= female) or seven (= male) lamellae. *M. hippocastani* has noticeably shorter elytra and a more elongated pygidium; further, the head, pronotum and elytra are more uniformly chestnut-brown in colour. **Egg:** 3 × 2 mm when laid, but then enlarging; oval, whitish to yellowish white. **Larva:** up to 50 mm long; body plump and white, with the enlarged, translucent anal segment somewhat darkened by the underlying gut contents; head brown and shiny; legs and mouthparts strong; anal slit transverse and wavy, surmounted by two parallel, longitudinal lines of spines. **Pupa:** 25–35 mm long; whitish to brown.

LIFE HISTORY

Adults of *M. melolontha* occur in late April, May and June, to feed on the foliage of various trees and shrubs. They are nocturnal and are frequently attracted to light. After a few weeks, mated females burrow into the soil for about 20 cm to lay eggs, each depositing up to about 70 in batches of 12–30. The eggs enlarge considerably whilst in the soil and hatch in 1–2 months, depending on temperature. Larvae feed on plant roots for an extended period, overwintering deeper in the ground and returning to more superficial layers each spring. They pass through three instars during the course of their development, usually pupating during the third (sometimes the fourth) summer in oval earthen cells formed 40 cm or more below the surface. Adults are produced about 6 weeks later, but they do not emerge from the soil until the following spring. The total life cycle, therefore, typically extends over 3 (occasionally 4) years. As a result, significant adult invasions tend to be cyclical, in the case of *M. melolontha* occurring every 3 or 4 years. The life cycle of *M. hippocastani* is broadly similar to that of *M. melolontha*, but usually extends over 4 or 5 years.

DAMAGE

Apple: adults feed on developing fruitlets, a single individual often taking one or more large bites out of several in a cluster. The holes sometimes extend to the core. Attacked fruitlets either drop prematurely or eventually mature into apples with one or more corky pits on the surface. Attacks on foliage are usually unimportant. Larvae sometimes attack the roots, weakening young trees. **Strawberry:** young plants are very susceptible to damage by chafer grubs, attacks on the rooting system often causing wilting and eventual death; maidens planted in recently broken-up pasture are particularly liable to be attacked, particularly in woodland areas.

Phyllopertha horticola (Linnaeus) (**244–245**)
Garden chafer

A polyphagous, minor pest of fruit crops, but associated mainly with pastures and grassland. Eurasiatic. Widely distributed in Europe; most numerous on light soils and particularly common in grassland areas.

244 Garden chafer (*Phyllopertha horticola*).

243 Cockchafer (*Melolontha melolontha*).

245 Garden chafer (*Phyllopertha horticola*) – larva.

DESCRIPTION
Adult: 7–11 mm long; head and thorax metallic bluish green; elytra shiny reddish brown; legs black. **Larva:** up to 15 mm long; white, with a light brown head; anal slit transverse and surmounted by two parallel longitudinal rows of spines; anal segment held close to head when in repose.

LIFE HISTORY
Adults fly in warm, sunny weather during May and June, and are active for about 3–4 weeks. Each female lays about 10–15 eggs in the soil from late May onwards. Larvae feed on plant roots (especially grasses) from July onwards and become fully fed by the autumn. They then overwinter in earthen cells and pupate in the spring, adults emerging about a month later.

DAMAGE
Adults frequently browse on the foliage and flowers of fruit trees, and will also bite into apple and pear fruitlets. Fruitlet damage is similar to that caused by adult cockchafers (*Melolontha* spp.), but usually far less extensive.

Serica brunnea (Linnaeus) (**246**)
Brown chafer
Larvae of this widely distributed, polyphagous species occasionally browse on the roots of fruit crops, particularly those planted in the vicinity of woodlands and forests. The larvae (up to 18 mm long) are creamy white, with a yellowish-brown head, numerous reddish body hairs and the anal slit surmounted by a horizontal arc of spines. Adults (7–11 mm long) are mainly reddish brown; they occur most frequently in July and August. The life cycle is normally completed in 2 years.

Family **BUPRESTIDAE**
(jewel beetles)

A large family of mainly tropical beetles, with relatively few European representatives. Adults (1–12 mm long) more or less elongate, and usually distinctly metallic in appearance. Larvae (sometimes known as 'flat-headed borers') are dorsoventrally flattened, often with a very large prothoracic segment into which much of the head is retracted. The larvae are mainly wood-borers and feed in galleries excavated just beneath the bark of host plants. Several species are of economic importance as forestry (timber) pests. Some species infest herbaceous plants, and several can induce the formation of abnormal swellings (often termed pseudo-galls). Just a few species are associated with fruit crops.

Agrilus aurichalceus Redtenbacher (**247**)
Raspberry jewel beetle
Associated with *Rosa* and *Rubus*, including cultivated raspberry. Widely distributed in mainland Europe from France eastwards. An introduced species in the USA.

DESCRIPTION
Adult: 4.5–7.0 mm long; mainly metallic olive green to metallic greenish blue; elytra elongate, with a sinuous outline, and distinctly swollen between the middle and apex.

LIFE HISTORY
Larvae feed singly within the stems of host plants, each burrowing upwards just beneath the epidermis from the point at which the egg was deposited. At first, a relatively tight spiral gallery is formed, which causes a localized swelling (pseudo-gall) about a centimetre long; later, the spiralling becomes more 'open' and the gallery straightens before terminating in an elongated,

246 Brown chafer (*Serica brunnea*) – larva.

247 Raspberry jewel beetle (*Agrilus aurichalceus*).

flask-shaped pupal chamber a few centimetres above the level of the pseudo-gall. Typically, the larval gallery remains filled with frass, as there are no external openings through which this might be expelled. Pupation occurs in the spring. Adults are active from May to July.

DAMAGE

Infested plants become progressively weakened and, in serious cases, fruiting is reduced. Adults browse on the leaves and can cause noticeable loss of tissue; on raspberry, however, such damage is of no significance.

Agrilus derasofasciatus Lacordaire (248)

Vine jewel beetle

A minor, secondary pest of grapevine in various parts of mainland Europe, including, for example, France, Germany and Spain. Also present in North Africa and an introduced pest in the USA.

DESCRIPTION

Adult: 4–6 mm long; olive green to greenish bronze (disc of pronotum darker); some forms are mainly black; tips of elytra dentate.

LIFE HISTORY

Adults occur from May to August. Larvae feed from summer onwards, forming spiral galleries in small to medium-sized stems. They eventually overwinter and pupate in the late spring, each in a small chamber formed beneath the epidermis.

DAMAGE

Larval infestations typically occur in grapevines that are already dead or are showing signs of decay. Although adults feed on the leaves of healthy vines, such damage is of no importance.

Agrilus sinuatus (Olivier) (249–250)

Pear jewel beetle

Associated mainly with pear, but also attacking apple; other hosts include *Crataegus monogyna* and *Sorbus*. Widely distributed in central Europe and currently (owing to warmer summers) extending its range as a pest into countries such as Belgium and the Netherlands. Recorded from the British Isles, but not reaching pest status; also present in North Africa and an introduced pest in the USA.

DESCRIPTION

Adult: 7–11 mm long; coppery red to purplish red; hind margin of pronotum strongly sinuate. **Larva:** up to 22 mm long; body dirty creamy white and dorsoventrally flattened, with the thoracic region noticeably enlarged; tip of abdomen with a pair of pointed, forceps-like projections.

LIFE HISTORY

Adults usually emerge in early June, but their appearance may be delayed by cool, wet weather. At first they graze on the foliage of pear trees, before

249 Pear jewel beetle (*Agrilus sinuatus*).

248 Vine jewel beetle (*Agrilus derasofasciatus*).

250 Pear jewel beetle (*Agrilus sinuatus*) – adult and typical damage to young leaf.

eventually mating. The adults may also feed on the foliage of other rosaceous trees, including almond. Larvae feed beneath the bark of host trees from July onwards, each forming a long, zigzag-shaped gallery in a small branch; galleries extending up to 1 m in length are reported in pear trees. Larvae usually pupate in the following May. However, unfavourable conditions may significantly delay their development, and individuals may then pass through two winters.

DAMAGE

Infestations are particularly damaging to young pear trees, causing premature fruit drop and rendering trees liable to attack by secondary organisms such as bark beetles (subfamily Scolytinae). Old, vacated galleries may become particularly noticeable several years after their formation, each developing into a large and distinctive scar on the bark, as the tree continues to grow. Adults feeding on foliage remove irregular sections from the edges, at least on young leaves the cut areas becoming necrotic, but such damage is of no significance.

Agrilus viridis (Linnaeus)
Beech jewel beetle

This widely distributed, Eurasiatic species is associated mainly with *Fagus sylvatica*, but will also attack many other trees and shrubs; occasionally, currant, gooseberry and grapevine are attacked, the larvae forming spiral galleries in the shoots and stems. Records of this species infesting raspberry, however, result from its confusion with raspberry jewel beetle (*Agrilus aurichalceus*). Adults (5–11 mm long) occur in various colour forms, including metallic blue, bluish black, bronze, green and coppery.

Capnodis tenebrionis (Linnaeus)
larva = flat-headed woodborer

A pest of stone fruit trees in Mediterranean regions and in central Europe, where it is associated mainly with almond, apricot, nectarine and peach.

DESCRIPTION

Adult: 15–25 mm long; mainly black and broad-bodied, with a paler black-marked, rounded pronotum; elytra parallel-sided, but narrowing abruptly to end in a blunt apex. **Larva:** up to 70 mm long; body white, flattened, clearly segmented and strongly tapered posteriorly beyond the pronotum; head armed with prominent black mandibles; pronotum very broad and rounded, with a brownish dorsal plate.

LIFE HISTORY

Adults occur at any time from spring to autumn. Those emerging from pupae in the spring live for only a few months and do not survive the winter. However, summer-emerging individuals survive for much longer, as they will hibernate during the first winter. Oviposition occurs only at high temperatures (over 25°C), each female being capable of laying several hundred eggs. The eggs hatch within 2–3 weeks. Larvae then bore within the main roots, hollowing out large, sawdust-filled chambers. Larval development lasts for up to a year. The fully fed larva typically pupates in an elongate chamber formed in the main rootstock immediately below the graft, and adults emerge about a month later.

DAMAGE

Infestations are of particular importance in young stone-fruit plantations. Adults graze on the leaves; they also damage the bark of the shoots and smaller branches. The larvae tunnel within the main roots and cause considerable damage; young trees may be killed and mature ones weakened.

Coroebus elatus (Fabricius)
Strawberry jewel beetle

At least in France, this southern European species is a sporadic pest of cultivated strawberry. Wild hosts (with which this species is more frequently associated) include *Agrimonia eupatoria*, *Filipendula vulgaris*, *Potentilla recta* and *Poterium sanguisorba*. Eurasiatic. Also found in North Africa.

DESCRIPTION

Adult: 6–8 mm long; greenish bronze and highly metallic; body elongate, but relatively broad. **Larva:** up to 15 mm long; body whitish, with a swollen thoracic region, and terminating in a pair of reddish-brown projections.

LIFE HISTORY

Adults occur from late May to July, most eggs being deposited in July. Larvae bore within crowns of plants from mid-summer onwards to form fine galleries that extend downwards below soil level. Pupation occurs in the following spring.

DAMAGE

Infested plants are weakened and cropping is reduced.

Family **ELATERIDAE**
(click beetles)

A small group of elongate beetles, with a hard exoskeleton and the head deeply retracted into the thorax. When disturbed, adults lie on their backs, feigning death, but then suddenly propel themselves into the air (with an audible click) by flexing the body between the prothorax and mesothorax, where a peg-like 'spring' (the prosternal process) is located between the first and second pair of legs. The soil-inhabiting larvae (commonly known as 'wireworms') are long, thin, cylindrical and tough-skinned, with well-developed thoracic legs and a ventral pseudopod on abdominal segment 10. A few species are very destructive agricultural pests.

Agriotes lineatus (Linnaeus) (**251–252**)
A common click beetle
A generally common agricultural pest, but sometimes also a problem on crops such as raspberry, strawberry and hop. Eurasiatic. Widely distributed in Europe, particularly in central and northern regions. An introduced pest in North America.

251 A common click beetle (*Agriotes lineatus*).

252 A common click beetle (*Agriotes lineatus*) – larva.

DESCRIPTION
Adult: 7–10 mm long; dark yellowish brown, the elytra striated alternately light and dark. **Egg:** 0.6 mm across; slightly oval and whitish. **Larva:** up to 25 mm long; body shiny yellowish brown, with a darker head; tip of body pointed; jaws powerful, but legs small; abdominal segment 9 helmet-shaped, with a pair of dark, sensory, dorsolateral pits.

LIFE HISTORY
Eggs are laid in groups in the soil during May and June, usually amongst grass or other vegetation. The larvae hatch about a month later and begin feeding on plant roots and other underground vegetable matter. They are most active in spring and autumn, but their development is slow and usually extends over 4 or 5 years. When fully grown, usually in July or August, they pupate in separate earthen cells formed 10–25 cm below the surface. Adults are produced after about a month, but they normally remain within their cells until the following spring.

DAMAGE
Wireworms bite through the roots and bore into the base of plants, sometimes causing them to wilt and die. Strawberry plants or hop sets planted in recently broken-up grassland are particularly liable to be attacked. Adults are generally harmless; however, they occasionally occur on fruit blossom, notably apple, and may then damage the floral parts. Damage to the inflorescences of red current is also reported.

Agriotes obscurus (Linnaeus) (**253**)
A common click beetle
This species is most abundant in more northerly areas and particularly favours light soils with a high organic content. The life cycle and habits are essentially similar to those of *Agriotes lineatus*, and the larvae also occasionally attack crops such as raspberry and strawberry. Adults (7–10 mm long) are dark brown in colour.

253 A common click beetle (*Agriotes obscurus*).

Agriotes sputator (Linnaeus) (**254**)

A common click beetle

Larvae of this relatively small species (adults 6–7 mm long) are recorded as minor pests of strawberry and various other crops. The distribution is more southerly than that of both *Agriotes lineatus* and *A. obscurus*.

Ctenicera cuprea (Fabricius) (**255**)

Upland click beetle

larva = an upland wireworm

Strawberry crops planted in hilly and montane regions are sometimes subject to attack by larvae of upland click beetles, especially *C. cuprea*. Adults (11–16 mm long) are dark-bodied, with a metallic green or coppery sheen, and the elytra sometimes partly tinged with yellowish; the antennae of males are long and each bears eight prong-like projections, whereas those of females are serrated, with mainly triangular flagellar segments. Larvae (up to 35 mm long) are light brown and stout-bodied; unlike species of *Agriotes*, abdominal segment 9 subdivided by a bulb-shaped notch and there are no sensory pits. The larvae take up to 5 years to complete their development, and cause similar damage to those of *Agriotes lineatus*.

254 A common click beetle (*Agriotes sputator*).

255 Upland click beetle (*Ctenicera cuprea*).

Family **CANTHARIDAE** (**256**)

Narrow, elongate, soft-bodied beetles with smooth, downy elytra, and long, filiform antennae. Adults and larvae are typically predacious, but adults of a few species are also at least partly phytophagous.

Cantharis obscura Linnaeus

A minor pest of apple, pear, quince, peach and other fruit trees in mainland Europe; also reported in Scotland as a pest of raspberry. Widely distributed in Europe.

DESCRIPTION

Adult: 9–15 mm long; head black; pronotum black with yellow or brownish-orange lateral margins; elytra black.

LIFE HISTORY

Adults occur from spring onwards and occasionally browse on the fruit buds and open blossoms of fruit trees. They are also associated with *Quercus*, feeding on the inflorescences and leaf petioles. Larvae are entirely predacious.

DAMAGE

Loss of tissue on fruit trees is usually of little or no significance. On raspberry, however, tissue of fruiting laterals may be removed down to the pith, resulting in the breakage or wilting of shoots. Adults of several other species of *Cantharis* (e.g. *C. fusca* Linnaeus, *C. livida* Linnaeus and *C. pellucida* Fabricius) also occur occasionally on fruit crops. They are most frequently found feeding on the open blossoms, but any damage caused is of little or no importance.

256 *Cantharis fusca* – adult.

Family **BOSTRYCHIDAE**

A mainly tropical family of cylindrical beetles, with the pronotum extending forward as a hood to cover much of the head; antennae 10-segmented, including a 3-segmented club. Larvae are more or less scarabaeiform and often wood borers.

Sinoxylon sexdentatum (Olivier)

A minor pest of grapevine; other hosts (although less frequently attacked) include mulberry, peach, pear, fig and olive. Widely distributed in the Mediterranean basin, extending into southern Europe.

DESCRIPTION

Adult: 4–6 mm long; robust and relatively broad; pronotum dull brownish black, subspherical and very large, armed with numerous short projections; elytra reddish brown and rectangular, each with three backwardly directed spines posteriorly; pronotum and elytra finely pubescent; antennal club broadly lamellate. **Larva:** up to 7 mm long; white.

LIFE HISTORY

In early spring, adults feed on the shoots, buds, leaves and flower clusters of grapevines; they may also invade a wide range of other plants. In May and June, each female beetle bores into a woody shoot, usually close to the base of a bud, to form a peripheral gallery beneath the rind, where mating takes place. The mated female then excavates a longitudinal burrow in which eggs are laid. She then abandons the shoot and crawls or flies away to infest another shoot or a nearby vine. Following invasion of the wood by the adult females, sawdust ejected from the mating and maternal galleries often accumulates on the ground beneath infested vines. Larvae feed throughout the summer months, forming long galleries, formed beneath the bark of the shoots and branches, and eventually pupate. New adults emerge in the spring through circular flight holes, there being just one generation annually.

DAMAGE

Feeding by adults in early spring weakens host plants. Later, the section of a shoot infested by larvae is often reduced to a mass of fine sawdust, enclosed by a thin outer layer of rind. Such shoots may subsequently break off. Attacks by this pest occur mainly on poorly growing, already debilitated grapevines.

Family **NITIDULIDAE**

A varied group of mainly small beetles, with clubbed antennae; club 3-segmented; elytra typically shorter than the abdomen, exposing the last one or two abdominal tergites.

Glischrochilus hortensis (Fourcroy) (**257**)

In dry weather, adults of this widely distributed and often common species sometimes tunnel into ripe strawberries, presumably in search of moisture. However, although the beetles tend to aggregate and are frequently attracted in large numbers to tree sap, it is unlikely that populations in strawberry beds are ever large enough to cause serious crop losses except, perhaps, in small, non-commercial plots. The beetles are 5–6 mm long and shiny black, with four orange-red dots on the elytra. In Canada, a related species (*Glischrochilus quadrisignatus* Say) is known to attack raspberries; this species is now established in mainland Europe (e.g. in Hungary), where it is attracted to damaged fruits and is implicated in the transmission of pathogens.

257 *Glischrochilus hortensis* – adult.

Meligethes aeneus (Fabricius) (**258**)
Pollen beetle

A generally abundant species, associated mainly with brassicaceous plants (including mustard, oilseed rape and vegetable brassicas, upon which it is a well-known pest). Adults often invade fruit blossom, notably in raspberry and strawberry plantations, sometimes in considerable numbers. Although usually harmless, the beetles may damage the petals and other floral parts in order to gain access to pollen, upon which they feed. Adults (1.5–2.7 mm long) are mainly black with a bronzy-green metallic sheen.

258 Pollen beetle (*Meligethes aeneus*) – adults on strawberry blossom.

Family **BYTURIDAE**

A small group of small, hairy beetles, with clubbed antennae and toothed tarsal claws. Larvae are cylindrical, with well-developed thoracic legs; the anal segment has a pair of dorsal processes and a ventral pseudopod.

Byturus tomentosus (Degeer) (**259–261**)
Raspberry beetle

A generally common and important pest of blackberry, loganberry and raspberry. Widely distributed in central and northern Europe.

DESCRIPTION

Adult: 3.5–4.5 mm long; brown, covered with decumbent, yellowish-brown hairs, which later fade to greyish brown or grey. **Egg:** 1.5 × 0.4 mm; shiny creamy white. **Larva:** up to 8 mm long; body pale yellowish brown, with darker brown plates arranged segmentally along the back and a pair of prominent dorsal cerci on the penultimate abdominal segment; head brown; thoracic legs well developed.

LIFE HISTORY

Adults appear from late April or early May onwards and are at once attracted to open blossoms of various rosaceous plants, including apple, pear and *Crataegus*. They fly readily and are particularly active in warm, sunny weather. Peak numbers are reached in early June, by which time most beetles will have migrated to loganberry or raspberry and begun to feed on the flowers and unopened flower buds. Attacks on blackberry occur somewhat later in the season. Mating occurs shortly after the adults have commenced feeding and the first eggs are laid about a week later, usually in blossoms that have set fruit. A female may lay 100 or more eggs, each being placed in a blossom along the filament of an anther or

259 Raspberry beetle (*Byturus tomentosus*).

260 Raspberry beetle (*Byturus tomentosus*) – larva.

style, usually one per flower. The period of egg-laying extends over 2 or 3 weeks, depending on weather conditions. Eggs hatch after about 10 days, generally from the green-fruit to early pink-fruit stages onwards. At first the small larvae feed on developing drupelets near their empty egg shells, but they soon crawl to the basal drupelets; at the pink-fruit stage, when the receptacles have softened, the fruit plugs are invaded. Feeding continues in the plug, and on the adjacent surfaces of ripening drupelets, each larva passing through four instars. When fully grown, after 5–7 weeks, the larvae vacate the fruits and drop to the ground. They then form earthen cells, usually within the top 5 cm of soil, but sometimes as deep as 20 cm or more. Pupation takes place after a delay of about a month. The adult stage is reached 5–6 weeks later. However, individuals remain in their pupal cells until the following spring, there being just one generation annually.

DAMAGE
Adult: the importance of feeding damage varies according to conditions. If large numbers of beetles migrate to loganberry or raspberry before blossom time, then damage can be very serious, resulting in the production of many small, malformed fruits and heavy crop losses. Buds can be destroyed completely; those at the centre of young bud clusters may be severed from the stalks and will then drop off. Open blossoms are also damaged, the stamens and nectaries being bitten and destroyed; in extreme cases the whole flower, with the exception of the sepals, may be devoured. **Larval:** damage caused by the larvae is usually more important. Attacked drupelets turn brown and hard, resulting in the development of shrivelled, misshapen fruits. In addition, the presence of the grubs in crops sent for processing may result in the whole consignment being rejected by the canners.

Family **TENEBRIONIDAE**

A varied family of often wingless, sombre-coloured beetles; elytra often fixed rigidly to the body; antennae thickened and often more or less clubbed. Larvae are elongate, with the posterior spiracles mounted on a pair of projections (respiratory processes). Several species are well-known pests of flour and other stored products in food warehouses.

Lagria hirta (Linnaeus) (**262**)
A polyphagous species on various shrubs, and a minor pest of blackberry, raspberry and strawberry. Widely distributed in Europe, but less abundant in the north.

262a *Lagria hirta* – larva.

261 Raspberry beetle (*Byturus tomentosus*) – larval damage to loganberry fruit.

262b *Lagria hirta* – adult.

DESCRIPTION
Adult: 7–10 mm long; head, legs and pronotum black; elytra yellowish brown, clothed in whitish hairs, elongate and noticeably broadened beyond the middle; antennae 11-segmented; hindwings fully developed. **Larva:** up to 12 mm long; body dark brown to blackish and hairy; head black and shiny.

LIFE HISTORY
Beetles are active from mid-May to the end of July, when they browse on the leaves of various plants. Eggs are laid in the soil and hatch about a month later. Larvae then begin to feed on vegetable debris around the base of plants; they may also remove tissue from foliage in contact with the ground. They eventually overwinter and complete their development in the spring. Fully fed larvae pupate in the soil without forming cocoons, adults emerging about 2 weeks later.

DAMAGE
Occasionally, adults bore into fruits. However, such attacks are rarely of significance.

Family **CERAMBYCIDAE**
(longhorn beetles)

A large family of small to very large, often brightly coloured beetles, with usually long, narrow elytra and very long antennae. The wood-boring larvae are soft and fleshy, with an enlarged prothorax and much reduced, non-functional thoracic legs. Longhorn beetles are sometimes associated with fruit trees; however, as larval feeding is restricted to unhealthy, dead or decaying wood they are of only secondary importance. Attacks on fruit trees sometimes follow infestations by primary pests. Apple-wood longhorn beetle (*Pogonocherus hispidulus* (Piller & Mitterpacher)) (5.5–7.0 mm long), for example, often invades trees attacked by apple clearwing moth (*Synanthedon myopaeformis*) (family Sesiidae) (Chapter 6).

Obera linearis (Linnaeus)
Hazel longhorn beetle
A pest of hazelnut and walnut in southern and eastern Europe, its range extending into northern Turkey. Although a potentially important pest, significant infestations are most often established on non-cultivated stands of *Corylus avellana*.

DESCRIPTION
Adult: 11–16 mm long; black with yellowish legs; elytra very long and narrow. **Larva:** up to 20 mm long; white.

LIFE HISTORY
Adults occur from May to July. Eggs are then laid singly, each at the base of a 1-year-old shoot. Following egg hatch, each larva bores upwards from the oviposition point to form a superficial gallery that partly encircles the shoot. The larva eventually hibernates and in the following year excavates a very long, downward-orientated borrow in the pith (that may extend for up to 60 cm), expelling frass at intervals through small holes made in the wall. Eventually, a small chamber is excavated at the end of the gallery. Here the larva overwinters for a second time. It then pupates in the chamber, 1–2 months before the appearance of the adult.

DAMAGE
On hazelnut, superficial burrows formed in the 1-year-old wood cause the shoot tips to desiccate and split, the affected tissue turning yellowish or bright red and keeling over. Tunnelling by older larvae weakens branches. Heavy infestations have an adverse effect on growth and may also reduce cropping potential.

Tetrops praeusta (Linnaeus) (**263**)
Little longhorn beetle

A minor pest of fruit trees, particularly apple, pear, cherry and plum, but usually absent from regularly pruned orchards and not important as a pest. Present throughout Europe. Also found in North Africa.

DESCRIPTION
Adult: 3–5 mm long; mainly black with hairy, dark brown to brownish-yellow elytra; legs mainly yellowish.

LIFE HISTORY
Adults feed on the young foliage of various broadleaved trees during April and May, and are often abundant in old, neglected orchards. The larvae feed in the twigs and stems of dead branches, completing their development within a year. Young beetles emerge in the spring through small, round flight holes, most often found in the tips of dead spurs.

DAMAGE
Adults eat only the epidermis of leaves and are not of economic importance. Similarly, the larvae are of no consequence in orchards as they attack only dead wood.

Trichoferus griseus (Fabricius)

In Mediterranean areas, a polyphagous pest of trees and shrubs, including cultivated fig and (to a lesser extent) pomegranate.

DESCRIPTION
Adult: 9–16 mm long; ash-grey, mottled with brown and slightly pubescent; elytra parallel-sided; antennae of male as long as body, those of female shorter.

LIFE HISTORY
Adults emerge in late July or early August. Eggs are then laid in sheltered situations on the bark of host plants, usually on small or medium-sized branches and often where mechanical damage has already occurred. Following egg hatch, larvae bore into the wood to feed. The larvae form long galleries up to 30 cm long, and usually complete their development in the following spring.

DAMAGE
On fig, infested branches are discoloured (the bark turning noticeably dark), fail to bear fruit and eventually die; subsequently, such branches often become invaded by the bark beetle *Hypoborus ficus* (subfamily Scolytinae).

Vesperus xatarti (Dufour-Mulsant)
Grapevine vesperus beetle

A potentially serious pest of grapevine in Mediterranean areas. Also attacks a wide range of woody and herbaceous crops, including fruit trees, fig, olive and various vegetable crops.

DESCRIPTION
Adult: 20–30 mm long; brownish and apterous; head subquadrate; thorax narrow and rounded; abdomen bulbous and elongate; elytra short and atrophied; antennae of male very long, those of female much shorter than body. **Egg:** 2.0 × 0.8 mm; white. **Larva:** very large, with a well-developed head, three pairs of thoracic legs; body brown, much wrinkled and sac-like, with numerous reddish, bristle-like hairs.

LIFE HISTORY
Adults appear from late December onwards, with a peak of emergence in February. After mating, eggs are laid in large groups in, for example, cracks or crevices on stakes or beneath the bark of old vine stumps. The eggs are then coated with a protective, yellowish secretion. A female is capable of laying up to a thousand eggs in her lifetime. Following egg hatch, larvae enter the soil to feed on the roots of host plants. The larvae occur relatively shallowly in spring and autumn, but much deeper (typically down to about 0.5 m) during summer and winter. There are five larval instars. Pupation occurs in an earthen cell during the late winter, but sometimes earlier. There is one generation annually.

DAMAGE
Larvae gnaw the roots of host plants, often causing extensive damage. Host plants may be weakened or even killed. Owing to their vigorous root growth, olive trees are able to resist attacks.

263 Little longhorn beetle (*Tetrops praeusta*).

Family **CHRYSOMELIDAE**
(leaf beetles)

A very large family of small to medium-sized, often brightly coloured or metallic-looking, rounded or oval-bodied beetles; antennae filiform; hind legs sometimes greatly enlarged and adapted for jumping, as in flea beetles. Adults and larvae are typically phytophagous, the former usually forming the overwintering stage; the latter are typically eruciform, with well-developed thoracic legs; larvae of most species live freely on leaves, but some mine within leaves or stems or attack the roots.

Agelastica alni (Linnaeus) (**264–267**)
Alder leaf beetle

An important and destructive pest of *Alnus* and various other broadleaved trees, particularly *Betula*, *Carpinus betulus* and *Tilia*; infestations may also occur on fruit trees (particularly apple) and hazelnut. Although widely distributed in mainland Europe, and well established in countries bordering the English Channel, this species is extinct in the British Isles.

DESCRIPTION

Adult: 6–8 mm long; bluish to violet with black antennae, tibiae and tarsi; body rather bulbous, with the elytra noticeably expanded towards the hind end. **Larva:** up to 11 mm long; cylindrical, mainly black and shiny.

LIFE HISTORY

Adults, which hibernate throughout the winter, emerge in the spring. After mating, the females eventually lay eggs in large, scattered groups on the fully expanded leaves of host plants. The eggs hatch in about 2 weeks. Larvae, which occur from June to July, feed gregariously on both sides of the leaves for about 3 weeks. At first they graze the surface, but later they bite out holes between the major veins. When fully grown, larvae drop to the ground to pupate, either on or just below the surface, and adults appear 1–2 weeks later. The young beetles graze on the foliage of host plants before overwintering. There is a single generation annually.

264 Alder leaf beetle (*Agelastica alni*) – female.

265 Alder leaf beetle (*Agelastica alni*) – egg batch.

266 Alder leaf beetle (*Agelastica alni*) – larvae.

267 Alder leaf beetle (*Agelastica alni*) – young larvae attacking apple leaf.

DAMAGE

Attacked foliage becomes peppered with large, irregular holes; heavily infested trees are weakened and their growth retarded. Direct damage may also be caused to fruit blossoms. Damage is of particular significance on young trees.

Altica ampelophaga Guérin-Méneville (268–271)
Vine leaf flea beetle

A sporadically important pest of grapevine in southern Europe; also present in North Africa.

DESCRIPTION

Adult: 4–5 mm long; mainly dark metallic green, sometimes with a bluish sheen; legs black; hind femora greatly enlarged (typical of a flea beetle); antennae 11-segmented. **Egg:** 1.0 × 0.4 mm; dull, orange yellow. **Larva:** up to 8 mm long; body dark yellow, with numerous black plates; head black; young larva mainly black.

LIFE HISTORY

Adults are active in the spring, when they attack the newly emerging growth of grapevines. Eggs are deposited in groups on the underside of leaves from early April onwards, each female being capable of laying several hundred eggs in her lifetime. Duration of the various developmental stages varies considerably. Typically, the eggs hatch 1–2 weeks after being laid. Larvae then browse on the foliage for about a month, passing through three instars. They then pupate in the soil at a depth of about 5 cm, and adults of the next generation appear 1–3 weeks later. There are usually two, and sometimes three, generations annually. Adults of the final generation hibernate amongst leaf litter and in other shelter, mating occurring in the following spring.

DAMAGE

Particularly in spring, adults and first-generation larvae can cause significant defoliation by attacking the freshly sprouting vines and damaging the leaves and flower clusters. In summer, attacks on the expanded foliage often result in the appearance of lace-like areas on the laminae where tissue has been removed, but the upper surface left largely intact.

268 Vine leaf flea beetle (*Altica ampelophaga*) – adults in copula.

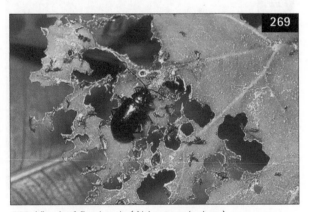

269 Vine leaf flea beetle (*Altica ampelophaga*).

270 Vine leaf flea beetle (*Altica ampelophaga*) – larvae.

271 Vine leaf flea beetle (*Altica ampelophaga*) – 'old' larval damage.

Aphthona euphorbiae (Schrank) (272–273)
Large flax flea beetle

This local, but widely distributed flea beetle is of greatest significance as a pest of flax and linseed; however, in spring, adults also feed on the foliage of a wide range of other plants, including apple and strawberry. Damage to fruit crops is restricted to slight loss of photosynthetic tissue and is of little or no importance. Adults are 1.5–2.0 mm long, mainly black, with a blue or greenish metallic sheen; the pronotum and elytra are finely punctured. Other flea beetles associated mainly with agricultural crops also sometimes browse on the foliage of cultivated fruits. These include mangold flea beetle (*Chaetocnema concinna* (Marsham)) and flax flea beetle (*Longitarsus parvulus* (Paykull)): the former on strawberry (usually in April and May) and the latter on apple (usually in May and June). Adults of *C. concinna* are 1.5–2.0 mm long, black and shiny, with each mid- and hind tibia bearing a pointed projection; those of *L. parvulus* are 1.2–1.7 mm long, black or brown, with a slight metallic lustre.

Batophila aerata (Marsham) (274–275)
A raspberry flea beetle

Adults of this widely distributed and often common European species feed on the foliage of various plants, particularly *Rubus*. Minor infestations of the adult beetles sometimes occur in late May and June, and again in the autumn, on cultivated blackberry, loganberry and raspberry. However, the beetles are usually present in only small numbers. Leaf damage is unimportant and restricted to the removal of tissue from the upper surface, the affected areas soon turning brown; the extent of damage, however, often belies the small size of the causal agents. Adults are 1.0–1.8 mm long and black, with a greenish-bronze sheen; the legs are reddish.

Batophila rubi (Paykull)
A raspberry flea beetle

This species, which occurs from southern Europe to Scandinavia, is often found on blackberry, loganberry, raspberry and strawberry. However, as in the case of *Batophila aerata*, this species is of only minor

272 Large flax flea beetle (*Aphthona euphorbiae*) – adults in copula.

274 A raspberry flea beetle (*Batophila aerata*).

273 Large flax flea beetle (*Aphthona euphorbiae*) – damage to strawberry leaflet.

275 A raspberry flea beetle (*Batophila aerata*) – damage to raspberry leaf.

importance. Adults are 1.5–2.0 mm long, black and shiny with red legs, and are somewhat plumper than those of the previous species.

Bromius obscurus (Linnaeus)
Vine leaf beetle

A pest of grapevine in south-central and southern Europe. An introduced pest in the USA.

DESCRIPTION
Adult: 5–6 mm long; head and pronotum black; elytra brown, brownish black or black, with a slight whitish pubescence, and noticeably broader than the pronotum. **Egg:** 1.0 × 0.5 mm; yellow. **Larva:** up to 8 mm long; body whitish and more or less C-shaped; head large and yellowish.

LIFE HISTORY
This species is entirely parthenogenetic. The adult females emerge in May or June and survive for up to 3 months. At first they feed on the foliage and green shoots; later in the season they also attack the rachides and developing grapes. Eggs are laid in the soil at the base of the vines, in groups of 20–30. Following egg hatch the larvae burrow down to graze on the roots. The larvae are fully grown by the following spring. They then pupate, and adults emerge 3–5 weeks later.

DAMAGE
Adults remove elongate sections of tissue from between the major veins on the expanded leaves, severely damaged leaves becoming tattered and torn. Damaged fruits develop darkened strips and are unmarketable as table grapes. Larvae graze away extensive areas of epidermal tissue from the roots and sometimes also burrow into the underlying wood; heavily infested vines may be weakened.

Galerucella sagittariae (Gyllenhal) (**276**)
Northern strawberry leaf beetle

An important pest of strawberry in parts of northern and north-eastern Europe, notably in Sweden and Finland. Loganberry and raspberry are also attacked.

DESCRIPTION
Adult: 4.0–5.5 mm long; yellowish brown to olive brown, with strongly punctured elytra; antennae dark-tipped and 11-segmented. **Egg:** 0.7 × 0.6 mm; elongate-oval, brownish white to pinkish beige. **Larva:** up to 10 mm long; body yellowish to greenish, with numerous black plates; head black.

LIFE HISTORY
Adults emerge from hibernation in the spring and often aggregate on the leaves of host plants. Eggs are laid singly or in small groups on either side of leaves, and also on the leaf petioles. Following egg hatch, larvae browse on the underside of the leaves, usually from early June onwards. Individuals pass through three instars and are fully grown within 3–4 weeks. They then pupate in the soil, close to the surface. New adults appear in the late summer and then feed briefly before eventually overwintering.

DAMAGE
The epidermis of foliage is rapidly grazed away and affected parts of leaves eventually turn brown. Heavy infestations lead to loss of plant vigour and can have an adverse effect on cropping.

276 Northern strawberry leaf beetles (*Galerucella sagittariae*).

Galerucella tenella Linnaeus (**277**)
Strawberry leaf beetle

A pest of strawberry and, to a lesser extent, loganberry and raspberry. Wild hosts include *Filipendula ulmaria*, *Geum rivale* and *Potentilla anserina*. Widely distributed in central and northern Europe.

DESCRIPTION
Adult: 3.0–4.5 mm long; mainly yellowish brown to reddish brown, with strongly punctured elytra; antennae 11-segmented. **Egg:** 0.6 × 0.4 mm; elongate-oval, brownish white to pinkish white. **Larva:** up to 8 mm long; body elongate, dull brownish white to yellowish, with numerous black plates; head black.

LIFE HISTORY & DAMAGE
Essentially similar to that of northern strawberry leaf weevil (*Galerucella sagittariae*).

Psylliodes attenuata (Koch) (**278**)
Hop flea beetle

A minor pest of hop; formerly of greater importance than nowadays. Widely distributed in many hop-growing districts of Europe.

DESCRIPTION
Adult: 2–3 mm long; greenish black with a bronzy sheen; legs, except hind femora, reddish; antennae 10-segmented; face with frontal lines forming an X-shaped pattern between the eyes. **Egg:** 0.50 × 0.25 mm; oval and pale yellow. **Larva:** up to 6 mm long; body whitish; head, prothoracic plate and anal plate reddish brown.

LIFE HISTORY
Adults hibernate in the soil or under vegetation, emerging in late spring. Eggs, up to 300 per female, are then laid in cracks in the soil around the base of host plants. The eggs hatch in about 7–10 days. Larvae feed on the roots of hop (and other plants) for about a month before finally pupating in earthen cells. New adults begin to appear from late July or early August onwards. They feed on the foliage in summer and may also attack the developing burrs (strobili). Activity ends in early autumn, as the beetles take up their winter quarters.

DAMAGE
Attacks are most harmful in nursery beds and on newly planted hops. By feeding on young bines and leaves in May and early June, adults (if numerous) have an adverse effect on plant development, particularly in cold weather when growth is slow. Damage to the developing burrs is most likely to occur in wet conditions.

277 Strawberry leaf beetle (*Galerucella tenella*).

278 Hop flea beetle (*Psylliodes attenuata*).

Family **ATTELABIDAE**
(weevils)

A small group of weevils, with straight antennae, often flattened, scale-less elytra, the tarsal claws fused basally and the external edge of the mandibles untoothed. Larvae apodous, with a well-developed head.

Apoderus coryli (Linnaeus) (**279–281**)
Hazel leaf-roller weevil

A minor pest of hazelnut, and often abundant on wild *Corylus avellana*; other hosts include *Carpinus betulus*, *Fagus sylvatica* and *Ostrya carpinifolia*. Eurasiatic. Present throughout Europe.

DESCRIPTION
Adult: 6–8 mm long; mainly red, with a black head and antennae and mainly black legs. **Egg:** 1.0–1.5 mm long; oval, orange. **Larva:** up to 10 mm long; body bright orange; head brown. **Pupa:** 6–8 mm long; orange.

LIFE HISTORY
Adults emerge in the spring to feed on the leaves of host plants. Eggs are laid in May and June, singly or in small groups, usually in the midrib towards the tip of an expanded leaf. The lamina is then severed near the base, the cut extending from one edge to or just beyond the midrib; the cut tissue then curls laterally to remain suspended from the unsevered part of the lamina as a stumpy, cigar-like leaf roll. The larvae develop singly within these leaf rolls and then pupate, new adults appearing at the end of July or in early August. Larvae of a second generation complete their development in the autumn; they eventually overwinter on the ground, within their fallen habitations, and pupate in the spring.

DAMAGE
Although larval habitations may be very numerous on wild hosts, damage caused in cultivated hazelnut plantations is usually unimportant.

279 Hazel leaf-roller weevil (*Apoderus coryli*).

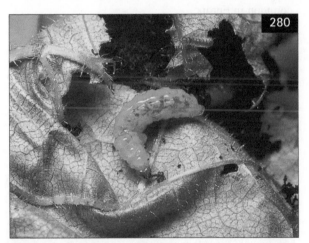

280 Hazel leaf-roller weevil (*Apoderus coryli*) – larva.

281 Hazel leaf-roller weevil (*Apoderus coryli*) – larval habitation.

Family **RHYNCHITIDAE**
(weevils)

A small group of weevils (formerly regarded as a subfamily within Attelabidae), with straight antennae, the tarsal claws separate (i.e. unfused) and the external edge of the mandibles dentate; rostrum usually elongated. Larvae apodous, with a well-developed head.

Byctiscus betulae (Linnaeus) (**282–285**)
Pear leaf-roller weevil
A potentially important pest of grapevine. Pear and various other broadleaved trees and shrubs are also attacked; in more northerly parts of its range associated mainly with *Betula* and *Corylus avellana* and of little or no pest status. Palaearctic. Widely distributed and often common in Europe.

DESCRIPTION
Adult: 4.5–7.0 mm long; brilliant metallic green, golden green through bluish green to blue or violet, often tinged with coppery red or reddish; male with a forward-projecting spine on each side of the pronotum. **Egg:** 0.8 × 0.6 mm; oval and white. **Larva:** up to 7 mm long; body white and plump; head light brown.

LIFE HISTORY
Overwintered adults appear from mid-May or early June onwards. They browse on the young leaves, forming characteristic channel-like markings or holes that often cover much of the laminae. After about a fortnight, the females deposit eggs in the midrib of expanded leaves of host plants, often 5–7 (occasionally more) per leaf. The leaf stalk (petiole) is then partly severed, causing the leaf to droop and to become folded into a tight, cigar-shaped cylinder that encloses the eggs and later becomes the larval habitation. These habitations desiccate and eventually turn black. Larvae feed for 3–4 weeks, usually completing their development by the end of August. They then pupate, new adults emerging about 10 days later. The young weevils browse on the leaves during the autumn and then hibernate.

282 Pear leaf-roller weevil (*Byctiscus betulae*).

283 Pear leaf-roller weevil (*Byctiscus betulae*) – larva.

284 Pear leaf-roller weevil (*Byctiscus betulae*) – larval habitation on pear.

DAMAGE

Grapevine: leaf browsing is often extensive and can have an adverse effect on grape development, as can loss of leaf tissue if larval habitations are abundant. **Pear:** larval habitations may be of significance on young trees, but infestations on older trees are of little or no importance.

Involvulus caeruleus (Degeer) (**286–287**)

Apple twig cutter

A local, sometimes common pest of apple. Pear, quince, medlar, almond, cherry, peach and plum are also attacked. Wild hosts include *Crataegus monogyna* and *Sorbus*. Palearctic. Present throughout Europe.

DESCRIPTION

Adult: 2.5–4.0 mm long; shiny metallic blue, clothed with black hairs. **Egg:** oval and translucent. **Larva:** up to 4 mm long; whitish with the brownish head retracted into the swollen prothoracic region.

LIFE HISTORY

Weevils appear from April or May onwards and are most active in warm, sunny weather. They feed on foliage and are often very numerous on young trees. Eggs are laid from early May onwards, a single egg being placed within the soft stem (about 10 to 15 cm from the tips of new vegetative shoots), which is then cut slightly below the oviposition point. Attacked shoots usually fall to the ground. However, if severed only partially they may remain attached and wither on the tree. Eggs hatch in about 2 weeks. Larvae then feed in the pith of the shoots for 3–4 weeks. They then enter the soil to pupate in earthen cells a few centimetres below the surface. Adults emerge in late summer and then hibernate under dead leaves or in other sheltered situations until the spring.

DAMAGE

The presence of this pest is betrayed in spring and early summer by the loss of young shoots or the sight of partially severed, withering ones. Later in the summer, lateral shoots develop below these damaged tips. Damage to established trees is rarely important, but attacks on young trees can be serious.

285 Pear leaf-roller weevil (*Byctiscus betulae*) – damage to grapevine leaf.

287 Apple twig cutter (*Involvulus caeruleus*) – larval habitation.

286 Apple twig cutter (*Involvulus caeruleus*).

148

Neocoenorrhinus aequatus (Linnaeus) (288–292)
Apple fruit rhynchites

A locally common pest of apple; other cultivated hosts include pear, quince, almond, apricot, cherry, plum and medlar. Widely distributed in Europe.

DESCRIPTION
Adult: 2.5–4.5 mm long; elytra reddish brown and hairy; head and thorax darker, strongly punctured and with a purplish to bronzy sheen. **Larva:** up to 4 mm long; whitish, with the brownish head retracted into the swollen prothoracic region.

LIFE HISTORY
Adult weevils occur from May to mid-July on various rosaceous trees and shrubs, including fruit trees. Here, they damage developing fruitlets by drilling small cylindrical holes into the flesh, at the base of which eggs (one per incision) may be laid. The eggs hatch after a week or so and the larvae then feed on the surrounding

288a, b Apple fruit rhynchites (*Neocoenorrhinus aequatus*).

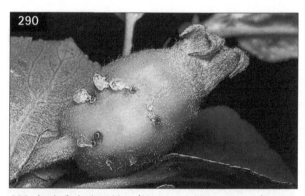

289 Apple fruit rhynchites (*Neocoenorrhinus aequatus*) – adult boring into fruitlet.

290 Apple fruit rhynchites (*Neocoenorrhinus aequatus*) – damage to fruitlet.

291 Apple fruit rhynchites (*Neocoenorrhinus aequatus*) – puncture scars on mature apple.

292 Apple fruit rhynchites (*Neocoenorrhinus aequatus*) – puncture scars on apricot, and onset of secondary rotting.

flesh, becoming fully grown in about 3 weeks. They then drop to the ground and eventually pupate in the soil, each within an earthen cell.

DAMAGE

Fruitlet damage can be serious and is very characteristic. On apple, for example, there may be 100 or more holes in one fruitlet, each about 1 mm across and up to 7 mm deep. Attacked apples remain marked and distorted, although the holes tend to close up as the fruitlets grow. Particularly in the case of plums, considerable quantities of gum issue from the sites of damage. Damaged fruitlets may also rot following invasion by secondary pathogens. Occasionally, the adult weevils cause harm in orchards before petal fall, biting holes into the buds and flowers.

Neocoenorrhinus cribripennis (Desbrochers)
Olive fruit rhynchites

A minor pest of olive in parts of the Mediterranean basin, including Corsica, Greece, southern Italy, Sardinia and Turkey. Attacks also occur on various other Oleaceae.

DESCRIPTION

Adult: 3.5–4.5 mm long; mainly red or yellow, with a greyish pubescence; prothorax finely punctured; elytra roundly quadrate, finely punctured and with distinct longitudinal striae. **Egg:** 0.55 × 0.40 mm; elliptical and yellowish. **Larva:** up to 4 mm long; body creamy white; head brown.

LIFE HISTORY

Adults emerge from hibernation from late April to late May. They then attack the young leaves, shoots and newly developing fruits. Eggs are laid singly within young olives. Larvae feed within the fruits and are fully grown by the autumn. They then vacate the fruit and enter the soil, where they form earthen cells several centimetres below the surface and eventually pupate. The adult stage is reached before the onset of winter, but the young weevils remain within their pupal chambers until the following spring.

DAMAGE

Adult feeding has an adverse effect on vegetative growth, and also causes malformation and distortion of developing fruitlets. Fruits infested by larvae are destroyed.

Neocoenorrhinus germanicus (Herbst)
Strawberry rhynchites

A somewhat local, but not infrequent, pest of strawberry. Blackberry, loganberry and raspberry are also attacked. Palaearctic. Widely distributed in Europe.

DESCRIPTION

Adult: 2–3 mm long; black with a greenish-blue sheen; at once distinguished from *Anthonomus rubi* (family Curculionidae) by body colour and by the straight antennae. **Egg:** 0.7 × 0.5 mm; whitish and translucent. **Larva:** up to 2.5 mm long; yellowish white, with the brownish head retracted into the swollen prothoracic region.

LIFE HISTORY

Adults emerge form hibernation from mid-March to early May. They then feed on strawberry and various kinds of *Rubus*. At first, they attack the young leaves. Later, they bite holes into blossom trusses, petioles and stolons. Egg-laying occurs from mid-April to August, reaching its height in May. Early in the season, eggs are laid in cavities bitten into leaf stalks (petioles); later, they are deposited in the petioles of young blossom trusses. On strawberry, tips of young stolons are also suitable oviposition sites. Eggs are always deposited in young, tender tissue and there may be from one to four in any one petiole, each placed within its own chamber. Once eggs are laid, the female encircles the stalk with a series of small punctures, just below the lowermost egg chamber. As a result, the leaf or blossom truss wilts and dies, either remaining attached to the plant by a small piece of rind or dropping to the ground. Eggs hatch 2–3 weeks later and the larvae burrow within the desiccated tissue, becoming fully grown within a few weeks. They then enter the soil to pupate, each in a loose, silken cocoon. The adult stage is reached 2–3 weeks later. However, the weevils remain in their cells throughout the winter, eventually appearing in the following spring.

DAMAGE

Strawberry: greatest damage occurs early in the season, destruction of young leaves weakening the plants and loss of flower trusses greatly reducing crop yields. Damage to the stolons, which occurs later in the summer, is rarely if ever severe enough to affect runner production. **Blackberry:** terminal blossom trusses are favourite oviposition sites, particularly whilst the buds are still tightly clustered together. Crop losses, however, are comparatively unimportant since it is usually the terminal cluster of an inflorescence that is destroyed. Nevertheless, such attacks can be of significance as these flowers produce the earliest fruits. Damage to plants may also cause lateral shoots to develop. **Loganberry and raspberry:** damage is usually limited to the tips of the new growth and typically results in the development of lateral shoots.

Neocoenorrhinus pauxillus (Germar) (**293–295**)

A pest of apple and pear in mainland Europe; occasionally, plum is also attacked. Widely distributed in Europe.

DESCRIPTION

Adult: 2–3 mm long; black, with a blue, dark blue or greenish-blue sheen.

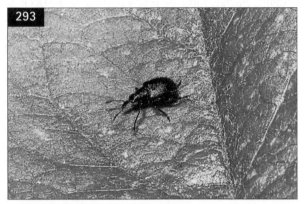

293 *Neocoenorrhinus pauxillus* – adult.

294 *Neocoenorrhinus pauxillus* – adult and feeding punctures on apple leaf.

295 *Neocoenorrhinus pauxillus* – typical damage to apple leaf.

LIFE HISTORY

Adults emerge from about mid-April onwards. They browse on the young leaves of host plants, forming numerous small holes in the laminae. Eggs are laid from early May onwards, each placed singly within the midrib on the underside of an expanded leaf, close to the point of attachment of the lamina to the petiole. The egg-laying female also bites into the midrib, causing the leaf to fold downwards at an angle of about 90°. The eggs hatch about a week later. Larvae feed for about 2 weeks before becoming fully grown. They then enter the soil and pupate. The adult stage is reached a few weeks later, but the weevils do not emerge from the soil until the following spring.

DAMAGE

Leaves folded over following damage to the midribs turn brown and fall prematurely. On mature trees, this is usually of no importance. However, heavy infestations on young trees may have an adverse effect on growth. Leaf browsing by adults is of no consequence.

Rhynchites auratus (Scopoli) (**296**)

A pest of cherry in mainland Europe; also present on other kinds of *Prunus*, including *P. padus*, *P. spinosa*, cultivated apricot and plum; walnut is also a host. Eurasiatic. Widely distributed in central and southern Europe.

DESCRIPTION

Adult: 5.5–9.5 mm long; metallic golden green to coppery, with upper surface of rostrum purplish; pronotum of male with a pair of anteriorly directed spines (cf. *Rhynchites baccus*).

LIFE HISTORY

Adults emerge in the spring from April onwards. They immediately invade host plants, where they graze on the blossoms and leaves. Later, they also probe the

296 *Rhynchites auratus* – adult.

developing fruitlets. From late May to early June, eggs are deposited singly in the fruitlets, each egg placed in a small cavity a few millimetres below the surface. The eggs hatch about 2 weeks later. Each larva then feeds within the pulp and when fully grown bores out of the fruitlet. The larva then enters the soil and forms a small chamber, 5–8 cm below the surface; the larva then pupates and the adult stage is reached about 2 months later. Adults remain within their pupal chambers and do not emerge until the following spring.

DAMAGE
Adults form holes in, or totally devour, the blossoms and leaves; they also form large cavities in the developing fruits. Small fruits, such as cherries, are often totally destroyed; attacks on apricot and plum fruits are less extensive, although often sufficiently serious to require crops to be downgraded.

Rhynchites baccus (Linnaeus)
A pest of apple, cherry and plum in mainland Europe; other hosts include pear, apricot and peach. Widely distributed in the warmer parts of Europe.

DESCRIPTION
Adult: 4.5–6.5 mm long; metallic golden red or purplish, with upper surface of rostrum mainly black; pronotum of male without spines (cf. *Rhynchites auratus*). **Larva:** up to 6.5 mm long; body white; head yellowish brown.

LIFE HISTORY
This species has a two-year life cycle, the details of which may vary from region to region. In warmer areas, young adults emerge in August and September. After a brief period of feeding on host plants they overwinter and reappear in the spring, usually from late March to mid-April. They then attack buds and, later, young fruitlets. Eggs are eventually laid within the fruitlets; they hatch a week or so later. Larvae feed for about a month and then enter the soil where they spend an extended period of up to 10 months in a state of diapause before eventually pupating. Adults emerge shortly afterwards.

DAMAGE
In spring, unopened buds are destroyed; later, developing fruits are peppered with small holes. Prior to egg-laying, fruit stalks (pedicels) may also be partly severed. Damaged fruits are often invaded by fungal pathogens, such as brown rot (*Sclerotinia fructigena*), of which the weevils may be accidental vectors.

Family **APIONIDAE** (weevils) (297)

A large family of small or very small, more or less pear-shaped weevils, with the basal antennal segment elongate and forming a scape; scape, at best, only slightly longer than antennal segments 2 and 3. Several species are damaging to agricultural crops such as field bean, clover and lucerne, and some are implicated in the spread of virus diseases. Although adults of several species are sometimes numerous on fruit trees and bushes, they do not breed upon them and are not important fruit pests.

Protapion apricans (Herbst)
Clover seed weevil
In May and June, adults of this generally distributed and common weevil sometimes feed on apple trees, cutting out characteristic small round holes in the young leaves; such damage, however, is of little or no consequence. The weevils are associated mainly with clover, their larvae feeding in the flower heads during the summer. Adults are 2.2–2.5 mm long and black bodied; the antennae are black, each with an orange scape; the legs are mainly orange, with black tarsi, mid- and hind tibiae. Other species to be found feeding on apple foliage in spring include the white clover seed weevil (*Protapion fulvipes* (Fourcroy)) and *Protapion nigritarse* Kirby.

297 *Protapion nigritarse* – adult.

Family **CURCULIONIDAE**
(true weevils and bark beetles)

The largest and most important family of weevils, with members characterized by their geniculate (elbowed) antennae which, unlike those of other weevil families, have a very long basal segment (scape); the rostrum (snout) is sometimes very long and the body often clothed in scale-like hairs. Larvae are apodous, with a well-developed head, and often adopt a C-shaped posture. Specific differences in the immature stages are usually slight, and larvae of closely allied species are often difficult if not impossible to separate with any degree of certainty.

In addition to those detailed below, many other weevils are also associated with fruit crops in Europe, usually as minor or occasional pests. In particular, these include a number of additional species within the genera *Otiorhynchus*, *Phyllobius* and *Polydrusus*. Their life cycles are essentially similar to those of their close relatives, as is the damage that such species may cause.

Bark beetles are now included within the Curculionidae, as a subfamily (see p.171).

Anthonomus piri Kollar (**298–299**)
Apple bud weevil
An important pest of pear in mainland Europe. In Britain, where this weevil is very local and rare, apple is attacked and pear is not a recorded host.

DESCRIPTION
Adult: 4.5–6.0 mm long; brownish or purplish red, clothed with golden, black, brown and white hairs, the latter forming a pale band across the elytra.

Egg: 0.8 × 0.6 mm; bean-shaped and creamy white. **Larva:** up to 7 mm long; creamy white, with a dark brown head; body fleshy and noticeably wrinkled, and more strongly C-shaped than that of *Anthonomus pomorum*.

LIFE HISTORY
Young adults are active in May and June. They feed on apple foliage for a few weeks and then seek shelter in which to aestivate during the summer months. The weevils reappear in September; eggs are then laid singly, each at the base of a small puncture made near the middle of a new fruit bud. The eggs hatch in October. Larvae then feed briefly on the fleshy tissue within the buds before hibernating; the larvae recommence feeding from late February or early March onwards, individuals becoming fully grown by the end of April. Pupation takes place in April and May within the shelter of an unopened bud, and the adult emerges about a month later.

DAMAGE
Adults make small punctures in leaf petioles, buds and spurs; this causes leaves to drop prematurely and may also result in bud death. Buds infested by larvae are hollowed out and fail to open, remaining as dead husks within which, in early spring, the immature stages of the weevil may be found. A circular exit hole, through which the young weevil emerged, is clearly visible in the side of a vacated bud.

298 Apple bud weevil (*Anthonomus piri*).

299 Apple bud weevil (*Anthonomus piri*) – damaged bud.

Anthonomus pomorum (Linnaeus) (**300–303**)
Apple blossom weevil

An important and potentially destructive pest of apple. Cultivated apple and *Malus sylvestris* are the normal hosts; however, pear, quince and, sometimes, medlar are also attacked. Present throughout Europe.

DESCRIPTION
Adult: 3.5–6.0 mm long; dark brown to black, covered with brown, greyish and whitish hairs, forming a whitish, and mottled, V-shaped mark across the elytra; a prominent whitish spot between the elytra and thorax. **Egg:** 0.7 × 0.5 mm; oval, white and translucent. **Larva:** up to 8 mm long; body mainly white, but tending to yellowish when fully grown; head dark brown. **Pupa:** 4–5 mm long; pale yellow.

LIFE HISTORY
Adults hibernate under tree bark, in crevices in posts, beneath debris and in other shelter, often migrating in numbers to suitable places in adjacent woods, hedgerows and ditches. They reappear from February onwards and are then active on warm days, flying about in search of host trees. If weevils arrive prior to bud burst, they wait for the bud scales to open before feeding. Eggs (about 40 or 50 per female) are laid singly from bud burst onwards, each inserted through a small puncture in the side of a flower bud and placed in a groove made by the female on an anther lobe. Eggs hatch within 10 days. The young larvae graze on the developing anthers and styles, and then attack the petals to form the familiar 'capped' blossom. The larvae are fully grown after about a month and then pupate within the shelter of the capped flowers. Adults emerge 2–3 weeks later, young weevils reaching their greatest numbers by mid-June. After feeding on the underside of apple leaves for about 3 weeks, the weevils disperse and take up their winter quarters. A few weevils from the parent generation may also survive the oncoming winter.

DAMAGE
The presence of brown, capped blossoms, formed after the larvae have nipped the petal bases to arrest their development, is characteristic of this pest. Light infestations are of little importance and in years of abundant fruit-set the pest may have a beneficial

300 Apple blossom weevil (*Anthonomus pomorum*).

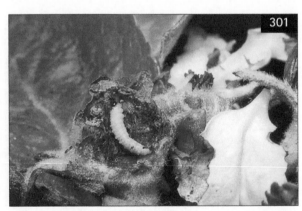

301 Apple blossom weevil (*Anthonomus pomorum*) – larva.

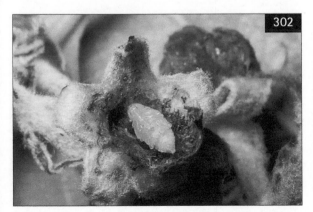

302 Apple blossom weevil (*Anthonomus pomorum*) – pupa.

303 Apple blossom weevil (*Anthonomus pomorum*) – 'capped' blossom.

154

thinning effect. However, when infestations are heavy, all except the most advanced flowers on the blossom clusters are destroyed and crops will be devastated. Adult feeding punctures in buds are distinctly larger than those drilled for egg-laying; they are rarely harmful on apple, but pierced pear buds tend to exude much sap, become distorted, and finally shrivel and die. Damage to foliage in early summer is restricted to the lower surface, the upper epidermis remaining intact. Occasionally, young adults drill shallow holes into the developing fruitlets; however, such damage is not serious.

Anthonomus rubi (Herbst) (304–306)
Strawberry blossom weevil

A locally important pest of strawberry. Blackberry and raspberry are also attacked. Eurasiatic. Widely distributed in Europe.

DESCRIPTION
Adult: 2–4 mm long; black, with a scattered greyish pubescence; at once distinguished from (*Neocoenorrhinus germanicus*) (family Rhynchitidae) by body colour

304 Strawberry blossom weevil (*Anthonomus rubi*).

and by the geniculate antennae. **Egg:** 0.5 × 0.4 mm; oval, white and translucent. **Larva:** 3.5 mm long; body dirty creamy white, noticeably C-shaped and wrinkled; head light brown.

LIFE HISTORY
Adults are active in warm, sunny weather from late April onwards, reaching peak numbers by late May. At first, they feed on strawberry foliage; later, flowers are also attacked. Eggs are deposited singly in unopened flower buds, mainly in June. As soon as an egg is laid, the female crawls a short distance along the flower stalk (pedicel) which she then girdles with several small punctures. Damaged buds cease to develop and either fall to the ground or remain dangling from the partially severed stalks. Eggs hatch in about 5–6 days and each larva (typically one per bud) feeds on the shrivelled receptacle and other floral parts beneath the sheltering canopy of withered sepals and petals. Larvae develop rapidly and are fully fed in about 2 weeks. Each then pupates in situ, a new generation of adults appearing about 2 weeks later. After feeding for a few weeks, the young weevils seek shelter amongst dead leaves and other debris, where they remain until the following spring. Attacks on blackberry and raspberry follow a similar pattern, but tend to occur somewhat later in the season.

DAMAGE
Adults make characteristic small round holes in the leaves and petals, but such damage is unimportant. Destruction of flower buds by egg-laying females, however, is more serious. **Strawberry:** early blossoms are particularly liable to be attacked and losses of 'king fruits' on cultivars such as Royal Sovereign may be significant; however, damage to more even-ripening cultivars is far less important. Strawberry cultivars producing few flowers (e.g. cv. Domanil) are most

305 Strawberry blossom weevil (*Anthonomus rubi*) – partly severed pedicel.

306 Strawberry blossom weevil (*Anthonomus rubi*) – petal damage.

seriously affected, more prolific-flowering cultivars tending to compensate for the loss of early blossom. **Blackberry and raspberry:** most damage occurs on the lateral spurs; primary blossoms tend to escape attack, as the weevils migrate to these hosts somewhat later in the season.

Barypeithes araneiformis (Schrank)
Smooth broad-nosed weevil
A minor pest of strawberry and, less frequently, raspberry. Widespread and locally common in central and western Europe.

DESCRIPTION
Adult: 3–4 mm long; shiny and virtually glabrous (cf. *Barypeithes pellucidus*), varying in colour from brownish yellow to black; body oval, with a short snout and a distinctly pointed abdomen.

LIFE HISTORY
Adults overwinter in the soil, appearing in strawberry plantations from February onwards. They then browse on the foliage. Eggs are laid in the soil and the larvae later feed on the roots of various weeds, including *Trifolium repens*. Second-generation adults appear in June and, in strawberry plantations, these often attack the developing fruits.

DAMAGE
Adult feeding on leaves is of little or no consequence. Attacks on the fruit, however, are of some importance. Green fruitlets are damaged, the weevils boring into the flesh to form small cavities characteristically wider than their entry hole. Although adults sometimes occur on apple trees, here they do no harm.

Barypeithes pellucidus (Boheman) (**307**)
Hairy broad-nosed weevil
This locally common species occurs on the foliage of raspberry and strawberry, and is sometimes numerous in plantations during May. Damage, however, is unimportant. Adults (3–4 mm long) are very similar to those of the previous species, but distinguished by the longer, denser and more upright pubescence. The larvae are associated with *Medicago lupulina*.

Curculio elephas (Gyllenhal)
A pest of chestnut in mainland Europe. Also associated with *Quercus*. Widely distributed in central and southern Europe; also present in North Africa.

DESCRIPTION
Adult: 6.0–10.5 mm long; ash-grey to yellowish grey; elongate-oval, with an extremely long, slender, curved rostrum. **Egg:** 0.45 × 0.35 mm; elliptical, white. **Larva:** up to 12 mm long; body white, plump and wrinkled; head brown.

LIFE HISTORY
Adults emerge from pupae in June, July or early August, but any overwintered adults appear somewhat earlier, usually in May. Eggs are deposited singly (rarely in twos or threes) in the fruits of chestnut from August to October, each egg being placed deeply within the tissue at the base of a tunnel-like hole bored into the developing fruit by the egg-laying female. On *Quercus*, the life cycle is somewhat different and egg-laying occurs from July onwards. Following egg hatch, the larvae feed for 4–6 weeks, typically in September and October. When fully grown they vacate the fruits and enter the soil, where they hibernate in earthen chambers formed several centimetres below the surface. Pupation occurs in the following June or July, but some larvae may remain in situ for a further one or two winters.

DAMAGE
Adults browse on the foliage of host plants, typically biting out holes in the major leaf veins and causing distortion; the adults also feed directly on the developing fruits. Larvae bore within the inner tissue of the fruits, forming frass-filled cavities; in the case of heavy infestations, crop losses are considerable.

307 Hairy broad-nosed weevil (*Barypeithes pellucidus*).

Curculio nucum Linnaeus (**308–311**)

Nut weevil

A pest of nut plantations, particularly filbert and hazelnut. Generally distributed, and locally common on *Corylus avellana*. Palaearctic. Also present in North Africa.

DESCRIPTION

Adult: 6–9 mm long; black and shiny, but covered with light brown or greyish-brown hair-like scales producing a mottled, gingery appearance; body diamond-shaped; rostrum very long, slender and curved. **Egg:** 0.8 × 0.5 mm; oval and glossy white. **Larva:** up to 10 mm long; body white and plump; head brown and relatively small.

308 Nut weevil (*Curculio nucum*).

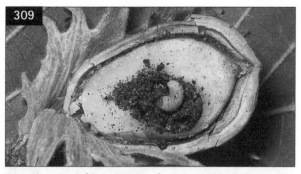

309 Nut weevil (*Curculio nucum*) – infested hazelnut.

310 Nut weevil (*Curculio nucum*) – fully fed larva.

LIFE HISTORY

Adults emerge in May and are particularly active on sunny days. Eggs are laid in June, when host nutlets have reached a diameter of about 10–12 mm (but are still soft-shelled), each inserted through a small hole bored by the female through the nutlet wall. The eggs hatch a week or so later and the larvae, typically one per infested nut, begin feeding on the kernels. Larvae pass through four instars, becoming fully grown in late July or August. Each then escapes from the nut, forcing its way through the by now much enlarged oviposition hole; this aperture also serves as an exit for frass during larval development. Once on the ground, the larvae burrow into the soil to form earthen cells 10 cm or more from the surface. They eventually pupate in the following spring.

DAMAGE

Although the kernels of infested nuts are often completely destroyed, attacks are frequently overlooked during the growing season because the shells develop and ripen more or less normally. At harvest, however, infested nuts are at once recognized by the presence of a circular (c. 2 mm diameter) hole in the side wall. Secondary crop losses may also occur, as weevil damage to nut walls, even when eggs are not then laid, may allow pathogenic fungi to infect the nutlets.

Furcipes rectirostris (Linnaeus) (**312–313**)

Cherry fruit weevil

In mainland Europe, a pest of cultivated cherry and, occasionally, plum. Wild hosts include *Prunus avium, P. padus* and, occasionally, *P. spinosa*. Eurasiatic. Widely distributed in central Europe.

DESCRIPTION

Adult: 3.5–4.5 mm long; reddish brown, with yellowish hairs forming two pale crossbands on the elytra; anterior femora each with two spines. **Larva:** up to 6 mm long; body whitish and cylindrical; head reddish brown.

311 Nut weevil (*Curculio nucum*) – larval exit hole.

LIFE HISTORY

Adults occur in the spring, from mid-April to mid-June. At the post-flowering stage, eggs are deposited singly in the developing fruitlets of host plants. The eggs hatch in 2–3 weeks and the larvae then feed within the developing stones for about a month before pupating. Young adults emerge from infested fruits from late July onwards and, after feeding on the foliage, eventually hibernate from about mid-September onwards.

DAMAGE

Egg-laying females probe developing fruitlets with their rostrum, forming distinctive necrotic spots on the surface. Later in the season, a fruit from which a young adult has emerged bears a dark, black-rimmed hole on the surface. Infested cherry stones are filled with brown frass and have a small, round exit hole in the wall.

Magdalis barbicornis (Latreille)
Pear weevil

A generally common, but minor, pest of pear and other fruit trees, especially in old, neglected orchards. Also associated with various other Rosaceae, including *Crataegus monogyna* and *Sorbus*. Present throughout Europe. An introduced pest in the USA.

312 Cherry fruit weevil (*Furcipus rectirostris*) – adult.

313 Cherry fruit weevil (*Furcipus rectirostris*) – adult damage.

DESCRIPTION

Adult: 2.5–4.0 mm long; dull black; antennae black, each with a red scape and (in male) a long, broad club; rostrum short and broad; female with decumbent body hairs.

LIFE HISTORY

Adults feed on the underside of pear leaves from late April to June. Eggs are laid in the bark, under which the larvae then develop in isolated chambers aligned parallel to the main axis of the branch or stem. The larvae feed from summer onwards, either eventually pupating in the following spring or, in warmer regions, completing their development and producing adults in late summer or autumn. In the latter case, the adults then hibernate and reappear in the following spring.

DAMAGE

Adult feeding is often concentrated within an area of leaf surface 1–2 cm across, which then becomes peppered with small, brown punctures. The holes sometimes extend through the lamina, but are of no consequence. Parts of wood infested by the larvae sometimes desiccate and die, but damage is of little or no importance.

Magdalis cerasi (Linnaeus)

A minor, sporadic pest of apple and pear. Other hosts include *Crataegus monogyna* and *Sorbus*. Present throughout Europe and particularly common in central and southern areas, including the Mediterranean basin.

DESCRIPTION

Adult: 3–4 mm long; dull black; rostrum relatively long; pronotum broad and swollen laterally; antennae with club normal in both sexes (cf. *Magdalis barbicornis*).

LIFE HISTORY

Adults feed in spring on the foliage of various rosaceous trees, and often occur in neglected orchards. Eggs are laid in small groups from late May onwards, rather later in the season than those of related species. The eggs are laid in small groups, usually in small cracks and crevices in the bark at the base of the buds and young shoots. Larvae feed in galleries which extend upwards for a few centimetres within the wood. Fully fed larvae pupate at the ends of their feeding galleries, and the adult stage is reached in the autumn. The young adults either remain in situ until the following spring or emerge and then hibernate in sheltered situations nearby.

DAMAGE

As for pear weevil (*M. barbicornis*).

Magdalis ruficornis (Linnaeus) (**314**)
Plum weevil

A minor pest of plum and, occasionally, other fruit trees such as apple, apricot, cherry and peach. Wild hosts include *Crataegus monogyna*, *Prunus spinosa*, *Sorbus aucuparia* and *Spiraea salicifolia*. Widely distributed in Europe, including Mediterranean areas.

DESCRIPTION
Adult: 2.5–3.0 mm long; dull black, with scape of antennae red; rostrum moderately short; pronotum relative narrow, but swollen laterally; antennae with club normal in both sexes (cf. *Magdalis barbicornis*).

LIFE HISTORY
Adult weevils feed on the underside of plum leaves during May and June. The larvae occur in individual cavities hollowed out beneath the bark of host trees, but they do not form extensive galleries or tunnels. Larval development is similar to that of pear weevil (*M. barbicornis*).

DAMAGE
Larvae feeding in shoots may cause nearby buds to desiccate and drop off, but this is of little or no importance as infestations are largely confined to older wood. Adult feeding is also unimportant.

Mecinus pyraster (Herbst)
Lesser apple foliage weevil

This small (3–4 mm long), cylindrical, shiny black weevil (with a slight grey pubescence) is often numerous on apple trees in June. Although feeding on foliage, forming small holes in the laminae, it is not harmful. The weevils have also been recorded on pear, cherry and plum. The larvae feed in the inflorescences of *Plantago*.

Orchestes fagi (Linnaeus) (**315**)
Beech leaf mining weevil

Although breeding only on *Fagus*, the adults of this widely distributed and often common weevil browse on the foliage of a wide range of broadleaved trees and shrubs, including (at least in mainland Europe) rosaceous orchard trees and also bush fruits. The young weevils are active in mid-summer and again in the spring, following hibernation. They sometimes invade fruit trees in large numbers, to attack the leaves and developing fruits. When feeding on foliage, they pepper the surface with small holes and, if the midrib is damaged, the attacked leaves may wilt; injury to fruits results in the development of small, corky surface scars. Adults (2.2–2.8 mm long) are elongate-oval, black, with a greyish-brown pubescence, brown robust legs, short antennae and close-set eyes.

Otiorhynchus clavipes (Bonsdorff) (**316**)
Red-legged weevil

A polyphagous pest of plum, raspberry and, occasionally, apple, currant, gooseberry, strawberry and grapevine. Widely distributed in western Europe, particularly on light soils.

DESCRIPTION
Adult: 9–13 mm long; blackish, with elytra elongate-oval and distinctly pointed posteriorly; legs long and reddish; sculpturing on thorax and elytra shallow; wingless. **Egg:** 0.6 × 0.5 mm across; more or less spherical; whitish when laid, but soon becoming blackish. **Larva:** up to 12 mm long; body creamy white, plump, wrinkled and strongly C-shaped; head brown. **Pupa:** 8–11 mm long; white, with antennae, legs and other appendages free; eyes purplish.

314 Plum weevil (*Magdalis ruficornis*).

315 Beech leaf mining weevil (*Orchestes fagi*).

LIFE HISTORY

Adults appear in late April and May or from mid-June to August, depending on the timing of pupation. They are active at night, feeding on leaves and other aerial parts of various host plants, but drop to the ground immediately if disturbed. By day, the weevils hide in grass tussocks, under stones and in other shelter. Reproduction in the species is either sexual or parthenogenetic. Eggs are laid in the soil, scattered at random near the surface beneath host plants, each female depositing up to 300. The eggs hatch in about 3 weeks. Larvae then feed on plant roots. They overwinter either as young individuals (and pupate in the following summer) or when fully fed (and then pupate earlier, in the late spring); overwintering in the pupal stage is also reported.

DAMAGE

Adults bite holes into leaves and destroy buds, blossoms, developing fruitlets and young shoots. The upper foliage on raspberry canes is often attacked; later in the season, the weevils weaken and check the growth of young canes by gnawing at the bases. Larval damage to the roots of raspberry and strawberry plants is sometimes serious, particularly in spring, attacked plants wilting and sometimes dying; roots of currant and gooseberry bushes may also be attacked, but effects on bushes are usually slight unless the plants are already under stress from other factors.

Otiorhynchus cribricollis Gyllenhal (**317**)

Olive weevil

A minor pest of olive. Adults also feed on various other plants, including apple, grapevine, citrus and fig. Widely distributed in southern Europe, including France, Italy and Spain; also present in North Africa and an accidentally introduced (and sometimes an important) pest in Australia, New Zealand and the USA.

DESCRIPTION

Adult: 6.5–8.5 mm long; brown and shiny, with the thorax and elytra deeply punctured; antennae and legs reddish.

LIFE HISTORY

Adults of this mainly parthenogenetic, nocturnal species occur from the end of May or early June onwards. Eggs are eventually laid in the soil. They hatch about 2 weeks later. Larvae then feed on the roots of various plants, particularly *Artemisia gallica*, becoming fully grown by the following spring. They then pupate, each in an earthen cell; adults appear about a month later.

DAMAGE

Adults form U-shaped notches in the margin of leaves, but effects on plant growth are usually insignificant.

Otiorhynchus meridionalis Gyllenhal

This minor pest occurs in the Mediterranean basin, where the adults feed on the leaves and flowers of various members of the Oleaceae, including cultivated olive. At least in Spain, damage is also reported on citrus (lemon). Adults are 7–10 mm long and brownish black, with the abdomen elongate-oval and somewhat pointed posteriorly. Larvae feed on the roots of various plants.

316 Red-legged weevil (*Otiorhynchus clavipes*).

317 Olive weevil (*Otiorhynchus cribricollis*) – adult damage to leaf.

Otiorhynchus ovatus (Linnaeus) (**318**)
Strawberry weevil

A polyphagous pest of various greenhouse and forestry plants; also, occasionally, a pest of strawberry and, less significantly, loganberry and raspberry. Palaearctic. Widely distributed and often common in Europe. Also an introduced pest in North America, where it is of particular importance on strawberry.

DESCRIPTION

Adult: 4.5–5.5 mm long; somewhat shiny, dark brown to black with a short, scattered, pale yellow pubescence; disc of thorax furrowed; transverse sculpturing of elytral interstices only slight; wingless. **Larva:** up to 6 mm long; body creamy white to brownish white; head brown.

LIFE HISTORY

This species is mainly parthenogenetic, with adult females appearing in late April and May. Eggs are laid about 10 days after spring feeding commences. They are usually deposited in the soil, but sometimes also on leaf petioles close to the ground. Productivity is relatively low, a weevil usually laying no more than 50 eggs. The eggs hatch in about 3 weeks. Larvae then feed on plant roots, pupating in the autumn within an earthen cell a few centimetres below the surface. The adult stage is reached about 3 weeks later, but individuals remain

318 Strawberry weevil (*Otiorhynchus ovatus*).

319 A lesser strawberry weevil (*Otiorhynchus rugifrons*).

inside their cells until the spring. Old adults still alive in the late summer and autumn may survive the winter in sheltered situations close to their breeding grounds.

DAMAGE

Similar to that caused by *Otiorhynchus rugosostriatus*, but usually not severe.

Otiorhynchus raucus (Fabricius)

A minor pest of fruit crops such as currant, gooseberry and strawberry; also polyphagous on many vegetable crops and ornamental plants. Widely distributed in Europe, particularly in warmer regions, and especially numerous in Mediterranean areas.

DESCRIPTION

Adult: 6–7 mm long; reddish brown, with lighter and darker markings on the elytra; elytra relatively broad.

LIFE HISTORY

Similar to that of *Otiorhynchus rugosostriatus*. Reproduction is often, but not exclusively, parthenogenetic. Adults are diurnal and graze on the leaves of a wide range of plants, including fruit crops. Larvae are polyphagous on the roots of various plants and are particularly damaging to cultivated *Paeonia*.

DAMAGE

Adult feeding on fruit crops is of significance only if the weevils invade crops in large numbers. The larvae are of little or no importance on fruit crops.

Otiorhynchus rugifrons (Gyllenhal) (**319**)

A lesser strawberry weevil

An occasional pest of strawberry, particularly on light soils. Widespread, but locally distributed in Europe.

DESCRIPTION

Adult: 4.5–6.0 mm long; dull black, with a yellowish pubescence; interstices on elytra with a double row of tubercles; sculpturing distinct; wingless. **Larva:** up to 7 mm long; body creamy white to brownish white; head brown.

LIFE HISTORY

Adults of this parthenogenetic species appear from mid-April onwards. Eggs are laid in the soil and, occasionally, on leaf petioles close to the ground. Larvae feed on the strawberry roots, although in their first instar they may mine within the petioles upon which the eggs were laid. Pupation usually takes place in the following spring, adults emerging 3–4 weeks later.

DAMAGE
Similar to that caused by *Otiorhynchus rugosostriatus*, but usually not severe.

Otiorhynchus rugosostriatus (Goeze) (320–322)
A lesser strawberry weevil

A potentially important pest of strawberry; other hosts include currant, gooseberry, raspberry and a wide range of ornamental plants. Widespread in Europe and the Mediterranean basin. An introduced pest in the USA and in some other parts of the New World.

DESCRIPTION
Adult: 5–7 mm long; blackish to reddish brown, with a yellowish pubescence; elytra parallel-sided and rounded posteriorly; thorax and elytra strongly sculptured; wingless. **Larva:** up to 8 mm long; body creamy white to brownish white; head brown.

LIFE HISTORY
This species is parthenogenetic, with adults active at night from May or June onwards. Eggs (several hundred per female) are laid in the soil and sometimes on the lowermost leaf petioles, mostly from late July to September. They hatch in about 3 weeks. Larvae then feed on the roots of host plants, causing most damage in the spring. Pupation occurs in an earthen cell several centimetres below the surface. Weevils emerge from the pupa a few weeks later. However, they remain within the pupal chamber for about a week while the body shell hardens and gradually takes on the typical dark adult coloration. Most adults die before the onset of winter, but some are able to survive for another season, hibernating in sheltered situations on the ground or in the soil.

DAMAGE
Adults bite large holes into strawberry leaves, but such damage is not serious. The larvae, however, are very damaging, attacked plants making poor growth, wilting and often dying, particularly during the fruiting period.

Otiorhynchus salicicola Heyden (323)
This European species occurs mainly in central and southern Europe and (at least in Italy) has been reported as a pest of kiwi fruit. The weevils also occur on certain other broadleaved plants, but are associated mainly with *Prunus laurocerasus*. Heavy infestations can occur locally, with the adults causing considerable damage to the foliage. Adults (11.0–13.5 mm long) are shiny black, with small patches of whitish-yellow hairs on the elytra and in fresh specimens a distinct, yellowish, bloom-like pulverescence.

320 A lesser strawberry weevil (*Otiorhynchus rugosostriatus*).

321 A lesser strawberry weevil (*Otiorhynchus rugosostriatus*) – larva.

322 A lesser strawberry weevil (*Otiorhynchus rugosostriatus*) – pupa.

323 *Otiorhynchus salicicola* – adult.

Otiorhynchus singularis (Linnaeus) (**324–326**)
Clay-coloured weevil

A pest of apple, pear, currant, gooseberry, raspberry, grapevine and hop. Widely distributed, particularly in western Europe.

DESCRIPTION
Adult: 6–7 mm long; shiny black and strongly sculptured, but covered with greyish-brown scales that give the body an overall dull, light and irregular pattern; body often encrusted with mud; wingless. **Larva:** up to 8 mm long; body creamy white, plump and wrinkled; head brown.

LIFE HISTORY
Eggs are laid in the soil during the summer. Larvae feed on the roots of a wide range of host plants from late summer onwards, eventually pupating in early February. Adults appear shortly afterwards. Males are very rare and reproduction is mainly parthenogenetic. The weevils feed voraciously at night on foliage, buds and bark, causing most harm from April to June. They hide by day in cracks in the soil, and under straw mulches and other shelter around the base of their foodplants. Some individuals may survive for several seasons, hibernating during the winter months.

324 Clay-coloured weevil (*Otiorhynchus singularis*).

DAMAGE
Although larval attacks on the roots can weaken host plants (in particular, raspberry), adults are usually more important, the type of damage they cause varying from crop to crop. **Apple:** attacks are often made in May and June, large irregular areas of bark being removed from young trees; buds of young grafts are also destroyed, producing blind stalks; on older trees, the weevils bite the leaves, fruitlets and stalks (petioles). **Pear:** most damage is caused in May and June to the base of fruitlets and their petioles, the fruitlets sometimes being decapitated; young vegetative shoots are attacked occasionally and are sometimes completely girdled. **Black currant:** buds on young bushes are frequently attacked, resulting in the development of misshapen plants with forked shoots; new growth is often girdled and killed; leaf petioles on young shoots are also gnawed, causing them to fold over in a characteristic manner. **Gooseberry:** in May and June, the weevils graze on the shoots causing them to wilt and break off. **Raspberry:** in April and May, the basal buds are often bitten out; later, from May onwards, the young spawn is attacked, causing death or excessive branching of the new growth; also, from April onwards, the young shoots, fruit blossom and unopened fruit buds on fruiting canes are attacked. **Grapevine:** weevils attack the buds, causing loss of inflorescences or vegetative shoots. **Hop:** the weevils are partial to new growth at the tips of shoots, severely checking growth and causing a proliferation of laterals; also, the bines are damaged near soil level, which may cause leaves to wilt and stems to break; most important damage occurs in May, and to growth less than 1 m tall.

325 Clay-coloured weevil (*Otiorhynchus singularis*) – damage to apple shoot.

326 Clay-coloured weevil (*Otiorhynchus singularis*) – damage to black currant.

Otiorhynchus sulcatus (Fabricius) (327–331)

Vine weevil

A very important pest of outdoor and greenhouse-grown ornamental plants, and a serious pest of strawberry. Highbush blueberry, currant, gooseberry, grapevine and other fruit crops are also attacked. Present, and often abundant, throughout much of Europe. Also introduced to many other parts of the world, including North America and Australasia.

DESCRIPTION

Adult: 8–10 mm long; black and shiny, the elytra with scattered patches of yellowish hairs; elytra parallel-sided; pronotum and elytra strongly sculptured; wingless. **Egg:** 0.7 mm across; more or less spherical; white at first, soon turning brown. **Larva:** up to 11 mm long; body creamy white to brownish white; head brown. **Pupa:** 7–9 mm long; white to creamy white, with purplish eyes.

LIFE HISTORY

Similar to that of *Otiorhynchus rugosostriatus*. Young adults often aggregate in large numbers in suitable shelter, such as under plastic mulches, and in infested black currant plantations will often roost in the bushes during the daytime.

DAMAGE

Although, in common with other wingless weevils, adults feed on foliage and often cause noticeable damage, the larval stage is most harmful. In particular, the larvae are often responsible for serious damage in strawberry beds, infested plants collapsing during the fruiting season and patches of poor growth often persisting into the following spring. Large numbers of

327 Vine weevil (*Otiorhynchus sulcatus*).

328 Vine weevil (*Otiorhynchus sulcatus*) – larva.

329 Vine weevil (*Otiorhynchus sulcatus*) – pupa.

330 Vine weevil (*Otiorhynchus sulcatus*) – adult damage to strawberry foliage.

331 Vine weevil (*Otiorhynchus sulcatus*) – patches of larval damage in strawberry field.

larvae feeding on the roots of gooseberry and other such crops can also cause decline, particularly under dry conditions. On black currant, adults roosting in bushes at harvest-time may become at least temporary contaminants in consignments of mechanically harvested fruit.

Peritelus noxius Boheman
Bulbous grapevine weevil

A polyphagous pest of grapevine; attacks may also occur on other fruit crops, particularly raspberry and strawberry. Wild hosts include *Artemisia vulgaris* and *Inula*. Widely distributed in mainland Europe, but most frequently reported as a pest in southern France and northern Italy.

DESCRIPTION
Adult: 3.5–5.5 mm long; brown to greyish; elytra distinctly globular; legs and antennae robust, the latter relatively long.

LIFE HISTORY
Similar to that of grey bud weevil (*Peritelus sphaeoides*), with overwintered adults present from late March to early July.

DAMAGE
Adults graze the edges of leaves and also destroy buds and flowers.

Peritelus sphaeroides Germar
Grey bud weevil

A pest of fruit trees, bush fruit, grapevine and walnut in mainland Europe. Also associated with various other trees and shrubs, including *Carpinus betulus*, *Fagus sylvatica* and *Juniperus*. Widely distributed in western and southern Europe, and particularly common in the western Mediterranean basin.

DESCRIPTION
Adult: 5–8 mm long; greyish brown, with scattered greyish-black markings on the elytra; rostrum broad and short.

LIFE HISTORY
Overwintered adults occur from early or mid-April to late June, and may often be found in orchards and vineyards, where they attack the buds, flowers and young foliage. Eggs are laid in the soil from early May onwards and hatch about a month later. Larvae feed on plant roots throughout the summer and eventually pupate, each in an earthen chamber. New adults appear from mid-August onwards, eventually hibernating in the soil from late September onwards.

DAMAGE
Adults destroy buds and flowers, and also cause minor damage to leaves. Larval damage to roots is of lesser importance.

Philopedon plagiatum (Schaller) (**332**)
Sand weevil

A local, minor pest of grapevine, particularly in weedy sites. Widely distributed in Europe and most numerous in sandy districts; also present in North Africa.

DESCRIPTION
Adult: 4–8 mm long; oval and stout-bodied, with protruding eyes and relatively thick legs; body black, but thickly clothed in brownish to brownish-grey scales; elytra with light and dark scales, producing a longitudinally striped pattern, and clothed with short upright hairs; antennae reddish. **Larva:** up to 8 mm long; body white; head light brown.

LIFE HISTORY
Adults overwinter in the soil and emerge from April onwards. They then immediately begin to feed on host plants. Eggs are deposited in the soil, usually from late May to mid-June. Larvae feed from June or July onwards, attacking the roots of *Carduus*, *Cirsium*, grasses and various other plants. Development is slow and larvae do not become fully grown until late in the following year; they then pupate and adults appear in the autumn.

DAMAGE
Adult weevils bite out large semicircular notches from around the margins of leaves. The larvae, which feed on plant roots, are of no economic importance.

332 Sand weevil (*Philopedon plagiatum*).

Phyllobius argentatus (Linnaeus) (**333–334**)
Silver-green leaf weevil

An often common, but minor and ephemeral, pest of fruit trees and hazelnut; generally abundant in spring on forest trees and shrubs, particularly *Betula*. Widely distributed in Europe.

DESCRIPTION
Adult: 4.5–6.0 mm long; body black, but appearing bright green owing to a covering of shiny, golden-green, disc-like scales; legs and antennae light brown, partially clothed in golden-green scales. In *Phyllobius*, unlike *Polydrusus*, the scapal grooves on the rostrum are directed towards the eyes, and the legs and antennae are somewhat thicker.

LIFE HISTORY
In common with those of several other species of leaf weevil (both *Phyllobius* and *Polydrusus* spp.), adults occur in considerable numbers in spring on broadleaved trees and shrubs, and often fruit trees, where they feed on the leaves and flowers. In dull conditions the weevils tend to remain motionless amongst the foliage, but they are readily detected if a branch is tapped over a cloth or tray. Later, the weevils migrate to various hedgerow shrubs and to ground herbage. Eggs of *P. argentatus* are laid in the soil during the late spring or early summer. Following egg hatch, the larvae feed on the roots of various weeds and grasses, including *Lamium*, *Rumex* and *Poa*. Pupation occurs in the early spring, within earthen cells, adults emerging shortly afterwards.

DAMAGE
In orchards, the adult weevils bite holes into the leaves and flower petals, typically around the edges. However, although such damage may be widespread it is rarely important except, perhaps, on very young trees and nursery stock. The extent of damage varies considerably from year to year and, apparently, is not related to the size of populations invading the trees.

Phyllobius glaucus (Scopoli) (**335**)
Associated with a wide range of trees and shrubs; a minor pest of fruit trees, particularly apple, pear, cherry and plum. Widely distributed in Europe.

DESCRIPTION
Adult: 8–12 mm long; body black and more or less coated in elongate, greenish-brown, hair-like scales (but coloration extremely variable); legs reddish brown.

LIFE HISTORY & DAMAGE
Similar to that of silver-green leaf weevil (*Phyllobius argentatus*).

333 Silver-green leaf weevil (*Phyllobius argentatus*).

334 *Phyllobius* sp. – adult damage to pear leaf.

335 *Phyllobius glaucus* – adult.

Phyllobius oblongus (Linnaeus) (336)

Brown leaf weevil

A very common, but minor and ephemeral, pest of fruit trees (including apple, pear, apricot, cherry, damson and plum) and hazelnut. Often abundant in spring on various other broadleaved trees and shrubs, particularly *Alnus, Betula* and *Fagus sylvatica*. Widely distributed in Europe; also present in North Africa and an introduced pest in the USA.

DESCRIPTION
Adult: 3.5–6.0 mm long; head and thorax black; elytra, legs and antennae brown; body hairs sparse and pale.

LIFE HISTORY
Similar to that of silver-green leaf weevil (*Phyllobius argentatus*).

DAMAGE
As for silver-green leaf weevil (*P. argentatus*).

Phyllobius pomaceus Gyllenhal (337–339)

Nettle leaf weevil

A locally important pest of strawberry, but associated mainly with *Urtica*. Present throughout Europe.

DESCRIPTION
Adult: 7–9 mm long: body black, coated with green, or sometimes yellowish-green to coppery-green, scales; antennae blackish, but reddish basally; legs black. **Larva:** up to 8 mm long; body creamy white; head brown.

LIFE HISTORY
Adults occur in the spring from May onwards and are often abundant on *Urtica dioica*. Eggs are laid in the soil during the early summer, the larvae eventually attacking the roots of weeds such as *Taraxacum officinale* and, occasionally, cultivated strawberry plants. The larvae are fully grown by the early spring. They then pupate, adults emerging several weeks later.

336 Brown leaf weevil (*Phyllobius oblongus*).

338 Nettle leaf weevil (*Phyllobius pomaceus*) – larva.

337 Nettle leaf weevil (*Phyllobius pomaceus*) – adult and damage to strawberry blossom.

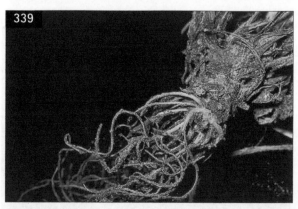

339 Nettle leaf weevil (*Phyllobius pomaceus*) – larval damage to strawberry roots.

DAMAGE

Larval attacks on the root system of strawberry plants are often severe. Infested plants wilt under stress and may die. Newly emerged young adults feed on open blossoms; however, damage is limited to the petals and is not important.

Phyllobius pyri (Linnaeus) (**340–341**)
Common leaf weevil

A very common, but minor and ephemeral, pest of fruit trees (especially apple, pear, quince, cherry, damson and plum) and hazelnut. Often abundant in spring on many other broadleaved trees and shrubs. Palaearctic. Present throughout Europe.

DESCRIPTION

Adult: 5–7 mm long; black, covered with elongate coppery, golden or greenish-bronze scales.

LIFE HISTORY

Similar to that of silver-green leaf weevil (*Phyllobius argentatus*). In early summer, the adults are particularly partial to *Urtica* and are often abundant in nettle beds. Occasionally, the larvae are sufficiently numerous in grassland and pastures to be considered pests.

DAMAGE

As for silver-green leaf weevil (*P. argentatus*).

Plintha caliginosus (Fabricius) (**342**)
Hop root weevil

A generally uncommon, but occasionally persistent, pest of hop. Raspberry and strawberry are also attacked. Widely distributed in central Europe and recorded from most, if not all, the main hop-growing areas.

DESCRIPTION

Adult: 5.5–9.0 mm long; dark brown to black and deeply sculptured; pronotum and elytra narrow and of similar width, and often coated with mud; wingless. **Larva:** up to 10 mm long; body creamy white; head brown; distinguished from that of *Otiorhynchus* by the presence of a median vertical line on the front of the head.

340 Common leaf weevil (*Phyllobius pyri*).

342a, b Hop root weevil (*Plintha caliginosus*).

341 Common leaf weevil (*Phyllobius pyri*) – larva.

LIFE HISTORY

Adults occur throughout the year, but are most numerous in late summer and autumn. They are very sluggish, rarely leaving the soil, and feed mainly by scraping the underground parts of hop bines. Eggs are laid singly in punctures made in rootstocks or underground portions of newly planted bines from late summer onwards, mainly from September to November. At first, larvae tunnel in the cortex; later, they enter the pith, and the burrows eventually become plugged with accumulated frass. Larvae are fully grown after 9–18 months; they then pupate within their galleries, mainly during the summer. Adults emerge 2–3 weeks later.

DAMAGE

Although feeding by adults may weaken the bines, attacks are of little or no importance as they are confined, largely, to the end of the growing season. Larvae, however, cause considerable harm, particularly in young hop gardens and to newly planted hop sets. The attacked bines are weakened and in severe cases killed; a single infested rootstock may contain 20 or more grubs. Young larvae tend to occur in the swollen basal portions of the bines and if these are used as sets or cuttings infestations are easily spread to previously clean areas.

Polydrusus formosus (Mayer) (343)
A green leaf weevil

A minor pest of fruit trees (including apple, pear, cherry, damson and plum) and hazelnut; more frequently associated with woodland trees and shrubs, particularly *Corylus avellana* and *Quercus*. Locally common in Europe. An introduced pest in the USA. Other species of *Polydrusus* associated with fruit crops in Europe include *P. inustus* Germar, *P. marginatus* Stephens, *P. mollis* (Ström) and *P. pterygomalis* Boheman.

DESCRIPTION

Adult: 5–8 mm long; body black, coated in green scales, often producing a coppery sheen; antennae with a dark club. In *Polydrusus*, unlike *Phyllobius*, the scapal grooves on the rostrum are directed below the eyes, and the legs and antennae are thinner. **Larva:** up to 7 mm long; body creamy white and C-shaped; head brown.

LIFE HISTORY

Adults are active in late spring and summer, usually from April or May onwards, when they browse on the buds, young leaves and open blossoms of a wide variety of trees and shrubs. The soil-inhabiting larvae feed on plant roots. They complete their development in the autumn and then overwinter. Pupation occurs in the spring.

DAMAGE

Adults are capable of causing extensive damage to fruit buds, shoots and open blossoms, but heavy infestations in orchards are infrequent. At least on apple, adults may also damage fruitlets, resulting in the development of corky scars on the skins of mature fruits.

Sciaphilus asperatus (Bonsdorff) (344)
Scaly strawberry weevil

An occasional pest of strawberry; raspberry and loganberry are also hosts. Widely distributed in central and northern Europe.

DESCRIPTION

Adult: 4–6 mm long; blackish, densely clothed with grey or yellowish-grey scales and often with a metallic sheen; thorax and elytra with strong, prominent, scale-like setae; antennae reddish and noticeably thinner than in *Otiorhynchus*.

343 A green leaf weevil (*Polydrusus formosus*).

344 Scaly strawberry weevil (*Sciaphilus asperatus*).

LIFE HISTORY
As for *Otiorhynchus ovatus*.

DAMAGE
Adults are capable of causing extensive damage to foliage; in common with those of *Otiorhynchus* spp., larvae attack the roots of plants.

Sitona lineatus (Linnaeus) (345–347)
Pea & bean weevil
A generally abundant and well-known pest of leguminous crops such as field bean and pea. The adults also browse on the leaves or open blossoms of various non-leguminous plants, including fruit trees and strawberry. However, although sometimes abundant in strawberry beds, they are not important pests of fruit crops. Widely distributed in Europe.

DESCRIPTION
Adult: 4–5 mm long; blackish, but covered with greyish or yellowish, hair-like scales which give the body a paler, longitudinally striped appearance; antennae reddish.

LIFE HISTORY
Adults emerge from hibernation from late March onwards. They then feed on a range of plants before eventually invading legumes (Fabaceae), notably crops of field beans. Eggs are usually laid in the soil close to host plants, and they hatch 2–3 weeks later. Larvae then feed on the root nodules. They pass through five instars and complete their development about 6 weeks later. Fully fed individuals pupate a few centimetres beneath the soil surface, each in an earthen cell, and new adults emerge 2–3 weeks later. These young adults are very active and migrate from host crops, usually by walking. They then browse on the leaves of many kinds of plant, before eventually hibernating.

DAMAGE
Adults typically make small, U-shaped notches along the edges of leaves. Such damage on fruit crops is usually caused by young weevils during the late summer and autumn, and attacks most frequently occur on plants growing in the vicinity of recently harvested legume crops. On peach, marginal leaf rolls caused earlier in the season by pests such as peach leaf-roll aphid (*Myzus varians*) (Chapter 3) afford suitable daytime shelter for adults; the weevils then emerge at night to browse on the leaves and, here again, cause the characteristic U-shaped notching. In spring, the overwintered weevils may visit orchards to feed on open fruit blossom, typically making holes in the petals. Although often very noticeable, damage caused is of no significance.

345 Pea & bean weevil (*Sitona lineatus*).

346 Pea & bean weevil (*Sitona lineatus*) – damage to strawberry leaf.

347 Pea & bean weevil (*Sitona lineatus*) – damage to peach leaf.

Stereonychus fraxini (Degeer) (**348–349**)

Olive leaf weevil

A minor pest of olive. Infestations also occur on other members of the Oleaceae, including *Fraxinus excelsior*, *Phillyrea latifolia* and *Syringa vulgaris*. Widespread in south-central and southern Europe, including many parts of France and southern Germany; also found in Asia Minor and North Africa.

DESCRIPTION

Adult: 3 mm long; greyish brown to reddish brown, with a blackish thoracic disc; eyes, unusually, located on top of head. **Larva:** up to 4 mm long; greenish yellow, with a black head.

LIFE HISTORY

Adults overwinter in debris on the ground, and in other shelter, emerging in the spring. They then feed on the buds and, later, on the leaves and leaf petioles of host plants. Eggs are deposited close to the veins on the underside of the expanded leaves. Following egg hatch, the larvae graze on the surface of the leaves, forming a series of window-like patches. When fully fed they pupate, each in an oval, yellowish-brown to brownish,

348 Olive leaf weevil (*Stereonychus fraxini*).

349 Olive leaf weevil (*Stereonychus fraxini*) – pupal cocoon.

parchment-like cocoon formed on the leaf surface. Young adults appear about 10 days later; they feed on the foliage before entering hibernation. There is one generation annually.

DAMAGE

Weevils feeding on unopened buds in the spring can delay the appearance of the new growth and, if infestations are heavy, adult and larval feeding on the foliage can reduce plant vigour.

Strophosoma melanogrammum (Förster) (**350**)

Nut leaf weevil

A minor and sporadic pest of hazelnut; generally abundant on a wide range of forest trees and shrubs, particularly *Corylus avellana*. Widely distributed in Europe. An introduced pest in North America.

DESCRIPTION

Adult: 4–6 mm long; robust, characteristically rounded, black or brown, covered with rounded, brownish or greyish scales; wingless; eyes angular and prominent.

LIFE HISTORY

Overwintered adults of this polyphagous species feed during the spring and early summer on the buds and leaves of various forest trees and shrubs. Males are known to occur in northern Europe, but elsewhere the weevil breeds entirely parthenogenetically. Eggs are laid in the soil and larvae attack the roots of plants such as *Deschampsia cespitosa* and *Rumex*. Pupation occurs in late summer, young weevils emerging in the autumn. They feed briefly and then enter hibernation.

DAMAGE

Adult feeding is limited to biting holes into buds and leaves and is likely to be important only on young trees.

350 Nut leaf weevil (*Strophosoma melanogrammum*).

Subfamily SCOLYTINAE
(bark beetles)

A subfamily, within the Curculionidae, of small, dull-coloured, more or less cylindrical, wood-boring beetles; antennae short, with a distinct scape and large, flattened club; elytra often concave apically. Eggs usually oval or spherical, pearly white and shiny, and laid in 'maternal' galleries formed within the host tree by the adults. Larvae are apodous, white (sometimes with a pinkish tinge when feeding), with a brown head and a very large prothoracic region that gives them a rather hunchback appearance. Development takes place entirely within the host, individuals passing through five instars. Larvae of some species are dependant upon the presence in their galleries of hyphae of so-called ambrosia fungi, upon which they feed. Bark beetles (some species also known as ambrosia beetles or as shot-hole borers) are of considerable economic importance, particularly to the timber trade. Several species attack fruit trees, but infestations usually occur only in dying branches or in sickly trees with root problems. Attacks are particularly common, for example, on plum trees infected with bacterial canker (*Pseudomonas syringiae* pv. *mors-prunorum*). Unlike many wood-boring insects, bark beetles do not invade dead, seasoned timber.

Hylesinus oleiperda (Fabricius)
A pest of olive. Other hosts include *Elaeagnus*, *Fagus sylvatica*, *Fraxinus excelsior*, *Ligustrum vulgare*, *Robinia pseudoacacia* and *Syringa vulgaris*. Widely distributed, particularly in central and southern Europe. In more northerly regions, including the British Isles (where the pest is known as the 'lesser ash bark beetle'), infestations most often occur on *Fraxinus excelsior*.

DESCRIPTION
Adult: 2.5–3.2 mm long; oval-bodied; mainly black, with red antennae and tarsi; hairs on upper surface of body often yellowish; elytra with distinct longitudinal striae; antennal club 4-segmented.

LIFE HISTORY
This species breeds in characteristic galleries formed in the sapwood immediately beneath the bark of host trees. The maternal gallery, in which from 30 to 40 eggs are laid, is relatively broad and more or less Y-shaped, with a short basal gallery and a pair of long 'arms'; the larval feeding galleries, which radiate from the maternal gallery, are several centimetres in length. In southern Europe there is typically just one generation annually. Adults occur in June and July, but sometimes later. Larvae usually form the overwintering stage.

DAMAGE
Attacks are of most importance on young trees. Infested branches are weakened and shoots may wilt and become desiccated. In addition, small branches may break off.

Hypoborus ficus (Erichson)
A minor pest of fig. Widely distributed in and around the Mediterranean basin.

DECRIPTION
Adult: 1.0–1.3 mm long; mainly matt black; prothorax and elytra with a sparse coating of short, stout hairs; body elongate-oval and parallel sided, with the small head hidden from above by the large prothoracic region.

LIFE HISTORY
A relatively broad, horizontal gallery (5–10 mm long) is formed beneath the bark by individuals of both sexes. Eventually, 30–40 eggs are laid. Following egg hatch, larvae form long feeding tunnels in the sapwood, each extending upwards or downwards immediately beneath the bark for about 20 mm and terminating in an elongate pupal chamber. There are three or more generations annually.

DAMAGE
Infested branches are weakened and may be killed, but damage is usually of little or no economic importance as the pest is most often present on weakened and abandoned trees, including those previously infested by the longhorn beetle *Trichoferus griseus* (family Cerambycidae).

Leperesinus varius (Fabricius)
A pest of olive; other hosts include *Acer*, *Carpinus betulus*, *Corylus avellana*, *Fraxinus excelsior*, *Juglans regia*, *Robinia pseudoacacia* and *Syringa vulgaris*. Widely distributed in Europe, including the British Isles where it occurs mainly on *Fraxinus excelsior* and is known as the 'ash bark beetle'.

DESCRIPTION
Adult: 2.5–3.5 mm long; body relatively narrow; mainly black, with reddish-brown antennae and tarsi; prothorax and elytra covered in grey scales, interspersed with irregular patches of darker ones; elytra dentate posteriorly; antennal club 3-segmented.

LIFE HISTORY
The insects breed mainly in felled or fallen trees and shrubs, but will also invade live ones previously weakened or damaged by other factors. Adults occur from early March to May, and typically appear in several distinct waves. The larvae develop beneath the

bark, forming irregular galleries (each about 20 mm long) that extend both upwards and downwards from a pair of broad, more or less horizontal maternal chambers. There is a single generation annually.

DAMAGE
The surface of infested wood becomes cracked and callosed, and hosts may be weakened.

Phloeotribus scarabaeoides (Bernard) (351)
Olive bark beetle
A pest of olive. Other hosts include *Nerium oleander* and, to a lesser extent, *Fraxinus excelsior* and *Syringa vulgaris*. Widely distributed in southern Europe, particularly in the Mediterranean basin, southern France and the Balkans; also found in Syria.

DESCRIPTION
Adult: 2.0–2.5 mm long; black, the elytra patterned with dark brown scale-like hairs and (especially posteriorly) coated in yellowish-grey hair-like scales and fine, erect hairs; front of pronotum with several dentate ridges; elytra with distinct striae and single rows of interstitial punctures; large, club-like tip of antenna formed from three lamellate segments.

LIFE HISTORY
When attacking a branch or trunk the adult female bores through the bark. She then excavates a more or less horizontal transverse tunnel on either side of the entry point, thereby forming a two-armed maternal gallery in which up to 60 eggs are laid. Following egg hatch, each larva bores vertically upwards or downwards from the maternal gallery, within the sapwood; in this way, more or less parallel feeding galleries are formed, each extending for several centimetres. When fully grown, each larva pupates at the end of its feeding gallery in a

351 Olive bark beetle (*Phloeotribus scarabaeoides*) – damage to shoot.

silken cocoon; young adults eventually emerge through small rounded holes bored through the bark. Following their emergence, the beetles often feed by boring into stems at the base of a shoot or in the axil between a shoot and a fruit peduncle. In southern Europe there are from two to four generations annually, with either adults or larvae overwintering. During spring and early summer, the adults tend to lay eggs in prunings, rather than in living trees; at that time they will also attack olive wood stacked as firewood.

DAMAGE
Adult feeding can result in heavy losses of young shoots, flowers and fruits. In addition, larval infestations cause direct damage to shoots, branches and trunks; growth may be stunted and all parts of the host above the larval feeding galleries may wilt.

Scolytus amygdali Guérin
Small almond-tree borer
A pest of almond; attacks also occur on apricot and peach. Present throughout the Mediterranean area.

DESCRIPTION
Adult: 2–3 mm long; head black, with a characteristic longitudinal facial keel (especially well developed in female); pronotum black, with longitudinal lines of punctures; elytra gingery brown, with darker patches dorsally.

LIFE HISTORY
In spring, adults invade host plants to bore within young shoots and buds, in order to feed on the sap contents. Later, they seek out unhealthy branches within which the females initiate elongate (20–30 mm long), longitudinal breeding galleries between the bark and sap wood. Mating takes place at the entrance to the galleries, after which eggs are laid. Larvae feed within side-branches excavated away from the maternal gallery, and eventually pupate. A further generation occurs during the summer and, at least in the most favourable of regions, a third generation in the autumn.

DAMAGE
Infestations on otherwise healthy trees are of no significance. However, branches or young trees weakened by other factors (including pests or pathogens) may be killed.

Scolytus mali (Bechstein) (352–353)
Large fruit bark beetle
A secondary pest of apple, plum and, less frequently, pear, cherry and other fruit trees. Palaearctic. Present throughout Europe.

DESCRIPTION
Adult: 3.5–4.5 mm long; head and thorax black; elytra dark brown; thorax and elytra shiny, the latter with smooth interstices.

352 Large fruit bark beetle (*Scolytus mali*) – female.

353 Large fruit bark beetle (*Scolytus mali*) – maternal gallery and larval feeding galleries.

LIFE HISTORY
Adults appear in late spring or early summer, flying in sunny weather. At first, they invade the new shoots and buds of host plants, in order to feed on the sap contents. Later, they burrow through the bark to the sap wood, where a longitudinal gallery is formed. Each chamber is about 5–12 cm long and has an irregular, slightly enlarged basal section in which mating takes place. Eggs (usually 50–60) are deposited in the gallery and after these hatch, the larvae tunnel away from the maternal gallery, between the bark and sap wood, to form long, perpendicular and more or less straight burrows. The larval galleries lie close together and each ends in a slight bulb in which pupation occurs. The larvae feed during the autumn and winter months, pupating in the spring. Young adults eventually emerge, biting their way through the bark to leave numerous, conspicuous flight (exit) holes.

DAMAGE
Attacks typically occur in already unhealthy trunks and large branches 20 cm or more in diameter, weakening or killing them. In addition, the adults may bore into shoots and buds, which then wither and die.

Scolytus rugulosus (Müller) (**354**)
Fruit bark beetle
A secondary pest of apple, pear, cherry, plum and other fruit trees. Widely distributed in Europe.

DESCRIPTION
Adult: 2.0–2.5 mm long; dark brownish to black, with brown legs and antennae; thorax shiny, longer than broad; elytra dull, the interstices wrinkled.

LIFE HISTORY
Adults appear in the spring and summer. When initiating an attack, they burrow into the bark of the

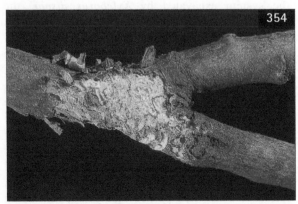

354 Fruit bark beetle (*Scolytus rugulosus*) – damage on plum.

host tree until they reach the sap wood. A narrow, longitudinal maternal gallery about 1.5–3.0 cm long is then excavated, within which the female lays about 50 eggs. Once eggs hatch, the larvae tunnel outwards from the initial gallery, between the bark and sap wood, to form a sinuous series of shallow tunnels that often criss-cross, particularly towards their extremities. At the end of its tunnel each larva forms a deeper cell in which pupation takes place. When adults eventually emerge they bore through the bark to the outside, forming the familiar shot-hole-like flight (exit) holes. There are usually two generations in a season, larvae of the second continuing their development during the winter months.

DAMAGE

This species attacks the trunks of small unhealthy trees, and branches up to about 6 cm in diameter; a serious infestation may kill the whole tree.

Xyleborinus saxeseni (Ratzeburg)

Fruit-tree wood ambrosia beetle

A secondary, polyphagous pest of a wide range of trees; sometimes associated with fruit trees, particularly apple and cherry. Palaearctic. Present throughout Europe; also present in North Africa and the USA.

DESCRIPTION

Adult: 2.0–2.5 mm long; dark brown, with yellowish body hairs; elytra elongate and parallel-sided, with the hind ends meeting in a more or less straight line.

LIFE HISTORY

Adults appear in the spring. After mating, each female bores into a host tree down to the heart wood, usually by way of dead tissue or a surface lesion. The tunnel is then extended transversely on one or both sides to form breeding galleries in which eggs are laid. Unlike the system excavated by *Xyleborus dispar*, the individual passages are narrow and flat. Once eggs have hatched, the larvae enlarge their chamber considerably to form a cavity 2 cm or more across. The faeces they produce form a brownish-yellow pulp, which lines the walls of the chamber and accumulates in the associated passages; considerable quantities are also ejected to the outside through the original entrance burrow. This pulp forms a substrate upon which ambrosia fungi can develop, lining the inner chamber and forming an essential part of the larval diet. All stages occur within the breeding chambers, the inhabitants often packed tightly together. Up to 300 individuals are reared in a colony, and there may be two generations annually.

DAMAGE

Infestations do not occur in healthy fruit trees (cf. *Xyleborus dispar*); damage, therefore, is of only secondary importance.

Xyleborus dispar (Fabricius)

larva = broadleaved pinhole borer

A primary and secondary pest of a wide variety of trees, including apple, pear, almond, apricot, cherry, peach, plum and walnut. Palaearctic. Widely distributed in Europe, including Mediterranean areas, but local and usually rare in more northerly regions.

DESCRIPTION

Adult female: 3.0–3.5 mm long; dark brown to black, with light brown legs and antennae; thorax and elytra with a sparse coating of yellowish hairs; thorax rounded and with numerous small projections anteriorly. **Adult male:** 1.5–2.0 mm long; thorax relatively small and abdomen short, giving the body a hunchbacked appearance.

LIFE HISTORY

Adults occur on the trunks of host trees from May to July; they are very active in hot, sunny weather. After mating, females bore into the bark of host trees, usually entering through a surface crack; they then burrow down to the sap or heartwood before excavating a transverse tunnel to either side. From each of these, cylindrical breeding galleries are produced, directed perpendicularly both upwards and downwards. Eggs are laid at the entrances to these passages, each female depositing up to 50 in her lifetime. Larvae occur from May or June onwards and, since eggs are laid over a period of several weeks, larvae of various sizes may be found within the colony at any one time. The larvae do not feed on the actual wood, but instead browse on ambrosia fungi which develop on the walls of the breeding passages. Adults occur in the galleries throughout the year, but are most numerous from mid-winter to early summer. Males are much rarer than females, particularly in the summer, usually forming no more than 10–20% of populations. Young adults may be found in the breeding galleries, before their emergence from infested trees, tightly packed one behind the other.

DAMAGE

Orchard infestations tend to be restricted to a small number of trees, often those growing in waterlogged conditions. However, relatively young, vigorous trees may also be colonized. The beetles attack both the trunks and larger branches and, in some cases, trees may become extensively riddled with galleries and will eventually die.

Chapter 5
True flies

Family **TIPULIDAE**
(crane flies)

Slow-flying insects (widely known as 'daddy-longlegs'), with elongate bodies, long and narrow wings, and very long, fragile legs. Larvae (colloquially known as 'leatherjackets') are elongate, more or less cylindrical and slightly tapered anteriorly, with a soft, but tough, leathery skin; the head is small and indistinct; the final abdominal (anal) segment bears numerous fleshy papillae, the features of which are often useful for distinguishing between species. Leatherjackets are typically soil inhabiting; they feed on the roots, and sometimes on the lowermost leaves, of a wide range of plants.

Nephrotoma appendiculata (Pierre) (**355**)
Spotted crane fly
A minor pest of strawberry and, occasionally, other fruit crops such as raspberry. A wide range of other low-growing plants are also attacked, particularly in allotments and gardens. Widely distributed in Europe.

DESCRIPTION
Adult: 12–20 mm long; body mainly yellow, golden yellow and black; wings 10–15 mm long, with a pale yellow or light brown stigma. **Egg:** 0.8 × 0.4 mm; oval and black. **Larva:** up to 30 mm long; body greyish brown; anal papillae rounded.

LIFE HISTORY
Adults are most numerous in May. Eggs are then deposited at random on the soil surface. The eggs do not hatch until the early autumn, owing to a period of summer diapause. Larvae feed briefly in the autumn on the subterranean parts of plants before overwintering. Activity is resumed in the early spring, with most larvae completing their development by the middle of April. They then pupate in the soil. Adults appear shortly afterwards, the pupae wriggling to the surface shortly before emergence.

DAMAGE
Particularly under dry conditions, attacked plants may wilt and this can be detrimental to the growth of young plants and to fruit production on older ones.

Tipula oleracea Linnaeus (**356**)
A common crane fly
A polyphagous and occasional pest of various crops, including cane fruit, strawberry and hop. Widely distributed in Europe, but generally less common than *Tipula paludosa*.

DESCRIPTION
Adult: 15–23 mm long; body mainly grey, the thorax with indistinct, light brown, longitudinal stripes; wings 18–28 mm long; legs brownish, with black tarsi, and

355 Spotted crane fly (*Nephrotoma appendiculata*) – female.

356 A common crane fly (*Tipula oleracea*) – male.

distinctly thinner than those of *T. paludosa*; antennae 13-segmented. **Larva:** up to 40 mm long; brownish grey, with a dull, dusty and wrinkled appearance; head black; anal segment with two pairs of elongate anal papillae.

LIFE HISTORY

This species has a similar biology to that of *T. paludosa*, but adults occur earlier in the year, typically from May onwards. Also, development is more rapid than is the case in *T. paludosa*, the larvae often pupating before the winter. Under particularly favourable conditions there may be two generations annually.

DAMAGE

Leatherjackets attack the underground parts of stems, roots and crowns of plants. At night, they also come to the surface and feed on the basal parts of stems; leaves in contact with the soil may also be holed or shredded. Damage is most evident in spring, when infested plants make poor growth and may begin to wilt, and is most likely to occur when young blackberry, loganberry or raspberry canes, strawberry runners or hop sets are planted in recently ploughed pastures or old hop gardens. Young strawberry plants are particularly liable to be attacked and if damage is severe they may have to be ploughed in. In general, attacks tend to be less severe in light, dry soils and in hot, dry years.

Tipula paludosa Meigen (357–358)

A common crane fly

A sometimes very common and important agricultural pest, particularly in pastures and grassland. The larvae sometimes also attack horticultural crops such as cane fruits, strawberry and hop. Widely distributed in Europe; also present in North America and Asia.

DESCRIPTION

Adult: 17–25 mm long; body grey, with a brownish or yellowish-red tinge, the thorax with indistinct longitudinal stripes; wings 13–23 mm long; legs brown and distinctly thicker than those of *Tipula oleracea*; antennae 14-segmented. **Egg:** 1.0 × 0.4 mm; oval, black and shiny. **Larva:** up to 45 mm long; similar to that of *T. oleracea*, but with just one pair of elongate anal papillae. **Pupa:** 20–30 mm long; brown and elongate, with paired respiratory horns on the head.

LIFE HISTORY

Adults emerge from June onwards, but are most abundant in late summer or early autumn. On average, a female deposits about 300 eggs; most are laid from mid-August to the end of September, and placed just below the soil surface in batches of five or six. Eggs hatch in about 14 days and the larvae then feed on plant roots near or at the soil surface. Feeding ceases during the winter to be resumed in the spring, the larvae becoming fully grown by about June and then pupating in the soil. In common with those of other species, the pupae wriggle upwards to protrude out of the ground shortly before adults emerge. There is just one generation annually.

DAMAGE

Apart from timing, similar to that caused by *Tipula oleracea*, with most significant damage occurring in the spring following the autumn in which eggs were laid.

357 A common crane fly (*Tipula paludosa*) – female.

358 A common crane fly (*Tipula paludosa*) – larva.

Family **BIBIONIDAE**
(St. Mark's flies) (**359**)

A small family of mainly black, hairy flies of robust appearance; males often hover rather sluggishly in the air in conspicuous groups. Larvae are cylindrical, with a prominent head, well-developed mouthparts and fleshy processes on each body segment. Bibionid larvae abound in soil that contains decaying vegetable matter, in manure heaps and the like, and sometimes cause damage to the roots of plants growing in soil with a high organic content.

Dilophus febrilis (Linnaeus) (**360**)
Fever fly

The subterranean larvae of this generally abundant species typically occur in gregarious masses, and sometimes attack the roots of strawberry plants and also the underground parts of hop plants. Attacks usually occur only locally, and are rarely of importance; however, damage may be of some significance on young strawberry plants or on recently planted hop sets. Eggs are most frequently deposited in well-manured soil or amongst decaying vegetable matter, and infestations most often follow the spreading of compost on the land or the planting of crops on sites of former manure heaps. There are usually three broods during the year, but adults are most often encountered in the early spring when, in common with other members of the family (genus *Bibio*), they are attracted into flowering orchards to forage for nectar; in such situations, they are often regarded as important pollinators. Fever flies (6 mm long) are characterized by a circlet of spurs on each front tibia and by a double series of spines at the front of the thorax. The larvae are about 10 mm long and dull brownish in colour; pupae are whitish, with a hunchbacked appearance.

Family **CECIDOMYIIDAE**
(gall midges)

Small to very small, delicate flies, with broad, often hairy wings and a much reduced venation; antennae long, with bead-like (moniliform) segments; legs long and thin; genitalia clearly visible. Larvae are short-bodied and narrowed at both ends, with the head small and inconspicuous; they usually possess a sternal spatula ('anchor-process' or 'breastbone'), which is often of characteristic shape for the genus or species.

Contarinia humuli (Theobald)
Hop strig midge
larva = hop strig maggot
An uncommon pest of hop in various hop-growing areas, including England and parts of mainland Europe.

DESCRIPTION
Adult: 1.5–2.0 mm long; body pale yellow; legs brown. **Egg:** very small, fusiform and translucent. **Larva:** up to 3 mm long; whitish, with a distinct spatula.

LIFE HISTORY
Adults are active in warm, sunny and calm conditions in late July and August, eggs being laid in developing hop burrs (strobili) from early August onwards. Larvae feed gregariously from mid-August to mid-September, there being up to 50 attacking a single cone. Most burrow into the central axis (strig) of the burr or into the base of the bracts, but some may feed externally. Fully grown larvae leave the burrs from late August onwards, and may 'jump' over the soil surface for some distance before burrowing into the ground to spin cocoons in which they will overwinter and, eventually, pupate – the ability to 'jump' (by flexing and suddenly straightening the body)

359 Bibionid larvae.

360 Fever fly (*Dilophus febrilis*) – adults in copula.

is a feature of *Contarinia* larvae, and helps them to disperse and then force their way into the ground prior to pupation. Larvae of some other dipterans, e.g. those of Mediterranean fruit fly (*Ceratitis capitata*), but not members of the genus *Dasineura*, share this ability.

DAMAGE

The strigs of infested burrs become blackened and riddled with tunnels. Heavily infested burrs turn brown and their growth is severely stunted. Damage is particularly serious on cvs Fuggle and Tutsham.

Contarinia pruniflorum Coutin & Rambier
Plum fruit-bud midge

Although associated mainly with wild *Prunus*, infestations of this southerly distributed midge occurred widely on cultivated plum in the Mediterranean region during the 1950s, and are currently causing concern in apricot orchards in southern France. Adults appear from early February or early March onwards, and deposit eggs in the outermost petals of the unopened flower buds. The eggs hatch a few days later. The translucent-whitish to lemon-yellow larvae (up to 2.5 mm long) feed gregariously within the flower buds for about 3 weeks. They then vacate the buds, drop to the ground and enter the soil, where they spin cocoons and eventually pupate. There is one generation annually, but some adults do not appear until after a second winter. Infested flower buds fail to open, turn brown and die, and heavy infestations can have a detrimental effect on cropping.

Contarinia pyrivora (Riley) (**361–364**)
Pear midge

An important pest of pear. Attacks, however, are usually most severe on garden trees. Fruits of ornamental *Pyrus*, e.g. *P. salicifolia*, are also attacked. Widely distributed in Europe; also present in North America and probably elsewhere in the world.

DESCRIPTION

Adult: 2.5–4.0 mm long; head and thorax black to greyish black, the latter with a pair of pale longitudinal stripes dorsally; abdomen greyish black, that of newly emerged female pale brownish yellow; wings dusky; antennae of male elongate, but those of female relatively short; female with a very long, extendible ovipositor. **Egg:** very small, fusiform, whitish and semitransparent. **Larva:** up to 5 mm long; yellowish white, with a conspicuous, brownish spatula.

362 Pear midge (*Contarinia pyrivora*) – larval damage to fruitlet.

363 Pear midge (*Contarinia pyrivora*) – larvae.

361 Pear midge (*Contarinia pyrivora*) – early larval damage to young fruitlet.

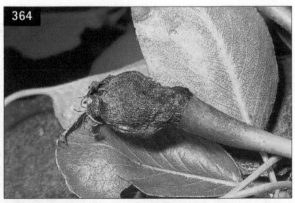

364 Pear midge (*Contarinia pyrivora*) – shrunken remains of attacked fruitlet.

LIFE HISTORY

Adult midges appear in April and May, when blossoms of mid-season pear cultivars are beginning to open. Eggs, usually from 10 to 30, are deposited in open blossoms or in those still at the white-bud stage, and are usually placed in groups on the anthers. If more than one female deposits eggs in the same flower, eventually up to 100 larvae may occur within a single fruitlet. Eggs hatch in 4–6 days and the larvae then feed within the flesh of the developing fruitlet, forming a black cavity. Larvae are fully fed in about 6 weeks. They then force their way out of the fruitlet (which may by then have dropped to the ground) and enter the soil to spin silken cocoons at a depth of about 5–8 cm. Here they overwinter, eventually pupating in the spring. There is one generation annually.

DAMAGE

For an initial period of about 2 weeks, infested fruitlets grow more rapidly than healthy ones and become either noticeably rounded or malformed. However, they cease development when about 15–20 mm in diameter and then usually shrivel, crack and decay, the outer skin turning black; finally, the damaged fruitlets drop to the ground. Even healthy fruitlets may drop off, following competition for assimilates (nutrients) from fast-growing infested fruitlets in the same cluster. Primary or secondary losses of fruit may be considerable, particularly on isolated garden trees.

Contarinia ribis (Kieffer)
Gooseberry flower midge

A minor pest of gooseberry in parts of mainland Europe, including the Czech Republic, Germany, France and the Netherlands.

DESCRIPTION
Larva: up to 2.5 mm long; yellowish white to orange.

LIFE HISTORY
Adults are active in early spring, when from one to five eggs are laid in the flower buds of host plants. Larvae feed within the buds in April and May, eventually dropping to the ground and entering the soil, where they overwinter. There is a single generation annually.

DAMAGE
Infested flower buds swell, fail to open and, eventually, desiccate. The calyx becomes thickened and fleshy, and turns whitish, yellowish or red.

Contarinia rubicola Kieffer
Blackberry flower midge

A minor and local pest of blackberry. Present in various parts of Europe, including the British Isles, France and Germany.

DESCRIPTION
Adult: 1.5–1.6 mm long; body dark grey to greyish brown; abdomen of female yellowish, with broad, greyish-brown crossbands. **Egg:** very small, fusiform and translucent. **Larva:** up to 3 mm long; whitish, with a distinct spatula.

LIFE HISTORY
Adults occur in June and July, eggs being laid within flower buds at the base of the stamens. The larvae feed gregariously within the unopened buds during July and August; as many as 27 larvae have been recorded together. When mature, the larvae drop to the ground. They overwinter in cocoons in the soil and pupate in the spring, there being one generation annually.

DAMAGE
Infested flower buds swell and usually fail to open; the expanded petals are wrinkled and the stamens wither and turn black. If only a few larvae are present in a bud, then a small, but severely distorted, fruit may be produced; more usually, however, an infestation leads to complete loss of the fruit.

Contarinia viticola (Rübsaamen)
Vine flower midge

A pest of grapevine in various parts of central Europe, including France, Germany and Luxembourg.

DESCRIPTION
Larva: up to 2.5 mm long; yellowish white.

LIFE HISTORY
Midges invade crops in about May and eggs are then laid in the swelling flower buds, either singly or in small groups. Infested buds continue to swell and the larvae feed on the contents. When fully grown, the larvae drop to the ground and enter the soil to spin silken cocoons in which to overwinter. Pupation takes place shortly before the appearance of the adults. There is just one generation annually.

DAMAGE
Infested flower buds swell and fail to open; they then wither and drop off. This results in the development of incomplete, irregular bunches of grapes and a reduction in crop yields. Infestations are restricted to cultivars at an appropriate growth stage when females are seeking suitable oviposition sites.

Dasineura mali (Keifer) (**365–366**)

Apple leaf midge

An increasingly common, but usually minor, pest of apple. Widely distributed in Europe; also present in North America.

DESCRIPTION

Adult: 1.5–2.5 mm long; body dark brown, marked above with black; abdomen in female red. **Egg:** very small, fusiform and reddish. **Larva:** up to 3 mm long; white or creamy white, becoming red or orange; spatula bilobed.

365 Apple leaf midge (*Dasineura mali*) – larvae.

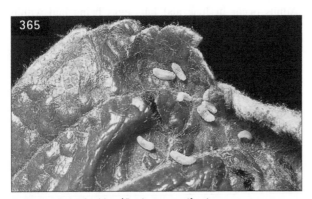

366a, b Apple leaf midge (*Dasineura mali*) – galled leaves.

LIFE HISTORY

Eggs are laid in the furled margins of young apple leaves. They hatch in 3–5 days and the larvae then feed on the upper epidermis, sheltered by the rolled-up leaf margins. Larvae are fully grown in 2–3 weeks and then drop to the ground to pupate in silken cocoons in the soil. Occasionally, pupation may also take place in the curled leaf or in other shelter. Adults emerge about 2 weeks later. Lack of rain may delay the exit of fully grown larvae from hardened leaves and, hence, lengthen the developmental period. There are three or more generations annually, larvae of the final brood overwintering in their cocoons and pupating in the spring.

DAMAGE

Infested foliage becomes distorted and the leaf margins remain tightly rolled; these galls are often tinged with red or purple; attacks are particularly prevalent on terminal growth and watershoots. Damage will stunt growth and although this is of little or no consequence in a fruiting orchard it can be serious on nursery stock, scions and young trees.

Dasineura oleae (Löw) (**367–368**)

Olive leaf gall midge

This minor pest of olive occurs in various parts of the Mediterranean basin. The yellow larvae (up to 3 mm long) cause elongate swellings on the leaves, within which development takes place. These unilocular galls, which measure up to 5 mm long and about 2 mm in width, each contain a single larva. The galls appear to have no adverse effect on plant growth or crop yields.

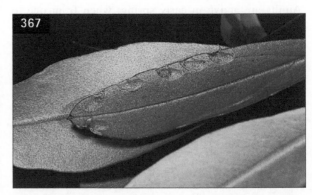

367 Olive leaf gall midge (*Dasineura oleae*) – galls, viewed from above.

Dasineura plicatrix (Loew) (369–370)

Blackberry leaf midge

A minor pest of blackberry, loganberry and raspberry; often common on wild *Rubus fruticosus*. Widely distributed in central and western Europe.

DESCRIPTION

Adult: 1.5–2.0 mm long; brownish or yellowish; abdomen of female noticeably pale. **Egg:** very small, fusiform and whitish. **Larva:** up to 2.5 mm long; whitish, with a bilobed spatula.

LIFE HISTORY

Adults appear in May and June. Eggs are laid in furled leaves and hatch a few days later. The larvae then feed gregariously on the upper surface of young unopened leaves, becoming fully grown in about 3 weeks. They then drop to the ground and eventually pupate in cocoons in the soil. The pupal stage lasts for about 3 weeks, a second generation of adults appearing in July and August, and larvae in August and September. The winter is spent as fully grown larvae in cocoons, pupation occurring in the spring.

DAMAGE

An attacked leaf tends to fold along the midrib, which then swells. Growth of infested shoots is usually checked, at least during the period of attack. Severely damaged leaves become twisted and distorted, and eventually turn black.

Dasineura pyri (Bouché) (371–372)

Pear leaf midge

An often common pest of pear. Widely distributed in Europe; also present elsewhere, including North America and New Zealand.

DESCRIPTION

Adult: 1.5–2.0 mm long; brown, with distinct black crossbands on the abdomen. **Egg:** very small, fusiform

369 Blackberry leaf midge (*Dasineura plicatrix*) – galled new growth.

370 Blackberry leaf midge (*Dasineura plicatrix*) – galled leaves.

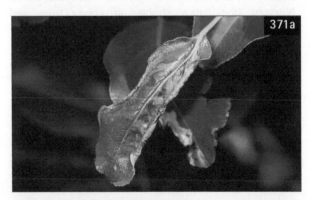

368 Olive leaf gall midge (*Dasineura oleae*) – galls, viewed from below.

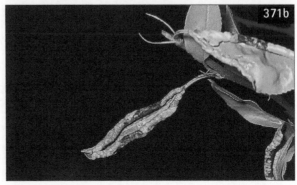

371a, b Pear leaf midge (*Dasineura pyri*) – galled leaves.

and whitish. **Larva:** up to 2.5 mm long; whitish, with a bilobed spatula.

LIFE HISTORY

Adults appear from mid- or late April onwards, but are most numerous in August or September. Eggs are laid within the rolled margins of young pear leaves, usually up to 35 per leaf. They hatch within a few days and the larvae then feed gregariously on the upper epidermis, preventing the leaf margin from unrolling. Larvae are fully grown in about 2 weeks. They then drop to the ground to spin cocoons and pupate a few centimetres beneath the surface. During very dry conditions, the larvae may be trapped within the tightly rolled leaves and, consequently, pupation may be delayed. A mass exodus of larvae from such leaves often then takes place after a shower of rain. The pupal stage lasts for about 2 weeks. The number of generations in a season (usually three) is dependent to some extent upon the availability of young foliage; broods may be restricted to two in very dry seasons.

DAMAGE

Leaves are much distorted by severe attacks and their margins remain tightly rolled, meeting along the mid-line. Such foliage turns red, and finally blackens and dies. Tightly rolled, blackened foliage can also be produced by tortrix moth larvae (family Tortricidae), notably by those of *Acleris rhombana* (see Chapter 6). Nursery stock, new shoots on young trees and watershoots on fruiting trees are most liable to be attacked. At the start of the season, damage appears only at the shoot tip. However, later, as shoots elongate, a succession of terminal leaves will become infested so that damage occurs at various levels.

Dasineura ribis (Barnes)
Black currant flower midge

A pest of black currant in northern Europe, particularly in Finland and Sweden.

DESCRIPTION

Adult: 2.0–2.5 mm long; grey, the abdomen tinged with red. **Larva:** up to 3 mm long; yellowish orange to reddish orange.

LIFE HISTORY

Adults appear in the spring and eggs are then laid in small numbers in the young flower buds. Larvae feed gregariously within the swelling buds and are fully grown in about a month. They then drop to the ground and enter the soil, where they overwinter and eventually pupate. There is one generation annually.

DAMAGE

Infested flower buds are malformed and swollen irregularly, particularly apically. The galled portions appear pale flecked, tinged more or less densely with reddish brown. Crop losses following heavy infestations can be considerable.

Dasineura tetensi (Rübsaamen) (**373**)
Black currant leaf midge

An important pest of black currant. First reported in 1931, in Kent, England; now recorded in most parts of the world where black currants are grown, and often a serious problem.

DESCRIPTION

Adult: 1–2 mm long; body brownish to yellowish; abdomen paler, with dark crossbands above. **Egg:** 0.70 × 0.27 mm; fusiform and whitish. **Larva:** up to 2.5 mm long; usually white, but sometimes becoming orange; spatula bilobed.

LIFE HISTORY

There are up to three main generations annually, adults appearing from April to August, but there may be a further generation on black currant cultivars (e.g. Goliath) still producing young leaves late in the summer. Eggs are laid

372 Pear leaf midge (*Dasineura pyri*) – severely galled shoot.

373 Black currant leaf midge (*Dasineura tetensi*) – severely galled shoot.

within the folds of young leaves, particularly in the tips of new growth, on either the upper or the lower surfaces. They hatch within a few days. The larvae then feed gregariously within the shelter of the furled leaves for about 10–14 days, before dropping to the soil to pupate in small, silken cocoons formed a few centimetres below the surface. The pupal stage lasts for about 2 weeks. There are usually four or five larvae on any one infested leaflet, but sometimes more. Individuals of the final generation overwinter in their cocoons, and pupate in the spring. The adult midges fly in calm conditions, but remain quiescent beneath the bushes if there is any wind.

DAMAGE
Infested leaves become discoloured, crumpled and distorted; they usually fail to expand normally and eventually turn black. In the early stages of attack, an infested leaf may twist characteristically to one side as it begins to unfurl. Damage to fruiting bushes is relatively unimportant and the quality and quantity of the crop is not affected. However, on nursery stock, distortion of the new growth results in unsightly plants and, often, misshapen bushes. Damage to cuttings can be particularly serious and may render them useless. Distortion of the foliage could mask signs of reversion virus disease in a plantation and may also prevent the recognition of rogue cultivars when inspecting nursery stock.

Dasineura tortrix (Löw)
Plum leaf-curling midge
Associated with various fruiting species of *Prunus*, including apricot, bullace, cherry, myrobalan, plum and *P. spinosa*, but not important as a pest except in nurseries. Widely distributed in central Europe (e.g. France, Germany and Italy); also present in parts of north-western Europe (including Denmark and England).

DESCRIPTION
Larva: up to 2.5 mm long; white. **Gall:** a fusiform bunch of distorted leaves and shortened internodes at the tip of a shoot.

LIFE HISTORY
Adults appear in the spring. Later, larvae may be found feeding gregariously in the loosely rolled margins of the galled leaves; such galls become particularly obvious towards the end of May and in June. Fully-fed larvae eventually drop to the ground to pupate, adults emerging shortly afterwards. There are up to three generations annually, with fully fed larvae overwintering in cocoons in the soil.

DAMAGE
Infestations disrupt the growth of terminal shoots, the shortened internodes causing the leaves to mass together; infested leaves also become interlocked and the leaf margins curled upwards. Such damage is usually transient; however, new growth arising from infested buds may turn black and die. Attacks are of greatest significance in nurseries.

Janetiella oenophila (Haimhoffer) (374–376)
Vine leaf gall midge
A common, but minor, pest of grapevine. Widely distributed in mainland Europe, including France, Germany, Hungary, Italy, Portugal and Switzerland.

DESCRIPTION
Larva: up to 2.5 mm long; salmon pink to orange. **Gall:** lenticular, more or less rounded in outline and about 3 mm in diameter; hard to the touch, yellowish to reddish, shiny above and hairy below – cf. galls formed by *Viteus vitifoliae* (Hemiptera: Phylloxeridae) (Chapter 3).

LIFE HISTORY
Adults occur in May. Eggs are then deposited in the leaves of grapevines and, sometimes, in the stalks (rachides) of

374 Vine leaf gall midge (*Janetiella oenophila*) – galled leaf.

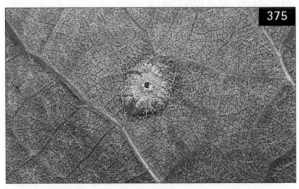

375 Vine leaf gall midge (*Janetiella oenophila*) – underside of gall.

376 Vine leaf gall midge (*Janetiella oenophila*) – larva.

377 Blackberry stem gall midge (*Lasioptera rubi*) – gall.

young fruit clusters. The eggs hatch in about 10 days. Larvae then feed within characteristic galls formed on the leaves (and sometimes on the young rachides), there usually being from five to 20 galls per infested leaf; much higher infestations, of several hundred galls per leaf, are also reported. The larvae are fully grown in about 3 weeks, usually in June. They then drop to the ground to form whitish cocoons in the soil. In many regions this species is single brooded, with the larvae overwintering and pupating in the spring. In other areas, two generations are reported, in which case the first-generation larvae pupate and give rise to a second generation of adults in the same summer. Larvae of the second brood eventually overwinter in the soil, pupating in the spring.

DAMAGE
Leaf galls are usually of little or no economic importance. However, when galls are present in very large numbers, young galled foliage may become distorted and its development retarded. In addition, galling of the young rachides can result in distortion of developing fruit clusters.

Lasioptera rubi Heeger (377)
Blackberry stem gall midge
A generally common, but minor, pest of blackberry, loganberry and raspberry. Often common on wild *Rubus fruticosus*. Widely distributed in central and western Europe.

DESCRIPTION
Adult: 2 mm long; head small, with large black eyes; thorax dark brown, with an anterior silver band; abdomen dark brown, with black and silvery-white crossbands; wing veins with black scales; a distinct clear area midway along the anterior margin of each wing. **Larva:** up to 3 mm long; orange red, with a deeply incised bifid spatula. **Pupa:** 2 mm long; orange red.

LIFE HISTORY
Adults occur in May, June or July and lay their eggs in batches of up to 15 or more at the base of buds and lateral shoots. Eggs hatch in about 10 days and the larvae then burrow through the epidermis into the shoot. The infested shoot swells and within 6 weeks a distinct, unilocular gall is formed – cf. galls formed on *Rubus* by rubus gall wasp (*Diastrophus rubi*) (Hymenoptera: Cynipidae) (see Chapter 7). The larvae continue feeding for several months and become fully grown by the following April or May; they then pupate. Adult midges appear about 3 weeks later, emerging through cracks or decaying parts of the gall.

DAMAGE
The galls are thick-walled, and become filled with blackish-brown frass and rotting tissue. They are sometimes as much as 40 mm long and are either walnut-shaped (completely encircling the stem) or asymmetrical (formed to one side of the stem); terminal galls may also occur. The galls stunt the growth of infested plants, and can have an adverse effect on both leaf and fruit production.

Prolasioptera berlesiana (Paoli)
Olive fruit midge
A secondary pest of olive throughout the Mediterranean area.

DESCRIPTION
Adult: 1.5–2.0 mm long; rusty red, with black markings dorsally on both thorax and abdomen; antennae black. **Egg:** 0.20–0.25 mm long; whitish. **Larva:** up to 3 mm long; reddish orange. **Pupa:** 1.5–2.0 mm long; brown.

LIFE HISTORY
Midges appear in late June and early July, coinciding with the initial period of oviposition in fruits by olive fruit fly (*Bactrocera oleae*) (family Tephritidae). Eggs

are then laid, usually singly, in puncture wounds formed in the developing fruits by olive fruit fly. The midge egg hatches in about a day, and the larva at first attacks the egg of the fruit fly, slowly imbibing the contents. After a few days, both the remains of the fruit fly egg and surrounding plant tissue become invaded by a fungus (*Sphaeropsis dalmatica* – anamorph: *Macrophoma dalmatica*), introduced by the egg-laying female midge. The fungal mycelium is, in turn, then used as food by the midge larva. Midge larvae are fully grown in 8 to 10 days. They then vacate the fruits and drop to the ground. Pupation occurs in silken cocoons in the soil, adult midges emerging about a week later. There are three or four generations annually.

DAMAGE
Midge-infested fruits develop characteristic oval spots on the surface (up to 7 mm long), and eventually turn black and drop to the ground. However, although damage to olive fruits (primarily due to the introduced fungus) is of some importance, crop losses are probably outweighed by the beneficial effects of the midge in reducing populations of olive fruit fly – a far more significant problem. Further, the midge larvae are not of primary importance, as they merely invade already pest-damaged olive fruits.

Putoniella pruni (Kaltenbach) (378–380)
Plum leaf gall midge
Associated mainly with *Prunus spinosa*; also reported on wild *P. insititia* and, at least in France and Italy, on cultivated plum. Widely distributed in Europe.

DESCRIPTION
Larva: up to 3 mm long; yellow to reddish orange. **Gall:** up to 40 mm long; a longitudinal, glabrous, strong-walled swelling on the underside of a leaf; upper surface with a narrow, longitudinal slit; usually green, but sometimes tinged with dark red.

LIFE HISTORY
Larvae feed gregariously in unilocular leaf galls during May and early June. They then drop to the ground and overwinter in the soil, eventually pupating in the following year a couple of weeks before the emergence of the adults. The midges occur mainly in April and early May.

DAMAGE
Galls cause distortion of leaves and may have a detrimental effect on photosynthesis.

Resseliella oculiperda (Rübsaamen)
larva = red bud borer
A local and sporadic pest of budded stocks and grafts,

378 Plum leaf gall midge (*Putoniella pruni*) – gall along midrib, viewed from above.

379 Plum leaf gall midge (*Putoniella pruni*) – gall along midrib, viewed from below.

380 Plum leaf gall midge (*Putoniella pruni*) – galls.

attacking apple, pear, plum and other fruit trees; particularly well known as a pest of cultivated *Rosa*. Distinct 'apple' and 'rose' strains occur, both (but particularly the latter) with flexible host preferences. Widely distributed in Europe, including the Czech Republic, Denmark, England, France and Germany.

DESCRIPTION

Adult: 1.4–2.1 mm long; body dark reddish brown. **Larva:** up to 3.5 mm long; salmon pink to red (colourless or whitish when young), with a bilobed spatula.

LIFE HISTORY

There are three generations in a year, adults occurring in May and June, in July and August and from mid-August to the end of September. Eggs are laid in graft slits or cuts made in the bark of newly budded stock. They hatch in about a week and the larvae then feed in small groups on sap in the cambium between scion and stock, becoming fully grown in 2–3 weeks. The larvae then drop to the ground and enter the soil to spin small cocoons a few centimetres below the surface. They then pupate and adults emerge 2–3 weeks later. Larvae of the autumn generation overwinter in their cocoons and pupate in the spring.

DAMAGE

Larval feeding prevents grafts or buds from taking and, as a result, the scion or buds wither and die. Even where part of a graft has successfully taken, an infestation will eventually lead to bud failure. Most damage is caused by larvae attacking grafts or buds from August onwards, when losses on unprotected nursery stock can be considerable.

Resseliella oleisuga (Targioni-Tozzetti)

Olive stem midge
Associated with Oleaceae, particularly olive, upon which it is a minor pest. Present in Mediterranean areas.

DESCRIPTION

Adult: 2.5–3.0 mm long; black, with an orange (= female) or grey (= male) abdomen. **Egg:** 0.25–0.30 mm long; sausage shaped. **Larva:** up to 3 mm long; whitish to orange. **Pupa:** reddish yellow to orange.

LIFE HISTORY

First-generation midges appear in the spring. Eggs are then laid in cracks in the bark of host plants, usually in groups of up to 30 at the base of branches. The eggs hatch within a few days and larvae then feed gregariously beneath the bark. After about 3 weeks, fully grown larvae vacate the host and enter the soil, where they pupate; adults of the next generation appear 7–10 days later. There are typically three generations annually, one each in spring, summer and autumn, larvae of the final generation overwintering.

DAMAGE

Leaves on infested shoots may wither and turn brown, and developing fruits may become desiccated. Bark overlying infested parts of host trees becomes necrotic and, in the wake of persistent attacks, infested branches are eventually killed.

Resseliella ribis (Marikovski)

Black currant stem midge
A pest of black currant in Finland and in some other parts of Europe, including Poland and parts of the former USSR. Also a minor pest on red currant and white currant.

DESCRIPTION

Adult: 1.8–2.0 mm long; grey to greyish orange. **Larva:** up to 3.5 mm long; orange.

LIFE HISTORY & DAMAGE

Apart from host plant, similar to that of raspberry cane midge (*Resseliella theobaldi*), but with just two generations annually.

Resseliella theobaldi (Barnes) (**381–382**)

Raspberry cane midge
An important pest of raspberry; occasionally,

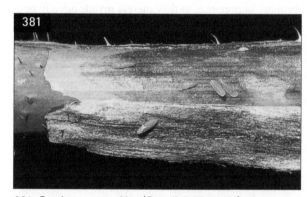

381 Raspberry cane midge (*Resseliella theobaldi*) – larvae.

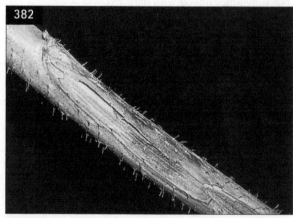

382 Raspberry cane midge (*Resseliella theobaldi*) – larval damage to stem.

loganberry is also attacked. A well-known pest in England, and also for many years a serious problem in the raspberry-growing areas of Scotland. Nowadays, also of importance in various parts of mainland Europe, including Scandinavia.

DESCRIPTION
Adult: 1.4–2.1 mm long; body dark reddish brown; long-legged and of delicate appearance. **Egg:** 0.94 × 0.33 mm; shiny and translucent. **Larva:** up to 3.5 mm long; salmon pink to yellowish, but colourless when young; spatula bilobed. **Pupa:** 1.3–2.0 mm long; dark reddish-brown, with prominent anterior and posterior respiratory processes.

LIFE HISTORY
The midges emerge in early May or later, depending on spring temperatures (midge appearance in Scotland, for example, tends to be about a month later than in southern England), and lay eggs in splits on young raspberry canes, usually when the spawn is about 20–30 cm high. Eggs hatch within a week and the larvae then feed in large clusters beneath the rind, close to longitudinal splits in the epidermis. Larvae feed for 2–3 weeks, passing through three instars. They then drop to the ground and enter the soil at the base of the canes to spin small cocoons in which they pupate. Adults emerge in 2–3 weeks and these produce a second brood of larvae in July and August. There is usually a third brood of larvae in September and these, when fully fed, overwinter in cocoons, pupating in the spring. Up to 100 first-brood larvae may infest a single cane; in later generations, there may be several hundred larvae per infested cane.

DAMAGE
Midge larvae cause only superficial damage to the canes, tissue in the feeding areas becoming discoloured, turning brown or black. However, midge-infested canes are particularly susceptible to certain fungal pathogens, such as raspberry cane blight (*Leptosphaeria coniothyrium*) and raspberry spur blight (*Didymella applanata*). Collectively, these fungi cause the disease known as 'midge blight', which may be sufficiently serious to kill the young canes and, consequently, reduce fruiting potential for the following year. There are noticeable differences in the susceptibility of different raspberry cultivars: those with readily splitting canes (e.g. cvs Glen Clova, Malling Enterprise and Malling Promise) are very prone to damage, whereas those whose canes are less liable to split (e.g. cvs Malling Landmark and Norfolk Giant) are usually infested only slightly.

Family **TEPHRITIDAE**
(large fruit flies)

Small to medium-sized flies, with relatively broad, patterned wings; eyes often large and colourful; females possess a distinct oviscapt.

Bactrocera oleae (Gmelin)
Olive fruit fly
The most important pest of olive. Widely distributed from the Canary Islands to India, and present throughout the Mediterranean basin.

DESCRIPTION
Adult: 4–5 mm long; head yellowish brown, with darker markings; thorax dark grey to blackish, with three incomplete black stripes dorsally; scutellum white and prominent; abdomen black, suffused with reddish brown dorsally; antennae relatively long; legs yellowish brown; wings transparent and elongate, with dark veins, and each with a brownish stigma-like mark on the anterior margin and a more diffuse brown mark at the apex; female with a distinct, retractile oviscapt. **Egg:** 0.7–0.8 mm long; white and opaque. **Larva:** up to 7 mm long; whitish and maggot-like. **Puparium:** 3.5–4.5 mm long; yellowish brown.

LIFE HISTORY
Adults appear from late March or early April onwards, depending on temperature. Mating, however, does not take place until later in the year. Eventually, eggs are eventually laid singly in developing olive fruits, each inserted just beneath the skin. The eggs hatch within a few days in high summer, but in up to 2 weeks or so under cooler conditions. Larvae then tunnel within the pulp for about 2 weeks before eventually pupating in situ. The period of pupation is relatively short (about 2 weeks) in summer, but greatly protracted (up to 3 months) in winter. Adults survive in the field for up to 6 months, and each female is capable of depositing several hundred eggs over a period of 1–3 months. Details of the life cycle vary from region to region, there being from two to four generations annually, depending on temperature. The winter is usually passed in the pupal stage, within puparia formed in the soil.

DAMAGE
The pest is capable of causing large reductions in crop yields, following infestation of fruits by the larvae. The pulp of infested fruits is partly or completely destroyed, and attacked fruits may also drop prematurely. In the case of harvested infested fruits, the pH of the pulp is altered (acidity increased), and this has a detrimental effect on oil quality. Increased oxidation, following damage to the protective outer skin of the fruit, also results. Further, pest-damaged fruits are subject to attack

by a range of secondary organisms (fungal pathogens), notably following secondary invasion of infested fruits by olive fruit midge (*Prolasioptera berlesiana*) (family Cecidomyiidae); this also has an adverse effect on oil production. Finally, consignments of harvested table olives may be rejected even when merely oviposition marks (stings) are found on the surface of just a small number of fruits.

Ceratitis capitata (Wiedemann) (**383–388**)
Mediterranean fruit fly

An important, southerly distributed pest of peach; infestations also occur on various other temperate fruit crops, including apple, pear, quince, apricot, cherry, plum, strawberry, grapevine and kiwi fruit. Other hosts include citrus, fig, olive and prickly pear. Present throughout the Mediterranean basin. Also established elsewhere, including much of Africa and parts of America, Asia and Western Australia. The pest is found, occasionally, in central and northern Europe (mainly as larvae in consignments of imported fruits), but is unable to survive cold conditions and cannot become established in such areas. Even in relatively mild regions, such as southern France, its survival from one year to the next is strictly limited.

DESCRIPTION
Adult: 4–5 mm long; eyes bright emerald green, clouded with reddish brown; thorax mainly black, suffused with greyish yellow; scutellum yellow and black; abdomen mainly yellowish orange, with two silvery crossbands; wings clear, with black veins and marked with black spots and brownish-yellow patches; legs yellowish brown. **Egg:** 1 mm long; white and narrowly fusiform. **Larva:** up to 8 mm long; white and translucent, with prominent black mouth-hooks, but on completion of feeding appearing dirty creamy white. **Puparium:** 4–5 mm long; yellowish brown to reddish brown and barrel-shaped.

LIFE HISTORY
Eggs are deposited a few millimetres beneath the surface of the fruit, usually in groups of up to nine, through a slit made with the female's oviscapt. Several females may lay eggs in the same fruit, so numbers of larvae eventually present in any one infested fruit may

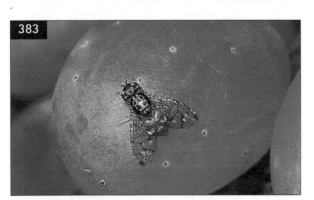

383 Mediterranean fruit fly (*Ceratitis capitata*).

384 Mediterranean fruit fly (*Ceratitis capitata*) – larvae.

385 Mediterranean fruit fly (*Ceratitis capitata*) – fully fed larva in soil.

386 Mediterranean fruit fly (*Ceratitis capitata*) – puparium.

be considerable. Also, adults are relatively long-lived and a female may deposit several hundred eggs in her lifetime. Eggs hatch within a few days of being laid and the larvae then feed on the flesh, causing extensive damage. Individuals are fully fed in about 2 weeks. They then leave the fruit (which, by then, may already have dropped to the ground) through small holes bored through the skin. The emerged larvae are capable of flexing their bodies and 'jumping', allowing them to disperse over the ground before eventually burrowing into the soil, where pupation takes place a few centimetres below the surface. There are usually two or more generations in a season, the number and rate of development depending on temperature. At an optimum temperature of 32°C, the complete life cycle from egg to adult occupies no more than 2 weeks. The winter is normally passed in the pupal stage in the soil.

DAMAGE

The oviposition hole in an infested fruit may be surrounded by a sunken and often discoloured area of rotting tissue, the punctures (stings) in themselves being sufficient in the case of high-quality produce to make

fruits unmarketable. Damage caused to the flesh by the larvae also renders the fruit useless. In addition to direct damage inflicted by the pest, infested fruits may also be invaded by secondary, pathogenic fungi.

Rhagoletis cerasi Linnaeus (389–390)
European cherry fruit fly

The 'southern' race of this insect is a serious pest of cherry in mainland Europe (e.g. in Austria, Italy, Portugal, southern France, southern Germany, Spain and Switzerland), especially sweet cherry. A so-called 'northern' race, associated with *Lonicera*, occurs in countries to the north and east of Switzerland; *Symphoricarpos rivularis* is also a host. Although established on cherry in parts of Scandinavia, this insect is not present in the British Isles.

DESCRIPTION

Adult: 3.5–5.0 mm long; body black, with yellow markings on the head and thorax; wings transparent, with a distinctive pattern of dark, bluish-black stripes. **Larva:** up to 6 mm long; body whitish and translucent. **Puparium:** 3–4 mm long; pale yellowish brown.

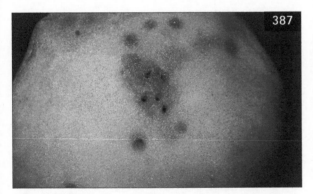

387 Mediterranean fruit fly (*Ceratitis capitata*) – external damage to peach.

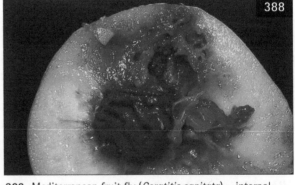

388 Mediterranean fruit fly (*Ceratitis capitata*) – internal damage to peach.

389 European cherry fruit fly (*Rhagoletis cerasi*).

390 European cherry fruit fly (*Rhagoletis cerasi*) – larva.

LIFE HISTORY

Adults occur from late May to early July and are particularly active in hot, dry conditions. They also frequently rest on leaves of the foodplant, imbibing honeydew and other sugary excretions. Eggs are laid from mid-June onwards, each being inserted singly beneath the epidermis of a ripening fruit. The eggs hatch within 1–2 weeks, and larvae feed on the flesh of the developing fruits for about 4 weeks. They then vacate the fruits and enter the soil, to pupate a few centimetres below the surface. There is a single generation annually; individuals may remain in the pupal stage over one, two or three winters.

DAMAGE

Attacked fruits often rot, and heavy infestations will reduce marketable yields. However, the extent of damage varies from year to year. Consignments of harvested cherries found to be infested may be rejected by fruit processors.

Rhagoletis cingulata (Loew)

American eastern cherry fruit fly

This North American species has been known to occur in Europe since 1983 (having been recorded from Germany, northern Italy, the Netherlands and Switzerland) and, at least in Germany, the pest now appears to be well established. Recorded foodplants include Morello (= sour) cherry, sweet cherry, plum and various wild *Prunus* species (e.g. *P. avium* and *P. padus*). Adults are most readily distinguished from those of *Rhagoletis cerasi* by differences in wing pattern. For example, in *R. cerasi* the dark banding extending around the wing tip is unbranched, but in *R. cingulata* it is distinctly bifid (giving the adult insect, when in repose and viewed from above, the superficial appearance of a pseudoscorpion).

Rhagoletis completa (Cresson)

Walnut husk fly

A minor pest of walnut; of North American origin, and now established in some parts of Europe (e.g. Italy and Switzerland).

DESCRIPTION

Adult: 4–5 mm long; reddish brown to yellowish brown, with a creamy-yellow scutellum; wings clear, marked with broad, blackish crossbands. **Larva:** up to 6 mm long; whitish to yellowish white, and translucent.

LIFE HISTORY

Adults occur from early June onwards, and eggs are eventually laid beneath the skin of developing walnuts. The eggs hatch within a week. Larvae feed inside the developing fruits for up to 5 weeks; they bore within the mesocarp and sometimes also attack the pericarp and kernel. Fully fed larvae drop to the ground and enter the soil to pupate, the puparia then overwinter.

DAMAGE

Infested fruits bear a small, black oviposition scar. Later the surface of such fruit blackens, as the underlying tissue is destroyed and also turns black.

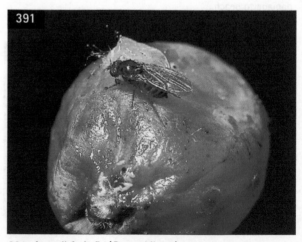

391 A small fruit fly (*Drosophila* sp.).

Family **DROSOPHILIDAE**

Small to very small flies, much attracted to fermenting juices; compound eyes bright red; antennal arista generally plumose. Larvae are translucent to whitish and elongate, with conspicuous anterior and posterior spiracles, and fleshy posterior papillae. Adults of several species are known as 'vinegar flies', and these are frequently abundant on dumped fruits, including the discarded remains of crushed grapes.

Drosophila spp. (**391**)
Small fruit flies

Several species occur in association with overripe or bruised apples, peaches, plums, raspberries, strawberries, grapes and other fruits. Generally abundant in Europe and in other temperate parts of the world.

DESCRIPTION
Adult: 2–3 mm long; thorax mainly yellowish brown; abdomen yellowish, with brown crossbands. **Egg:** 0.5 mm long; white and elongate, with a pair of oar-like respiratory processes. **Larva:** up to 4.5 mm long; whitish, with distinct black mouth-hooks and a pair of prominent anterior spiracular processes and a pair of long posterior respiratory processes. **Puparium:** 3–4 mm long; brown, with short anterior respiratory processes and long posterior respiratory processes.

LIFE HISTORY
The winter is passed within the puparium, adults occurring outdoors from May to October. Eggs are laid in decaying fruit and other similar situations where the larvae will eventually feed. Development is rapid, and there are several generations annually.

DAMAGE
The flies usually breed in fermenting or decaying matter and merely aid the breakdown of the organic material. In vineyards, however, larvae feeding on rotting tissue within ripening bunches of grapes may spread from unsound to sound fruits, eating away the flesh to leave only the empty skins. Vinegar flies are also implicated in the spread of the spores of fungal pathogens.

NOTE
The Asiatic species *Drosophila suzukii* (Matsumura) has recently spread to many parts of the world, including North America and southern Europe (e.g. France, Italy and Spain). Unlike other species, adults are able to deposit eggs beneath the skins of healthy fruits, and larvae then cause extensive damage and crop losses. Recorded hosts in Europe include cherry, peach, blackberry, blueberry, raspberry and strawberry.

Family **AGROMYZIDAE**

Small to very small, mainly black or grey, often yellow-marked, flies with a distinct break on the costal vein. Females of some species pierce the leaves of host plants and lap up the exuding sap. Larvae possess distinctive black mouthparts (the cephalopharyngeal skeleton), and paired anterior and posterior spiracles, details of which are often used to distinguish between species. The larvae of most species mine within leaves, to form serpentine or blotch-like galleries in which two discontinuous trails of frass accumulate (cf. galleries formed by leaf-mining lepidopterous larvae, Chapter 6).

Agromyza flaviceps Fallén
Agromyza igniceps Hendel (**392**)
larvae = hop leaf miners

These two widely distributed European species are associated with *Humulus lupulus*; occasionally, minor infestations are also found on cultivated hop. The larvae of both species form irregular linear mines in the leaves: mines of the former containing diffused accumulations of greenish frass, and those of the latter with the frass distributed in distinct black strips. Mines of *A. igniceps* (which occurs early in the season) tend to split open before the larvae have completed their development. At least in the case of *A. flaviceps* there can be two generations annually. Adults of these species are separable as follows: *A. flaviceps* – legs bright yellow, wing length c. 3.2 mm; *A. igniceps* – legs black, wing length c. 2.5 mm.

392 *Agromyza flaviceps* – leaf mine on hop.

Agromyza potentillae (Kaltenbach) (**393–394**)

larva = strawberry leaf miner

A local pest of raspberry and strawberry; also attacks various ornamental and wild herbaceous Rosaceae, including *Potentilla*. Holarctic. Widely distributed in Europe.

DESCRIPTION

Adult: wing length 2.0–2.6 mm; body black to greyish black; legs somewhat yellowish, with black femora. **Larva:** up to 3 mm long; whitish, with distinct, black mouth-hooks; posterior spiracles each with three bulbs. **Puparium:** 2–3 mm long; yellowish brown, with prominent (horn-like) anterior respiratory processes. **Leaf mine:** a light brown blotch, often tinged with red, visible from above.

LIFE HISTORY

Adults of the first generation occur in May or June and lay eggs singly on the foliage of host plants. The larvae feed inside the leaves, each forming a linear mine that later develops into an elongated blotch. When fully fed, the larva vacates its mine through a small slit made in the underside of the leaf and then pupates amongst shelter on the ground. In temperate regions there are two generations annually, adults of the second appearing in August. Occupied mines occur from June to July and again from September to October.

DAMAGE

Larval feeding is extensive, but damage is limited to the destruction of tissue within the galleries. There may be one or more mines in a single leaf and, if numerous, the pest can retard the growth of young strawberry plants; attacks on older strawberry plants and on raspberry, however, are of little or no consequence.

Phytomyza heringiana Hendel (**395–397**)

A usually minor pest of apple. Widely distributed in Europe.

DESCRIPTION

Adult: wing length 1.5–1.8 mm; mainly dull black to greyish; fore knees yellowish. **Larva:** up to 2.5 mm

393 *Agromyza potentillae* – leaf mine on strawberry.

395 *Phytomyza heringiana* – leaf mine with protruding puparium.

394 *Agromyza potentillae* – leaf mines on raspberry.

396 *Phytomyza heringiana* – puparium.

long; whitish and translucent. **Puparium:** 2.0–2.5 mm long; light brown; anterior respiratory processes brownish black and broadly bifid, each forming an outwardly curved, horn-like projection; posterior respiratory processes brownish black, bifid and divergent. **Leaf mine:** an irregular, linear to blotch-like gallery, with a wavy outline; visible from above.

LIFE HISTORY

Adults of the first generation emerge from late March or early April onwards. Eggs are then laid singly on the expanded leaves of host plants. First-brood larvae, which mine within the leaves, occur in May and June. They eventually pupate within the feeding gallery, adults of the second generation emerging in late June or early July. The puparium remains protruding from the mine following emergence of the adult. Second-brood larvae mine within apple leaves from early August onwards. They pupate in the autumn, individuals overwintering in the pupal stage.

DAMAGE

Larval mines can cause slight distortion of foliage, but have little or no adverse effect on tree growth.

Family **MUSCIDAE**

Small to large flies, with large squamae covering the halteres. Larvae are maggot-like.

Mesembrina meridiana (Linnaeus) (**398**)

Adults of this widely distributed species occur from May to September and are locally common in rural areas. In summer and autumn they are frequently attracted to mechanically damaged (e.g. bird-pecked) orchard fruits, particularly apples and pears, where (along with wasps and other insects) they feed avidly on the exuding juices. The flies do not damage sound fruits. However, they can contaminate the skins with their faeces, thereby requiring harvested fruits to be washed prior to sale or consumption. Adults (15–16 mm long) are mainly black, with a distinctive golden-metallic sheen on the face (between the compound eyes and the frons) and prominent golden-yellow squamae and wing bases. The larvae most frequently feed in cow dung and horse dung, where they are facultative (opportunist) carnivores.

397 *Phytomyza heringiana* – old, vacated leaf mine.

398 *Mesembrina meridiana* – adult.

194

Family **ANTHOMYIIDAE**

Small to medium-sized, often greyish flies, typically similar in appearance to house flies. Larvae are maggot-like and usually phytophagous, attacking plant stems and roots. Several species are important pests of agricultural or horticultural crops.

Pegomya rubivora (Coquillett) (**399–400**)
Loganberry cane fly
A minor pest of blackberry, loganberry and raspberry. Widely distributed in northern Europe, but of local occurrence.

DESCRIPTION
Adult: 6 mm long; greyish black. **Larva:** up to 8 mm long; white, with black mouth-hooks. **Puparium:** 6 mm long; light brown.

LIFE HISTORY
Adults emerge in April or May and lay their eggs in the leaf axils on the terminal shoots of the primocanes (those representing the initial, usually non-fruiting, first-year growth). Following egg hatch, the larvae burrow into the pith of the canes, tunnelling downwards for about 10 cm before girdling the stem just beneath the epidermis. The larvae feed gregariously and continue to tunnel down to the base of the canes, where they eventually pupate from late July onwards. The winter is spent in the pupal stage, within the puparium.

DAMAGE
The tips of girdled canes wither and die, but attacks are sporadic and not serious.

399 Loganberry cane fly (*Pegomya rubivora*) – infested shoots.

400 Loganberry cane fly (*Pegomya rubivora*) – damage to base of young stem.

Chapter 6
Butterflies and moths

Family **HEPIALIDAE**
(swift moths)

Primitive, medium-sized moths, with vestigial mouthparts and short antennae. Larvae have well-developed legs, with crochets on the abdominal prolegs forming several concentric circles. Swift moth larvae are soil-inhabiting; they feed on the roots of various wild and cultivated plants.

Hepialus humuli (Linnaeus) (**401–403**)
Ghost swift moth
A frequent pest of grassland and other crops, including raspberry, strawberry and hop. Widely distributed in central and northern Europe.

401 Ghost swift moth (*Hepialus humuli*) – female.

DESCRIPTION
Adult female: 50–70 mm wingspan; forewings greyish ochreous, with a pinkish tinge at the apex. **Adult male:** 46–50 mm wingspan; wings silvery white; body yellowish brown. **Egg:** 0.7 × 0.5 mm; white when laid, soon becoming shiny black. **Larva:** up to 50 mm long; body whitish and more or less opaque; pinacula reddish brown; head and prothoracic plate reddish brown. **Pupa:** 30–45 mm long; dark brown; abdomen with several transverse series of dorsal spines.

LIFE HISTORY
Adults occur in June and July, sometimes later. They are active at dusk, each female broadcasting several hundred eggs whilst hovering over grasses and other low-growing vegetation. Following egg hatch, the young larvae burrow into the soil to feed on plant rootlets. Later, they attack larger roots and may also bite into stolons and the lowermost parts of stems. The larvae construct silk-lined feeding tunnels in the soil, often retreating rapidly into them when disturbed. The larvae consume large quantites of food. However, growth is slow and individuals pass through 12 or more instars. Development usually extends over 2 (but occasionally 3) years. Pupation normally occurs in late April or May, in an earthen cell formed 10–25 cm beneath the surface. Pupae of swift moths typically wriggle up to the surface shortly before adult moths emerge.

402 Ghost swift moth (*Hepialus humuli*) – male.

403 Ghost swift moth (*Hepialus humuli*) – larva.

DAMAGE

Although frequently established in hop gardens, this pest usually causes little serious harm, at least to older plants. However, if infested hops are grubbed, the roots of following crops (such as raspberry or strawberry) may be severely damaged, plants being weakened and perhaps killed. Crops planted into recently ploughed, infested grassland or pasture are also likely to be attacked. Root damage is particularly severe during the second summer of larval development.

Hepialus lupulinus (Linnaeus) (**404–407**)
Garden swift moth

An often common pest of grassland and a wide range of crops, including gooseberry, raspberry, strawberry and hop. Widely distributed and often common in central, northern and south-eastern Europe.

DESCRIPTION

Adult: 25–40 mm wingspan; forewings yellowish brown, variably marked with white (especially in the male); hindwings yellowish grey, darker in the male. **Egg:** 0.5 mm across; almost spherical; whitish when laid, but soon becoming black. **Larva:** up to 35 mm long; body whitish, shiny and partly translucent, the dark gut clearly visible; pinacula ochreous; head and prothoracic plate light brown. **Pupa:** 20 mm long; reddish brown; abdominal segments with ventral projections and dentate dorsal ridges.

LIFE HISTORY

Adults fly at dusk in May and June, and occasionally in August or September, often skimming over grassland in considerable numbers. Up to 300 eggs are broadcast at random by each female, whilst in flight. Larvae feed in the soil on roots of grasses and many other plants,

404 Garden swift moth (*Hepialus lupulinus*) – female.

405 Garden swift moth (*Hepialus lupulinus*) – male.

406 Garden swift moth (*Hepialus lupulinus*) – larva.

407 Garden swift moth (*Hepialus lupulinus*) – pupa.

including those with fleshy or woody underground systems. They inhabit narrow feeding burrows within the soil, down which they retreat if disturbed; they are, however, often unearthed during soil cultivation. The larvae feed throughout the winter, and usually pupate in April. Occasionally, however, development extends over another year. Pupation occurs in a loosely woven silken tunnel among the root system of the host. About 6 weeks later, pupae wriggle to the soil surface where they remain protruding after emergence of the adults. Flushes of newly emerged moths tend to appear after rainfall.

DAMAGE
Strawberry plants are frequently attacked, particularly if planted in recently ploughed grassland or pasture. Large tunnels are bored into the crowns, retarding growth and often causing plants to wilt; badly damaged plants may be killed. Most serious damage is caused in autumn, winter and early spring. If unchecked, infestations in established strawberry plantations may persist and increase in importance from year to year. As with ghost swift moth (*Hepialus humuli*), subterranean parts of other soft-fruit crops and hop plants may also be attacked, the larvae of both species sometimes occurring together.

Family **NEPTICULIDAE**

Very small moths, with small 'eye-caps' (scales at the base of each antenna, forming a hood-like structure that partly covers the eye); forewings usually dark (often with metallic markings) and the venation reduced. Larvae are unicolourous, semitransparent and almost apodous, the thoracic legs being reduced to short, extendible lobes and the abdominal prolegs to fleshy humps without hooked (crocheted) terminal pads; the head is wedge-shaped. The larvae feed in leaves, forming sinuous mines or blotches. Pupation usually occurs outside the mine in a small, flattened, parchment-like cocoon.

Stigmella aurella (Fabricius) (**408**)
Blackberry pygmy moth
A minor pest of blackberry. Generally abundant in the wild on *Rubus fruticosus*. Widely distributed in Europe, but usually absent from regions with severe winters; also present in North Africa.

DESCRIPTION
Adult: 6–7 mm wingspan; head orange; forewings dark purple (the basal half with a coppery sheen), with a broad, golden-metallic, almost median, crossband; hindwings dark grey. **Larva:** up to 6 mm long; body yellowish, with a greenish-brown gut visible; head yellowish brown. **Leaf mine:** a long, winding and gradually widening gallery, visible from above.

LIFE HISTORY
Adults occur in May, with a second generation appearing in August. Eggs are laid singly, usually one per leaf. Larvae feed in June and July, and those of the

408 Blackberry pygmy moth (*Stigmella aurella*) – leaf mine on blackberry.

second brood from September to March or April. The precise emergence times of adults vary and, perhaps owing to confusion with its close relatives or biological races, breeding often appears to be continuous. Pupation occurs amongst debris on the ground, in a greenish or pale ochreous cocoon.

DAMAGE

Mines are relatively large and may cause some leaf discoloration, but attacks are not of economic importance.

Stigmella fragariella (Heyden) (**409–410**)

A minor pest of loganberry and raspberry; at least in mainland Europe, infestations are also reported on cultivated strawberry. Wild hosts include *Agrimonia eupatoria*, *Fragaria vesca* and *Rubus idaeus*. Elongate mines are formed on the leaves in July and again in the autumn. Adults are essentially similar in appearance to those of blackberry pygmy moth (*Stigmella aurella*), but the head is black; also, the forewings are basally metallic green and have a paler crossband. Unlike blackberry pygmy moth (of which, according to some authorities, *S. fragariella* may be merely a biological race), the winter is passed in the pupal stage.

Stigmella malella (Stainton) (**411–412**)
Apple pygmy moth

A locally common pest of apple. Widely distributed in Europe.

DESCRIPTION

Adult: 4–5 mm wingspan; head dark ochreous; forewings almost black, each with a narrow, shiny, whitish crossband beyond the middle; hindwings grey. **Larva:** up to 4 mm long; body pale yellow; gut reddish brown; head brown. **Leaf mine:** commencing as a narrow, sinuous gallery on the upper surface, which soon becomes distinctly broadened; the dark, reddish-brown or black frass forms a more or less complete line.

LIFE HISTORY

Eggs are laid in May, and again in August, on the underside of apple leaves, usually close to a vein. First-generation larvae feed in June and July, and those of the second generation in September and October. When fully fed, each larva escapes through the upper leaf surface and pupates in a yellowish-orange cocoon constructed at the base of a leaf or amongst other shelter on the tree; the cocoons are sometimes formed on the ground.

409 *Stigmella fragariella* – leaf mine on loganberry.

411a, b Apple pygmy moth (*Stigmella malella*) – leaf mines.

410 *Stigmella fragariella* – larva.

411a, b Apple pygmy moth (*Stigmella malella*) – leaf mines.

DAMAGE
Several mines may occur in a leaf, but they cause little or no distortion. Attacks are rarely important on established trees but, if mines are numerous, they can be detrimental on young trees and nursery stock.

Stigmella plagicolella Stainton (**413**)
Although associated mainly with wild *Prunus*, notably *P. spinosa*, mines are found occasionally on cultivated plum, including damson. Present throughout Europe.

DESCRIPTION
Adult: 4–5 mm wingspan; head orange; forewings brownish, with a purplish or purplish-bronze sheen and a silvery-white to gold-tinted crossband; hindwings grey. **Larva:** up to 4 mm long; body whitish yellow to yellow and shiny; head brown. **Leaf mine:** commencing as a very narrow, contorted gallery that abruptly changes to a broad blotch in which the frass is distributed more or less centrally.

LIFE HISTORY
Eggs are laid on the underside of leaves, usually close to the midrib. Adults occur in May and June, with those of a second generation appearing in August. Larvae mine the leaves in July and from September onwards, eventually pupating in pale, yellowish-brown cocoons. Occasionally, the initial narrow galleries of two or more larvae may become united into a common blotch. Fully-grown larvae overwinter in cocoons formed on the ground; they pupate in the spring.

DAMAGE
As for apple pygmy moth (*Stigmella malella*).

Stigmella pomella (Vaughan) (**414–415**)
A minor pest of apple. Widely distributed in Europe.

DESCRIPTION
Adult: 5–6 mm wingspan; head orange yellow; eye-caps whitish; forewings entirely dark greyish brown, with a slight purplish sheen; hindwings grey. **Larva:** up to 5 mm long; body orange yellow; head light brown. **Leaf mine:** commencing as a narrow, contorted gallery, but soon developing into a semi-linear, brownish blotch containing irregular lines of frass.

LIFE HISTORY
Adults occur in two generations, from early May to mid-June and from late July to early September. Eggs

412 Apple pygmy moth (*Stigmella malella*) – puparium.

414 *Stigmella pomella* – adult.

413 *Stigmella plagicolella* – leaf mine on plum.

415 *Stigmella pomella* – leaf mine on apple.

are laid mainly on the underside of leaves, close to a major vein, and hatch 2–4 weeks later. Larvae feed mainly in June and July, and from late September onwards, those of the autumn brood being more numerous. The winter is passed in the pupal stage.

DAMAGE
As for apple pygmy moth (*Stigmella malella*).

Stigmella pyri (Glitz) (**416**)
Pear pygmy moth
A pest of pear in various parts of mainland Europe, including France, Germany, the Netherlands and Switzerland; also associated with plum and other kinds of *Prunus*.

DESCRIPTION
Adult: 4–5 mm wingspan; head reddish ochreous; forewings shiny golden brown, tinged apically with purple; hindwings grey. **Larva:** up to 4 mm long; body bluish green and translucent; gut reddish; head brownish to virtually colourless. **Leaf mine:** a narrow, tightly coiled gallery, becoming greatly broadened and superficially blotch-like, visible from above; dense black frass clearly visible.

LIFE HISTORY
First-generation adults occur from late May to early July and those of the second from late July to early September. Eggs are laid mainly on the upper side of leaves and hatch about 3 weeks later. Larvae mine within the leaves from late June to the end of July, with those of the second brood feeding from late August to the end of October. Pupation occurs in reddish-brown cocoons formed on the ground. The winter is passed in the pupal stage.

DAMAGE
Attacks are rarely important on established trees. However, if mines are numerous, infestations can be harmful to young trees and nursery stock.

Family **TISCHERIIDAE**

A small family of very small moths, with narrow, pointed wings; head with forward-projecting scales. The leaf-mining larvae are very flat bodied, with a flat, wedge-shaped head, broad thoracic segments and the abdomen tapering posteriorly; the thoracic legs are much reduced and abdominal prolegs absent.

Tischeria ekabladella (Bjerkander) (**417–418**)
A minor pest of chestnut, but more frequently found on wild *Castanea sativa*; *Quercus* is also a host. Present throughout Europe.

DESCRIPTION
Adult: 8–11 mm wingspan; forewings deep ochreous yellow, speckled apically with blackish scales; hindwings grey. **Larva:** up to 7 mm long; body pale yellow and translucent, with the darker gut partly visible; head light brown; prothoracic plate brown. **Leaf mine:** a whitish, initially shell-shaped, blotch on the upper surface.

417 *Tischeria ekabladella* – leaf mines on chestnut.

416 Pear pygmy moth (*Stigmella pyri*) – leaf mines.

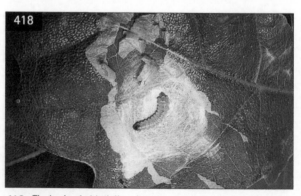

418 *Tischeria ekabladella* – larva.

LIFE HISTORY

Adults occur in June, depositing eggs on the leaves of host trees. The eggs hatch in the early autumn and the larvae feed within the leaves from about September to November. Unlike many leaf-mining insect larvae, a slit is formed at the edge of the mine through which frass is ejected. When fully fed, each larva forms a circular chamber in the middle of its mine, within which it overwinters and, eventually, pupates.

DAMAGE

Although several mines may be present in a single leaf, they do not cause distortion and adverse effects on hosts are minimal, especially as attacks occur rather late in the season.

Tischeria marginea (Haworth) (**419–422**)

Larvae mine within the leaves of wild *Rubus*; infestations are also found, occasionally, on cultivated blackberry and raspberry. Widely distributed in central and south-eastern Europe.

DESCRIPTION

Adult: 7–8 mm wingspan; forewings brownish yellow, the leading edge (costal margin) dark purplish; hindwings dark grey. **Larva:** up to 6 mm long; body yellowish green; head, prothoracic plate and anal plate greyish brown to black; young larva translucent, with the greenish gut contents clearly visible, and with a black head. **Pupa:** 5 mm long; brownish black. **Leaf mine:** an elongate, somewhat wrinkled, light brown blotch on the upper surface.

LIFE HISTORY

Adults occur in May and June, and in July or August. Eggs are laid singly on the upper surface of leaves. Larvae of the first brood occur from June to July, and those of the second from September to March. Pupation takes place within the mine and the pupa protrudes from the leaf surface on the emergence of the adult.

DAMAGE

The larval mines cause slight distortion of leaves, but are otherwise harmless.

419 *Tischeria marginea* – adult.

420 *Tischeria marginea* – larva.

421 *Tischeria marginea* – pupa.

422 *Tischeria marginea* – leaf mine on blackberry.

Family **INCURVARIIDAE**

Small, mainly day-flying, metallic-looking moths, with well-developed, often very long, antennae. Larvae are frequently leaf miners when young and some are persistent shoot borers; the abdominal prolegs have rudimentary crochets arranged in a single transverse band.

Lampronia capitella (Clerck)

larva = currant shoot borer
A local and generally uncommon pest of currant and gooseberry, particularly red currant and white currant; infestations are most frequently found in gardens. Eurasiatic. Widely distributed in central and northern Europe.

DESCRIPTION
Adult: 14–18 mm wingspan; head ochreous yellow; forewings rather shiny, dark purplish brown, each with three more or less triangular, creamy-white markings and white terminal cilia; hindwings light purplish grey. **Egg:** 0.7 mm long; lemon-shaped and almost colourless. **Larva:** up to 11 mm long; body dull olive green; head and prothoracic plate black; young larva white, but reddish when hibernating. **Pupa:** 8 mm long; brown.

LIFE HISTORY
Adults appear in late May or June and frequently bask in sunshine. Eggs are laid in currant or gooseberry fruitlets (often more than one per fruitlet) through a slit made by the female ovipositor. Larvae feed on the developing seeds from June to July. However, when still about 2 mm long, they vacate the fruits and spin cocoons low down on the bushes beneath dead bud scales or in bark crevices. The larvae reappear in the early spring and then invade the new buds or shoots. Individual larvae may move from shoot to shoot, attacking several buds, or they may burrow into the pith of a shoot or twig. Pupation takes place in April or May, usually in a flimsy cocoon formed within a mined shoot. The empty pupa remains protruding from the shoot after emergence of the adult moth.

DAMAGE
Young larvae cause fruits to ripen prematurely, their presence being indicated by the accumulation of frass which is extruded through the tiny entry hole. Fruit losses are of little or no importance. However, in spring, the larvae may kill buds and this can lead to significant crop losses, as can removal of infested buds by chaffinches, sparrows and other insectivorous birds. Mined shoots may wilt and die, and damage to heavily infested bushes can be considerable.

Lampronia rubiella (Bjerkander) (**423–424**)

Raspberry moth
larva = raspberry shoot borer
An occasionally serious pest of loganberry and raspberry, particularly in northerly districts. Widely distributed and locally common in Europe and Asia Minor. Also now established in Canada.

DESCRIPTION
Adult: 9–12 mm wingspan; head yellow to yellowish brown; forewings shiny, dark golden brown, with numerous creamy-yellow spots of various sizes; hindwings purplish grey. **Egg:** 0.35 mm long; oval, colourless and almost transparent. **Larva:** up to 8 mm long; body bright red; head, prothoracic plate and anal plate black; body of young larva, before hibernation, more or less colourless. **Pupa:** 5–7 mm long; brown.

423 Raspberry moth (*Lampronia rubiella*).

424 Raspberry moth (*Lampronia rubiella*) – bud damage.

LIFE HISTORY

Adults are active both at night and in sunny weather during May and June, although somewhat later in more northerly areas. Eggs are laid in tissue at the base of the stamens or petals, usually one per flower. Immediately after egg hatch the larvae feed on the surface of the developing fruitlets, but soon tunnel into the receptacles. When about 3 weeks old, and almost 4 mm long, they vacate the fruit to hibernate in the soil near the roots of host plants. The larvae reappear in spring, at bud burst, and crawl up the canes to continue feeding. Each bores into a bud or young shoot, making an obvious entry hole at the base, and then consumes the contents. In some cases, an individual completes its development in a single bud or shoot, eventually spinning up and pupating within it. Other larvae may attack several buds or shoots; these wandering larvae eventually pupate in white, silken cocoons spun on leaves, canes, supporting posts or other structures. Adults emerge about 3 weeks later.

DAMAGE

Young larvae do little or no harm to the developing fruits and attacks are rarely noticed at this stage. In spring, however, infestations cause withering or death of buds and young shoots, especially towards the tips of canes; if damaged portions are broken open, the larvae or pupae may be found inside. Although infested primary buds can be replaced, if secondary ones are also attacked there is no replacement and re-growth cannot occur. Heavy infestations seriously reduce fruit yields.

Family **HELIOZELIDAE**

A small family of very small, often metallic-scaled, day-flying moths. Larvae are virtually apodous, with a cylindrical, but posteriorly tapered, body. They mine within the leaves or petioles of host plants.

Antispila rivillei Stainton (**425–426**)

A southerly distributed pest of grapevine. Widely distributed in the Mediterranean basin, its range including parts of southern Europe, Asia and the Near East.

425a, b *Antispila rivillei* – adult.

426 *Antispila rivillei* – vacated leaf mines on grapevine.

DESCRIPTION

Adult: 6 mm wingspan; forewings metallic black, with a purplish sheen, and with golden-metallic markings forming a subbasal crossband, two triangular wedges and a large subterminal spot. **Egg:** 0.1 mm long; oval and whitish. **Larva:** up to 4 mm long; body yellowish and translucent; head brownish. **Leaf mine:** a short linear gallery that terminates in a broad blotch that is then excised to form a portable case.

LIFE HISTORY

Adults occur in the spring, and eggs are then deposited on the upper side of leaves, often several on a leaf. Larvae mine within the leaves, forming galleries up to 20 mm in length. When fully grown each larva cuts out a small, oval section of the upper and lower epidermis at the end of the mine, to form a flat, portable case. The encased larva then crawls away and usually settles on the bark of the vine, where it eventually pupates in a strong, silk-lined cocoon formed within the case. The by now light-brown cases are often likened to grains of rice. There are three generations annually, individuals of the final brood overwintering in the pupal stage.

DAMAGE

The mines formed by this species may be very numerous, which can result in extensive loss of leaf tissue.

Family **COSSIDAE**

Large or very large, stout-bodied, relatively primitive moths that usually lack a proboscis; antennae usually pectinate in male, simple in female. Larvae are wood-borers in tree trunks or branches. They are stout-bodied, with a well-developed prothoracic plate and large mandibles; crochets on the abdominal prolegs are biordinal or triordinal, arranged in a complete circle.

Cossus cossus (Linnaeus) (**427–428**)
Goat moth

A generally uncommon pest of broadleaved trees; the host range includes, for example, apple, pear, cherry, plum, chestnut, walnut, citrus and olive, but attacks are found most often in *Betula*, *Fraxinus excelsior*, *Salix* and *Ulmus*. Eurasiatic. Widely distributed in Europe.

DESCRIPTION

Adult: 70–100 mm wingspan; body and wings dull greyish brown, partly suffused with whitish grey and marked irregularly with black. **Egg:** 1.7 mm long; oval and reddish brown. **Larva:** up to 100 mm long; body pink when young, later with yellowish sides and a dark red

427 Goat moth (*Cossus cossus*).

428 Goat moth (*Cossus cossus*) – damage at base of host tree.

back; head and prothoracic plate black. **Pupa:** 50–60 mm long; dark brown, with double transverse rows of spines on the abdominal segments.

LIFE HISTORY
The moths fly at night in June or July, resting during the daytime on tree trunks, fence posts and similar situations. Eggs are laid on tree trunks in batches of up to 50, a female laying several hundred eggs during her lifetime. After eggs hatch, the larvae bore into the tree bark and down to the sap wood. Development usually takes 3 or 4 years, the larvae feeding on both heart and sap wood for much of the time, but hibernating in their galleries during the winter months. Fully grown larvae pupate just beneath the bark in a very strong silken cocoon incorporating chips of wood. Some larvae crawl away from the host tree and spin cocoons in mounds of soil; pupae have sometimes been found in ants' nests.

DAMAGE
Attacks in orchards are uncommon and are usually confined to one or two trees, which are eventually killed; cherry is most often attacked. Feeding galleries are very extensive, and the tunnels are considerably larger than those formed by larvae of leopard moth (*Zeuzera pyrina*), the larvae burrowing in all directions in both sap and heart wood. Also, goat moth larvae normally confine their attacks to the trunks of mature trees, rarely infesting branches or shoots. Sap frequently exudes from the bark of host trees, a useful (although not diagnostic) clue to the presence of the pest. The larvae emit an unpleasant goat-like smell that is easily detected if the galleries are opened up.

Zeuzera pyrina (Linnaeus) (**429–431**)
Leopard moth
A pest of a wide range of trees and shrubs, including apple, pear, quince, cherry, plum, black currant, grapevine and walnut. also, in southern Europe, an important pest of citrus and olive trees. Eurasiatic. Widely distributed in Europe, but less common in more northerly regions. An introduced pest in North America.

DESCRIPTION
Adult: 45–65 mm wingspan: wings white and translucent, with black or blue-black spots; body similarly coloured; male considerably smaller than female and with antennae strongly pectinate basally. **Egg:** 1 mm long; oval and pinkish orange. **Larva:** up to 60 mm long; body yellow, with large, black pinacula; head, prothoracic plate and anal plate brownish black; young larva pinkish, the prothoracic plate with a frilly hind margin. **Pupa:** 25–35 mm long; reddish brown.

LIFE HISTORY
Eggs are laid in June or July in wounds or cracks in the bark, including adult exit holes; each female deposits several hundred eggs, usually in batches. The newly emerged larvae disperse rapidly from the egg mass and each bores into the tree to begin feeding. Very young larvae may attack leaf petioles and main veins, buds and shoots, but later they feed in the larger twigs and

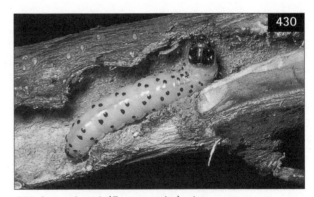

430 Leopard moth (*Zeuzera pyrina*) – larva.

429 Leopard moth (*Zeuzera pyrina*) – male.

431 Leopard moth (*Zeuzera pyrina*) – pupa.

branches. They tunnel up the heart wood, forming a frass-filled gallery 40 cm or more long. Development is completed in 2 or 3 years, the larva pupating within its feeding gallery in a silken cocoon into which particles of wood are incorporated. In early summer, the pupa wriggles out of the cocoon to the surface of the branch where it remains protruding after emergence of the adult.

DAMAGE
Young trees up to 15 years of age are most frequently attacked, and the larvae usually occur in stems or branches less than 10 cm across. Their presence is indicated initially by the accumulation of frass and particles of wood that are forced out of the entry holes, and later by the withering of leaves and die-back of shoots. Infested branches may die and if a young main stem is invaded the whole tree can be killed. Attacks in more northerly areas are heaviest in years following hot summers favourable for egg-laying.

Family ZYGAENIDAE

A family of small to medium-sized, often brightly-coloured, day-flying moths, with either bipectinate or clubbed antennae. Larvae have a small, retractile head and are often extremely colourful, with clusters of body hairs arising from large verrucae.

Aglaope infausta Linnaeus (432–435)
Almond leaf skeletonizer moth
A southerly-distributed and important pest of almond, especially in Portugal and Spain; also a locally significant pest on fruit trees, including apple, pear, quince, apricot, cherry, peach and plum. Widely distributed in southern Europe.

DESCRIPTION
Adult: 20 mm wingspan; head and thorax black, the latter with a red crossband anteriorly; forewings mainly greyish black, suffused with red basally, especially along the wing margins; hindwings greyish black, extensively suffused with red, especially on the apical half; antennae bipectinate in both sexes. **Egg:** 0.7 × 0.4 mm; pale

432 Almond leaf skeletonizer moth (*Aglaope infausta*) – female.

433 Almond leaf skeletonizer moth (*Aglaope infausta*) – male.

yellow. **Larva:** up to 15 mm long; body mainly purple, black and yellow dorsally, creamy white ventrally, with clusters of hairs arising from distinct verrucae; head small and inconspicuous; prothoracic plate relatively large and fleshy, mainly black, marked with purple and yellow. **Pupa:** 10 mm long; light brown.

LIFE HISTORY

Adults occur, and eggs are laid, in July. Larvae then feed for a short time before spinning silken, scale-like hibernacula on the bark of host trees. They then remain in diapause from mid-summer onwards and do not recommence feeding until the following spring. In wetter years, however, when host trees are able to produce late new growth, feeding will be protracted and the larvae will not form their hibernacula until late summer. In spring, larvae again become active. They feed fully exposed on the leaves, either drop to the ground or remain temporarily suspended on a thread of silk when disturbed. Larvae complete their development in about 3 weeks. They then pupate, each in a broad (8 × 4 mm), pinkish or yellowish-white cocoon spun either singly or in batches on the foliage and developing fruits, or amongst withered leaves. Adults appear about a month later.

DAMAGE

Feeding by overwintered larvae in the spring has an adverse effect on the development of host trees, with leaves grazed extensively; the foliage of almond is often reduced to a skeletal collection of midribs. In wetter summers, damage caused by young larvae can be particularly severe and can lead to a reduction in cropping.

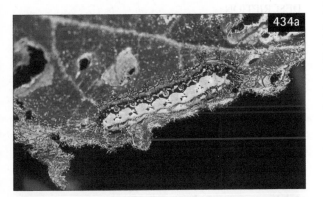

434a, b Almond leaf skeletonizer moth (*Aglaope infausta*) – larva.

435 Almond leaf skeletonizer moth (*Aglaope infausta*) – pupal cocoons.

Theresimima ampelophaga (Bayle-Barelle)
Vine leaf skeletonizer moth

A local pest of grapevine. Widely distributed in the Mediterranean basin, Hungary and the former Yugoslavia; also found further east through the Crimea to the Caspian Sea.

DESCRIPTION

Adult: 20–25 mm wingspan; wings greyish brownish; antennae black to bluish-metallic, very long and bipectinate in both sexes; head and thorax bluish-metallic. **Larva:** up to 16 mm long; body partly yellowish white, yellowish grey, brownish red and blackish, with distinct verrucae bearing clumps of light brown or pale hairs.

LIFE HISTORY

In regions where there are two generations annually this species overwinters in the pupal stage and adults of the first generation appear in April or May. Eggs are then laid in groups on the underside of leaves. The eggs hatch within about 2 weeks. Larvae feed for a few weeks and then pupate, each in a silken cocoon formed on the bark of the foodplant. Second-generation adults appear in August. Larvae of the second brood complete their development in the autumn and then pupate. In the northern parts of its range (e.g. in Hungary), where there is just one generation annually, the winter is passed as second- or third-instar larvae and adults occur from June to August.

DAMAGE

Larvae feeding on the leaves cause significant defoliation, which can reduce crop yields.

Family **PSYCHIDAE**

A family of case-inhabiting larvae (often called 'bagworms'). Pupation takes place within the larval case. Adults lack a proboscis and are short-lived. Females are wingless and usually remain within their larval/pupal cases throughout their lives. Males are active, day-flying moths, with strongly bipectinate antennae.

Pachythelia unicolor (Hufnagel) (**436**)
Associated mainly with grasses, although the larvae will also browse on the leaves of strawberry, grapevine and other cultivated plants. Widely distributed in mainland Europe.

DESCRIPTION
Adult male: 22–25 mm wingspan; wings expansive, mainly smoky black. **Larva:** up to 35 mm long; greyish brown to brownish black, with conspicuous creamy-white markings on the head and thoracic segments. **Case:** up to 40 mm long; composed of pieces of grass and other vegetative material.

LIFE HISTORY
Larvae of this case-inhabiting species feed mainly on grasses from late summer onwards, completing their development in the following spring. The case, within which pupation takes place, is then attached firmly to a suitable support such as a stake or fence or tree trunk. Adults occur in June and July.

DAMAGE
In dry years, when grass growth is poor, the larvae may cause noticeable damage to the leaves of cultivated plants, particularly grapevine.

Family **LYONETIIDAE**

A family of small moths, with narrow, elongate wings and long hair fringes. Larvae are mainly leaf miners, with three pairs of thoracic legs and a complete circle of crochets on each abdominal proleg.

Bucculatrix crataegi Zeller (**437**)
A minor pest of apple and pear; at least in mainland Europe, plum is also attacked. Wild hosts include *Crataegus monogyna*, *Sorbus aucuparia* and *S. torminalis*. Widely distributed in Europe; also present in North Africa and parts of Asia.

DESCRIPTION
Adult: 7–9 mm wingspan; head with a scale tuft between the antennae; forewings whitish, patterned by yellowish, blackish-tipped scales; hindwings grey. **Larva:** up to 8 mm long; body dull greyish green, with a yellowish-white spiracular line and yellowish-white pinacula; early (leaf-mining) larva pale greenish yellow, with a darker head and prothoracic plate; thoracic legs and five pairs of abdominal prolegs fully developed in all instars.

LIFE HISTORY
Adults occur in May and early June. Eggs are laid singly on the upper side of leaves, usually close to the major veins. Young larvae inhabit short, irregular mines; older larvae feed externally, grazing away small patches of the upper epidermal leaf tissue. Larval development is usually completed by the end of August. Individuals then pupate in cocoons formed on the foodplant or amongst debris on the ground. In northern Europe, adults usually emerge in the following year. However, if conditions are favourable, there may be a partial second

436 *Pachythelia unicolor* – case.

437 *Bucculatrix crataegi* – damage to apple leaf.

generation in September. In warmer parts of Europe, including Mediterranean regions, this pest is typically double brooded.

DAMAGE

Larval mining and 'windowing' of foliage is of little or no economic importance; the 'windows' may subsequently split and develop into a series of holes.

Leucoptera malifoliella (Costa) (**438–439**)
Pear leaf blister moth
An often common, but local, pest of apple and pear. Wild hosts include *Crataegus monogyna*, *Malus sylvestris*, *Prunus spinosa* and *Sorbus aucuparia*. Eurasiatic. Widely distributed in Europe.

DESCRIPTION
Adult: 6–8 mm wingspan; forewings shiny metallic grey, marked with orange, white and black, each with an apical black-edged, purplish-gold spot; hindwings lead grey. **Egg:** 0.3 mm across; discoid and brownish. **Larva:** up to 4 mm long; light green; plump and tapering slightly posteriorly; five pairs of abdominal prolegs; lateral tubercles present on the last thoracic and the first two abdominal segments. **Pupa:** 3 mm long; dark brown. **Leaf mine:** a more or less circular brownish blotch, with darker spiral markings.

LIFE HISTORY
Adults occur from late April or May to June or July. Eggs are laid singly on the underside of leaves, sometimes more than one per leaf. Each larva feeds in a blotch formed on the upper side of the leaf. In August or September, the fully grown larvae vacate their mines; each then spins a white, boat-shaped cocoon in a crevice on the bark or amongst rubbish on the ground, and eventually pupates. There is one generation annually.

DAMAGE
Attacked leaves sometimes contain several mines which are at first brown, but later become purplish brown, then black. Infested leaves are not distorted, but they may drop prematurely. Established trees are rarely harmed, but attacks on young trees or nursery stock may be of some importance.

Lyonetia clerkella (Linnaeus) (**440–445**)
larva = apple leaf miner
Associated mainly with rosaceous trees and shrubs, and often a common pest of apple and cherry; at least in mainland Europe, also a pest of pear, quince, peach and plum. Palaearctic. Also present in Madagascar.

DESCRIPTION
Adult: 8–9 mm wingspan; forewings shiny white (but often partly or entirely suffused with brown), marked apically with a darker spot and by several black streaks which extend through the fringe of cilia; hindwings dark

438

439

438 Pear leaf blister moth (*Leucoptera malifoliella*) – leaf mines on apple.

439 Pear leaf blister moth (*Leucoptera malifoliella*) – larva.

440

440 *Lyonetia clerkella* – adult.

grey. **Larva:** up to 8 mm long; body green, translucent and moniliform; head and legs brown. **Pupa:** 3.5 mm long; light green, with yellowish-brown wing cases. **Leaf mine:** a very long gallery, widening gradually throughout its length and ending in a relatively long final chamber; pale coloured, with a central line of dark frass.

LIFE HISTORY

Adults hibernate in suitable shelter under loose bark, amongst thatch and in out-buildings, reappearing in April. Eggs are usually laid singly, each in a small hole made by the ovipositor in the underside of a leaf. They hatch in about 2 weeks and each larva then commences to mine towards the upper surface. Feeding is completed in 3–4 weeks. The larvae then bite their way out of their galleries and wander about on the foliage for a few hours before spinning flimsy, hammock-like cocoons about 6–7 mm long. These are suspended by strands of silk attached to leaves or to rough bark. Pupation then takes place, the adult moths appearing about 2 weeks later. There are usually three generations annually, adults occurring in June and in August, and from October to April.

DAMAGE

Tissue surrounding or isolated by the mines becomes discoloured; if several mines are present, an infested leaf may eventually shrivel and die.

441 *Lyonetia clerkella* – larva.

442 *Lyonetia clerkella* – leaf mines on apple.

443 *Lyonetia clerkella* – leaf mines on cherry.

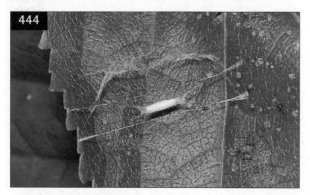

444 *Lyonetia clerkella* – pupal cocoon.

445 *Lyonetia clerkella* – pupal cocoons on bark of cherry tree.

Lyonetia prunifoliella (Hübner)

Associated in mainland Europe with various rosaceous trees including, occasionally, apple, pear, quince, peach and plum; *Betula* is also a host. Palaearctic. Often common, particularly in the warmer parts of Europe.

DESCRIPTION

Adult: 9–10 mm wingspan; head and thorax usually white; forewings mainly brown to reddish brown, marked with numerous white wedges and strigulae; hindwings brownish. **Larva:** up to 9 mm long; light green. **Leaf mine:** initially, a very narrow gallery that then leads into a broad blotch, often located at the leaf margin. The gallery and blotch are sometimes quite separated.

LIFE HISTORY

Adults emerge in the autumn and then overwinter. In spring, they again become active, depositing eggs on the underside of the leaves of host plants. Larvae mine within the leaves during the summer before eventually pupating in a white, hammock-like cocoon, usually slung beneath a leaf. At least in some regions, there may be a second brood of larvae in the autumn.

DAMAGE

As for *Lyonetia clerkella*.

Family **GRACILLARIIDAE**

A very large family of small moths, with very narrow wings bearing long hair fringes. Markings on the front (costal) margin of the wing (e.g. strigulae or wedges) or on the hind margin (dorsum) are often characteristic. Adults often rest with the anterior part of the body raised and the wings held roof-like over the body, with their tips touching the substratum. Larvae lack prolegs on abdominal segment 6; crochets on the abdominal prolegs are either scattered or may be arranged as a lateral penellipse surrounding several larger crochets. The larvae are mainly leaf miners and in their young stages are sap-imbibers, with a peculiar, flattened head. Pupation occurs within the larval habitation, and the pupa remains protruding from the mine following the emergence of the adult.

Callisto denticulella (Thunberg) (**446–448**)

An often abundant pest of apple, especially on unsprayed garden trees. Widely distributed in Europe; also present in North America.

446 *Callisto denticulella* – adult.

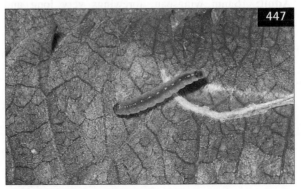

447 *Callisto denticulella* – larva.

448 *Callisto denticulella* – larval habitation on apple leaf.

449 *Caloptilia roscipennella* – adult.

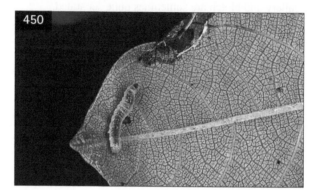

450 *Caloptilia roscipennella* – larva.

451 *Caloptilia roscipennella* – pupal cocoon.

DESCRIPTION
Adult: 10–12 mm wingspan; forewings blackish, with four triangular, creamy-white marks on the costal margin and two on the dorsum; hindwings grey. **Larva:** up to 7 mm long; body ochreous; head black. **Pupa:** 5 mm long; light brown.

LIFE HISTORY
Adults occur in May and June. Eggs are then laid singly on the leaves. Larvae, which feed mainly in July and August, at first mine within the leaves, forming irregular blotches; later, they feed externally in the shelter of a folded leaf edge held tightly down with silk. Pupation occurs in a dense brown cocoon formed amongst rubbish on the ground.

DAMAGE
Leaf edges folded over by this species are very characteristic, but damage caused is of little or no importance.

Caloptilia roscipennella (Hübner) (**449–451**)
larva = walnut leaf miner
A minor pest of walnut in central Europe and in Mediterranean areas.

DESCRIPTION
Adult: 16 mm wingspan; forewings very long and narrow, greyish white to light brownish grey or ochreous to reddish brown, with scattered blackish dots and a somewhat silvery sheen; hair fringes ochreous to black; femur and apical half of tibia of each foreleg and mid-leg with conspicuous patches of projecting black scales. **Larva:** up to 9 mm long; body pale yellowish green and translucent, with the gut contents clearly visible; head pale yellowish green.

LIFE HISTORY
Overwintered adults occur in May, depositing eggs on the leaves of host plants, especially on young walnut trees 2–3 m in height. For their first two instars, larvae mine within the leaves; subsequent instars feed within cone-like habitations, each formed by folding over the tip of a leaflet. Fully grown larvae pupate in elongate, shiny, membranous cocoons, adults emerging about 2 weeks later. A second brood of larvae feed in late summer. Second-generation adults emerge in September and then hibernate, reappearing in the following spring. Moths of the genus *Caloptilia* typically rest with the anterior part of the body raised, supported by the forelegs and mid-legs, and the hind end of the body touching the substratum.

DAMAGE

Larvae cause noticeable damage to foliage, but infestations are of little or no significance.

Metriochroa latifoliella (Millière) (452–454)

Olive leaf miner moth

A minor pest of olive in the Mediterranean basin; also associated with *Phillyrea angustifolia* and *P. media*.

DESCRIPTION

Adult: 8 mm wingspan; body and wings mainly grey, with a violet sheen. **Larva:** up to 6 mm long; body yellow and dorsoventrally flattened; abdomen terminating in a pair of divergent, conical processes; legs poorly developed; head flattened and wedge-shaped. **Pupa:** 4.5 mm long; brown; head with a pointed appendage dorsally; cremaster with a bifid tip. **Mine:** an elongate gallery, up to 6 cm long.

LIFE HISTORY

There are three generations annually, with adult active from late March to late May, from late June to early August, and in late September and October. Eggs are laid on the underside of leaves. Following egg hatch, larvae burrow into the leaves where they form elongate mines. Each mine commences as a linear gallery directed towards the tip of the leaf; it then turns back abruptly and continues towards the petiole, becoming gradually wider throughout its length. Larvae of the spring and summer generation are fully fed in 1–3 months, depending on temperature, but those of the final generation do not complete their development until the following spring. Pupation occurs in a small cocoon spun at the end of the larval feeding gallery, adults emerging 2–3 weeks later.

DAMAGE

Although mines are often numerous, and on an infested leaf may extend throughout most of the lamina, the impact on crop yield or tree growth is rarely significant.

Parornix devoniella (Stainton) (455)

A minor pest of hazelnut, and abundant in woodlands on wild *Corylus avellana*. Present throughout central and northern Europe.

DESCRIPTION

Adult: 9–10 mm wingspan; forewings greyish brown, with numerous lighter costal strigulae and several blackish elongated markings; hindwings grey. **Larva:** up to 6 mm long; body greenish yellow and translucent; head yellowish brown. **Pupa:** 2.0–2.5 mm long; dark brown.

452 Olive leaf miner moth (*Metriochroa latifoliella*).

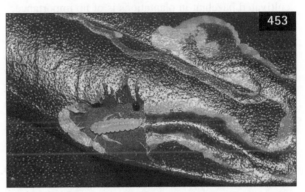

453 Olive leaf miner moth (*Metriochroa latifoliella*) – larva.

454 Olive leaf miner moth (*Metriochroa latifoliella*) – mine.

455 *Parornix devoniella* – larval habitation on hazelnut.

LIFE HISTORY

Adults occur in May and August, with two broods of larvae feeding in July and September. The larvae initially produce small blotches in leaves; later, they feed externally, each folding part of the leaf edge over as a shelter. Pupation also occurs in a folded leaf edge. Pupae of the first generation occur in July or August in folded leaves on host trees; those of the second occur from September onwards, typically in fallen leaves.

DAMAGE

The presence of folded leaf edges immediately indicates the presence of this insect, but damage caused to cultivated hazelnut in plantations is of no importance.

Parornix torquillella (Zeller) (**456**)

In mainland Europe (e.g. the Netherlands) reported as an occasional pest of cultivated plum, but usually associated with wild hosts, especially *Prunus spinosa*. Present and often numerous in much of Europe and Asia Minor, but less common or absent in the north.

DESCRIPTION

Adult: 9–10 mm wingspan; forewings purplish brown to brownish grey, with whitish markings and usually numerous whitish costal strigulae; hindwings grey. **Larva:** up to 6 mm long; body whitish yellow to yellowish green; head yellowish brown to brown; prothoracic plate with four black spots.

LIFE HISTORY

Adults occur from mid-April to mid-June, eggs then being laid on the underside of leaves of host plants. The eggs hatch about 3 weeks later, and the young larvae mine within the leaves. These larvae form small, brownish-white blotch mines on the underside of leaves, often in the apex of a major vein. Later, each larva feeds at the edge of a leaf, in a folded-over section held down tightly with silk. Pupation occurs either in the shelter of a narrowly folded leaf margin or amongst leaf litter on the ground. Second-generation adults occur from mid-June to September, producing larvae that complete their development and then pupate from mid-October onwards. Owing to confusion with the closely related species *Parornix finitimella* (Zeller), some doubt exists concerning the life cycle of *P. torquillella*. At least in some northerly regions, including the British Isles, this species is probably single-brooded.

DAMAGE

Photosynthetic activity of heavily infested leaves is reduced, but minor infestations are of little or no significance.

Phyllonorycter blancardella (Fabricius) (**457–460**)

Apple leaf blister moth

A usually minor pest of apple. Generally common throughout Europe. An accidentally introduced pest in North America.

DESCRIPTION

Adult: 6–9 mm wingspan; forewings dark coppery brown (often marked with blackish scales), with shiny white, dark-edged markings, including four costal and three dorsal strigulae; antennae white, ringed with brown; thorax mainly coppery brown, partly marked with white. **Larva:** up to 5 mm long; body yellow, with the reddish gut visible; head and prothoracic plate light brown; younger larva with a greyish body and a black head. **Pupa:** 4 mm long; dark brown. **Leaf mine:** a blotch on the underside, with a characteristic mosaic pattern in the epidermal tissue visible from above.

LIFE HISTORY

In northern Europe, adults first appear in May. Eggs are then laid singly on the underside of leaves. They hatch 2–3 weeks later, and the young larvae immediately

456 *Parornix torquillella* – larval habitation on plum.

457 Apple leaf blister moth (*Phyllonorycter blancardella*).

burrow into the leaf to begin feeding. At first, the larva (which at the early stage of development has a very large thorax and a large, wedge-shaped head with suctorial mouthparts) feeds by imbibing sap; later, the body becomes more even-shaped and the larva feeds directly on the plant cells, rapidly enlarging the mine. The larva is fully fed in 4–6 weeks and then pupates within the mine. The adult emerges 2–3 weeks later, usually in August, leaving the remains of the pupa protruding from the mine. Larvae of a second brood

feed during September and October. The winter is then spent as pupae in mined, fallen leaves. In warmer regions there are three or four generations annually.

DAMAGE

Strands of silk spun within the mine causes some distortion of the leaf lamina, which is especially pronounced when several blotches occur in the same leaf. When numerous, mines have an adverse effect on the growth of host trees. Particularly on unsprayed trees, mines of *P. blancardella* and *P. corylifoliella* often occur in the same leaf.

Phyllonorycter cerasicolella (Herrich-Schäffer) (461–462)

Cherry leaf blister moth

A minor pest of cherry; in mainland Europe also present on almond, peach and plum. Wild hosts include *Prunus avium* and *P. cerasus*. Widely distributed in Europe.

DESCRIPTION

Adult: 7–8 mm wingspan; forewings orange brown, with white, partly dark-edged, markings. **Larva:** up to

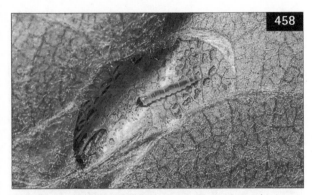

458 Apple leaf blister moth (*Phyllonorycter blancardella*) – larva.

459 Apple leaf blister moth (*Phyllonorycter blancardella*) – mine viewed from above.

461 Cherry leaf blister moth (*Phyllonorycter cerasicolella*).

460 Apple leaf blister moth (*Phyllonorycter blancardella*) – mine viewed from below.

462 Cherry leaf blister moth (*Phyllonorycter cerasicolella*) – mine viewed from below.

5 mm long; body pale yellow, with a dark green gut; abdominal segment 6 with a distinct orange spot; head dark brown. **Leaf mine:** a pale blotch formed between two veins on the underside, often at the base or leaf margin; on cherry and plum, never located against the midrib.

LIFE HISTORY
Eggs are laid on the underside of leaves, often near the base. Larvae then mine singly within the leaves in June or July, each eventually pupating in a tough whitish cocoon formed within the mine. Larvae of a second brood feed from September onwards and eventually pupate in the spring. Adults occur in May and in August.

DAMAGE
Infested leaves are distorted, but adverse effects on hosts are negligible unless mines are numerous.

Phyllonorycter coryli (Nicelli) (**463**)
Nut leaf blister moth
Generally common on wild *Corylus avellana* throughout central and northern Europe, and also sometimes present as a minor pest on cultivated hazelnut.

DESCRIPTION
Adult: 7.0–8.5 mm wingspan; forewings dark brown, with white, dark-edged markings, including four costal and three dorsal strigulae; hindwings grey. **Larva:** up to 5 mm long; body yellowish to greenish, with the green gut visible; abdominal segment 6 with an orange patch; head light brown.

LIFE HISTORY
At least in northern Europe, adults typically occur in May and in August, depositing eggs on the upper side of hazel leaves. Larvae mine within the leaves from June to July and from September to October.

DAMAGE
Although there are often several mines in a leaf, causing considerable distortion, the pest is not of economic importance.

Phyllonorycter corylifoliella (Hübner) (**464–466**)
Associated mainly with rosaceous trees, including *Crataegus monogyna*, *Sorbus aria* and *S. aucuparia*, and both wild and cultivated apple and pear. Often numerous in gardens, but not an important fruit pest. Common and widely distributed in central and northern Europe.

464 *Phyllonorycter corylifoliella* – adult.

465 *Phyllonorycter corylifoliella* – larva.

463 Nut leaf blister moth (*Phyllonorycter coryli*) – leaf mines on hazelnut.

466 *Phyllonorycter corylifoliella* – leaf mine on apple.

DESCRIPTION

Adult: 8–9 mm wingspan; forewings chestnut-brown, partly suffused with black, and with relatively indistinct whitish or light brown markings, including one costal and one or two dorsal strigulae; basal streak distinctly angulated; thorax brown, fringed with white and with a white central line. **Larva:** up to 5 mm long; body pale dirty white to yellowish, shiny and transparent; head blackish brown. **Pupa:** 4 mm long; dark brown. **Leaf mine:** a large, pale-buff blotch on the upper side, characteristically flecked with reddish brown.

LIFE HISTORY

In temperate parts of Europe there are two generations annually. Adults typically occur in May to June and in July or August, with larvae feeding from May or June onwards and from August or September onwards. Unlike apple leaf blister moth (*Phyllonorycter blancardella*), eggs are laid on the upper surface of leaves; also, although second-brood larvae are fully fed in the autumn, pupation may not occur until the following spring.

DAMAGE

As for apple leaf blister moth (*Phyllonorycter blancardella*).

Phyllonorycter leucographella (Zeller) (467–470)

Firethorn leaf blister moth

A Mediterranean species that, in recent years, has extended its range into more northerly parts of Europe, including Austria, Belgium, England, France, Germany, the Netherlands and Switzerland. Although breeding mainly on *Pyracantha coccinea* (and, in Mediterranean areas, on *Calycotoma spinosa*), leaf mines often also occur on apple and cherry; other rosaceous hosts include *Chaenomeles*, *Cotoneaster*, *Crataegus* and *Sorbus*.

DESCRIPTION

Adult: 8–9 mm wingspan; head with a crest of white scales; forewings golden brown, with two white costal strigulae, four narrow, white dorsal strigulae and a long, narrow, white basal streak; hindwings grey. **Larva:** up to 5 mm long; body yellow, with the red gut partly visible; head brown. **Pupa:** 4 mm long; light brown. **Leaf mine:** an expansive silvery blotch on the upper surface.

468 Firethorn leaf blister moth (*Phyllonorycter leucographella*) – larva.

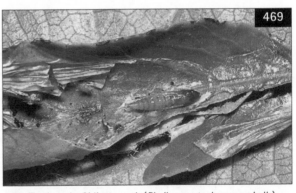

469 Firethorn leaf blister moth (*Phyllonorycter leucographella*) – pupa.

467 Firethorn leaf blister moth (*Phyllonorycter leucographella*).

470 Firethorn leaf blister moth (*Phyllonorycter leucographella*) – leaf mine on cherry.

LIFE HISTORY

Adults occur from May to June and from August to mid-September, eggs being deposited singly on the leaves of various host plants. The larvae mine within the leaves, each forming a large blotch that initially extends along the midrib. Later, the mine broadens and may then extend over much of the upper surface. On *Pyracantha coccinea*, an affected leaf eventually folds upwards longitudinally, to form a protective, pod-like envelope within which larval development (still within the mine) is completed and pupation occurs. Mines on apple and cherry, however, often split open, leading to the death of the occupants before they can pupate. Larvae occur during the summer months and from September onwards. The latter, on evergreen hosts such as *P. coccinea*, continue to feed throughout the winter and eventually pupate in the spring; larvae feeding on deciduous hosts such as apple and cherry, however, are unable to survive the winter as they cannot continue their development in fallen leaves. Larvae in fallen leaves of evergreen hosts also die.

DAMAGE

Mines reduce photosynthetic activity and, being very obvious, are unsightly. On young trees, heavy infestations are likely to have an adverse effect on growth.

Phyllonorycter messaniella (Zeller) (**471**)

Although associated mainly with *Quercus ilex*, mines also occur on leaves of various other trees, including *Castanea sativa* and deciduous *Quercus*; more rarely, mines occur on cultivated fruit trees, including apple, peach and plum. Present throughout central and southern Europe; also, an introduced and harmful pest in Australia and New Zealand.

DESCRIPTION

Adult: 7–9 mm wingspan; forewings golden ochreous and shiny, with narrow, white, dark-edged markings, including four costal and four dorsal stigulae, and a long, narrow, basal streak; hindwings greyish brown. **Larva:** up to 5 mm long; body yellow to whitish yellow; head brown. **Leaf mine:** a pale blotch on the underside, between two major veins, with a strong central fold.

LIFE HISTORY

This species is said to breed more or less continuously, so long as conditions remain favourable. However, details of the life cycle are not fully documented. On *Quercus ilex*, fully fed larvae overwinter in their mines, and then pupate in the spring. At least on peach, larvae also appear to overwinter (in the fallen leaves).

DAMAGE

Leaf mines cause slight distortion, but adverse effects on hosts are negligible unless mines are numerous.

Phyllonorycter spinicolella (Zeller) (**472–474**)

Widely distributed in mainland Europe on cultivated damson and plum, particularly young trees. In the British Isles associated most frequently with *Prunus spinosa*.

472 *Phyllonorycter spinicolella* – adult.

471 *Phyllonorycter messaniella* – leaf mine on peach, viewed from below.

473 *Phyllonorycter spinicolella* – leaf mine on plum, viewed from above.

DESCRIPTION
Adult: 6.5–7.5 mm wingspan; forewings golden orange, with four white costal, three white dorsal strigulae and a white basal streak; most strigulae strongly dark edged inwardly; thorax golden orange, edged with white, with a white central line. **Larva:** up to 5 mm long; body greenish white, with the dark gut visible; abdominal segment 6 with an orange spot; head greenish yellow. **Mine:** an elongate blotch on the underside of a leaf, with the lower surface strongly creased and causing the upper surface to arch.

LIFE HISTORY
Adults of the first generation occur in May, depositing eggs singly on the underside of leaves. Larvae mine within the leaves in July and then pupate, adults of the second generation appearing in August. Larvae of the second brood feed from September onwards, overwintering in their mines and pupating in the spring.

DAMAGE
Mines sometimes cause noticeable distortion of leaves, but their presence on fruit trees is usually of only minor significance.

Family **PHYLLOCNISTIDAE**

Very small, mainly white moths, characterized by the presence of 'eye-caps' (scales at the base of each antenna, forming a hood-like structure that partly covers the eye); forewings narrow and elongate, with long hair fringes. The leaf-mining larvae are apodous, and sap-imbibers throughout their development.

Phyllocnistis citrella (Stainton) (**475–480**)
Citrus leaf miner moth
An important pest of citrus, especially in south-east Asia, its distribution extending southwards to Australia and, in recent years, westwards through Turkey into southern Europe. Also now established in Africa and South America.

DESCRIPTION
Adult: 5–8 mm wingspan; forewings mainly white, partly suffused with yellowish orange, and marked with black striae and a prominent black pre-apical

475 Citrus leaf miner moth (*Phyllocnistis citrella*) – adult.

474 *Phyllonorycter spinicolella* – leaf mine on plum, viewed from below.

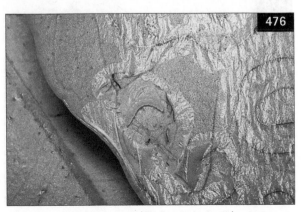

476 Citrus leaf miner moth (*Phyllocnistis citrella*) – larva.

spot. **Larva:** up to 3.5 mm long; body translucent, partly suffused with yellow. **Pupa:** 2.5–3.0 mm long; golden brown; head with a distinctive horn-like projection ('cocoon-piercer'). **Mine:** an elongate, sinuous gallery, on either the upper or the lower surface of a leaf.

LIFE HISTORY

Adults overwinter and reappear in the spring. Eggs are then laid singly (and occasionally in twos or threes) on the expanded leaves of host plants, typically close to the midrib. Eggs may also be deposited on succulent green shoots. Larvae form very long, strongly serpentine mines that usually (although not invariably) terminate at the leaf margins. Pupation occurs at the end of the mine, typically with the leaf margin slightly folded over the pupal chamber. There are several generations annually, and development from egg to adult takes about 6 weeks.

DAMAGE

Infestations are of greatest significance on young trees, when distortion and loss of photosynthetic leaf area can have a detrimental effect on plant growth. Tissue adjacent to the mines may also become necrotic. When mines are present on the underside of leaves the upper epidermis turns partly yellowish, betraying their presence. Heavy infestations can lead to premature leaf fall and death of young growth. Mines formed in young shoots are very superficial and appear as black tracks that terminate in a small rounded chamber.

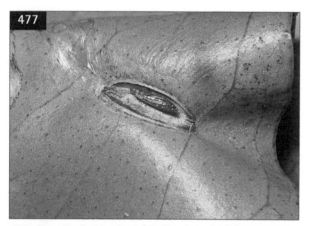

479 Citrus leaf miner moth (*Phyllocnistis citrella*) – pupa.

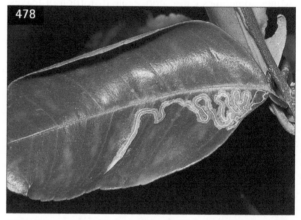

478 Citrus leaf miner moth (*Phyllocnistis citrella*) – fresh leaf mine.

479 Citrus leaf miner moth (*Phyllocnistis citrella*) – vacated leaf mine.

480 Citrus leaf miner moth (*Phyllocnistis citrella*) – vacated mine in young shoot.

Phyllocnistis vitegenella Clemens (**481–482**)
American grapevine leaf miner moth

An American pest of grapevine. Also found in northern Italy, near Vicenza, in 1995, where it is now established.

DESCRIPTION
Adult: 5.5 mm wingspan; forewings brilliant white, marked apically with brown strigulae (that extend through the hair fringes) and a prominent black terminal spot. **Larva:** up to 4 mm long; whitish. **Pupa:** 3 mm long; brown; head with a horn-like projection ('cocoon-piercer'). **Mine:** a very long, contorted gallery on the upper side of a leaf.

LIFE HISTORY
Adults overwinter and reappear in the spring. Eggs are then laid on, and larvae then mine within, the expanded leaves. Pupation occurs in a distinct chamber formed at the end of the feeding gallery, often (but not exclusively) in a slight upwardly directed fold at the leaf margin. Adults of the first generation appear in early June, the remains of the pupa protruding from the mine following emergence. There are several generations (in central Italy four or five) annually.

DAMAGE
The leaf mines are conspicuous, although (apart from curling at the leaf margins, where the mines often terminate) infestations cause little or no distortion. Nevertheless, leaf damage becomes more and more extensive as the season progresses and, in the case of persistent infestations within one and the same leaf, there can be significant loss of photosynthetic activity.

481 *Phyllocnistis vitegenella* – adult.

482 *Phyllocnistis vitegenella* – leaf mines on grapevine.

Family **SESIIDAE**
(clearwing moths)

A distinctive family of unusual, day-flying, wasp-like moths, with much of the wing area between the veins devoid of scales and with the abdomen terminating in a fan-like tuft of scales. Larvae are stem borers in trees and shrubs. Crochets on the abdominal prolegs are uniordinal and typically arranged in two transverse bands, but in a single band on the anal prolegs. The pupa has a heavily sclerotized head, used to cut its way out of the larval gallery prior to the emergence of the adult.

Pennisetia hylaeiformis (Laspeyres)
Raspberry clearwing moth
larva = raspberry crown borer
A local pest of *Rubus*, particularly raspberry. Present in central and north-eastern Europe, including Scandinavia. Absent from the British Isles and the Netherlands.

DESCRIPTION
Adult: 20–25 mm wingspan; wings mainly clear, marked with black; body mainly black, with narrow transverse bands on the abdomen. **Larva:** up to 20 mm long; body creamy white, with the central blood vessel clearly visible dorsally; head reddish brown; prothoracic plate light brown. **Pupa:** 12–15 mm long; yellowish brown.

LIFE HISTORY
Adults are active from June onwards, and often settle on the leaves of host plants to bask in sunshine. Eggs are laid in the soil, close to host plants. They hatch from July onwards, and larvae then invade the root systems, eventually boring into the taproots. The larvae overwinter within their hosts, and recommence activity in the early spring, each larva boring upwards and feeding within the pith of a second-year cane. Larvae are fully grown by about May. They then pupate in situ, and adults emerge a few weeks later.

DAMAGE
Leaves on infested fruiting canes wilt and fruit trusses fail to develop, resulting in crop losses.

Synanthedon myopaeformis (Borkhausen) (**483–485**)
Apple clearwing moth
A local and generally uncommon pest of apple, usually attacking only old and neglected trees; other hosts include pear, almond, peach, cherry and plum. Widely distributed, particularly in central and southern Europe.

DESCRIPTION
Adult: 20–25 mm wingspan; wings mainly clear, with brownish-black veins and borders; body black, with a red crossband on the abdomen. **Larva:** up to 20 mm long; body somewhat flattened dorsoventrally, dull whitish brown, with the dark central blood vessel clearly visible dorsally; abdominal prolegs short; head shiny reddish brown and retractile; young larva whitish, with a light brown head. **Pupa:** 10–13 mm long; light brown, with a dark brown head and cremaster.

483 Apple clearwing moth (*Synanthedon myopaeformis*).

484 Apple clearwing moth (*Synanthedon myopaeformis*) – larva.

485 Apple clearwing moth (*Synanthedon myopaeformis*) – pupa.

LIFE HISTORY

Adults appear in June or July, but sometimes earlier, and are very active in warm, sunny conditions. Eggs are laid in crevices in the trunks of suitable host trees or at the base of the main branches, often close to wounds or areas of existing infestations. Attacks may also be initiated in parts of trees damaged by pests such as cherry bark tortrix moth (*Enarmonia formosana*) (family Tortricidae). As soon as the eggs hatch, the young larvae burrow into the host tree and tunnel within or just beneath the bark, to form irregular, winding galleries. Development lasts for about 20 months, first- and second-year larvae often occurring together. When fully grown, usually in April, May or June of the second year, each larva spins a tough, greyish cocoon and pupates. These cocoons occur close to the surface amongst decayed wood and masses of frass. About 2–4 weeks later, the pupa breaks out of the cocoon and wriggles to the surface of the bark, where it remains protruding from the exit hole after the adult moth has emerged.

DAMAGE

Bark covering infested parts of a tree peels off readily and the presence of the pest encourages invasion by fungal pathogens. Severe attacks lead to softened, blackish patches on the bark, from which masses of sticky sap may exude. However, the pest usually limits its attention to sickly trees and is, therefore, mainly of secondary importance. On suitable host trees, infestations often persist from year to year.

Synanthedon tipuliformis (Clerck) (486–488)

Currant clearwing moth

larva = currant borer

A locally common pest of black currant; red currant and gooseberry are also attacked. Widely distributed in Europe. Accidentally introduced into North America, Australia and New Zealand.

DESCRIPTION

Adult: 17–21 mm wingspan; wings mainly clear, with brownish-black veins and borders; forewings partly orange brown apically; body metallic bluish black or purplish black when freshly emerged, with three (= female) or four (= male) narrow yellow crossbands. **Egg:** 0.6 × 0.4 mm; oval and yellowish white. **Larva:** up to 18 mm long; body creamy white, with the dark central blood vessel clearly visible dorsally; head light brown and retractile. **Pupa:** 10–12 mm long; light brown.

LIFE HISTORY

Adults occur from late May to July. They fly in sunny weather, especially during the early morning. They also bask in sunshine on leaves of the foodplant and visit nearby flowers such as those of *Rubus fruticosus*. Eggs are laid singly on the bark close to a bud or side shoot, usually one per stem. They hatch in about 10 days and

487 Currant clearwing moth (*Synanthedon tipuliformis*) – larva.

486 Currant clearwing moth (*Synanthedon tipuliformis*).

488 Currant clearwing moth (*Synanthedon tipuliformis*) – larval exit hole in branch.

the young larvae immediately bore down to the pith and begin feeding. A larva normally works along the pith towards the younger wood, burrowing in young shoots or in the older stems. Feeding continues throughout the summer and autumn, growth being virtually completed by the onset of winter. In April or May, the larva eventually bites a channel to the surface, but keeps this tunnel capped by a thin window of rind. A tough, silken cocoon, incorporating particles of dark brown frass, is then spun in the hollowed-out pith, within which the larva pupates. Sometimes, pupal chambers are formed near to old exit holes made by members of previous generations. Several weeks later, just before the adult emerges, the pupa breaks out of its cocoon, wriggles to the surface and breaks through the rind. The pupa remains protruding from the exit hole after emergence of the adult, with the empty antennal sheaths characteristically swept backwards.

DAMAGE
Leaves on infested young shoots may wilt and fruit trusses may fail to develop. However, apart from the presence of exit holes, bushes often show little outward sign of attack. Infested shoots and stems have dark brown or blackened piths, and are weakened and easily broken if bent; during spring frost-protection measures, for example, branches on attacked bushes may snap when weighed down with ice. Damage may also predispose bushes to attack by fungal pathogens, such as *Phomopsis* spp. The pest is often overlooked and most likely to be discovered during winter pruning.

Family **CHOREUTIDAE**

A distinctive group of small, day-flying moths, with broad wings that are held flat (i.e. horizontally) over the body in repose. Larvae are usually web-forming, with prominent pinacula and long, pencil-like abdominal prolegs.

Choreutis nemorana (Hübner) (**489–494**)
Fig-tree skeletonizer moth
A generally common, but minor, pest of fig. Widely

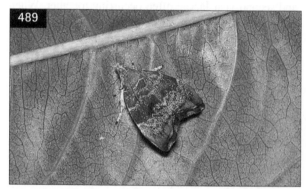

489 Fig-tree skeletonizer moth (*Choreutis nemorana*).

490 Fig-tree skeletonizer moth (*Choreutis nemorana*) – larva.

491 Fig-tree skeletonizer moth (*Choreutis nemorana*) – pupal cocoon.

distributed in the Mediterranean area, including southern Europe and North Africa; also present in parts of Asia, and in the Canary Islands and Madeira.

DESCRIPTION
Adult: 16–20 mm wingspan; forewings mainly reddish brown to ochreous brown, suffused with black and marked extensively with white to grey scales; hindwings brownish, each with a pair of pale spots towards the margin. **Egg:** 0.5 mm across; creamy white

492 Fig-tree skeletonizer moth (*Choreutis nemorana*) – pupa.

493 Fig-tree skeletonizer moth (*Choreutis nemorana*) – larval web.

494 Fig-tree skeletonizer moth (*Choreutis nemorana*) – leaf damage.

and spherical. **Larva:** up to 20 mm long; body light green, shiny and semitransparent, with a pale dorsal line and numerous large, black verrucae; head ochreous yellow and shiny, marked with black. **Pupa:** 7–8 mm long; brown.

LIFE HISTORY
Overwintered adults appear in the early spring, and eventually deposit eggs in groups on the leaves of fig. Larvae feed from mid-May onwards, each protected by a thick web of silken threads. Larvae are fully grown a few weeks later. They then pupate, each in a dense, white, boat-shaped cocoon spun on the leaves or elsewhere on the foodplant. Adults of the summer generation appear in July, and second-brood larvae feed from the end of July to early October. Adults emerge in the autumn and then hibernate.

DAMAGE
Larvae cause noticeable distortion of leaves and also extensive discoloration, scarification and tattering of the laminae.

Choreutis pariana (Clerck) (**495–499**)
Apple leaf skeletonizer moth
larva = apple leaf skeletonizer
Locally common, attacking apple, pear and cherry, but rarely found on sprayed trees. Palaearctic. Widely distributed in Europe. An introduced pest in North America.

DESCRIPTION
Adult: 11–13 mm wingspan; forewings reddish brown or yellowish brown, suffused with greyish, and with dark wavy crosslines; hindwings dark greyish brown. **Larva:** up to 14 mm long; body yellowish to light green, with conspicuous black pinacula on each segment; head yellowish brown; legs relatively long. **Pupa:** 5–6 mm long; brown.

495 Apple leaf skeletonizer moth (*Choreutis pariana*).

LIFE HISTORY

Adults hibernate in thatch, amongst dead leaves, in hedgerows and so on, resuming activity in the spring at about bud burst. Eggs are laid on leaves during April. Larvae then feed from May onwards. At first, they graze the upper surface of leaves close to the midrib; later, they roll the edges of the leaves and feed in the shelter of slight webs. The larvae wriggle violently if disturbed and fall to the ground. Individuals are fully grown in about 3 weeks. Each then pupates in a dense, white, cigar-shaped cocoon about 9 mm long, spun under a leaf, amongst withered leaves or in rubbish on the ground. Adults emerge a few weeks later, usually in July and early August. Larvae of a second brood appear in August. These eventually pupate to produce the overwintering adult generation in September or October.

DAMAGE

Larvae typically remove tissue from the upper surface of foliage, leaving the lower surface intact. Such damage is often conspicuous, but rarely sufficiently extensive to cause major harm. Larvae of the second generation are more important as they may graze the surfaces of fruits, causing distinct russeting.

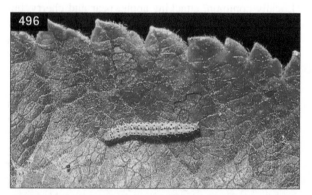

496 Apple leaf skeletonizer moth (*Choreutis pariana*) – larva.

497 Apple leaf skeletonizer moth (*Choreutis pariana*) – larval web.

498 Apple leaf skeletonizer moth (*Choreutis pariana*) – pupal cocoon.

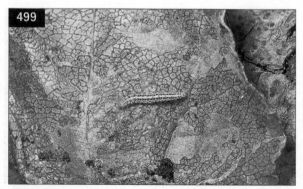

499 Apple leaf skeletonizer moth (*Choreutis pariana*) – larval damage.

Family **YPONOMEUTIDAE**

A variable and diverse family of small to medium-sized moths, with well-developed, projecting maxillary palps; wings long and narrow, held in a steeply sloping, roof-like posture when in repose. Larvae are of various forms; often, the abdominal prolegs have crochets arranged in several concentric circles, as in the genus *Yponomeuta* (subfamily Yponomeutinae), or have a circle of small crochets surrounding a penellipse of larger ones, as in the genus *Ypsolopha* (subfamily Plutellinae). The larvae of many species feed within buds, fruits or shoots; some are gregarious and web-inhabiting.

Argyresthia albistria (Haworth) (**500**)

Although associated mainly with *Prunus insititia* and *P. spinosa* this species also breeds, occasionally, on cultivated damson, myrobalan and plum. Widely distributed in Europe.

DESCRIPTION
Adult: 9–11 mm wingspan; head and thorax brilliant white; forewings mainly brown, with a wide, brilliant-white streak dorsally, interrupted by a brown fascia.

500 *Argyresthia albistria* – adult (left), larva (right).

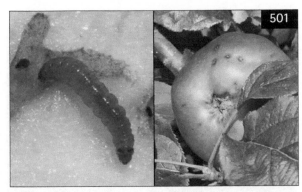

501 Apple fruit moth (*Argyresthia conjugella*) – larva (left), infested apple (right).

Larva: up to 7 mm long; body whitish to light green, and reddish intersegmentally; head dark brown; prothoracic and anal plates olive brown.

LIFE HISTORY
Adults occur from mid-June to September. When fully at rest, in common with other members of the genus *Argyresthia*, the adults remain with the head close to the leaf or other surface (and the antennae held sideways), and the body, wings and hind legs held upwards at an angle of about 45°. Eggs are laid singly on the shoots of host plants, either in small crevices or under flakes of bark; they are also stated to be deposited amongst the galls formed on shoot tips by plum spur mite (*Acalitus phloeocoptes*) (Chapter 8). The eggs hatch in the autumn. Young larvae then invade the buds, feeding superficially on two or three before eventually overwintering. In spring, the larvae mine within the flowering shoots and become fully grown in late April or May. Each then descends to the ground on a strand of silk and pupates on the ground in a flimsy cocoon.

DAMAGE
Infestations in spring can result in the death of many buds, flowers and shoots, particularly on the uppermost parts of host trees.

Argyresthia conjugella Zeller (**501**)
Apple fruit moth
Locally common in association with *Sorbus aucuparia*, and sometimes an important pest of apple. Eurasiatic. Widely distributed in Europe, but absent from Portugal, Spain and the Balkans; also present in North America.

DESCRIPTION
Adult: 10–12 mm wingspan; forewings dark purplish brown, each with whitish markings on the costal margin, a broad white streak towards the base and a dark oblique crossband; hindwings grey. **Egg:** 0.5 × 0.3 mm; oval, flat and whitish. **Larva:** up to 7 mm long; body dull, pale dirty yellow to pinkish; head, prothoracic plate and anal plate light brown. **Pupa:** 5 mm long; brown.

LIFE HISTORY
Moths occur in June. Eggs are laid on the fruitlets, each female depositing about 20–30. They hatch in about 10–14 days and the larvae immediately burrow into the flesh to feed. There are usually several larvae in an infested fruit, and they become fully grown by late September or October. They then vacate the fruit to pupate, either dropping to the ground with the fruit or lowering themselves on silken threads. Each larva spins a silken cocoon (with a flimsy, net-like outer layer and a

dense, white inner layer) and then pupates. Cocoons are usually formed in the soil, at a depth of about 3–5cm, but they may occur amongst rubbish on the surface or in bark crevices on the trunks of host trees.

DAMAGE

Infested fruits are riddled with tunnels and the skin becomes marked with small, discoloured, sunken blotches. Later, during the emergence of the fully fed larvae, the fruit skin is pierced by numerous holes, each 1–2mm in diameter. Damaged fruits soon rot, and heavy crop losses can occur. Attacks by this pest are often overlooked until close to harvest; they are most likely to occur in orchards close to woodlands and in seasons when rowan berries are scarce.

Argyresthia curvella (Linnaeus) (502)

A pest of apple, especially in neglected, unsprayed orchards. Widely distributed in central and northern Europe. Most often reported in the Czech Republic, England, Finland, Germany, northern Italy and Switzerland.

DESCRIPTION

Adult: 10–12 mm wingspan; forewings mainly shiny white, each with bronzy-brown markings apically and,

usually, a similarly coloured, oblique, tapering crossband; hindwings grey. **Larva:** up to 9mm long; body yellowish white; head brown; prothoracic and anal plates brownish.

LIFE HISTORY

Adults occur in June and July. Eggs are laid on the shoots and spurs, where they overwinter. The eggs hatch in April and the larvae then invade the opening flower buds to feed on the ovaries and other floral parts. In May, when fully grown, they descend to the ground. Each spins a cocoon in the soil and then pupates, adults emerging a few weeks later.

DAMAGE

Infested flower buds fail to produce viable blossoms and so do not set fruit. Attacks are most serious on trees with relatively few flowers.

Argyresthia pruniella (Clerck) (503)

Cherry fruit moth

A generally common pest of cherry and, less frequently, plum. Widely distributed in Europe and Asia Minor. An introduced pest in North America.

DESCRIPTION

Adult: 10–12mm wingspan; forewings light brown and each more or less extensively marked with white on the dorsal half, the pale area interrupted by a broad, pale brownish-yellow crossband that terminates well before the wing tip; hindwings grey. **Egg:** 0.5 × 0.3mm; flat, pear-shaped and reticulate; brown to grey or olive green. **Larva:** up to 10mm long; body light green to greenish yellow; head, prothoracic plate and anal plate brown. **Pupa:** 5mm long; brown, with a greenish tinge.

LIFE HISTORY

Adults occur in late June and July. They rest on the foliage and tree trunks, adopting the characteristic pose of other argyresthids, and are readily disturbed during the daytime.

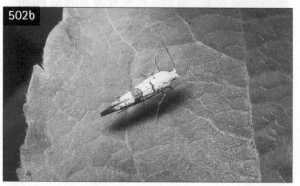

502a, b *Argyresthia curvella* – adult.

503 Cherry fruit moth (*Argyresthia pruniella*).

Eggs are laid on leaf scars, beneath bud scales and in other sheltered situations on the shoots and spurs, mainly about 2–3 m above ground level. The eggs usually do not hatch until the following spring; however, some may hatch earlier, in which case the larvae hibernate beneath the old egg shells in cocoons of silk. Larval feeding commences in the early spring, individuals entering the fruit buds and often attacking them before bud burst. Here, the larvae feed on the developing bud tissue; later, they occur in the flowers, usually hidden within the tunnelled-out ovary or young fruitlet. When fully grown, in mid- to late May, the larvae descend to the ground on silken threads and enter the soil to pupate at a depth of a few centimetres. The pupal cocoons are double-walled, with a dense inner wall and a flimsy, net-like outer one. Adults emerge a few weeks later.

DAMAGE
At first, the stamens and unexpanded petals are eaten; later, the ovaries and developing fruitlets are hollowed out. Crop losses, especially in unsprayed cherry orchards, may be considerable.

Argyresthia spinosella Stainton (**504–506**)
In common with *Argyresthia albistria*, this widely distributed species is also associated mainly with wild *Prunus* (especially *P. spinosa*) and is an occasional, minor pest on damson, myrobalan and plum. Larvae feed within flowering shoots in the spring, causing these to droop and the young leaves to turn black. Unlike *A. albistria*, however, eggs are the overwintering stage and hatch in March. The larvae (up to 7 mm long) are whitish, with the head, prothoracic plate and anal plate black. Adults (9–11 mm wingspan) are mainly white and purplish grey, with the tegulae, sides of head and base of the forewings ochreous. They occur from June to early July.

Euhyponomeutoides albithoracellus
(Tengström) (**507**)
Currant bud moth
A major pest of currant in Finland and northern Sweden; attacks also occur on gooseberry.

DESCRIPTION
Adult: 12–16 mm wingspan; forewings ochreous, with the costal margin broadly white; hindwings brownish grey; head, thorax and antennae brilliant white. **Larva:** up to 15 mm long; body mainly light green; head yellowish green.

LIFE HISTORY
Adults occur from late June to mid-July. Eggs are then laid on the upper surface of expanded leaves, typically

504 *Argyresthia spinosella* – adult.

505 *Argyresthia spinosella* – shoot damage.

506 *Argyresthia spinosella* – larva.

507 Currant bud moth (*Euhyponomeutoides albithoracellus*).

in depressions along the major veins. Young larvae, which form the overwintering stage, become active in the early spring. At first, they feed within the buds. Later, they feed externally on the emerging shoots, each larva spinning silken threads over the damaged leaflets. Fully fed larvae eventually pupate in silken boat-shaped cocoons, and adults emerge a few weeks later.

DAMAGE
In spring, buds invaded by larvae fail to open fully, the partially emerging leaf tips dying and eventually turning black. Young growth is also destroyed.

Prays citri Millière (508)
Citrus flower moth
A pest of citrus, particularly lemon, pomelo and sweet orange. Widely distributed in the Mediterranean area; also present in Asia and Australasia.

DESCRIPTION
Adult: 10–12 mm wingspan; forewings grey, with blackish markings; hindwings greyish brown, with smoky tips. **Egg:** 0.20 × 0.15 mm; oval and opalescent. **Larva:** up to 7 mm long; body whitish to light brown; head and prothoracic plate marked with brown; crochets on abdominal prolegs arranged in several concentric circles. **Pupa:** 5–6 mm long; brown.

LIFE HISTORY
Development of this pest is extremely rapid, allowing the completion of up to 11 overlapping generations annually in the Mediterranean basin (typically, a winter generation, three each in spring and autumn, and four in summer). All stages of the pest may often be found together and, at the height of summer, development from egg to adult can be completed in less than a month. From early spring onwards, eggs are laid singly, or in two or threes, on the flower buds. They hatch shortly afterwards. Larvae then feed singly within the buds, each destroying the innermost tissue and then moving on to attack another bud. Expelled frass and silken webbing are associated with the feeding sites, the larval webs often covering the infested inflorescences. Larvae also attack the developing fruits – hence the vernacular name 'citrus young-fruit borers'. Such attacks are launched from the very early stages of fruit development onwards, larvae typically entering via the side of the receptacle. Larvae may also feed on young foliage and young shoots. Pupation occurs in a flimsy, net-like cocoon spun in suitable shelter on the host plant or amongst rubbish on the ground.

DAMAGE
Attacked buds desiccate and die. Infested young fruits become filled with frass and drop prematurely; infested older fruits remain small and become distorted. Attacks on leaves and shoots are less significant.

Prays oleae Bernard (509–513)
Olive moth
A notorious pest of olive; also associated with other Oleaceae, including *Jasminum*, *Ligustrum vulgare* and *Phillyrea*. Widely distributed in the Mediterranean area; also present in South Africa.

508a, b Citrus flower moth (*Prays citri*).

509 Olive moth (*Prays oleae*).

DESCRIPTION

Adult: 13–14 mm wingspan; forewings mainly grey, tinged with silver and speckled with black; hindwings grey. **Egg:** 0.5 × 0.4 mm; whitish at first, later darkening to brown. **Larva:** up to 8 mm long; body mainly light green, marked dorsally with reddish brown, and with a pair of irregular, brownish or reddish-brown subdorsal lines; prothoracic plate light green, marked with reddish brown; crochets on abdominal prolegs arranged in several concentric circles. **Pupa:** 6 mm long; yellowish brown to dark brown.

LIFE HISTORY

This pest has three biologically distinct generations: anthophagous (flower feeding), carpophagous (stone feeding) and phyllophagous (leaf feeding). Larvae of the first feed in the spring, within and on the flowers. Those of the second bore into the stones of the fruits during the summer and autumn, and those of the third mine within the leaves over the winter period. Overwintered larvae complete their development in the spring, pupation taking place in cocoons spun in bark crevices, amongst spun leaves or amongst shelter on the ground. Adults appear in April or May and eggs are then deposited on the flower buds. The eggs hatch in about a week. Larvae then attack the developing flowers. Pupation occurs in the larval habitation, the next generation of adults appearing from May to July. Eggs of this generation are laid singly on or close to the developing fruits; the resulting larvae (typically one per fruit) then bore into the fruits, each burrowing down to the base of the stone; they feed for about 3 weeks before emerging and pupating. Adults of the final generation appear from September onwards; their eggs are deposited on the leaves. The larvae of this generation mine the underside of leaves, forming small, blotch-like galleries, clearly visible from above. Larvae pass through five instars, those of the leaf-mining generation vacating their mines and attacking new leaves at each moult.

DAMAGE

Larvae of the anthophagous generation destroy large numbers of flowers, reducing fruit set; those of the carpophagous generation cause premature fruit drop and significant yield loss. The extent of damage caused by leaf-mining larvae (the phyllophagous generation) is rarely of economic importance.

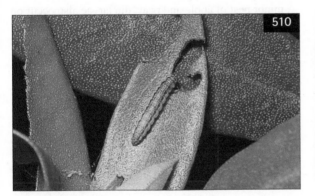

510 Olive moth (*Prays oleae*) – final-instar larva.

511 Olive moth (*Prays oleae*) – young larva.

512 Olive moth (*Prays oleae*) – pupa and pupal cocoon.

513 Olive moth (*Prays oleae*) – old leaf damage.

Scythropia crataegella (Linnaeus) (**514–516**)
Hawthorn webber moth

Although associated mainly with *Crataegus* and, to a lesser extent, *Cotoneaster*, minor infestations of this pest are occasionally found on damson. Widespread in central and southern Europe; also found in the more southerly parts of northern Europe.

514 Hawthorn webber moth (*Scythropia crataegella*).

515 Hawthorn webber moth (*Scythropia crataegella*) – larvae.

516 Hawthorn webber moth (*Scythropia crataegella*) – pupa.

DESCRIPTION
Adult: 13–14 mm wingspan; forewings silvery white, mottled with purplish grey, each with two purplish-grey or brownish-grey crossbands; hindwings greyish brown. **Larva:** up to 15 mm long; body reddish brown, the thoracic segments marked dorsally with yellowish orange; body hairs whitish and relatively long; abdominal prolegs with crochets arranged in several concentric circles; head black. **Pupa:** 7–8 mm long; black to reddish brown, banded with creamy white.

LIFE HISTORY
Infestations are usually first noticed during the spring, when the overwintered larvae feed gregariously on the foliage of host plants, protected by a thin, expansive communal web. Larvae are fully grown from mid-June onwards. They then pupate within the web, and moths emerge in late June and July. Eggs, deposited on the twigs of host plants, hatch in the late summer. The young larvae then mine briefly in the leaves before hibernating.

DAMAGE
Larvae cause significant defoliation; infestations can affect the development of fruitlets and new shoots.

Swammerdamia pyrella (Villers) (**517–519**)
A minor pest of apple and pear; the larvae also occur on *Crataegus*. Generally common in Europe; also present in North America and Japan.

DESCRIPTION
Adult: 10–13 mm wingspan; head white or pale ochreous; thorax greyish brown; forewings greyish

517 *Swammerdamia pyrella* – adult.

brown, with a darker (but indistinct), oblique crossband, a white subapical spot on the costal margin and coppery terminal cilia; hindwings mainly grey, but whitish basally. **Larva:** up to 14 mm long; body yellowish to whitish, marked extensively with dark reddish brown; head brownish black.

LIFE HISTORY

First-generation adults occur in May, sometimes earlier, and those of the second in July and August. The larvae, which are solitary feeders, attack the upper epidermis of expanded leaves, each sheltered beneath a slight web formed close to the tip of a leaf. Larvae of the first brood feed from May or June to July, and those of the second from August or September to October. Pupation occurs in dense, whitish cocoons secreted on the foodplant or amongst debris on the ground.

DAMAGE

Larvae typically graze away the epidermis to leave irregular patches of brownish or whitish tissue. Such damage, however, although very obvious, is of little or no importance.

Yponomeuta evonymella (Linnaeus) (**520–521**)
Cherry small ermine moth

This widely distributed, and often abundant, European species is associated mainly with wild *Prunus padus*; other hosts include *P. cerasus*, *P. domestica* and *Sorbus aucuparia*. At least in mainland Europe, cultivated cherry is also a host. Affected trees are often completely defoliated, and coated in dense, polythene-like sheets of webbing. However, compared with certain other related species of small ermine moth, this species is usually of minor significance as a fruit pest. Adults (23–25 mm wingspan) are larger than those of common small ermine moth (*Y. padella*) (another *Prunus*-infesting species of *Yponomeuta*), and the forewings are pure white, with five or six longitudinal rows of black dots (including a row of 9–11 dots towards the lower margin); the hind wings are dark grey. Also, pupation occurs in white, opaque cocoons formed in clusters within the larval web.

518 *Swammerdamia pyrella* – larva.

520 Cherry small ermine moth (*Yponomeuta evonymella*) – adult and pupal cocoons.

519 *Swammerdamia pyrella* – larval web on apple leaf.

521 Cherry small ermine moth (*Yponomeuta evonymella*) – web.

234

Yponomeuta malinellus Zeller (**522–527**)

Apple small ermine moth

A pest of apple, but rarely present in regularly sprayed orchards. Widely distributed in Europe; also present in Asia and the USA.

DESCRIPTION

Adult: 18–22 mm wingspan; forewings greyish white to white, with several black dots; terminal cilia grey; hindwings dark grey. **Egg:** flat, yellow to dark purplish grey. **Larva:** up to 20 mm long; body dirty yellowish grey, with black spots; head black; prothoracic and anal plates blackish; body of young larvae whitish. **Pupa:** 10 mm long; yellowish brown, with a darker head, thorax, wing cases and tip; cremaster with 6 long, filamentous setae.

522 Apple small ermine moth (*Yponomeuta malinellus*).

523 Apple small ermine moth (*Yponomeuta malinellus*) – final-instar larva.

524 Apple small ermine moth (*Yponomeuta malinellus*) – young larvae in web.

525 Apple small ermine moth (*Yponomeuta malinellus*) – cluster of pupae.

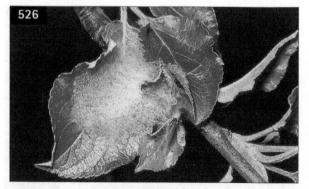

526 Apple small ermine moth (*Yponomeuta malinellus*) – young web.

527 Apple small ermine moth (*Yponomeuta malinellus*) – older web.

LIFE HISTORY

Adults occur in July and August and are very conspicuous when settled on or flitting around the foodplant. Eggs are laid on the bark of twigs and shoots in rafts several millimetres long. There are often 40–80 eggs in a batch, the eggs being placed in overlapping tile-like rows and then coated with a gelatinous layer that soon hardens. Newly laid egg batches are pale yellow and conspicuous, but they soon darken and then come to match the colour of the bark. The eggs hatch in the autumn and the small larvae hibernate in groups beneath the protective remains of their egg shells. At bud burst, the larvae crawl forth and burrow into the young bud tissue. They also burrow into the leaves, to form blotch-like mines in which they feed gregariously. Later, the larvae feed externally, attacking leaves and blossoms, sheltered by dense, greyish or whitish communal webs. Pupation occurs in late June, within dense, white, opaque cocoons placed in groups amongst the protective webbing (cf. common small ermine moth, *Yponomeuta padella*).

DAMAGE

Larvae mining within the young leaves produce reddish or brownish blotches, but such damage is of little or no importance. Older larvae, feeding gregariously in their webs, may cause considerable defoliation. Heavy infestations will reduce tree vigour and, perhaps, have an adverse effect on cropping.

Yponomeuta padella (Linnaeus) (528–530)

Common small ermine moth

Although found mainly on *Crataegus monogyna* and *Prunus spinosa*, colonies also occur occasionally on cultivated cherry and plum. Palaearctic. Present throughout Europe. An introduced pest in North America.

DESCRIPTION

Adult: 19–22 mm wingspan; forewings grey or greyish white to white, with several black dots forming four longitudinal rows (including a row of 4–7 dots towards the lower margin); hindwings dark grey. **Larva:** up to 22 mm long; body greenish grey, with black spots; head black; prothoracic and anal plates blackish. **Pupa:** 7–8 mm long; light brown to dark brown; cremaster with six long, posteriorly directed setae.

LIFE HISTORY

Adults occur in July and August. Eggs are laid on host trees in batches of about 50, and then coated with a yellowish gelatinous material that soon hardens. The eggs hatch about 3 weeks later, although some may not do so until the following spring. Larvae remain clustered beneath the protective shell, where they

overwinter. In the spring, the larvae become active and attack buds and young leaves, feeding gregariously beneath protective webs. When larval numbers are large, these webs soon become extensive and will cover whole shoots and branches. Pupation occurs in about June, in flimsy, semitransparent, greyish cocoons spun randomly within the larval webbing (cf. apple small ermine moth, *Yponomeuta malinellus*). Adults emerge about 2 weeks later.

528 Common small ermine moth (*Yponomeuta padella*).

529 Common small ermine moth (*Yponomeuta padella*) – larvae.

530 Common small ermine moth (*Yponomeuta padella*) – pupae.

DAMAGE

When numerous, larvae are capable of causing considerable defoliation, restricting the growth of trees and reducing cropping potential.

Ypsolopha persicella (Fabricius)

A southerly distributed pest of almond, apricot, peach and, less frequently, plum; grapevine is also reported as a host. Widely distributed in the warmer parts of Europe, especially in the Mediterranean basin.

DESCRIPTION

Adult: 20–22 mm wingspan; forewings hook-tipped (falcate), mainly yellowish white, but also with indistinct darker markings (including an incomplete diagonal line and a broader curved mark arising from the hind margin); hindwings grey to dark grey. **Larva:** up to 22 mm long; body green and fusiform, with whitish dorsal and subdorsal lines; head green; anal prolegs very well developed.

LIFE HISTORY

This species overwinters in the egg stage on the shoots of host plants, and the eggs hatch in April or early May. Larvae then feed for 1–2 months, and eventually pupate on the leaves in white, parchment-like, boat-shaped cocoons. Adults emerge about 3 weeks later, with a second generation of larvae feeding during the summer. These eventually pupate, and new adults are produced from mid-August onwards. In the more northerly parts of its range this species may have just one generation annually.

DAMAGE

Larvae destroy the young leaves and are of greatest significance in spring.

Ypsolopha scabrella (Linnaeus) (**531–532**)

Associated mainly with *Crataegus monogyna* and *Malus sylvestris*, but also present, occasionally, as a minor pest on cultivated apple, pear, cherry and plum trees. Locally common and widely distributed in Europe.

DESCRIPTION

Adult: 19–22 mm wingspan; white or greyish white to black, resembling a flake of dead wood. **Larva:** up to 20 mm long; body bright green and fusiform, with a white dorsal stripe, a pair of pale subdorsal lines, and several rows of small, black, seta-bearing spots; head greenish brown; anal prolegs very well developed.

LIFE HISTORY

Larvae feed on the leaves of host plants in May and June, sheltered beneath slight silken webs. If disturbed, they become extremely active and wriggle rapidly

531 *Ypsolopha scabrella* – adult.

532 *Ypsolopha scabrella* – larva.

backwards and drop to the ground. Pupation occurs in white, boat-shaped cocoons formed on the ground or beneath loose bark. Adults are present from July to late August or early September.

DAMAGE

Loss of leaf tissue is not extensive, and usually unimportant.

Zelleria oleastrella Millière (**533–535**)

A minor pest of olive in the Mediterranean basin, particularly in southern France, Italy and Spain.

DESCRIPTION

Adult: 14–15 mm wingspan; wings mainly grey, with a silvery sheen, marked with black and partly suffused with light brown. **Egg:** 0.8 × 0.7 mm; oval, yellowish green, with a reticulate pattern. **Larva:** up to 14 mm long; dark green to whitish green, with paler pinacula; head brownish; crochets on abdominal prolegs arranged in several concentric circles. **Pupa:** 4 mm long; brown; cremaster with four long spines.

533 *Zelleria oleastrella* – adult.

534 *Zelleria oleastrella* – larva.

535 *Zelleria oleastrella* – pupal cocoon.

They then pupate, each in a white, boat-shaped cocoon (6.0 × 2.5 mm) spun beneath a slight silken web. Adults emerge 2–3 weeks later. There are five overlapping generations annually. When at rest, the adults remain with the head close to the leaf or other surface, and the body, wings and hind legs held upwards at an angle of about 45°, reminiscent of members of the genus *Argyresthia* (see Chapter 6).

DAMAGE
Larvae destroy buds, leaves and shoots; they are particularly damaging to the terminal shoots of new grafts and can have an adverse effect on the subsequent structure of young trees.

LIFE HISTORY
Adults occur throughout the year, but in greatest numbers from May to November. From spring onwards, eggs are laid on the upper and lower surfaces of leaves; they hatch 4–10 days later, depending on temperature. Although eggs are laid on the leaves, the neonate larvae usually bore into a terminal bud, where they begin to feed. Occasionally, however, larvae may inhabit the tips of young shoots, bunching the tissue together with threads of silk. Larvae are fully grown in 2–4 weeks.

Family SCHRECKENSTEINIIDAE

A small family of small moths, with long, narrow wings; hind tibia with four spurs of equal length, a single row of short bristle-like spines and longer spines forming a small whorl at the apex; in repose, the hind legs are raised and project obliquely over the abdomen. Larvae have four pairs of abdominal prolegs, and specialized body hairs that include an internal tube through which fluid rises to form external beads of liquid; the spiracles are borne on chitinous projections known as 'mammillae'.

Schreckensteinia festaliella (Hübner) (536–537)

An often common species, the larvae feeding in the wild on the leaves of *Rubus fruticosus* and *R. idaeus*. Attacks are sometimes very damaging, but there is only an isolated report (on raspberry in Scotland) of the larvae causing harm to cultivated plants. Widely distributed in Europe; also present in North America.

DESCRIPTION
Adult: 10–12 mm wingspan; body mainly brownish to ochreous; forewings ochreous, marked with dark brown streaks; hindwings dark brown. **Larva:** up to 10 mm long; light green and bearing conspicuous whitish hairs. **Pupa:** 5–6 mm long; green; cremaster with about 10 brown spines.

LIFE HISTORY
Larvae feed beneath slight webs formed on the upper surface of leaves of the young, non-fruiting canes, where they devour the upper cuticle and remove the green tissue between the veins. Breeding occurs throughout the spring and summer. Pupation occurs in a net-like cocoon formed on the stems of the foodplant.

DAMAGE
Damaged leaves, which become flecked or entirely white in colour, may be unable to contribute to photosynthesis.

536 *Schreckensteinia festaliella* – adult.

537 *Schreckensteinia festaliella* – larva.

Family **COLEOPHORIDAE**
(casebearer moths)

A distinctive family of small moths, with very long, narrow and generally unicolourous wings; most species (those within the genus *Coleophora*) have paired patches of short spines on abdominal tergites 1–6 or 1–7. Larvae have a short, stubby abdomen; the abdominal prolegs, apart from the anal pair, are vestigial, with those on abdominal segment 6 often absent; the thoracic segments usually have a dorsal plate (with that on the prothorax usually covering the whole dorsal surface) and a well-developed anal plate. The larvae are mostly leaf-miners and, except when very young, inhabit characteristic portable cases. The mines are characteristically blotch-like, with a small circular hole at the point of attachment of the case. In the vast majority of species, pupation occurs within the larval case.

Coleophora anatipennella (Hübner) (**538–539**)
Cherry pistol casebearer moth
A local pest of apple and, especially, cherry. Wild hosts

include *Crataegus monogyna* and *Prunus spinosa*. Widely distributed in Europe and Asia Minor.

DESCRIPTION
Adult: 13–17 mm wingspan; forewings white, suffused with yellowish grey; hindwings darkish grey. **Egg:** minute, rounded and ribbed. **Larva:** up to 10 mm long; body reddish orange; head and thoracic plates blackish; three pairs of abdominal prolegs. **Case:** 8–10 mm long; pistol-shaped; brown, grey or black, and somewhat shiny.

LIFE HISTORY
Adults occur in late June and July. Larvae initially mine the leaves; they may also, occasionally, mine within the fruits. Later, the larvae inhabit small, portable, more or less pistol-shaped cases. In autumn, the case-inhabiting larvae feed on the underside of leaves, which then become peppered with minute mines. In early autumn, the cases are attached to twigs and branches, where the larvae overwinter. Activity is resumed in the spring, at about bud burst. The larvae then attack the leaves, typically grazing away the upper surface and sometimes biting completely through the lamina. Buds are also attacked. The protective case is constantly enlarged during larval growth, larvae completing their development by about mid-May. Larvae then pupate within their cases, adults appearing 3–4 weeks later.

DAMAGE
Leaf damage by young larvae in autumn is insignificant. Also, loss of tissue during the period of spring feeding on buds and leaves is rarely extensive on fruit crops.

Coleophora hemerobiella (Scopoli) (**540–542**)
Fruit tree casebearer moth
A pest of apple, cherry and plum. Wild hosts include *Crataegus monogyna*, *Prunus spinosa* and *Sorbus aria*.

538 Cherry pistol casebearer moth (*Coleophora anatipennella*).

539 Cherry pistol casebearer moth (*Coleophora anatipennella*) – case.

540 Fruit tree casebearer moth (*Coleophora hemerobiella*).

Widely distributed in central Europe and in parts of northern Europe; also present in Asia Minor.

DESCRIPTION

Adult: 12–15 mm wingspan; forewings whitish to greyish, speckled with dark scales; hindwings mainly grey. **Larva:** up to 9 mm long; body brown; head and thoracic plates brownish black; three pairs of abdominal prolegs. **Case:** up to 11 mm long; dark purplish brown or greyish brown, long and slender; young case pistol-shaped and strongly reflexed.

LIFE HISTORY

This species has a two-year life cycle. Adults occur in June and July. Eggs are laid singly on the base of a leaf or against the midrib. Larvae feed briefly in July or August (but later in more northerly regions), initially inhabiting a silk-lined blotch mine within which they pass their first autumn and winter. Larvae resume feeding in the following spring, converting their shelter into a portable, pistol-shaped case. Feeding continues for about 2 months, until June. The larvae then enter diapause and eventually overwinter for a second time. Overwintered larvae become active at bud burst, usually in April and again attack the leaves. After about a month the pistol-shaped cases are abandoned, each larva then inhabiting a new, elongate, tubular case. Larvae, which may feed on developing fruitlets as well as on foliage, are fully fed in late May or June. They then pupate within their tubular cases, adults emerging shortly afterwards.

DAMAGE

Leaf mining, especially by older larvae in their second summer, is often extensive, causing leaves to pucker; owing to loss of tissue, photosynthetic activity may be reduced significantly. Attacked fruitlets develop characteristic round holes, coinciding with the attachment point of the cases and entry points of larvae, and usually drop prematurely.

Coleophora potentillae Elisha (**543–545**)

Strawberry casebearer moth

Polyphagous on a wide range of (mainly rosaceous)

541 Fruit tree casebearer moth (*Coleophora hemerobiella*) – case, alongside abandoned inital case.

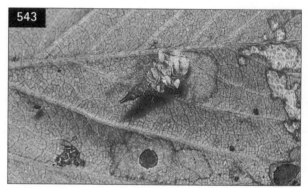

543 Strawberry casebearer moth (*Coleophora potentillae*) – case.

542 Fruit tree casebearer moth (*Coleophora hemerobiella*) – leaf mines on apple.

544 Strawberry casebearer moth (*Coleophora potentillae*) – leaf mines on strawberry leaflet, viewed from above.

herbaceous plants and shrubs, including *Rubus*; occasionally found on cultivated strawberry. Locally distributed in the British Isles (mainly in England), France, Germany and Scandinavia, usually in damp situations.

DESCRIPTION
Adult: 8–10 mm wingspan; forewings shiny brownish grey to bronzy brown, with greyish-brown cilia; hindwings grey. **Larva:** body dull yellowish brown, with mainly black prothoracic and mesothoracic plates; no metathoracic plate; anal plate very large; head blackish brown. **Case:** 6–7 mm long; light brown, with a ventral keel and a series of frilly leaf fragments adorning the basal half; 'mouth' angled at about 30–50° to the main axis.

LIFE HISTORY
Adults occur in late June and July, and eggs are then laid on the underside of a wide range of host plants. The case-inhabiting larvae feed from August onwards, typically forming a series of blotches in the underside of the leaves. Fragments of epidermal tissue are frequently removed from the leaves and added to the larval case, typically forming a series of skirt-like frills around the base. In the autumn, fully fed larvae attach themselves to the stems of woody plants, upon which they overwinter. These individuals then pupate in May. In some districts, larvae overwinter prior to becoming fully grown; they complete their development in the spring and then pupate, usually in June.

DAMAGE
Leaf mines may be very numerous and conspicuous. However, as they occur rather late in the season, their presence on strawberry plants is of little or no importance.

Coleophora serratella (Linnaeus) (**546–548**)
Hazel casebearer moth

A polyphagous species on various broadleaved trees and shrubs, including *Alnus*, *Betula*, *Carpinus betulus*, *Corylus avellana*, *Sorbus* and *Ulmus*. Sometimes a minor pest on cultivated hazelnut; also

546 Hazel casebearer moth (*Coleophora serratella*) – adult.

547 Hazel casebearer moth (*Coleophora serratella*) – case.

545 Strawberry casebearer moth (*Coleophora potentillae*) – leaf mines on strawberry leaflet, viewed from below.

548 Hazel casebearer moth (*Coleophora serratella*) – leaf mines on hazelnut.

recorded on apple. Widely distributed and generally common in Europe; also present in North America and Asia.

DESCRIPTION
Adult: 11–14 mm wingspan; forewings yellowish brown to dark brown; hindwings dark greyish brown. **Larva:** body dark brown; head and thoracic plates black. **Case:** 7 mm long; greyish brown to dark brown and cylindrical, with a serrated dorsal keel; 'mouth' angled at about 30° to the main axis.

LIFE HISTORY
Adults occur in July and early August. Eggs are laid on the underside of leaves, usually close to a major vein, and hatch within 3 weeks. At first the young larvae mine within the leaves, each forming a small, pear-shaped blotch. On moulting to the second instar, the larva forms a small case within which to continue its development. Feeding continues into October. Individuals then move onto the twigs and settle down to overwinter. Activity is resumed after bud-burst. The larvae then mine the leaves from the underside, gradually enlarging their protective cases as development continues. The larval feeding galleries soon become obvious from above, as brown, semitransparent, blotches. Pupation occurs within the case from early June onwards, usually fully exposed on the upper surface of an expanded leaf.

DAMAGE
Leaf mines reduce photosynthetic activity and, if abundant, may have an adverse effect on growth.

Coleophora spinella (Schrank) (**549–551**)
Apple & plum casebearer moth
A generally common, but minor, pest of apple and, less frequently, cherry and plum. Often present on *Crataegus*

monogyna; other hosts include *Prunus spinosa* and *Sorbus aria*. Widely distributed in Europe; also present in North America, Asia Minor and Japan. Recent revision of the genus *Coleophora* has separated *C. spinella* from *C. coracipennella* (Hübner) and *C. prunifoliae* Doets on the basis of differences in the genitalia; reliable external characters to distinguish between these species at any stage in the life cycle, however, are wanting; the biology of all three species is similar. Only *C. spinella* is considered here.

DESCRIPTION
Adult: 10–12 mm wingspan; forewings grey to brownish grey and shiny; hindwings greyish brown; antennae white, ringed with dark brown. **Larva:** body reddish brown; head, thoracic plates and anal plate black; three pairs of abdominal prolegs. **Case:** 5–6 mm long; more or less cylindrical, light to dark brown; young case pistol-shaped, brown to greyish.

LIFE HISTORY
Adults occur in June and July. Eggs are then laid singly, mainly on the underside of leaves. They hatch in

550 Apple & plum casebearer moth (*Coleophora spinella*) – abandoned primary case.

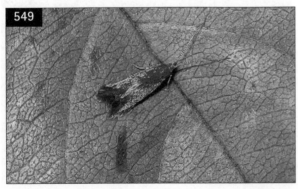

549 Apple & plum casebearer moth (*Coleophora spinella*).

551 Apple & plum casebearer moth (*Coleophora spinella*) – case and leaf mine.

September. Initially, larvae mine within the leaves. However, each soon encloses itself within a small portable case made of silk and excised pieces of leaf tissue. The case-inhabiting larvae then form small blotch mines in the leaves before eventually migrating to the twigs or stems. Here they anchor themselves to the bark and hibernate until the following spring. Activity is resumed at bud burst, the larvae initially attacking the opening buds and young foliage. Material is added to the case, gradually increasing its size. Eventually, however, usually 2–3 weeks later, a new, larger, cigar-shaped case is formed from a fresh leaf fragment, the smaller pistol-shaped case being abandoned nearby. Feeding continues until May, each larva protruding out of its case and eating out a series of blotch mines in the leaves. Buds and, at least on apple, fruitlets are also attacked. When fully fed, the larva attaches its case to a branch or to the tree trunk and pupates, the adult moth emerging about 2 weeks later.

DAMAGE
If numerous, leaf blotches can distort leaves and reduce photosynthetic activity, but attacks on fruit crops are rarely extensive; as in the case of fruit tree casebearer moth (*Coleophora hemerobiella*), attacked fruitlets develop characteristic round holes, coinciding with the attachment point of the cases and entry points of larvae, and usually drop prematurely.

Coleophora violacea (Ström) (**552**)
In mainland Europe, a minor pest of apple and certain other fruit trees. Polyphagous on various rosaceous trees and shrubs, including apple, cherry, pear, plum, *Crataegus monogyna*, *Prunus spinosa* and *Rosa*; non-rosaceous hosts include *Betula, Carpinus betulus, Castanea sativa, Corylus avellana* and *Tilia*. Widely distributed in Europe; also present in North Africa. Prior to the mid-1950s (and also, often, beyond) this species was widely known as *Coleophora paripennella* Zeller. The 'true' *C. paripennella*, however, is associated with herbaceous plants, especially *Centaurea nigra* and *Cirsium arvense*.

DESCRIPTION
Adult: 9–11 mm wingspan; dark greyish brown and shiny. **Larva:** yellowish brown; head, thoracic plates and anal plate black; four pairs of abdominal prolegs. **Case:** 5.5–6.5 mm long; blackish brown, with a paler ventral keel; 'mouth' at 0° (i.e. parallel) to the main axis; anal end abruptly tapered and curved downwards dorsally; leaf fragments often remain attached to the lateral and dorsal surfaces.

LIFE HISTORY
Adults occur in May and June. Eggs are then deposited on the leaves of a wide range of host plants. Larvae feed from August onwards, each inhabiting an elongate case that is held prostrate to the substratum. The case is enlarged as the larva grows. However, unlike that of apple & plum casebearer moth (*Coleophora spinella*), is never changed. When fully grown, usually in October, the larva attaches its case to a suitable surface, such as a nearby branch, and then hibernates. Pupation occurs in the following spring.

DAMAGE
Larvae cause slight damage to foliage, but attacks are usually of no economic importance.

552 *Coleophora violacea* – case.

Family **OECOPHORIDAE**

Moderately small to medium-sized moths, with prominent labial palps and broadly elongate forewings; antennae often very long, and usually with a pronounced basal tuft of hair (the pecten). Larvae have a well-developed prothoracic plate and are tortrix-like (see family Tortricidae), but the anal prolegs often have the semicircle of crochets broken into two bands and the long subdorsal (SD) seta on abdominal segment 8 is often placed more dorsally than the spiracle.

Batia unitella (Hübner) (**553**)

This species occurs widely in central and northern Europe, where it is associated with old apple trees. Larvae are mainly brown, with white dorsal and subdorsal lines along the back. They feed in the wood of apple trees from late summer or autumn onwards, and complete their development in the following spring. Adults (14–17 mm wingspan) are mainly golden brown, with brownish-black hindwings. They occur in June and July, and may often be beaten from the foliage of host trees. Since larvae confine their feeding to already dead, decaying or dying wood they are not of importance – cf. those of cherry bark tortrix moth (*Enarmonia formosana*) (family Tortricidae).

Carcina quercana (Fabricius) (**554–555**)

A minor pest of fruit trees, particularly apple, pear, cherry and plum; blackberry and raspberry are also attacked. Wild hosts, with which this species is most frequently associated, include *Fagus sylvatica* and *Quercus*. Widely distributed, particularly in central and southern Europe. An introduced pest in Canada.

DESCRIPTION

Adult: 15–20 mm wingspan; forewings bright yellow to ochreous yellow, mainly suffused with greyish pink or purplish pink; hindwings pale; antennae longer than wings. **Egg:** yellow, with a warted surface. **Larva:** up to 15 mm long; body bright apple-green, with two yellowish lines along the back and somewhat flattened; head rather large and quadrate. **Pupa:** 8 mm long; brown.

LIFE HISTORY

Larvae feed singly on the underside of leaves in May and June, each sheltering beneath a transparent web of transverse threads. When fully grown, typically from mid-June onwards, each larva pupates beneath an opaque, sheet-like web. Adults appear in July and August.

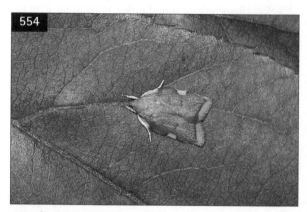

554 *Carcina quercana* – adult.

553 *Batia unitella* – adult.

555 *Carcina quercana* – larva.

DAMAGE

Young larvae bite out patches in the underside of leaves, often leaving the upper surface intact. Later, complete holes are made in the lamina near the web, either in the centre or at the margin of the leaf. The larval webs do not noticeably deform the leaves and damage is not serious.

Diurnea fagella (Denis & Schiffermüller) (556–559)

A minor pest of apple and hazelnut, but associated mainly with woodland broadleaved trees and shrubs, including *Betula*, *Fagus sylvatica*, *Quercus* and *Salix*. Widely distributed in Europe.

DESCRIPTION

Adult female: 18–20 mm wingspan, but wings much reduced and pointed; forewings whitish, suffused to a greater or lesser degree with brownish black, and bearing blackish and buff scale tufts; hindwings greyish to brownish black; palps long. **Adult male:** 25–28 mm wingspan; coloration similar to female.

Larva: up to 18 mm long; body pale, dull yellowish green, with yellowish intersegmental markings; head light brown; prothoracic plate mainly yellowish, with a pair of brown lateral markings; third pair of thoracic legs fleshy and projecting well beyond width of body; young larva dirty greenish grey and translucent. **Pupa:** 10 mm long; brown; cremaster with a cluster of long, hooked setae.

LIFE HISTORY

Adults appear in March and April. The females, although incapable of flight, are very active and crawl away with considerable speed if disturbed. Eggs are laid on various hosts, the larvae then feeding from June to October in spun or, occasionally, rolled leaves. Fully fed larvae spin cocoons amongst debris on the ground. The winter is passed in the pupal stage.

DAMAGE

Feeding is rarely extensive and damage usually limited to slight loss of leaf tissue.

556 *Diurnea fagella* – adult female (dark form).

557 *Diurnea fagella* – adult male (light form).

558 *Diurnea fagella* – larva.

559 *Diurnea fagella* – young larva.

Diurnea lipsiella (Denis & Schiffermüller) (560–561)

Associated with various trees and shrubs, including *Acer campestre*, *Prunus spinosa*, *Quercus*, *Rosa* and *Rubus fruticosus*; in mainland Europe, occasionally a minor pest on fruit trees and cane fruits. Widely distributed in Europe.

DESCRIPTION

Adult female: 23–25 mm wingspan; forewings grey, with darker markings, pointed and broadly elongate; hindwings light brownish grey. **Adult male:** 23–25 mm wingspan; forewings light brown, with a darker tip; hindwings greyish brown. **Larva:** up to 24 mm long; body creamy white; head and prothoracic plate brownish black; thoracic legs black, the third pair enlarged and projecting well beyond width of body. **Pupa:** 10–11 mm long; reddish brown, with tip of abdomen darker dorsally; cremaster with a cluster of long, hooked setae.

LIFE HISTORY

Adults occur in October and November. Eggs are then laid on various host plants and hatch in the following spring. Larvae feed within a bunch of strongly spun leaves from April to July. When fully grown they enter diapause in a slight silken cocoon, formed amongst debris on the ground; pupation occurs in late summer.

DAMAGE

Larvae cause slight defoliation and distortion of shoots, but are usually present on fruit crops in only small numbers.

Esperia sulphurella (Fabricius) (562–563)

This generally common and widely distributed European species breeds in dead, decaying or dying wood. The larvae often occur on fruit trees and are sometimes mistaken for those of cherry bark tortrix moth (*Enarmonia formosana*) (family Tortricidae). The black and yellow adults (12–16 mm wingspan) occur in April and May, flying in sunshine. The relatively slender, light brownish-grey, translucent larvae (up to 15 mm long) feed throughout the summer, autumn and winter. They produce masses of dark brown frass, mixed with strands of silk, that accumulates on the bark of infested trees and often betrays their presence. Pupation occurs in a silken cocoon, formed within the feeding burrow during March or April. The pupa

560 *Diurnea lipsiella* – adult female.

562 *Esperia sulphurella* – adult.

561 *Diurnea lipsiella* – larva.

563 *Esperia sulphurella* – larva.

(8–12 mm long) is brown and has an elongate, 8-spined cremaster. Unlike larvae of cherry bark tortrix moth, this species does not attack healthy trees and is not a cause of tree death.

Oecophora bractella (Linnaeus)

Reported in the Netherlands as a minor, secondary pest of currant, but associated mainly with mulberry trees, the larvae feeding in dead or decaying wood. Present mainly in central and southern Europe; also established in Asia Minor.

DESCRIPTION

Adult: 14–17 mm wingspan; forewings brownish black, with a bronzy sheen and metallic-blue markings, mainly bright yellow basally, with a distinct bright yellow spot on the front margin towards the apex; hindwings brownish black. **Larva:** body brownish grey; head light brown; prothoracic and anal plates brown.

LIFE HISTORY

Larvae feed from July to October and then hibernate; they resume activity in the spring and complete their development in April or May. Adults occur from late May to the end of July. There is one generation annually.

DAMAGE

On currant, larvae extend primary damage caused by currant clearwing moth (*Synanthedon tipuliformis*) (family Sesiidae). They are otherwise of no importance.

Family **GELECHIIDAE**

A distinctive group of small moths, with prominent, upwardly curved labial palps and long, narrow forewings; the margin of each hindwing is usually distinctly sinuous. Larvae have small abdominal spiracles and relatively narrow abdominal prolegs, the latter bearing few crochets arranged as a mesal penellipse.

Anarsia lineatella Zeller (**564**)
Peach twig borer moth

An often common pest of peach in mainland Europe; infestations also occur on almond, apricot and, occasionally, apple, cherry and plum (in the case of plum, especially on yellow-fruiting cultivars). Widely distributed in central and southern Europe; also present in North Africa, North America and Asia. Occasionally found in more northerly parts of Europe (including Scandinavia), but not as an established pest.

DESCRIPTION

Adult: 12–16 mm wingspan; forewings dark brown to greyish black, with irregular black markings and a black patch on the costal margin; hindwings light brownish grey. **Egg:** 0.5 × 0.3 mm, elongate-oval and orange yellow. **Larva:** up to 16 mm long; body dark brown to reddish brown, paler intersegmentally; head, prothoracic plate and anal plate black. **Pupa:** 6–8 mm long; dark brown.

LIFE HISTORY

First-generation adults typically appear in May and early June and those of the second in late July and August. Eggs are laid singly at the base of leaves and hatch about 2 weeks later. The larvae then bore within the young shoots and will often damage several during the course of their development. Larvae may also attack developing fruits, tunnelling within the flesh and also boring into the stone. Fully grown larvae pupate in silken cocoons,

564 Peach twig borer moth (*Anarsia lineatella*) – damage to shoot.

adults emerging 10–12 days later. Second-brood larvae feed from mid-August to early September and then hibernate, each sheltering in a silken cocoon formed under the bark of host trees. They complete their development in the following spring, attacking the blossoms and also boring into the young shoots. Under favourable conditions there are three or four generations annually.

DAMAGE

The tips of infested shoots wither and die, but such damage is usually of importance only in nurseries or in young orchards. Attacks on developing fruits, however, can lead to significant crop losses. The infested fruits develop large necrotic areas, become distorted and are quite unmarketable. Attacked apricots exude considerable quantities of gum.

Gelechia rhombella (Denis & Schiffermüller) (565–567)

A minor pest of apple and pear. Widely distributed across most of Europe to Russia.

DESCRIPTION

Adult: 12–16 mm wingspan; forewings ochreous grey, peppered with black and marked with several small, black patches; hindwings grey. **Larva:** up to 14 mm long; body dark green, purplish-tinged, and with a pair of bluish-white stripes along the back and a whitish stripe along each side; head dark brown; prothoracic plate brown, with a white anterior border. **Pupa:** 6 mm long; reddish brown.

LIFE HISTORY

Larvae feed in the spring, each attacking a single leaf. The infested leaves are folded upwards along the midrib to form a characteristic, pouch-like habitation, within which the larva then grazes on the epidermal tissue. Sometimes a larva will inhabit a folded-down leaf edge. If disturbed, the larvae wriggle violently backwards and drop to the ground. Larvae are fully grown in late May or June. They then pupate in flimsy silken cocoons. Adults emerge shortly afterwards.

DAMAGE

Damaged leaves turn partly brown, and are then very noticeable; however, except occasionally on very young trees, infestations are of little or no importance.

Recurvaria leucatella (Clerck) (568)

A minor pest of apple and pear, but associated mainly with *Crataegus monogyna*, *Malus sylvestris* and *Sorbus*. Eurasiatic. Widespread in central and northern Europe.

565 *Gelechia rhombella* – adult.

566 *Gelechia rhombella* – larva.

567 *Gelechia rhombella* – larval habitation on apple.

568 *Recurvaria leucatella* – adult.

DESCRIPTION
Adult: 13–14 mm wingspan; head and thorax creamy; forewings blackish to brown, each with several raised black scale tufts, a broad white patch towards the base and an irregular white crossline towards the apex. **Larva:** up to 11 mm long; body brownish, with a pinkish-red tinge; head, prothoracic plate and anal plate dark brown to black. **Pupa:** 5 mm long; brown.

LIFE HISTORY
Adults appear in July, but 1–2 months earlier in warmer regions. Eggs are then laid singly or in small batches on the underside of leaves, usually close to a major vein. Young larvae mine the leaves in late summer and then overwinter in silken cocoons spun amongst dead foliage. In spring, the larvae attack flowers and leaves, sheltering during the daytime within small portions of webbed tissue. The larvae are very active when disturbed. Pupation occurs in June, within a small cocoon hidden amongst dead leaves or debris. The pupa is not protruded from the cocoon following emergence of the adult.

DAMAGE
Attacked flowers may fail to set fruit, but the pest is normally present in only small numbers; damage to foliage is of little or no consequence.

Recurvaria nanella (Denis & Schiffermüller) (569–570)
A destructive pest of apple, pear, almond, apricot, peach, plum and other fruit trees. Widely distributed in central and southern Europe, but local and of little or no importance in more northerly regions. An introduced pest in North America (e.g. in Canada, where it is known as the 'lesser bud moth').

DESCRIPTION
Adult female: 12 mm wingspan; forewings mainly whitish, marked with grey, especially basally; hindwings grey; antennae, palps and tarsi white, ringed with black. **Adult male:** similar to female, but forewings mainly dark grey to blackish, marked with white. **Egg:** 0.45 × 0.30 mm; suboval and somewhat flattened. **Larva:** up to 10 mm long; body yellowish or yellowish green; head, prothoracic plate and anal plate black; anal comb robust and 6-pronged; young larva dull reddish orange. **Pupa:** 5 mm long; dark green to brown.

LIFE HISTORY
Adults occur in June and July, but later in cooler areas. Eggs are then deposited either singly or in small batches on the underside of expanded leaves, typically close to a major vein. The eggs hatch about 2 weeks later. Larvae then mine within the leaves, each forming a small linear gallery with numerous blind-ending side branches. Larvae enter diapause whilst in their second instar, and recommence feeding in the following spring. They then invade the buds and also attack the blossoms and young leaves. Larvae are fully grown within 2 months. They then pupate, each in a white cocoon formed amongst debris on the ground. Adults emerge about 2 months later, although emergence may be more protracted under cool conditions.

DAMAGE
Leaf mining by larvae prior to overwintering is of no significance. In spring, however, larvae destroy buds and flowers, and also cause considerable leaf distortion.

569 *Recurvaria nanella* – early mine in almond leaf.

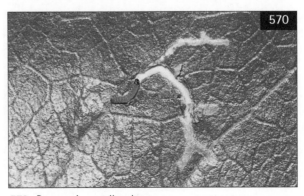

570 *Recurvaria nanella* – larva.

Family **BLASTOBASIDAE**

A small group of small moths, with elongate wings. Larvae are tortrix-like (see family Tortricidae), but have relatively small abdominal spiracles; also, on abdominal segments 1–7, the large pinaculum above each spiracle is characteristically crescent-shaped. The larvae usually feed amongst debris and other dry material.

Blastobasis decolorella (Wollaston) (**571–574**)
Straw-coloured apple moth

This insect, a native of Madeira, was found in Britain in 1946. At that time it was restricted to the London area; however, it is now very widely distributed and locally common, particularly in the southern half of England. Although typically breeding in vegetable rubbish, and normally harmless, serious attacks can occur on apple trees (notably on cv. Egremont Russet and other cultivars that bear short-stalked fruits) in both gardens and commercial orchards. Attacks also occur on almond and peach.

DESCRIPTION
Adult: 18–21 mm wingspan; forewings pale ochreous yellow, each with four darker (but often inconspicuous) spots and scattered darker scales; hindwings paler. **Egg:** 0.7 × 0.6 mm; oval and whitish. **Larva:** up to 13 mm long; body rather plump, purplish brown and shiny; head and prothoracic plate dark brown.

LIFE HISTORY
Adults occur from June to late autumn. On fruit trees, and on almond, larvae have been found feeding from July onwards; they are often concealed beneath dead leaves or amongst other shelter, but are most frequently found within damaged tissue around the stalks of developing fruits. Pupation occurs in small cocoons amongst debris on the ground.

DAMAGE
Apples and other tree fruits (notably peaches) growing in tight clusters or set against the shoots are especially liable to be attacked. The larvae bore into the flesh around the stalks or where adjacent fruits are touching. Large areas of skin and associated flesh are removed. Wounds tend to weep and become covered by a sticky mass of black frass. Attacked fruits are unfit for storage or for marketing, and crop losses may be considerable. Although larvae may feed inside mummified fruitlets, they do not penetrate to the core of maturing fruits. Similarly, on almond, although the larvae burrow extensively within the exocarp, they do not cause harm to the nuts. Larvae also damage the bark of shoots adjacent to infested fruits.

571 Straw-coloured apple moth (*Blastobasis decolorella*).

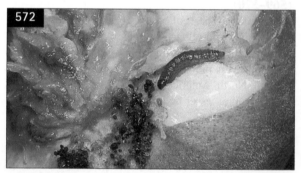

572 Straw-coloured apple moth (*Blastobasis decolorella*) – larva.

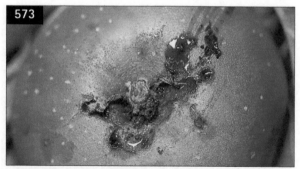

573 Straw-coloured apple moth (*Blastobasis decolorella*) – damage to apple.

574 Straw-coloured apple moth (*Blastobasis decolorella*) – damage to peach.

Family **MOMPHIDAE**

A small group of small moths, with long, narrow, strongly fringed wings. Body of larvae covered in small setae; caudal plates present on the caudal segments. The larvae are often leaf miners, seed-borers or shoot-borers.

Spuleria atra (Haworth) (**575–578**)
Pith moth
A persistent pest of apple, particularly in young orchards, but rarely found on regularly sprayed trees. Eurasiatic. Widely distributed in Europe.

575 Pith moth (*Spuleria atra*).

DESCRIPTION
Adult: 10–14 mm wingspan; forewings narrow, mainly brownish, suffused with black, and each with two black scale tufts; hindwings grey. **Egg:** 0.6 × 0.3 mm; oval, pale yellowish to brownish. **Larva:** up to 10 mm long; body brownish pink, with three caudal plates; head and prothoracic plate dark brown. **Pupa:** 5–7 mm long; golden brown; slender, with long wing cases; cremaster forming two long, tubular ventral projections directed obliquely forwards.

576 Pith moth (*Spuleria atra*) – larva.

LIFE HISTORY
Adults are most numerous in late July and early August. Eggs are laid in leaf axils and at the base of leaf petioles or small flower buds. They hatch about 2 weeks later and the tiny larvae immediately bore into the bark, usually close to a bud, to leave a small entry hole surrounded by yellowish-brown frass. Each larva tunnels within a shoot or spur throughout the autumn and winter months. At this stage there are few outward signs of their presence. However, in spring, quantities of frass appear at the surface, forced out through cracks which appear in the damaged tissue as the shoots swell. In June, when fully grown, the larvae burrow to the surface and break through the bark, usually close to a lateral or terminal bud, or just below a blossom truss. Pupation then occurs, still within the larval burrows, adult moths emerging about a month later.

577 Pith moth (*Spuleria atra*) – pupa.

DAMAGE
Leaves and blossom trusses on infested shoots and spurs wilt and turn brown, usually just before the flowers are fully open. Terminal buds may also be killed and weak lateral shoots will then develop. With heavy infestations, fruit set is seriously affected. In addition, damage to the bark allows easy access for pathogenic fungi, especially apple canker (*Nectria galligena*). Rarely, young larvae may burrow into maturing fruits, forming galleries in the flesh.

578 Pith moth (*Spuleria atra*) – damage to apple shoot.

Family **COCHYLIDAE**

A group of moths essentially similar to members of the family Tortricidae, of which they are sometimes considered merely a subfamily. Adults often have prominent, beak-like labial palps; the forewings are typically elongate, often rather angular, and held in a steep, roof-like posture when in repose. Larvae typically feed within flowerheads, seedheads, fruit clusters, roots or stems of host plants.

Eupoecilia ambiguella (Hübner) (**579–581**)
Vine moth

An important pest of grapevine; other hosts include a wide range of plants such as *Frangula alnus*, *Hedera helix*, *Ligustrum vulgare*, *Lonicera*, *Ribes*, *Symphoricarpos rivularis* and *Viburnum*. Palaearctic. Widely distributed in central and southern Europe, but with a less southerly distribution than European vine moth (*Lobesia botrana*) (family Tortricidae).

DESCRIPTION

Adult 12–15 mm wingspan; forewings whitish ochreous, marked with yellow ochreous and with a dark, brownish to black median fascia; hindwings grey. **Egg:** 0.8 × 0.6 mm; greyish brown when laid, later becoming speckled with orange. **Larva** up to 12 mm long; body purplish brown or yellowish brown to olive green, with large, brown and moderately conspicuous pinacula; head and prothoracic plate dark brown or black; anal plate brownish to yellowish; abdominal prolegs each with 25–30 crochets (cf. *Lobesia botrana*, family Tortricidae). **Pupa:** 5–8 mm long; reddish brown; cremaster with a ring of 16 hook-tipped bristles and a pair of short horns dorsally.

LIFE HISTORY

In central Europe there are usually two, and in southern Europe usually three, generations annually. Adults of the first appear from about mid-April or May onwards, the period of activity varying from region to region. On

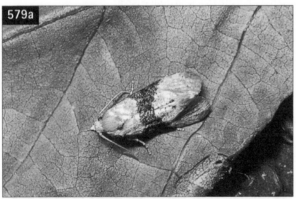

579a, b Vine moth (*Eupoecilia ambiguella*) – male.

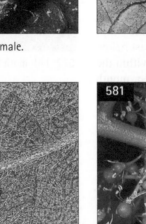

580 Vine moth (*Eupoecilia ambiguella*) – larva.

581 Vine moth (*Eupoecilia ambiguella*) – larval habitation.

grapevine, eggs are laid on the unopened flower buds and hatch 1–2 weeks later. Larvae then commence feeding and spin several buds together, the silken shelters becoming more dense and larger as larval development progresses. Larvae eventually pupate in their habitations, or in folded leaves or other situations, usually after feeding for about 3–4 weeks. Adults of the next generation appear about 10 days later. Larvae that develop in summer and autumn feed on immature or maturing fruits. When fully grown, larvae of the final generation spin overwintering cocoons, hidden in or beneath the bark of older vine stems, or in cracks on supporting posts, and eventually pupate in the spring (cf. European vine moth, *Lobesia botrana*, family Tortricidae).

DAMAGE
As for European vine moth (*Lobesia botrana*) (family Tortricidae).

Family **TORTRICIDAE**
(tortrix moths)

A large and important group of mainly small moths, including many well-known pests, with typically bluntly rounded and rectangular forewings and short hair fringes. The forewings often have distinctive basal, subbasal and median fasciae, a pre-apical spot or a costal blotch (as in the subfamily Tortricinae), or costal strigulae, a submarginal ocellus and a dorsal patch (as in the subfamily Olethreutinae). Eggs are typically flattened, scale-like or lenticular, and are laid singly, in pairs or in batches; when deposited in batches the eggs often overlap (reminiscent of roof tiles or fish scales). Larvae have five pairs of abdominal prolegs, each leg bearing a complete circle of similarly sized crochets, and usually well-developed prothoracic and anal plates; many species have an anal comb, with which particles of frass are ejected from the anus. The larvae often feed in spun or rolled leaves, and wriggle backwards rapidly if disturbed. Pupation may take place in the larval habitation or elsewhere, the pupa being protruded from the cocoon following emergence of the adult.

Acleris comariana (Lienig & Zeller) (**582–584**)
Strawberry tortrix moth
A pest of strawberry, particularly in low-lying (e.g. fenland) areas; also associated with *Fragaria vesca*, *Geum rivale* and *Potentilla palustris*. Widely distributed in Europe; also present in North America, China and Japan. Records of this species damaging raspberry are the result of confusion with broad-barred button moth (*Acleris laterana*).

DESCRIPTION
Adult: 13–18 mm wingspan; forewings greyish brown to yellowish brown or reddish brown, often marked with whitish, sometimes with blackish scale tufts and often

582 Strawberry tortrix moth (*Acleris comariana*).

with a conspicuous costal blotch; hindwings grey. **Egg:** 0.8 × 0.6 mm; pale creamy white (overwintering eggs darkening to brick-red). **Larva:** up to 15 mm long; body whitish green to dull green, somewhat darker along the back; pinacula brown; head yellowish brown, marked with blackish; prothoracic plate ochreous, with a dark brown or blackish hind margin; anal plate green; early instars pale cream to light brown, with a black or brown head. **Pupa:** 6.0–7.5 mm long; light brown; cremaster short and broad, with short lateral horns.

LIFE HISTORY

This double-brooded species overwinters as eggs laid singly on mature leaves, mainly on the underside. Descriptions of petioles and stipules as oviposition sites do not hold true for present-day strawberry cultivars. Eggs hatch in April or early May and the larvae then attack the young, unfurling leaves. Later, they feed amongst spun leaves which may sometimes be webbed to blossom trusses or small fruitlets. Larvae also feed directly in the flowers, sheltering beneath lightly webbed-down petals. First-brood larvae are fully grown by early to mid-June, having passed through six instars.

Each then pupates in a silken cocoon formed in the shelter of a folded leaf or between spun leaves. Adults appear from mid-June to July; they also lay eggs singly on the foliage. These eggs hatch in 10–14 days and a second generation of larvae then feeds from late July to September, eventually pupating to produce adults from September to October or early November.

DAMAGE

Larvae feed mainly on the leaves, and heavy infestations can lead to severe defoliation, especially when members of the second brood are present on the new summer growth arising from recently mown-off or burnt-off plants. Damage to flowers and young fruitlets will result in the loss or malformation of fruits. However, compared with other strawberry-feeding tortricids, the proportion of the larval population attacking blossoms and fruitlets is usually small.

Acleris cristana (Denis & Schiffermüller) (585–586)

A minor pest of fruit trees in mainland Europe, particularly apple and plum. Wild hosts include

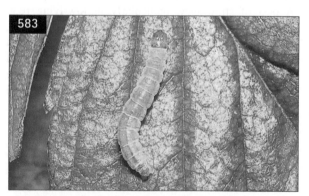

583 Strawberry tortrix moth (*Acleris comariana*) – larva.

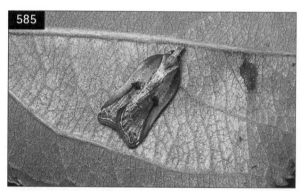

585 *Acleris cristana* – adult.

584 Strawberry tortrix moth (*Acleris comariana*) – larval habitation.

586 *Acleris cristana* – larva.

Crataegus monogyna, *Prunus spinosa* and *Sorbus aria*. Eurasiatic. Widespread in central Europe.

DESCRIPTION
Adult: 18–22 mm wingspan; an extremely variable species; forewings ranging from blackish through reddish brown to ochreous, with a very large central tuft of scales (ranging in colour from white to black) and usually with a distinct ochreous or white band along the hind margin (dorsum); costal margin of forewings noticeably indented; hindwings brownish grey. **Egg:** greyish and opalescent when laid, but then becoming reddish brown. **Larva:** up to 14 mm long; body green to yellowish green; head and prothoracic plate chestnut-brown to blackish brown. **Pupa:** 10 mm long; brown; cremaster short and broad, with pointed lateral horns.

LIFE HISTORY
Hibernated adults are active in the early spring and deposit eggs singly or in small batches on the twigs of host plants. Larvae feed on the young foliage from April onwards and eventually pupate to produce adults in late May in regions where there are two generations annually. In more northerly regions, where the pest is single-brooded, adults do not appear until August.

DAMAGE
Larvae contribute to damage caused in spring and summer by those of various other lepidopteran species, but numbers on fruit trees are usually small.

Acleris laterana (Fabricius) (587–588)
Broad-barred button moth
A minor pest of blackberry, loganberry and raspberry. Wild hosts include *Crataegus monogyna*, *Rosa*, *Rubus fruticosus*, *Salix* and *Vaccinium*. Widely distributed in central and northern Europe and Asia; also present in North America.

DESCRIPTION
Adult: 15–20 mm wingspan; forewings varying from silvery white to ochreous or olive yellow, suffused with grey and black, marked with black, brown or red, and usually with a large costal blotch; hindwings grey. Forms of this variable species are similar in appearance to *Acleris comariana* – the two species are best distinguished by differences in host plant and by details of their biology. **Egg:** greenish grey. **Larva:** up to 15 mm long; body whitish green; head brownish yellow; prothoracic plate light brown or greenish brown. **Pupa:** 7–8 mm long; reddish brown; cremaster short and broad, with short lateral horns.

LIFE HISTORY
In northern Europe, this species is single-brooded. Eggs are laid singly at the base of buds in late August and September. They hatch in the spring. The larvae then feed between two spun leaves during May and June, and may also attack the flowers. Pupation occurs in July, within a silken cocoon spun in a folded leaf; adults emerge about 3 weeks later. In warmer regions there are two generations annually, adults of the first occurring in July and those of the second appearing in September.

DAMAGE
Attacks on blossom trusses may result in malformation or loss of fruits, but foliage damage is of no consequence.

Acleris lipsiana (Denis & Schiffermüller)
In northern Europe, a local pest of fruit crops such as apple and strawberry. Usually most numerous in moorland areas, where it is associated mainly with

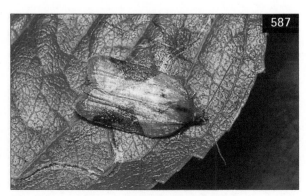

587 Broad-barred button moth (*Acleris laterana*).

588 Broad-barred button moth (*Acleris laterana*) – larva.

Myrica gale, *Vaccinium myrtillus* and *V. vitis-idaea*, but also with *Betula* and *Malus sylvestris*.

DESCRIPTION
Adult: 17–23 mm wingspan; forewings reddish brown to ochreous, variably suffused with grey, bluish grey and black; hindwings light grey. **Larva:** up to 15 mm long; body light green to yellowish green; head light reddish brown; prothoracic plate brown to blackish, with a whitish anterior border and a thin median line. **Pupa:** 8–9 mm long; brown; cremaster short and broad, with prominent, pointed lateral horns.

LIFE HISTORY
This species is single-brooded. Eggs are laid in the spring in small batches on the leaves of host plants. Larvae then feed in June and July, usually inhabiting a shelter of folded leaves or a spun leaf. Fully grown larvae pupate in the larval habitation or amongst other suitable shelter, adults emerging from August onwards. They then overwinter.

DAMAGE
Larvae feed voraciously on foliage. At least on apple, they also attack the developing fruitlets, causing extensive surface damage; considerable weeping occurs from the wounds.

Acleris rhombana (Denis & Schiffermüller) (589–593)
A generally common, but minor, pest of apple, pear and plum; the larvae also feed on *Crataegus monogyna* and *Sorbus aucuparia*, and on certain other trees and shrubs. Present across central and northern Europe to Asia Minor; also established in North America.

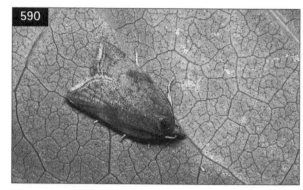

590 *Acleris rhombana* – adult (dark form).

591 *Acleris rhombana* – larva.

592 *Acleris rhombana* – early larval habitations on pear.

589 *Acleris rhombana* – adult (light form).

593 *Acleris rhombana* – larval habitation on plum.

DESCRIPTION

Adult: 13–19 mm wingspan; forewings with a subfalcate tip, and varying from dark reddish brown to ochreous, with dark markings forming a reticulate pattern (sometimes also with a distinct central scale tuft); hindwings light grey, darker in female. **Egg:** 0.8 × 0.5 mm; yellowish green. **Larva:** up to 14 mm long; body light green or yellowish green, with inconspicuous pinacula; head reddish brown to black; prothoracic plate reddish brown or greenish to black; anal plate light green; anal comb yellowish. **Pupa:** 7–9 mm long; reddish brown to blackish brown; cremaster short and broad, with pointed lateral horns.

594 *Acleris variegana* – adult.

595 *Acleris variegana* – larva.

LIFE HISTORY

Adults occur from August or September to October. Eggs are laid singly or in small batches on the bark of trunks and branches. They hatch in spring and the larvae then invade the opening buds. Later, larvae feed in webbed leaves, usually at the tips of the young shoots. They may also feed on blossom trusses. Pupation takes place in June or early July, usually in a cocoon formed in the soil. The pupal stage is protracted, lasting for 6–8 weeks or even more. Although typically single-brooded, in the most southerly parts of its range there are two generations annually.

DAMAGE

Larval damage is restricted mainly to the loss or distortion of younger leaf tissue, but with little or no effect on tree growth. On pear, early-instar damage on young leaves can cause the tissue to become tightly rolled and to turn black, reminiscent of galls formed by pear leaf midge (*Dasineura pyri*) (Chapter 5).

Acleris variegana (Denis & Schiffermüller) (594–595)

Associated with a wide range of rosaceous plants, including apple, pear, cherry, blackberry and raspberry, but of only minor pest status. Present and often common in various parts of Europe; also established in north-western Africa and Asia, and an introduced pest in North America.

DESCRIPTION

Adult: 14–18 mm wingspan; forewings usually whitish ochreous or purplish, variably suffused with grey, with much of the basal half white and with distinct, often black, subcentral scale tufts; hindwings grey. **Larva:** up to 14 mm long; body light green or yellowish green; head and prothoracic plate yellowish brown or greenish brown; anal plate green. **Pupa:** 7–8 mm long; light brown; cremaster broad, with prominent, pointed lateral horns.

LIFE HISTORY

Adults appear over an extended period, from late June or early July to September. Eggs are then laid in batches on either side of expanded leaves, usually along the midrib. Larvae feed from May to late June or early July, sheltered within loosely spun leaves or in folded leaf edges, eventually pupating in the larval habitation or amongst fallen leaves. There are conflicting opinions as to the overwintering stage of this insect. In some parts of its range, as in North America, two generations are completed annually.

DAMAGE

Young larvae skeletonize the leaves, but later instars graze more extensively; however, unless larvae are numerous, growth of the foodplant is not affected.

Adoxophyes orana (Fischer von Röslerstamm) (596–598)

Summer fruit tortrix moth

An important pest of apple, apricot, cherry, peach, pear, plum and other fruit trees; infestations also occur on currant and gooseberry. Also associated with many other broadleaved trees and shrubs. Eurasiatic. Widely distributed in Europe.

DESCRIPTION

Adult female: 18–22 mm wingspan; forewings greyish brown to orange brown, with dark markings and often noticeably reticulate (but less strongly patterned than in the male); hindwings grey. **Adult male:** 15–19 mm wingspan; forewings light greyish brown to orange brown, distinctly marked with dark brown and suffused with ochreous; hindwings grey. **Egg:** lemon-yellow. **Larva:** up to 20 mm long; body yellowish green, olive green or dark green, with small yellowish pinacula; head and prothoracic plate ochreous; thoracic legs light brown. **Pupa:** 10–11 mm long; dark brown.

596 Summer fruit tortrix moth (*Adoxophyes orana*) – female.

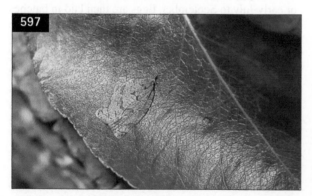

597 Summer fruit tortrix moth (*Adoxophyes orana*) – male.

LIFE HISTORY

Adults of the first generation occur from late May or June to mid-July. Eggs are then laid on the leaves of host plants in batches of up to 160 or more. The eggs hatch in about 1–3 weeks, depending on temperature. The young larvae then feed beneath silken webs spun on the underside of leaves, usually removing the lower epidermal tissue close to a midrib. Later, the larvae shelter within webbed foliage, especially at the tips of the shoots, and then feed more extensively. Foliage is often also webbed down against the fruits and fruitlets, which are also then attacked. Fully fed larvae pupate in the larval habitation, adults of the second generation emerging from late July or early August onwards. Larvae of the second brood appear in mid-August or early September. They feed for several weeks before overwintering, as second- or third-instar larvae, in silken hibernacula formed between dead leaves or mummified fruits and branches. The larvae reappear in late March or April and begin to feed on the opening buds. Later, they attack the blossom trusses and, finally, the leaves of the young shoots. Development is usually completed in late May, the larvae then pupating in a shelter of webbed leaves. Although typically double-brooded, under particularly favourable conditions second-brood larvae may complete their development in the autumn and give rise to a partial third generation of adults.

DAMAGE

Loss of buds may be of significance, but damage to foliage is usually relatively unimportant. Larvae feeding on mature fruits, notably on apples and pears, remove the skin and graze shallowly into the flesh to form large, irregular patches. These grazed areas become evenly russeted and the fruits are likely to be unmarketable. Damage to the surface of developing plums results in considerable weeping; the blemished fruits are also unmarketable.

598 Summer fruit tortrix moth (*Adoxophyes orana*) – larva and damage to apple leaf.

Ancylis achatana (Denis & Schiffermüller) (599–600)

A minor pest of fruit trees in mainland Europe, particularly apple and plum, but associated mainly with *Crataegus monogyna*, *Prunus spinosa* and, to a lesser extent, *Cotoneaster*. Widely distributed in Europe.

DESCRIPTION

Adult: 15–18 mm wingspan; forewings whitish to silvery grey, extensively marked with dark brown to yellowish brown, including a quadrate or triangular blotch on the hind margin; hindwings grey. **Larva:** up to 15 mm long; body reddish brown to dark brown, with pale pinacula; head and prothoracic plate brownish black to black; anal plate beige, with dark markings. **Pupa:** 10–12 mm long; brownish black.

LIFE HISTORY

Larvae feed mainly in May, each inhabiting a folded leaf. Individuals (unlike those of many other tortricids) are relatively inactive, even when disturbed. Pupation occurs in late May, and adults emerge in June and July.

DAMAGE

Larvae contribute to defoliation caused by those of various other torticid species, but are usually present on fruit trees in only small numbers.

Ancylis comptana (Frölich)

Southern strawberry tortrix moth

Although associated mainly with *Fragaria vesca*, *Poterium sanguisorba*, *Potentilla* and *Thymus*, larvae of this locally common species are also reported infesting cultivated strawberry, particularly in warmer regions. Widely distributed in Europe, particularly in coastal areas and on chalk and limestone soils, its range extending to China and Korea; also present in North America, as the subspecies *Ancylis comptana fragariae* (Walsh & Riley), where it is a well-known pest of strawberry and the larvae are known as 'strawberry leaf rollers'.

DESCRIPTION

Adult: 10–12 mm wingspan; forewings silvery white, suffused and marked with dull brown and grey, the latter forming distinct costal strigulae; markings variable, but usually including a distinct dark brown basal patch; cilia white, marked apically with brown; tip of forewing noticeably hooked (falcate); hindwings grey, with paler cilia. **Larva:** up to 12 mm long; body greyish green, with pale, relatively large, pinacula; head mainly light yellowish brown to dark brown; prothoracic plate yellowish brown; anal plate pale yellow, speckled with blackish brown; anal comb well developed.

LIFE HISTORY

Adults occur from April to June, with those of a second generation active from about mid-July to September. Eggs are laid singly on the underside of leaves and hatch 2–3 weeks later. The larvae from the first generation feed in June and July, usually pupating in July; those of the second brood feed in the late summer or early autumn and either complete their development and overwinter as fully grown individuals or feed intermittently during the winter; in either case, pupation occurs in the spring. Young larvae typically feed on the underside of leaves, sheltered by flimsy silken webs; later, each larva forms a tube-like habitation from a folded leaf, and feeds on the upper surface amongst silken webbing and accumulated frass.

DAMAGE

Heavy infestations on strawberry cause significant defoliation, and this can have an adverse effect on crop yields.

599 *Ancylis achatana* – adult.

600 *Ancylis achatana* – larva.

Ancylis selenana (Guenée)

A pest of apple and pear. Also associated with *Crataegus monogyna* and *Prunus spinosa*. Widely distributed in mainland Europe.

DESCRIPTION
Adult: 13–15 mm wingspan; forewings mainly reddish brown, with a distinct greyish-yellow submarginal ocellus and a strongly hooked (falcate) tip; hindwings greyish brown. **Larva:** up to 8 mm long; body greenish to yellowish; head ochreous; prothoracic plate yellowish, marked with black posteriorly. **Pupa:** 5 mm long; pale yellow to dark yellow.

LIFE HISTORY
This species usually overwinters as fully fed larvae, typically in a cocoon spun between two dead leaves. Pupation occurs in the early spring, and first-generation adults then emerge in early April. Eggs are laid singly and hatch about 10 days later (taking slightly longer than eggs laid in summer). Larvae feed on the underside of leaves, usually drawing two expanded leaves together as a shelter and grazing away the epidermal tissue. Pupation occurs in the larval habitation, and adults emerge shortly afterwards. Development of this pest is rapid, with up to five completed generations annually; larvae of the final brood feed well into the autumn.

DAMAGE
This species contributes to defoliation caused by larvae of many other tortricids; attacks are particularly serious on young trees.

Ancylis tineana (Hübner)

A minor pest of pear and plum in various parts of mainland Europe, particularly in central and southern regions. Wild hosts include *Betula*, *Crataegus monogyna*, *Populus tremula* and *Prunus spinosa*; also present in North America.

DESCRIPTION
Adult: 12–13 mm wingspan; forewings mainly grey, with a distinct greyish-white submarginal ocellus and a strongly hooked (falcate) tip; hindwings grey. **Larva:** up to 11 mm long; body pale yellowish grey to greyish green, with brownish pinacula; head yellowish brown; prothoracic plate ochreous; anal plate brown. **Pupa:** 5–6 mm long; light brown.

LIFE HISTORY
This species has two generations annually. Adults of the first generation occur in May, and their larvae feed from June to July; fully-fed larvae then pupate, each in a silken cocoon spun in the larval habitation. Adults of the summer generation appear from early July onwards, and these give rise to second-brood larvae that feed from September to October. These larvae overwinter and then pupate shortly before the appearance of adults.

DAMAGE
This species contributes to defoliation caused by larvae of various other tortricids.

Aphelia paleana (Hübner) (**601–602**)

A minor pest of strawberry, but associated mainly with grasses and herbaceous plants such as *Centaurea scabiosa*, *Filipendula ulmaria*, *Plantago* and *Scabiosa*. Widely distributed in central and northern Europe, particularly in coastal and lowland sites.

DESCRIPTION
Adult: 17–24 mm wingspan; forewings mainly creamy ochreous to whitish ochreous and unmarked; hindwings

601 *Aphelia paleana* – adult.

602 *Aphelia paleana* – larva.

dark grey, with pale cilia. **Larva:** up to 20 mm long; body dark greyish green to black, with prominent white pinacula; head and anal plate mainly black or brownish black; prothoracic plate black or brownish black, but paler anteriorly. **Pupa:** 8–10 mm long; dark brown; cremaster elongate, with four terminal and four lateral, hook-tipped bristles.

LIFE HISTORY

Adults occur in July and August, depositing eggs in batches on the leaves of a wide range of low-growing plants. Larvae feed in late summer or early autumn and then overwinter. They complete their development in the spring, eventually pupating in a folded-down leaf edge.

DAMAGE

Larvae cause minor damage to foliage and in spring also attack the open blossoms, webbing the petals down with silk. Numbers of larvae in strawberry plantations, however, are usually small.

Archips crataegana (Hübner) (603–604)

A minor pest of fruit trees, particularly apple, pear, cherry and plum. Also polyphagous on various broadleaved trees and shrubs including, *Fraxinus excelsior*, *Populus*, *Quercus*, *Salix*, *Sorbus*, *Tilia* and *Ulmus*. Palaearctic. Widely distributed in Europe.

DESCRIPTION

Adult female: 23–27 mm wingspan; forewings light brown to purplish brown, with chocolate-brown markings; hindwings brownish grey. **Adult male:** 19–22 mm wingspan; forewings lighter than in female; hindwings brownish grey. **Larva:** up to 23 mm long; body dull greenish black to dark olive green, with black pinacula; head shiny black; prothoracic and anal plates black; anal comb black, with 6–8 prongs. **Pupa:** 9–12 mm long; dark brown to black; cremaster elongate.

LIFE HISTORY

Adults of this single-brooded species occur from late June to August. Eggs are laid in conspicuous batches of about 30 on the trunks and main branches, and then coated with a hard, white substance which disguises them as bird-droppings. The eggs hatch in April or early May. The tiny larvae are very active and rapidly climb the tree to begin feeding on the underside of the leaves. Later, each feeds inside a tightly rolled leaf edge, usually on fully expanded foliage at the shoot tips. Pupation occurs in a rolled leaf or between two spun leaves, adults emerging a few weeks later.

DAMAGE

Feeding is restricted mainly to the expanded leaves and unless larvae are very numerous damage is insignificant.

603 *Archips crataegana* – female.

604 *Archips crataegana* – larva.

Archips podana (Scopoli) (**605–610**)
Fruit tree tortrix moth

An often common pest of apple. Other cultivated hosts include pear, apricot, cherry, plum, currant, gooseberry, blackberry, raspberry and hop. Also present on a wide range of ornamental and wild trees and shrubs. Eurasiatic. Widely distributed and often abundant in Europe. An introduced pest in North America.

DESCRIPTION

Adult female: 20–28 mm wingspan; forewing creamy ochreous to purplish brown, with a brown, reticulate pattern and darker markings, and a dark apical spot; hindwings brownish grey, suffused extensively with orange. **Adult male:** 19–23 mm wingspan; forewings purplish to chestnut-brown or purplish ochreous, with yellowish and dark brown, velvety markings and a dark

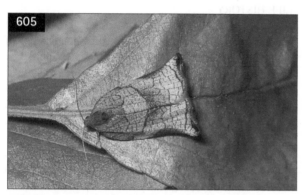

605 Fruit tree tortrix moth (*Archips podana*) – female.

606 Fruit tree tortrix moth (*Archips podana*) – male.

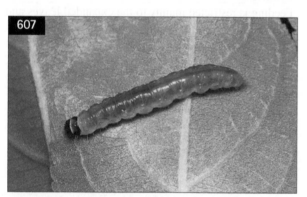

607 Fruit tree tortrix moth (*Archips podana*) – final-instar larva.

608 Fruit tree tortrix moth (*Archips podana*) – young larva.

609 Fruit tree tortrix moth (*Archips podana*) – damage to apple, caused in spring.

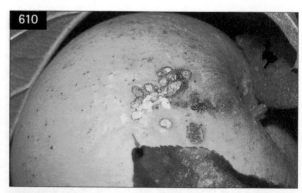

610 Fruit tree tortrix moth (*Archips podana*) – damage to apple, caused in autumn.

apical spot; hindwings greyish, partly tinged with orange. **Egg:** green. **Larva:** up to 22 mm long; body light green to greyish green, darker above, with pale pinacula; head chestnut-brown or black; prothoracic plate chestnut-brown, with darker lateral and hind margins, a pale anterior margin and a pale, narrow mid-line; anal plate green or grey; first thoracic spiracle elliptical and last abdominal spiracle distinctly larger than the rest. **Pupa:** 9–14 mm long; dark yellowish brown to blackish brown; cremaster elongate and with four pairs of strong, curved, hook-tipped bristles.

LIFE HISTORY

Adults occur from June to September, but are usually most abundant in July. Eggs are deposited on the leaves in flat, oval batches of about 50 or more; they are then covered with a protective, waxy secretion that soon hardens. These egg masses are extremely difficult to find as they closely match the colour of the leaves. The eggs hatch in about 3 weeks. The tiny, first-instar larvae then spin individual webs of silk, usually beneath leaves and close to the main veins, and begin feeding. After a few days, they moult and each then lives between two spun leaves. Particularly on apple, some larvae may occur beneath slight webs spun where two or more fruits are touching; larvae may also feed within the calyx cavities of the fruits. However, a greater proportion of larvae attack fruits only after moulting to the third instar. Most larvae overwinter in this growth stage, each sheltering in a dense, silken hibernaculum spun beneath a bud scale, beneath a dead leaf and a twig or spur, or within other shelter. Occasionally, second-instar larvae will also hibernate; in favourable seasons, when adult emergence and egg-laying is particularly advanced, some larvae may feed up and pupate to produce a partial second generation of moths in the second half of August and in September. However, most larvae, and any resulting from eggs laid by second-generation adults, complete their development in the spring. Overwintered larvae become active in late March or April, burrowing into the opening buds. They soon attain the fourth instar and, after yet another moult, then attack the flowers. Fifth-instar larvae may also feed on the young fruitlets. Sixth- and seventh-instar larvae tend to live on foliage, each webbing two or more leaves together and sheltering between them, or spinning a dead leaf to a healthy one or to a twig as a retreat. These final larval stages occur from early May to early June. Individuals then pupate within the larval habitation or within freshly spun leaves nearby. Adults emerge 3 or more weeks later.

DAMAGE

In autumn, the young larvae often damage mature or maturing apples, biting irregular pits in the skin. Such blemishes, although superficial, extending no more than 1–2 mm into the flesh, can seriously affect the marketability of the crop; they also allow easy access of fungal pathogens during storage. Similar damage is caused to pears, apricots, cherries, damsons and plums, but less frequently. At harvest, young larvae are sometimes carried into fruit stores and packing sheds where, if conditions are suitable, feeding (and, hence, fruit damage) may continue. In spring, attacks on buds can be serious, the overwintered larvae often totally destroying them; feeding by larvae or the presence of their webbing may also affect the development of young leaves and flowers. Attacks on fully expanded foliage of fruit trees and bushes are usually of little or no consequence. However, bitten fruitlets either drop prematurely or develop corky patches or pimples that affect the quality of the mature fruit.

Archips rosana (Linnaeus) (**611–613**)
Rose tortrix moth

A pest of apple, pear and plum; occasionally, black currant, gooseberry, raspberry and hop are also attacked. Also associated with a wide variety of other trees and

611 Rose tortrix moth (*Archips rosana*) – female.

612 Rose tortrix moth (*Archips rosana*) – larva.

shrubs. Palaearctic. Widely distributed in Europe and particularly common in orchards and gardens. An introduced pest in North America.

DESCRIPTION

Adult female: 17–24 mm wingspan; forewings reddish brown, with diffuse, somewhat reticulate, darker markings; hindwings grey, tinged apically with orange yellow. **Adult male:** 15–18 mm wingspan; forewings light brown to purplish brown, with dark brown to blackish, often reddish-tinged markings; hindwings grey. **Egg:** greenish grey. **Larva:** up to 22 mm long; body light green, but darker dorsally and paler ventrally; head and prothoracic plate light brown to black; anal plate green or light brown. **Pupa:** 9–11 mm long; yellowish brown.

LIFE HISTORY

Eggs are laid during the summer in batches on the bark of host plants. They hatch in April. The larvae then feed in the buds and, later, within spun leaves or a rolled leaf. Blossoms and young fruitlets are also attacked. Pupation occurs in the larval habitation from mid-May or June onwards, adults appearing from late June or July to early September.

DAMAGE

Leaf-feeding is of little consequence. However, attacks on buds, flowers and fruitlets can be serious, causing direct crop losses or, as on apple, the formation of russeted, badly misshapen and unmarketable fruit.

Archips xylosteana (Linnaeus) (**614–616**)
Brown oak tortrix moth

A minor pest of fruit trees, particularly apple, pear, cherry and plum. Also polyphagous on various wild trees, shrubs and woody plants, including *Abies*, *Acer*, *Betula*, *Fraxinus*, *Populus*, *Salix*, *Sorbus*, *Tilia*, *Ulmus* and, especially, *Quercus*. Eurasiatic. Widely distributed in Europe.

614 Brown oak tortrix moth (*Archips xylosteana*) – male.

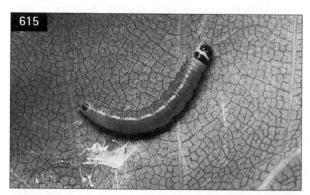

615 Brown oak tortrix moth (*Archips xylosteana*) – larva.

613 Rose tortrix moth (*Archips rosana*) – pupa.

616 Brown oak tortrix moth (*Archips xylosteana*) – larval habitation on apple.

DESCRIPTION

Adult: 15–23 mm wingspan; forewings whitish ochreous, suffused with olive grey, and with reddish-brown, pale-edged, variegated markings; hindwings greyish brown. **Larva:** up to 22 mm long; body whitish grey, sometimes grey or dark bluish grey, with paler sides; head shiny black; prothoracic plate black or dark brown, with a whitish mid-line and anterior border; anal plate black or brownish black; anal comb present. **Pupa:** 9–12 mm long; dark brown or black; cremaster elongate.

LIFE HISTORY

Adults are most numerous in July. Eggs are laid in batches on the trunks or branches and then coated with a brownish secretion that camouflages them against the bark. They hatch in late March, April or early May. Larvae feed in a similar way to those of *Archips crataegana*, each typically inhabiting either a tube-like fold along the edge of a leaf or a completely rolled leaf. Pupation takes place in May or June.

DAMAGE

Feeding is confined mainly to fully expanded leaves and is not of major importance; however, young fruitlets may also be attacked, resulting in the development of corky blemishes on the skin.

Argyrotaenia pulchellana (Haworth) (**617–620**)

In mainland Europe a pest of apple and grapevine. Also associated with a wide range of wild trees, shrubs and herbaceous plants. Widely distributed in Europe; also present in North America.

DESCRIPTION

Adult: 12–16 mm wingspan; forewings silvery grey to light brown, with dark reddish-brown markings; hindwings grey, with whitish cilia. **Egg:** lemon-yellow when laid, later becoming light brown. **Larva:** up to 18 mm long; body yellowish green, with pale and indistinct pinacula; head yellowish green or light green; prothoracic plate mainly light green; anal plate light green, often speckled with black; anal comb well

617 *Argyrotaenia pulchellana* – female.

618 *Argyrotaenia pulchellana* – larva.

619 *Argyrotaenia pulchellana* – larva and habitation.

620 *Argyrotaenia pulchellana* – pupa.

developed, with 6–8 prongs. **Pupa:** 8–9 mm long; yellowish brown to dark brown; cremaster elongate and robust, armed with several thick, curved bristles.

LIFE HISTORY

In Europe, there are up to three generations annually, with two being most frequently encountered. Adults of the first generation occur from April to May and those of the second from June to July. Eggs are laid in batches on the leaves of host plants and hatch within 2–3 weeks. At first, the larvae graze on the lower surface of leaves, often close to the midrib, but later they inhabit spun leaves. The larvae also feed on the developing inflorescences and fruitlets. Pupation takes place in the larval habitation, adults of the second generation emerging about 7–10 days later. Second-brood larvae attack leaves and maturing fruits. Pupae resulting from second-brood larvae usually overwinter. However, under favourable conditions, they may give rise to a third or at least a partial third generation.

DAMAGE

Apple: larval damage to maturing fruits is particularly serious, with large areas of skin and the underlying tissue removed, especially around the stalk; such fruits are unmarketable. **Grapevine:** early infestations reduce cropping potential by aborting flowers and destroying young fruitlets. Damage to older fruits results in darkened patches on the skin, of particular importance on table grapes; the presence of silk webbing in bunches of grapes is also a problem.

Cacoecimorpha pronubana (Hübner) (**621–627**)
Carnation tortrix moth

Highly polyphagous and a locally important pest of strawberry, both in the open and under protection. Other hosts include almond, cherry, raspberry, grapevine, citrus, olive and a wide range of other trees, shrubs and herbaceous plants. Of South African origin. Now widely distributed and locally common across Europe to Asia Minor; also present in North Africa and North America.

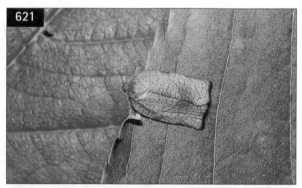

621 Carnation tortrix moth (*Cacoecimorpha pronubana*) – female.

622 Carnation tortrix moth (*Cacoecimorpha pronubana*) – male.

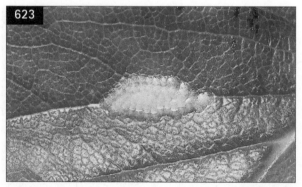

623 Carnation tortrix moth (*Cacoecimorpha pronubana*) – egg batch.

624 Carnation tortrix moth (*Cacoecimorpha pronubana*) – young larva.

DESCRIPTION

Adult female: 18–22 mm wingspan; forewings pale orange brown, reticulated with darker brown; hindwings mainly orange. **Adult male:** 12–17 mm wingspan; forewings orange brown, with variable reddish-brown and purplish-black markings; hindwings bright orange, each with a blackish border. **Egg:** light green. **Larva:** up to 20 mm long; body olive green to bright green, paler below, and with slightly paler pinacula; head greenish yellow or yellowish brown, marked with dark brown; prothoracic and anal plates green, marked with dark brown; anal comb green, usually 6-pronged. **Pupa:** 9–12 mm long; brownish black to black; cremaster elongate and tapered, with 8 strong, hooked bristles.

LIFE HISTORY

In Europe, this species has from two to five generations annually, but there is considerable overlap and all stages of the insect may occur together. Adults appear from April to October, or even later, but in the more northerly parts of its range are most numerous from May to June, and from late August onwards. The males have a characteristic, erratic flight and are very active in sunny weather. Eggs are laid on leaves in batches of up to 200 and they hatch 2–3 weeks later. Larvae then develop through seven instars, inhabiting spun leaves and feeding on the foliage, blossom trusses or developing fruits. Pupation occurs within the larval habitation, or in a freshly folded leaf or amongst webbed foliage, adults emerging shortly afterwards. The winter is usually passed as young larvae sheltering on the foodplant in silken hibernacula. Under favourable conditions, breeding is continuous and all stages of the pest may be found together.

DAMAGE

The larvae are voracious feeders and are capable of causing considerable harm to foliage, blossoms and fruits. Larvae may also mine within young shoots, causing noticeable distortion. Strawberries are often attacked close to harvest, the larvae burrowing into the flesh beneath the shelter of the calyx. In more northerly parts of its range, attacks on strawberry tend to be most severe on protected crops. On citrus, in spring and summer, larvae often feed on the outer tissue (mesocarp) of developing fruits. At first, such fruits are attacked close to the base. Later, the larvae browse more extensively, sheltered by webbed leaves. Feeding is restricted to the mesocarp, and the damaged areas soon become corky and discoloured; such fruits may also become distorted. Later in the year, ripe fruits are attacked, usually around the base; these damaged areas do not heal and, following invasion by secondary pathogens, may eventually rot.

625 Carnation tortrix moth (*Cacoecimorpha pronubana*) – early larval habitation on olive.

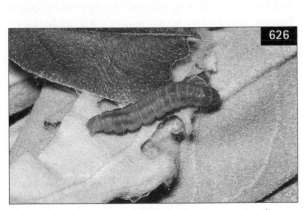

626 Carnation tortrix moth (*Cacoecimorpha pronubana*) – final-instar larva.

627 Carnation tortrix moth (*Cacoecimorpha pronubana*) – pupa.

Celypha lacunana (Denis & Schiffermüller) (628–629)

Dark strawberry tortrix moth

A pest of strawberry and especially important on crops grown under cloches. Occasionally also found on raspberry and other fruit crops. In the wild, associated with a wide range of herbaceous plants, trees and shrubs. Eurasiatic. Widely distributed in Europe.

DESCRIPTION

Adult: 14–18 mm wingspan; forewings pale ochreous or greyish ochreous, with greenish-black and scattered blue-grey markings; hindwings dark brownish grey. **Egg:** pale yellowish. **Larva:** up to 15 mm long; body rather slender, usually dark purplish brown, with blackish pinacula and with a greenish tinge apparent between the segments and below; head, prothoracic plate, anal plate and anal comb blackish brown or dark purplish brown; anal comb prominent, with five long prongs. **Pupa:** 8–10 mm long; dark brown; cremaster blunt, with eight long, hook-tipped bristles.

LIFE HISTORY

This species is double-brooded. Adults of the first generation occur in May and June. Eggs are then laid on the upper side of leaves in batches of two to three, usually close to the major veins. They hatch in about 10–21 days. Larvae feed amongst spun leaves, passing through five instars. They are very active if disturbed and wriggle rapidly backwards, thrashing the body violently from side to side. Pupation occurs in webbed leaves in late July or August and adults of the second generation appear in late July or August. Larvae from eggs laid by these moths feed briefly in September and October and then enter hibernation, usually in their second and third instars, each spinning up within a folded leaf edge. Feeding is resumed in the early spring, the larvae initially browsing on the foliage. During the blossom period they also attack the flowers, sheltering beneath a canopy of webbed petals. Pupation occurs in late April or May. Adults appear shortly afterwards.

DAMAGE

In common with other strawberry-feeding tortricids, this species contributes to loss of foliage, and larvae attacking the flowers cause abortion or malformation of the fruitlets. Damage tends to be particularly severe on protected crops.

628a, b Dark strawberry tortrix moth (*Celypha lacunana*).

629 Dark strawberry tortrix moth (*Celypha lacunana*) – larva.

Choristoneura diversana (Hübner)

A minor pest of apple, pear and plum. Other hosts include *Betula*, *Lonicera*, *Quercus*, *Rhamnus*, *Salix* and *Syringa*. Widely distributed in central and northern Europe; also present across northern Asia to Japan.

DESCRIPTION

Adult: 15–23 mm wingspan; forewings light brown, with darker markings; hindwings grey to brownish grey. **Larva:** body green or greyish green, with yellow pinacula; head and prothoracic plate dark brown or reddish brown.

LIFE HISTORY

At least in northern Europe there is just one generation annually, with adults active in July. Eggs are then laid in large batches on the leaves of host plants; they hatch in August and September. Larvae feed briefly and then hibernate. In spring, the larvae feed in spun leaves, becoming fully grown by late May or early June. Pupation then takes place in the larval habitation or amongst freshly webbed leaves.

DAMAGE
Before hibernating, young larvae may graze on the surface of maturing fruits. However, the insect is rarely found on sprayed fruit trees. Leaf damage caused in the spring is not important.

Choristoneura hebenstreitella (Müller)
Larvae of this generally distributed and locally common woodland species occur occasionally on apple, pear and plum, but they are not important fruit pests. Adults are similar to those of *Choristoneura diversana*, but much larger (20–29 mm wingspan); also, the forewing markings are distinctively olive brown and those of males have a short, but distinct, costal fold. The moths occur in June and July. Larvae feed briefly before hibernation, and complete their development in the following spring. Fully grown larvae are dark greyish green, with pale pinacula; the head is black or dark brown, the prothoracic plate brown (with a straight, white leading edge and partly black sides) and the anal plate yellowish brown.

Clepsis spectrana (Treitschke) (**630–631**)
Straw-coloured tortrix moth
A pest of black currant, blackberry, strawberry, grapevine and hop. Also associated with a wide range of wild herbaceous plants. Generally common in northern and central Europe, particularly in fenland and coastal habitats.

DESCRIPTION
Adult: 15–24 mm wingspan (male usually noticeably smaller than female); forewings pale ochreous to yellowish, with variable dark brown to blackish markings; hindwings light grey. **Larva:** up to 25 mm long; body brown to greyish olive green, paler laterally and ventrally, with whitish pinacula; head and prothoracic plate shiny black or blackish brown; anal plate whitish, marked with black or brown; anal comb with 6–8 long prongs. **Pupa:** 10–14 mm long; dull black; cremaster stout and elongate.

LIFE HISTORY
This species is double-brooded. Adults of the first generation occur from early June to July, sometimes earlier. Eggs are laid in small batches on the foodplant, hatching 2–3 weeks later. The larvae then feed beneath a web or in young, webbed leaves and 'capped' flowers. When fully grown, each pupates in a white, silken cocoon

in the larval habitation, in webbed leaves or amongst dead leaves. A second generation of adults appears in August and September. Larvae from these moths feed for a short time before hibernating in silken retreats, usually spun on the foodplant. The larvae reappear in early spring to continue feeding and complete their development, usually pupating in May or June.

DAMAGE
On black currant, blackberry, grapevine and hop, larvae may cause considerable damage to foliage, particularly on the young shoots. New growth arising on mechanically harvested black currant bushes is most likely to be attacked. On strawberry, damage in spring develops rapidly, particularly on protected crops; larvae are especially attracted to the flowers and, by feeding on the stamens and receptacle beneath the shelter of webbed-down petals, often cause loss or malformation of fruits.

630 Straw-coloured tortrix moth (*Clepsis spectrana*) – female.

631 Straw-coloured tortrix moth (*Clepsis spectrana*) – larva.

Cnephasia asseclana (Dennis & Schiffermüller) (632–634)

Flax tortrix moth

A sometimes common pest of currant, loganberry, raspberry and strawberry. Also associated with a wide range of wild herbaceous plants, including *Chrysanthemum leucanthemum*, *Lathyrus pratensis*, *Plantago*, *Ranunculus* and *Rumex*. Widely distributed and generally common in Europe; also present in Canada.

DESCRIPTION

Adult: 15–18 mm wingspan; forewings whitish grey, more or less suffused with black and ochreous, and with dark, greyish-brown, blackish-edged markings; hindwings greyish brown. **Egg:** greenish yellow. **Larva:** up to 15 mm long; body grey or bluish white to dark cream or greyish green, with large, black pinacula; head light or dark yellowish brown, marked with black; prothoracic plate light brown to dark brown, marked with black, and with a fine, whitish midline and a partly whitish border; anal plate yellowish brown, marked with black; anal comb dark brown or blackish, with about 6 long prongs. **Pupa:** 7–9 mm long; light brown; tip with several hook-tipped bristles; the pair of dorsal cremastal projections of moderate size (cf. *Cnephasia incertana*).

LIFE HISTORY

Moths occur from June to August. Eggs are deposited, either singly or in small batches, on herbaceous plants or on tree trunks and other rough surfaces. They hatch in about 3 weeks and the larvae then spin small cocoons in suitable shelter nearby, having fed only on their egg shells. They then overwinter within these hibernacula. Activity is resumed in the spring, the first-instar larvae mining leaves to feed in irregular, usually blotch-like mines. Later, each larva feeds amongst spun leaves or on a flower or flowerhead, on strawberry spinning the petals down with silk to form a 'capped' blossom. If disturbed, the larva usually rolls into a tight 'C' and drops to the ground. Pupation takes place in late May or early June in the folded edge of a leaf or amongst debris on the ground.

DAMAGE

Attacks on foliage are relatively unimportant, but damage to stamens and receptacles of flowers may result in the loss or malformation of fruits. On strawberry, the larvae can also cause direct damage to the developing fruits by tunnelling into the flesh beneath the calyx.

Cnephasia incertana (Treitschke) (635–636)

Light grey tortrix moth

A pest of strawberry, particularly in gardens. In mainland Europe also found (for example) on apple, pear, cherry and grapevine. A very polyphagous species, especially on herbaceous plants. Generally distributed and common in Europe. Also found in Asia Minor and North Africa.

632 Flax tortrix moth (*Cnephasia asseclana*).

633 Flax tortrix moth (*Cnephasia asseclana*) – larva.

634 Flax tortrix moth (*Cnephasia asseclana*) – larval habitation on strawberry.

DESCRIPTION

Adult: 14–18 mm wingspan; forewings greyish white, suffused with greyish brown, and with greyish-brown, blackish-suffused markings; hindwings mainly grey. **Egg:** olive brown or greenish. **Larva:** up to 15 mm long; body dull, dark green to brownish, with black or dark brown pinacula; head light brown, with a black hind margin; prothoracic plate black, with a fine, white medial line and a white front margin; anal plate large and black; unlike *Cnephasia asseclana*, no anal comb. **Pupa:** 7–9 mm long; dark brown to blackish; tip with several hook-tipped bristles; the pair of dorsal cremastal projections small and inconspicuous (cf. *C. asseclana*).

LIFE HISTORY

Similar to that of flax tortrix moth (*C. asseclana*), but larvae usually complete their development slightly earlier in the year; also, moths occur in late May, June and July.

DAMAGE

As for flax tortrix moth (*C. asseclana*).

Cnephasia longana (Haworth) (**637–638**)

larva = omnivorous leaf tier

A minor pest of apple, pear, black currant, raspberry, strawberry, hazelnut and hop, but associated mainly with herbaceous plants such as *Achillea millefolium*, *Aster tripolium*, *Chrysanthemum leucanthemum*, *Hypochoeris radicata* and *Lychnis flos-cuculi*. Widely distributed and locally common in Europe, particularly in coastal and chalkland areas; also present in the Canary Islands, north-west Africa and North America.

DESCRIPTION

Adult female: 15–22 mm wingspan; forewings whitish ochreous or greyish white, with yellowish-brown or greyish-brown markings; hindwings light grey. **Adult male:** 15–22 mm wingspan; forewings uniformly pale ochreous to brownish ochreous; hindwings light grey. **Egg:** salmon-pink. **Larva:** up to 15 mm long; body rather plump, greenish grey to yellowish grey, with pale longitudinal bands along the back and sides; pinacula small and blackish; head light brown; prothoracic plate shiny, light green to brownish or yellowish grey, marked with brown; anal plate light brown; anal comb light brown, with 6–8 long prongs. **Pupa:** 7–9 mm long; light brown; tip with numerous fine, hook-tipped bristles.

635 Light grey tortrix moth (*Cnephasia incertana*).

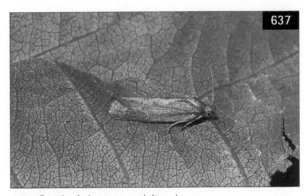

637 *Cnephasia longana* – adult male.

636 Light grey tortrix moth (*Cnephasia incertana*) – larva.

638 *Cnephasia longana* – larva.

LIFE HISTORY

Adults occur mainly in July. Eggs are then laid singly or in small batches on the foodplant or on rough posts and tree trunks. They hatch in late July or in August. After devouring its egg shell, each tiny larva crawls away and then spins a silken cocoon in which to overwinter. In early spring, the larvae leave their hibernacula to begin feeding on a wide range of host plants, often being dispersed by wind after producing a fine silken strand. The larvae feed in spun leaves, especially on terminal shoots, webbing the foliage tightly together. They also attack flowers and blossom trusses, and (as on apple) may also bore into the shoot tips. Pupation occurs in June, within a flimsy cocoon spun amongst debris on the ground or amongst dead leaves. Moths emerge 2–3 weeks later. There is just one generation annually.

DAMAGE

Direct damage to foliage is unimportant. However, on apple, pear and black currant, larvae tunnelling into shoots cause considerable disruption of growth, and are especially damaging in nurseries. When on strawberry, larvae commonly attack the flowers, causing death or the development of malformed fruits. Infestations are most likely to occur in weedy orchards and plantations, particularly in chalkland or coastal sites.

Croesia holmiana (Linnaeus) (**639–640**)

A polyphagous species on rosaceous trees and shrubs, including fruit trees (particularly apple and pear). Widely distributed in northern and central Europe; also present in Asia Minor.

DESCRIPTION

Adult: 12–15 mm wingspan; forewings yellowish orange to reddish brown, with a white costal blotch; hindwings light grey. **Larva:** up to 12 mm long; body yellowish green to ochreous; head brown and shiny; prothoracic plate brown, marked with black laterally. **Pupa:** 5–6 mm long; yellowish brown to orange yellow; cremaster broad, and bearing several short, hook-tipped bristles.

LIFE HISTORY

Adults occur in July and August, but sometimes earlier. Eggs are laid singly on the bark of host trees, where they overwinter. Larvae occur mainly in May and June, spinning the margins of two leaves together as a shelter. Pupation occurs in a freshly folded leaf or in the larval habitation.

DAMAGE

Larvae graze on the leaves and young fruitlets, contributing to damage caused by various other species.

Cydia funebrana (Treitschke) (**641–642**)

Plum fruit moth

larva = red plum maggot

A locally important pest of damson and plum; other recorded hosts include apricot and peach. Generally common and sometimes abundant on *Prunus spinosa*. Eurasiatic. Widely distributed in Europe.

DESCRIPTION

Adult: 11–15 mm wingspan; forewings dull purplish grey, with darker (but obscure) markings, and irregularly suffused with ash-grey; hindwings brownish grey. **Egg:** 0.7 × 0.6 mm; whitish and translucent when laid, later becoming yellowish. **Larva:** up to 12 mm long; body mainly bright pinkish red, with brownish, inconspicuous pinacula; head dark brown to black; prothoracic plate light brown and translucent, with darker markings; anal plate light brown and relatively

639 *Croesia holmiana* – adult female.

640 *Croesia holmiana* – larva.

narrow; anal comb with a weak base; young larva whitish and translucent. **Pupa:** 6–7 mm long; light brown; abdomen with dentate ridges dorsally; tip with several fine, hook-tipped bristles.

LIFE HISTORY

Adults occur from late April or May onwards. In cooler regions, where there is usually just one generation annually, they are most numerous from mid-June to mid-July; in more favourable districts, however, where the pest is double-brooded, peak numbers occur somewhat earlier in the season. Eggs are laid on the fruitlets from May or June onwards, often close to the depression which passes from fruit stalk to fruit tip, usually just one per fruitlet. The eggs hatch in about 2 weeks. Each larva then bites into the fruitlet a short distance from its empty egg shell. Larvae feed for several weeks, passing through five instars. When fully grown (in warmer regions usually from late June onwards, but under cooler conditions from July or early August onwards), they bore out of the fruits and spin silken cocoons under loose bark or in other suitable situations. In regions where there are two generations annually, larvae then pupate to produce a second generation of adults in July and August. Elsewhere, however, larvae usually remain in diapause until the following spring, although early-developing larvae may pupate and give rise to a partial second generation of adults. Second-generation adults tend to lay their eggs on the lower parts of maturing fruits; also, they often deposit more than one egg per fruit, but then only one of the larvae normally survives. Second-brood larvae vacate the fruits in late summer or autumn. They then overwinter and pupate in the spring.

DAMAGE

Larvae feed within relatively large fruitlets, attacks being initiated considerably later in the season than are those of plum-feeding sawflies (*Hoplocampa flava* and *H. minuta*) (see Chapter 7). On entering the flesh, the young larva forms a narrow, winding mine directed from the point of entry towards the fruit stalk. The mine soon turns brown and is then clearly visible through the skin. From near the stalk, the mine is extended to the centre of the fruit and as the larva grows the flesh around the stone is eaten and replaced by masses of wet, brown frass. When fully fed, the larva escapes through the side of the fruit, leaving a small (c. 2 mm) circular hole in the skin. Tell-tale particles of frass are not ejected from the larval galleries, but since attacked fruit ripen prematurely they are easily recognized amongst the developing crop. When infestations are initiated in nearly ripe fruits (typical of second-generation attacks) the larval mine runs directly from the point of entry to the stone. Fruit losses attributable to red plum maggot

641a, b Plum fruit moth (*Cydia funebrana*).

642 Plum fruit moth (*Cydia funebrana*) – larva.

are especially severe in years with a light fruit set and when conditions favour production of a marked second generation of the pest. Even low pest levels can be important in crops sent for processing, since the presence of only a few infested fruits may lead to the rejection of complete consignments.

Cydia janthinana (Duponchel) (643–644)

A pest of apple in parts of mainland Europe (notably southern Germany and Switzerland, where significant attacks have been reported), but the normal host is *Crataegus monogyna*. Widely distributed in Europe and Asia Minor.

DESCRIPTION

Adult: 9–11 mm wingspan; forewings mainly brown, with irregular, lighter and darker markings; hindwings mainly greyish brown. **Larva:** up to 10 mm long; body dull reddish to pinkish; head and prothoracic plate yellowish brown; anal plate yellowish brown, spotted with brownish black.

LIFE HISTORY

Adults occur in July and are often seen flying around hawthorn bushes in afternoon sunshine. Eggs are laid singly on the fruits of host plants and hatch several weeks later. Larvae, which occur from late August to late September or October, feed within the fruits, often spinning adjacent ones together lightly with silk. Ejected pellets of frass soon accumulate amongst slight webbing between adjacent fruits and this is a useful clue to the presence of a larva. Fully-fed larvae vacate infested fruits and then overwinter in cocoons formed under the bark of host trees or amongst leaf litter on the ground. Pupation occurs in the spring, shortly before the appearance of the adults.

DAMAGE

On apple, larvae destroy part of the flesh of infested fruits. However, the galleries, unlike those formed by larvae of codling moth (*Cydia pomonella*), remain largely free of frass. The larvae may also form irregular cavities and channel-like scars on the skins of attacked fruits.

Cydia lobarzewskii (Nowicki)

A locally important pest of apple, damson and plum in mainland Europe; cherry is also a host. Currently, of particular significance on apple in Switzerland. Although found in plum orchards in south-east England (Kent) during the mid-1900s, the pest is otherwise virtually unknown in the British Isles.

DESCRIPTION

Adult: 13–14 mm wingspan; forewings mainly greyish brown, dark brown to blackish; hindwings dark greyish brown. **Larva:** up to 12 mm long; body pinkish above, whitish yellow below; head reddish brown; prothoracic plate brownish and translucent.

LIFE HISTORY

Adults occur in May and June. Eggs are then laid singly on the developing fruitlets. Larvae feed within the fruitlets throughout July. When fully grown they emerge and spin cocoons in sheltered situations on the trunks of host plants. Here they overwinter and eventually pupate in the spring. Under suitable conditions, however, larvae may pupate the summer before overwintering and give rise to at least a partial second generation.

DAMAGE

Similar to that caused on apple and plum by codling moth (*Cydia pomonella*) and plum fruit moth (*C. funebrana*), respectively.

Cydia molesta (Busck) (645–650)

Oriental fruit moth

An important pest of peach; apple, pear, quince and almond are also attacked. Originating in eastern Asia (China and Japan) and now established in the warmer parts of Europe, including eastern Austria, the Czech Republic, France, Greece, Hungary, Italy and the former Yugoslavia. Sometimes found on imported fruit in more northerly countries, where it fails to become established. Also present in North Africa, North America, South America, western Asia and south-western Australia.

643 *Cydia janthinana* – larva.

644 *Cydia janthinana* – infested fruits of *Crataegus.*

DESCRIPTION

Adult: 11–14 mm wingspan; forewings mainly grey to blackish, with a small, distinctive, whitish spot; hindwings brownish, but paler basally. **Egg:** whitish and translucent when laid, later becoming yellowish and the developing embryo then visible as a red ring. **Larva:** up to 12 mm long; body pinkish brown; head, prothoracic plate and anal plate yellowish brown. **Pupa:** 6 mm long; yellowish brown; abdomen with reddish-brown dentate ridges dorsally.

LIFE HISTORY

In mainland Europe, adults appear from March onwards. Eggs are eventually laid singly on leaves and twigs of host plants; they hatch 1–4 weeks later, depending on temperature. Larvae bore into buds and shoots, and are fully grown within about 2–3 weeks. They then pupate, each in a whitish cocoon spun on the surface of the foodplant, under bark or, occasionally, in the soil. Adults emerge 1–2 weeks later. There are up to five generations annually, larvae of the later generations

645 Oriental fruit moth (*Cydia molesta*).

646 Oriental fruit moth (*Cydia molesta*) – fully fed larva.

647 Oriental fruit moth (*Cydia molesta*) – young larva in quince fruit.

648 Oriental fruit moth (*Cydia molesta*) – pupa.

649 Oriental fruit moth (*Cydia molesta*) – damage to pear shoot.

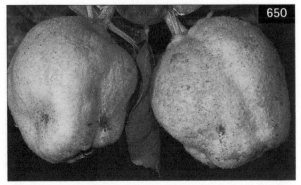

650 Oriental fruit moth (*Cydia molesta*) – damage to quinces.

feeding within the developing fruits. Such fruits have typical frass-filled holes on the surface, similar to those caused on pome fruits by codling moth (*Cydia pomonella*) larvae. Fully-fed larvae of the final generation overwinter in cocoons and pupate in the early spring.

DAMAGE

Infestations are particularly severe on peach, the larvae destroying buds in the spring and also disrupting the growth of young shoots, causing them to wilt. Later in the season, infested fruitlets may drop prematurely. Also, damage to older fruits causes distortion, and the pulp is partly destroyed and filled with brownish frass. Such fruits are unmarketable and heavy infestations result in significant crop losses.

Cydia pomonella (Linnaeus) (651–654)
Codling moth
larva = apple maggot
A notorious, worldwide pest of apple and, occasionally, pear, quince, peach, chestnut and walnut. Generally common, and often abundant, in Europe.

DESCRIPTION
Adult: 15–22mm wingspan; forewings blackish brown, more or less suffused with ash-grey and with a large, metallic, bronzy-black submarginal ocellus; hindwings brown. **Egg:** 1.3 × 1.0mm; whitish and opalescent when laid, later with the developing embryo visible as a red ring. **Larva:** up to 20mm long; body pale pinkish white (younger instars whitish), with dark pinacula; head and prothoracic plate brown; anal plate pale; no anal comb. **Pupa:** 8–10mm long; yellowish brown to dark brown; abdomen with dentate ridges dorsally; tip with several, hook-tipped bristles.

LIFE HISTORY
First-generation adults occur from mid- or late May onwards, earlier or later depending on temperature. Eggs are laid singly on developing fruitlets and on foliage, usually during warm evenings with temperatures above 15°C. They hatch in about 10–14 days, the small larvae then burrowing into the fruitlets. Fruitlets are often entered through the calyx ('eye'), but larger fruits are usually invaded through the side or near the stalk. Larvae are fully grown after about 4 weeks,

651 Codling moth (*Cydia pomonella*).

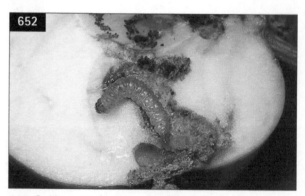

652 Codling moth (*Cydia pomonella*) – larva.

653 Codling moth (*Cydia pomonella*) – external damage to apple.

654 Codling moth (*Cydia pomonella*) – external damage to pear and secondary rotting.

passing through five instars. They then vacate the fruit, which may or may not have dropped to the ground, and spin cocoons under loose bark, in cracks in tree trunks and supporting stakes, and so on. In northern Europe, where this species is mainly single-brooded, fully-fed larvae usually overwinter and pupate in the spring; however, under favourable conditions, those spinning up by the end of July may pupate and produce a partial second generation of adults in August and early September. Larvae arising from eggs laid by these individuals complete their feeding in the autumn and then overwinter. Late-developing larvae still inside the fruits after harvest often form their cocoons in cracks in apple boxes or elsewhere in fruit stores and packhouses. In warmer parts of Europe there are regularly two or more generations annually.

DAMAGE
Apple: at first, an invading larva (typically one per infested fruit) forms a small cavity just beneath the skin and, after feeding for a few days, burrows down to the core, leaving a prominent, red-ringed entry hole in the side or near the eye, characteristically blocked by dry frass. Within the fruit, the larva eats away a large proportion of the flesh and also attacks the pips, the cavity becoming filled with brown frass. The entry point at the surface is greatly enlarged as tissue beneath is eaten away and the larva eventually escapes leaving a small, unplugged exit hole. Sometimes a larva will attack another fruit in the same cluster before becoming fully grown; damaged fruits tend to ripen and drop prematurely. Attacks can be very serious, particularly following the appearance of a second generation of larvae, and are often most severe close to stores and packhouses and on trees near piles of empty boxes that held the previous season's crop. Infestations of codling moth occur later in the year and in larger (older) fruits than those of, for example, fruitlet-mining tortrix moth (*Pammene rhediella*) and apple sawfly (*Hoplocampa testudinea*) (Chapter 7); also, there is little or no external accumulation of frass, most remaining within the larval gallery. **Pear:** larvae cause similar damage on pear and wounds are often an entry point for secondary fungal rots.

Cydia pyrivora (Danilevsky)
A pest of pear in eastern Europe; also reported from Austria and northern Italy (South Tyrol), and from various Mediterranean islands (e.g. Corsica, Crete, Sardinia and Sicily).

DESCRIPTION
Adult: 20–22 mm wingspan; essentially similar in appearance to that of *Cydia pomonella*; the two species are most reliably separated by genitalial differences.

Larva: similar to that of *C. pomonella*, but body greyish white and head dark brown to black; also distinguished by the much smaller number of crochets (approximately six) on the anal claspers (>20 in the case of *C. pomonella*).

LIFE HISTORY
There is just one generation annually, and adults appear somewhat later than those of codling moth (*C. pomonella*).

DAMAGE
Similar to that caused on pear by codling moth (*C. pomonella*).

Cydia splendana (Hübner) (**655–656**)
Acorn moth
A pest of chestnut and, occasionally, walnut. Associated in the wild with *Castanea sativa* and *Quercus*. Eurasiatic. Widely distributed in Europe.

DESCRIPTION
Adult: 14–22 mm wingspan; forewings mainly greyish white, suffused with darker grey and brown; costal strigulae usually pronounced; submarginal ocellus

655 Acorn moth (*Cydia splendana*).

656 Acorn moth (*Cydia splendana*) – larva.

large, with several black dashes; hindwings greyish brown. **Egg:** 0.72 × 0.55 mm; whitish when laid, later with the developing embryo visible as a purplish-red ring. **Larva:** up to 15 mm long; body greyish green to yellowish white and translucent; head yellowish brown; prothoracic plate and anal pale poorly sclerotized; pinacula inconspicuous; no anal comb.

LIFE HISTORY
Fully fed larvae overwinter in cocoons spun in the soil. They have an extended period of diapause and usually

657 Vine tortrix moth (*Ditula angustiorana*) – female.

658 Vine tortrix moth (*Ditula angustiorana*) – male.

659 Vine tortrix moth (*Ditula angustiorana*) – larva.

do not pupate until the summer, usually in June or July. Adults appear about 3–4 weeks later, with the main flight period extending either from July to August or from August to September. Some larvae may remain in diapause over a second winter. Eggs are laid singly along the veins of leaves close to the developing nuts. The eggs hatch 10–12 days later. Larvae then bore into the developing nuts to feed, becoming fully grown in the autumn.

DAMAGE
Part of the kernel of an infested nut is destroyed and replaced by a mass of blackish frass. Occasionally, heavy infestations occur and may result in considerable crop losses.

Ditula angustiorana (Haworth) (657–659)
Vine tortrix moth
A polyphagous and often common pest of fruit crops, particularly apple, pear, apricot, plum, raspberry and grapevine; also occurs on a wide range of forest trees and shrubs, including conifers. Widespread across Europe to Asia Minor; also present in parts of North Africa and North America.

DESCRIPTION
Adult female: 14–18 mm wingspan; forewings pale ochreous brown to whitish ochreous, with chestnut-brown, dark brown, blackish and bluish markings; hindwings brown. **Adult male:** 12–15 mm wingspan; forewings greyish brown to ochreous brown, distinctly marked with dark purplish brown, blue and black; hindwings dark brown. **Egg:** pale yellow. **Larva:** up to 18 mm long; body slender, pale yellowish green to brownish green or greyish green, and darker above, with light green pinacula; head greenish yellow or yellowish brown, marked with blackish brown; prothoracic plate yellowish green, light brown or dark brown; anal comb greenish or brownish, with four prongs; thoracic legs green, tipped with blackish brown; spiracles small, the last pair twice the diameter of the others. **Pupa:** 8 mm long; light brown; cremaster elongate, with eight tightly hooked bristles.

LIFE HISTORY
Adults occur mainly in June and July, the males often flying in sunshine. Eggs are laid on the leaves in moderately large batches. The newly emerged larvae feed on the foliage and after moulting will also attack the fruits. In autumn, whilst still small, they spin silken hibernacula on buds or spurs, where they overwinter until the early spring. They then attack buds, young leaves and, later, blossom trusses and fruitlets, often sheltering in spun leaves and becoming very active if

disturbed. Pupation occurs in May or June in a cocoon spun in a folded leaf, in webbed foliage or amongst dead leaves on the ground.

DAMAGE

The larvae are rarely sufficiently numerous to cause economic damage to leaves, flowers or developing fruitlets, and grazing of mature apple and pear fruits by young larvae in the autumn is superficial. The pest does, however, contribute to overall damage inflicted at that time by other lepidopterous larvae.

Enarmonia formosana (Scopoli) (**660–661**)
Cherry bark tortrix moth

A locally common pest of apple, pear, almond, apricot, cherry, peach and plum; usually associated only with older trees. Often also numerous on ornamental *Prunus* and *Sorbus*. Eurasiatic. Widely distributed in Europe. Also recently established as a pest in North America.

DESCRIPTION

Adult: 15–18 mm wingspan; forewings more or less brown to black, with a purplish sheen, and with irregular, yellowish-orange markings and silvery-white costal strigulae; hindwings dark brown. **Egg:** 0.7 × 0.6 mm; creamy white when laid, becoming reddish. **Larva:** up to 11 mm long; body translucent greyish white, with brownish-grey pinacula; head light brown; prothoracic and anal plates light greyish brown. **Pupa:** 7–9 mm long; light brown; cremaster broad and blunt (cf. *Esperia sulphurella*, family Oecophoridae).

LIFE HISTORY

Adults appear from May or early June to September, the extended emergence and flight periods giving the impression of two generations. The moths are active in sunshine and often make repeated short flights to and from the branches or trunks of infested trees, but they are well camouflaged and difficult to see when settled on the bark. Eggs are laid singly or in batches of two or three, usually on previously infested or otherwise damaged parts of host trees. They hatch in 2–3 weeks and the larvae then attack the bark to feed beneath the surface. Brown silk-lined tubes of frass often protrude from the bark of infested trees, at once indicating the presence of larvae. The larvae are usually fully grown by the following spring or early summer, passing through five instars. Each then pupates in a silken cocoon formed within the larval feeding gallery. Pupae remain protruding from the bark of infested trees after the adult moths have emerged.

DAMAGE

Larvae excavate irregular, often deep, galleries in the bark, members of successive generations feeding within and considerably enlarging the same burrows. Most feeding occurs in the bark tissue; the underlying cambium may also be damaged, but the feeding tunnels do not extend into the wood. Attacks are normally established only on older and, often, previously damaged trees. For example, they are often initiated in frost-damaged bark or adjacent to pruning wounds and other mechanically damaged areas; also, on apple, larvae sometimes occur in association with those of apple clearwing moth (*Synanthodon myopaeformis*) (family Sesiidae). Particularly on cherry and plum, a considerable quantity of gum may exude from infested parts of host trees and this, along with accumulations of light brown frass and silken webbing forced out of cracks in the bark, may be one of the first indications of an attack. Cherry trees are usually infested near the base of the trunks. However, apple trees, particularly large ones, are frequently damaged on the underside of the

660 Cherry bark tortrix moth (*Enarmonia formosana*).

661 Cherry bark tortrix moth (*Enarmonia formosana*) – larval frass tubes.

main branches, close to the trunk, and on the trunk near such areas. Infestations on apple result in the production of cracks, swellings and cankers, and are particularly serious; eventually, branches or even whole trees may be killed.

Epiblema uddmanniana (Linnaeus) (662–664)
Bramble shoot moth
larva = bramble shoot webber
An often important pest of blackberry, boysenberry, loganberry and tayberry. Generally common and widely distributed in Europe; also present in Asia Minor and North Africa.

DESCRIPTION
Adult: 15–20 mm wingspan; forewings light brownish grey, tinged with olive green, and each with a subtriangular, reddish-brown blotch on the trailing edge (dorsum); hindwings grey. **Egg:** 0.65 × 0.50 mm; translucent whitish when laid, but soon becoming creamy white. **Larva:** up to 15 mm long; body rather plump, dull darkish brown, with black or brownish-black pinacula; head, prothoracic plate and anal plate black or brownish black. **Pupa:** 6–10 mm long; reddish brown to blackish brown; tip of abdomen blunt.

LIFE HISTORY
Moths occur from late June to the end of July. Eggs are laid singly on leaves close to the tips of young canes, each female depositing up to 300. Eggs hatch in about 2 weeks and the young larvae immediately invade the growing points. Here, each larva (usually no more than one per cane) webs together the two halves of a partly opened leaf, moving up the cane to younger leaves as the inhabited leaf becomes tougher. Larvae also burrow into the actual growing points of the canes. However, when about 3 weeks old and still small, each spins a tough cocoon on the lower half of the plant, between a leaf base and the cane. Here they overwinter, most individuals entering hibernation before the end of August as third-instar larvae. Activity is resumed in late March or April. The larvae then web together leaves on the fruiting laterals or at the cane tips, or burrow directly into flower buds. Larvae also invade the new canes,

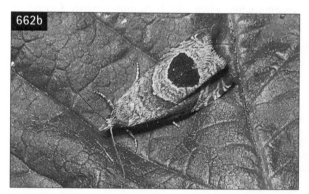

662a, b Bramble shoot moth (*Epiblema uddmanniana*).

663 Bramble shoot moth (*Epiblema uddmanniana*) – larva.

664 Bramble shoot moth (*Epiblema uddmanniana*) – larval habitation.

webbing together bunches of apical leaves to form tough, tent-like shelters. Masses of frass accumulate in the silken strands of the larval habitation, each web usually being occupied by a single larva. Larvae on fruiting canes migrate to the young vegetative growth after the blossom period, where they also form the characteristic tents of webbed leaves. Larvae spend much of their time in the shelter of their tents, but may vacate them at night to feed on younger foliage elsewhere on the canes. Larvae pass through five instars and then pupate from late May onwards, usually within the larval habitation. Adults emerge about 3 weeks later.

DAMAGE

The young larvae cause no harm before hibernation. In spring, however, attacks can be serious. Infested flower buds are hollowed out, leaving an outer shell of petals and sepals. Later, the terminal buds of webbed canes are killed, growth checked and canes distorted. Lateral shoots then develop from normally dormant buds. As a consequence, weakened canes are produced and cropping potential for the following year is greatly reduced.

Epinotia tenerana (Denis & Schiffermüller) (665)
Nut bud moth

Larvae of this often very common species feed on, for example, *Alnus*, *Corylus avellana* and *Crataegus*, and are sometimes a problem on cultivated hazelnut. Widely distributed in Europe.

DESCRIPTION

Adult: 12–16 mm wingspan; forewings brownish grey towards the base and brownish orange towards the apex, with a dark, angular basal patch, lead-grey markings and white costal strigulae; hindwings brownish grey, paler basally. **Egg:** 0.6 × 0.5 mm; pale yellow. **Larva:** up to 10 mm long; body greyish green or yellowish green; head and prothoracic plate dark brown; pinacula light brown; anal comb blackish; early instars with head and prothoracic plate black. **Pupa:** 5 mm long; dark yellowish brown; cremaster with six hook-tipped bristles.

LIFE HISTORY

Adults emerge in the summer, but are long-lived and usually do not deposit eggs until about September, placing them on the buds either singly or in batches of two or three. Eggs hatch about a month later and the larvae immediately burrow into catkin buds. Feeding continues throughout the autumn and winter, within developing catkins; in early spring, leaf buds and unfurling leaves are also attacked. Larvae continue to feed within spun leaves and are fully grown by the end of May. They then enter the soil and pupate. There is one generation annually.

DAMAGE

Larvae cause death of developing catkins and leaf buds, which results in considerable distortion of growth. Heavy infestations may reduce crop yields.

Epiphyas postvittana (Walker) (666–668)
Light brown apple moth

Apart from specimens imported in consignments of New Zealand apples, this important Australasian apple pest was unknown in Europe until 1936, when it was found breeding on ornamental *Euonymus* at Newquay, Cornwall, England. Although having become well established on a wide variety of ornamental plants in South West England, and also since having greatly extended its range eastwards into various parts of southern England, including East Anglia, it has not yet been reported as a pest of apple or other fruit crops. Evidence from Australasia suggests that this species is

665 Nut bud moth (*Epinotia tenerana*) – bud damage.

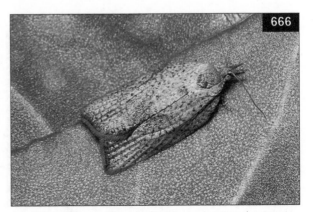

666 Light brown apple moth (*Epiphyas postvittana*) – female.

unlikely to become an economic pest of fruit crops in Britain, except possibly in the mildest parts of Cornwall and South Devon. Parts of mainland Europe, however, would certainly appear favourable for its development. In England, there are two generations annually, the yellowish-brown to reddish-brown adults (16–25 mm wingspan) occurring from May to October, and the yellowish-green larvae (up to 18 mm long) occurring in June and July and from September to April.

667 Light brown apple moth (*Epiphyas postvittana*) – male.

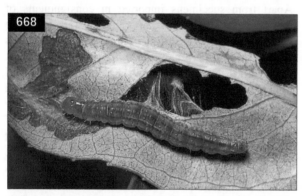

668 Light brown apple moth (*Epiphyas postvittana*) – larva.

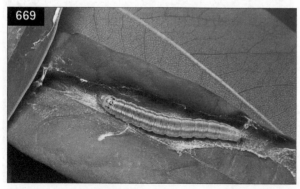

669 *Exapate congelatella* – larva.

Exapate congelatella (Clerk) (**669**)

In mainland Europe, a minor pest of apple, plum and other fruit trees; larvae are also reported on gooseberry, red currant and strawberry. Wild hosts include *Crataegus monogyna*, *Ligustrum vulgare* and *Prunus spinosa*. Widely distributed in northern and central Europe. Also found in western China.

DESCRIPTION
Adult female: 10–11 mm wingspan; brachypterous – forewings narrow and elongate, whitish grey, with dark markings; hindwings vestigial. **Adult male:** 20–23 mm wingspan; forewings elongate, mainly grey to greyish brown, speckled with black and suffused with whitish or silvery-white markings; hindwings brownish grey. **Larva:** up to 16 mm long; body light green to whitish green, with whitish subdorsal bands; head yellowish green, marked with brown; prothoracic plate light green marked with brownish black.

LIFE HISTORY
Adults occur from mid-September or October to December. Males are active fliers, but the females are incapable of flight and merely crawl over the foodplant. Eggs are laid either singly or in small batches on the bark of shoots and twigs of host plants. They hatch in the spring. Larvae, which usually inhabit spun shoots, attack the buds, leaves and developing fruitlets. They pupate amongst debris on the ground from about June onwards.

DAMAGE
On fruit trees and bushes, larvae bore into the buds and also attack the foliage of the young shoots. Notably on gooseberry, in addition to browsing on the leaves, the larvae devour the floral parts of the open blossoms, preventing them from setting fruit; they also bore into developing fruitlets, which then desiccate and drop off.

Gypsonoma dealbana (Frölich) (**670–672**)

A pest of hazelnut; larvae also occur on various wild hosts, including *Corylus avellana*, *Crataegus*, *Populus*, *Quercus* and *Salix*. Palaearctic. Widely distributed in central and northern Europe.

DESCRIPTION
Adult: 11–15 mm wingspan; forewings white to creamy white, marked more or less extensively with greyish brown, yellowish brown and dark brown, and each with a distinctive black dash subcentrally; hindwings grey. **Larva:** up to 9 mm long; body yellowish to pinkish white, with greyish-brown pinacula; head and prothoracic plate yellowish brown to black; anal plate and thoracic legs dark brown. **Pupa:** 6–7 mm long; brown; cremaster with 10 hook-tipped bristles.

LIFE HISTORY

In warmer regions, adults appear in May and June, but not until July in more northerly areas. After mating, eggs are laid on the underside of leaves, usually singly and close to a major vein. They hatch 1–2 weeks later. Larvae then feed on the leaves, again close to a major vein. They graze away the lower epidermis and underlying mesophyll, but leave the upper epidermis intact. Whilst still small, the larvae form cocoons on the shoots, between a new bud and a shoot, in which they then aestivate and eventually overwinter; larvae can also hibernate within galls formed on *Corylus* by filbert bud mite (*Phytoptus avellanae*) (Chapter 8). The larvae reappear very early in the spring to feed on developing buds, young shoots and expanding leaves. Fully grown larvae pupate in cocoons spun amongst debris on the ground or in the uppermost layer of the soil. There is one generation annually.

DAMAGE

In spring, larvae cause considerable disruption and deformation of the new growth, and this can have a detrimental effect on cropping.

Hedya dimidioalba (Haworth) (**673–674**)

Marbled orchard tortrix moth
larva = spotted apple budworm
A usually minor pest of apple, pear, almond, apricot, cherry and plum, but sometimes abundant in unsprayed orchards. Also associated with various non-cultivated hosts, including *Alnus*, *Crataegus monogyna*, *Fraxinus excelsior*, *Prunus spinosa*, *Quercus*, *Rosa* and *Sorbus*. Generally distributed and common in much of Europe; also present in North America.

670 *Gypsonoma dealbana* – adult.

671 *Gypsonoma dealbana* – larva.

673 Marbled orchard tortrix moth (*Hedya dimidioalba*).

672 *Gypsonoma dealbana* – larval habitation on hazelnut.

674 Marbled orchard tortrix moth (*Hedya dimidioalba*) – larva.

284

DESCRIPTION

Adult: 15–21 mm wingspan; forewings ochreous white apically, suffused with silver and ochreous grey, the remainder marbled extensively with dark brown, bluish grey and black; hindwings brownish grey. **Larva:** up to 20 mm long; body olive green to dark green, with black pinacula; head, prothoracic plate, anal plate and anal comb dark brown or black; thoracic legs black. **Pupa:** 8–10 mm long; dull black; cremaster tapered, with an apical tuft of eight hook-tipped bristles.

LIFE HISTORY

Adults occur in June and July. When at rest, in common with other species of the genus *Hedya*, they closely resemble bird-droppings. Eggs are laid singly or in small batches, mainly on the underside of leaves. They hatch in about 2 weeks and the larvae feed for several weeks before hibernating whilst still small, each in a dense cocoon spun within a bark crevice or beneath an old bud scale. Activity is resumed in late March or April, the larvae attacking the opening buds, blossom trusses, foliage and young shoots, often sheltering between two leaves webbed together with silk. Pupation occurs in spun leaves in late May or June, adults emerging 2–4 weeks later.

DAMAGE

In summer, larvae may cause superficial damage to apples or pears and, in spring, they contribute to that done to leaves and blossoms by larvae of other lepidopteran species. The larvae may also tunnel into the young shoots and cause wilting or death of the tips.

Hedya ochroleucana (Frölich) (**675–676**)

Although associated mainly with *Rosa*, and often a pest on cultivated rose bushes, the larvae are sometimes found on apple. At least in the northerly parts of its range there is just one generation annually, with larvae feeding mainly from April to mid-June, each inhabiting a bunch of spun leaves. Pupation occurs within the larval habitation in May or June, and adults fly in June and July. In warmer parts of Europe there may be two generations annually, with first-generation adults active in May and June and those of the second generation appearing in the late summer and early autumn. Fully fed larvae (up to 18 mm long) are dull greyish green to olive green, with inconspicuous pinacula and the head and prothoracic plate brownish black to black. Adults (16–23 mm wingspan) are mainly blackish to bluish grey and brownish grey, with the forewings partly creamy white and often tinged with pink.

675 *Hedya ochroleucana* – adult.

676 *Hedya ochroleucana* – larva.

677 Plum tortrix moth (*Hedya pruniana*).

678 Plum tortrix moth (*Hedya pruniana*) – larva.

Hedya pruniana (Hübner) (677–678)

Plum tortrix moth

An often abundant pest of damson and plum; also found, occasionally, on apple, cherry and pear. Wild hosts include *Corylus avellana*, *Prunus avium* and *P. spinosa*. Widely distributed in Europe, but with a less northerly range than that of marbled orchard tortrix moth (*Hedya dimidioalba*).

DESCRIPTION

Adult: 14–18 mm wingspan; forewings mainly white apically, the remainder mottled with dark brown, brownish grey and blackish; hindwings grey. **Larva:** up to 18 mm long; body bright green to olive green, with black pinacula; head, prothoracic plate, anal plate and anal comb black or brownish black. **Pupa:** 8–9 mm long; brownish black to black; cremaster with a cluster of strong, hook-tipped bristles.

LIFE HISTORY & DAMAGE

As for marbled orchard tortrix moth (*Hedya dimidioalba*). However, larvae cause most harm to the foliage and young shoots of damson and plum, and the adults appear slightly earlier in the season.

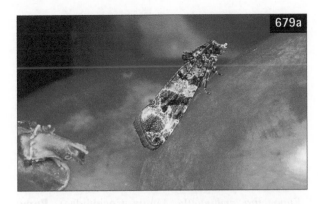

679a, b European vine moth (*Lobesia botrana*).

Lobesia botrana (Denis & Schiffermüller) (679–682)

European vine moth

An important pest of grapevine in central and southern Europe; in Spain, also associated with highbush blueberry. Wild hosts include *Berberis vulgaris*, *Clematis vitalba*, *Hedera helix*, *Ligustrum vulgare*, *Lonicera* and *Viburnum*. Widely distributed in central and southern Europe; also present in parts of Africa and Asia.

680 European vine moth (*Lobesia botrana*) – larva.

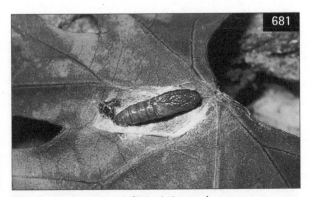

681 European vine moth (*Lobesia botrana*) – pupa.

682 European vine moth (*Lobesia botrana*) – larval habitation.

DESCRIPTION

Adult: 10–15 mm wingspan; forewings creamy white, with brown, olive brown, blackish and bluish-grey markings; hindwings dark brownish grey (= female) or whitish grey (= male). **Egg:** 0.7 × 0.6 mm; yellowish when laid, later becoming greyish and opalescent. **Larva** up to 11 mm long; body greyish green or yellowish green to brown, and often translucent; head yellowish brown; prothoracic plate brown; anal plate ochreous; anal comb with 6–8 prongs; abdominal prolegs each with 30–40 crochets (cf. *Eupoecilia ambiguella*, family Cochylidae). **Pupa:** 4.5–7.0 mm long; yellowish brown to dark brown; cremaster fan-shaped, with eight hook-tipped bristles.

LIFE HISTORY

Adults emerge in late April or early May. Eggs are laid mainly on flower buds and hatch a week or so later. Larvae then attack the flower buds and may also burrow into the pedicels, and typically inhabit silken webs spun within the developing inflorescences. Pupation occurs in the larval habitation, or in suitable shelter nearby, moths of the next generation appearing in about a week. Development of the second-brood larvae, which feed amongst the developing clusters of grapes, is often very rapid and, if conditions are favourable, a further one or two generations may be completed in the same season. The winter is passed in the pupal stage, within cocoons spun under bark, in cracks on supporting posts and in other sheltered situations (cf. vine moth, *Eupoecilia ambiguella*, family Cochylidae).

DAMAGE

Grapevine: larvae destroy flower buds and also cause the death of open flowers and young fruitlets; later in the season, larvae damage the developing or maturing grapes. Fungal pathogens often invade the damaged tissue, which can lead to additional problems and further crop losses. Infestations in spring are sometimes considered beneficial, as they may have a thinning effect and result in improved fruit quality at harvest. Attacks later in the season, however, are potentially very damaging, and particularly serious on table grapes. The silken larval habitations and sticky exudations from damaged surfaces are also unwelcome contaminants in bunches of grapes.

Lozotaenia forsterana (Fabricius) (683–684)

A pest of raspberry and strawberry, at least in England, but usually of only minor importance; larvae also feed on various other plants, including *Hedera helix* and *Ribes*. Widely distributed in Europe.

DESCRIPTION

Adult: 20–29 mm wingspan, female larger than male; forewings light greyish brown, with dark brown to blackish markings; hindwings grey. **Larva:** up to 25 mm long; body dull greyish green, darker above; head dark brown and shiny, with black markings; prothoracic plate yellowish brown, extensively marked with dark brown; anal plate yellowish green, marked with black; anal comb present; thoracic legs brown. **Pupa:** 12–14 mm long; dark brown.

LIFE HISTORY

In northern Europe, adults occur in late June and July. Eggs, which are laid on the foliage, hatch in September. The larvae then feed during the autumn before hibernating. Activity is resumed in April, each larva living between two or more leaves strongly spun together with silk. During the blossom period, they will also attack flowers. Pupation occurs in June between spun leaves.

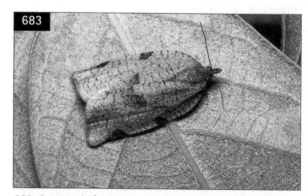

683 *Lozotaenia forsterana* – adult female.

684 *Lozotaenia forsterana* – larva.

DAMAGE

If numerous, larvae can cause considerable defoliation, but serious attacks are rare. In common with other strawberry-feeding tortricids, larvae feeding on flowers may cause fruit abortion or malformation.

Neosphaleroptera nubilana (Hübner)

A minor pest of plum; also reported on apple, apricot and pear. Wild hosts include *Crataegus monogyna* and *Prunus spinosa*. Widely distributed in Europe, but absent from south-western areas.

DESCRIPTION

Adult: 12–14 mm wingspan; forewings mainly dark greyish brown (in female strongly suffused with greyish white and then superficially similarly patterned to those of *Cnephasia asseclana*); hindwings greyish brown. **Larva:** up to 12 mm long; body light green and translucent; pinacula light green; head yellowish brown; prothoracic plate mainly brownish green; anal plate green.

LIFE HISTORY

Adults occur in June or July. They are mainly active at dusk. However, in sunny weather, males are often also on the wing during the daytime. Larvae inhabit spun shoots, within which they feed in the late summer or early autumn before overwintering. In spring, they again become active and usually complete their development in June. Pupation takes place in the larval habitation, adults emerging shortly afterwards. There is just one generation annually.

DAMAGE

Larvae contribute to defoliation caused by those of other spring-feeding species of Lepidoptera.

Pammene argyrana (Hübner) (**685**)

Reported in the Netherlands and Switzerland as a pest of apple, at least in old orchards. Also found in association with old apple trees in England, but not as a significant pest. Normally, this species breeds in galls formed on *Quercus* by oak-apple gall wasp (*Biorhiza pallida* (Olivier)) (Hymenoptera: Cynipidae). Widely distributed in Europe.

DESCRIPTION

Adult female: 10–13 mm wingspan; forewings white or creamy white, marked with black, grey and pale ochreous; hindwings mainly greyish brown. **Adult male:** similar to female, but hindwings white basally and with a broad, blackish-brown margin. **Larva:** up to 10 mm long; body pinkish white and translucent, with reddish-brown pinacula and crossbands of whitish spicules; head brown; prothoracic plate whitish grey,

685a, b *Pammene argyrana* – adult male.

with mainly blackish lateral and posterior borders; anal plate dark brown anteriorly, white posteriorly.

LIFE HISTORY

Adults appear in April and May. Eggs, which are usually laid singly or in pairs, hatch in about 2 weeks. On apple, larvae then invade the developing fruitlets, usually one per infested fruitlet, burrowing through the flesh to the core. When fully grown, usually in August, the larva emerges through a small hole bored through the skin of the fruit and eventually overwinters in a silken hibernaculum spun on the trunk of the host tree beneath bark flakes, lichen or moss. Pupation occurs in the early spring, shortly before the emergence of the adults.

DAMAGE

On apple, infestations tend to occur on early-flowering cultivars. Infested fruitlets contain a long, frass-filled gallery and, eventually, a circular exit hole on the surface. Such fruits often drop prematurely; however, if continuing to develop to maturity on the tree they are unmarketable.

Pammene fasciana (Linnaeus)

A pest of cultivated chestnut, and often common on wild *Castanea sativa* and *Quercus*. Widely distributed in Europe.

DESCRIPTION

Adult: 14–17 mm wingspan; forewings white, more or less suffused with grey, brown and ochreous, and with dark brown, blue-black or blackish markings forming costal strigulae and a distinct submarginal ocellus, the latter containing 6–7 black dashes; hindwings greyish brown, darker in female. **Egg:** 0.7 × 0.6 mm; creamy white, later with the developing embryo visible as a purplish-red ring. **Larva:** up to 13 mm long; body creamy white, with large brownish pinacula; head yellowish brown; prothoracic plate yellowish brown, marked with black; anal plate light brown, speckled with black; anal comb present. **Pupa:** 7–9 mm long; brown.

LIFE HISTORY

Adults occur from early June onwards, with peak activity in late June and early July. Eggs are laid in small batches, typically along the lateral veins on the upper side of the expanded leaves close to clusters of developing nuts. The eggs hatch within 2 weeks and larvae then feed within the developing nuts, typically from August to October. Fully grown larvae, which usually emerge only after the infested nuts have fallen to the ground, spin strong silken cocoons in crevices or beneath flakes of bark on the trunks of trees, or in rotten wood or other shelter, in which they then overwinter. Pupation occurs in May, and adults emerge about 3–4 weeks later.

DAMAGE

Much of the inner tissue of the nuts is consumed, and infested nuts often drop prematurely. Infestations can lead to significant crop losses.

Pammene rhediella (Clerck) (**686–690**)

Fruitlet-mining tortrix moth

An important pest of apple and plum. Pear and cherry are also attacked, but less frequently. In the wild, associated mainly with *Crataegus monogyna*. Common and widely distributed in Europe; also present in Central Asia.

DESCRIPTION

Adult: 9–12 mm wingspan; forewings dark purplish and metallic, with an orange tip; hindwings dark brown. **Egg:** 0.7 × 0.6 mm; whitish and translucent. **Larva:** up to 10 mm long; body rather plump, creamy white to greyish white, with moderately large, light brown or greyish pinacula; head, prothoracic plate and anal plate mainly brown; anal comb with a strong base; early instars whitish and translucent, with head and anal plate black. **Pupa:** 4–5 mm long; light brown.

LIFE HISTORY

Adults occur in late April, May and June. They are active in sunshine, and often fly in orchards at or above tree height. Eggs are laid singly on the underside of leaves close to flowers or clusters of young fruitlets. They hatch in about 2 weeks. The larvae then attack the

686 Fruitlet-mining tortrix moth (*Pammene rhediella*).

687 Fruitlet-mining tortrix moth (*Pammene rhediella*) – larva in apple.

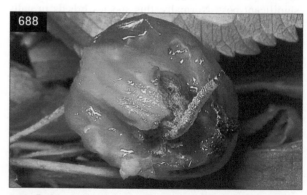

688 Fruitlet-mining tortrix moth (*Pammene rhediella*) – larva in cherry.

fruitlets, although they may firstly feed on the stamens and calyxes of the flowers, webbing them together with silk. If isolated fruitlets are attacked, the larva attaches a leaf to the surface as a shelter. Larvae feed within the fruitlets until late June or early July and then, on becoming fully grown, wander away to spin cocoons under loose bark on the trunks or larger branches. Here, they overwinter and eventually pupate in the spring, shortly before the emergence of the adults.

DAMAGE

Apple: fruitlets are usually attacked when about 2 cm in diameter. Initially, the young larvae feed within the eye (calyx), forming a strong, external web in which brown or black pellets of frass soon accumulate. Feeding also occurs around the stalk and at the sides where adjacent fruitlets are touching. One larva may damage several fruitlets in a cluster. The stalk of an attacked fruitlet is often loosened, following the loss of flesh around the base. Such fruitlets, and those with major flesh wounds, will drop prematurely; however, they sometimes remain in situ and subsequently wither, held in place by the silken web. The main gallery formed in the flesh of a mined fruit characteristically runs from the stalk to the core and then outwards to the shoulder; the mine is usually no more than 2 mm wide, virtually frass-free and is often lined with a whitish, sugary secretion upon which yeasts develop. Superficially damaged fruitlets often possess holes in the sides, each up to 2 mm deep; these eventually heal, and at maturity the apples bear one or more irregular pits with corky scar tissue at their base. Such fruits also tend to become misshapen. **Plum:** larvae form galleries in the flesh of fruitlets, causing considerable weeping of gum which then accumulates in the surrounding webbing. As on apple, very little frass remains within the larval gallery (cf. damage caused by larvae of plum fruit moth, *Cydia funebrana*), and attacked fruitlets either drop prematurely or, if superficially damaged, survive to maturity with the wounds healed over. **Cherry:** wounds in cherries often weep, and most of the flesh is destroyed.

Pandemis cerasana (Hübner) (**691–692**)
Barred fruit tree tortrix moth

A polyphagous pest of trees and shrubs, including apple, pear, cherry, plum, currant, blackberry, raspberry and hazelnut. Palaearctic. Widely distributed and often common in Europe.

689 Fruitlet-mining tortrix moth (*Pammene rhediella*) – infested apple fruitlets.

691 Barred fruit tree tortrix moth (*Pandemis cerasana*).

690 Fruitlet-mining tortrix moth (*Pammene rhediella*) – infested cherry.

692 Barred fruit tree tortrix moth (*Pandemis cerasana*) – larva.

DESCRIPTION

Adult: 16–24 mm wingspan; forewings pale yellowish to ochreous brown, with light brown, dark-edged markings; hindwings greyish brown; antennae of male each with a basal notch. **Larva:** up to 20 mm long; body rather thin and flattened, light green, but darker above, with pale pinacula; head light green to brownish green, marked with black; prothoracic plate light green or yellowish green, with sides and hind edge dark; anal plate green, dotted with black; anal comb with 6–8 prongs; first thoracic and last abdominal spiracles elliptical and much larger than the others. **Pupa:** 8–13 mm long; brown to brownish black; cremaster longer than wide, bearing four lateral and four terminal hook-tipped bristles.

LIFE HISTORY

In the more northerly parts of its range, this species has one generation annually, with moths occurring over an extended period from June to August or early September. The eggs are deposited in batches on the leaves or branches. Most hatch after a few weeks, but some not until the following spring. Young larvae in summer or autumn feed on the foliage and then, usually when in their third instar, move onto the twigs to spin silken retreats in which to overwinter. Activity recommences at bud burst, when any overwintered eggs also hatch. Larvae feed until May or early June, each inhabiting a rolled or folded leaf. Pupation occurs in a whitish cocoon spun in the larval habitation or in a newly folded leaf. Under favourable conditions there may be two generations annually, with adult activity peaking in June to July and again in September.

DAMAGE

Leaf damage is usually unimportant, as larvae rarely feed together in large numbers, but attacks on young blossoms and fruitlets in the spring may result in crop losses and in blemished fruit at harvest.

Pandemis corylana (Fabricius) (**693**)

This widely distributed and locally common species occasionally attacks cultivated hazelnut and fruit trees; wild hosts include *Corylus avellana, Fagus sylvatica, Frangula alnus, Larix* and *Pinus*. The adults occur from July to September. They resemble those of barred fruit tree tortrix moth (*Pandemis cerasana*), but are larger (18–24 mm wingspan) and have more strongly reticulated, brownish-orange forewings and paler apices to the hindwings. The green, rather slender larvae are up to 25 mm long, with a well-developed anal comb with up to nine prongs; they feed mainly from May to June or July between spun leaves or in longitudinally folded leaves.

Pandemis dumetata (Treitschke)

A minor pest of strawberry in various parts of mainland Europe; attacks are also reported (e.g. in Italy) on fruit trees, including apple, pear and peach. Eurasiatic. Widely distributed in Europe.

DESCRIPTION

Adult: 18–22 mm wingspan; forewings reddish ochreous to reddish brown, with a darker brown basal fascia, median fascia and reticulate pattern; inner margin of median fascia more or less straight (cf. *Pandemis heparana*); hindwings whitish, suffused with brownish grey. **Larva:** up to 25 mm long; body green, darker dorsally, with green pinacula; head green, marked with black; prothoracic and anal plates green.

LIFE HISTORY

On strawberry, larvae overwinter in silken cocoons spun on the leaves of the foodplant. In spring, they again become active, grazing on the young leaves and inhabiting folded leaves or folded-over leaf edges. When fully grown they pupate, each in a folded leaf or folded-over leaf edge, usually by the end of May or beginning of June. Adults, which represent a first generation of moths, appear in about mid-June. Eggs are then deposited in large batches on the leaves and hatch 7–10 days later. Larvae of the summer generation also feed on the leaves, eventually completing their development and producing a second-generation of adults in August. Second-brood larvae hibernate from late August or early September onwards. The life cycle on fruit trees is similar.

DAMAGE

Larvae contribute to defoliation caused by those of other tortricid species, such as strawberry tortrix moth (*Acleris comariana*) and *Ancylis comptana*.

693 *Pandemis corylana* – adult.

Pandemis heparana (Denis & Schiffermüller) (694–695)

Dark fruit tree tortrix moth

A polyphagous pest of trees and shrubs, including apple, pear, plum, currant and raspberry. Eurasiatic. Widely distributed in Europe and often common.

DESCRIPTION

Adult: 16–24 mm wingspan; forewings reddish ochreous to reddish brown, with a darker basal fascia, median fascia and reticulate pattern; inner margin of median fascia more or less toothed (cf. *Pandemis dumetana*); hindwings dark brownish grey; antennae of male each with a basal notch; labial palps long and prominent, especially in female. **Larva:** up to 25 mm long; body bright green, with pale sides; head yellowish green to ochreous; prothoracic plate green to yellowish green; anal plate light green, sometimes with dark speckles; anal comb whitish and usually with 6–8 prongs. **Pupa:** 10–12 mm long; brown to brownish black; cremaster fan-shaped, about as long as wide, with eight hook-tipped bristles.

694 Dark fruit tree tortrix moth (*Pandemis heparana*) – female.

695 Dark fruit tree tortrix moth (*Pandemis heparana*) – larva.

LIFE HISTORY

Adults appear from late June onwards. In cooler regions, where there is just one generation annually, eggs are then laid on the upper surface of leaves, usually in batches of 30–50. They hatch in 2–3 weeks and the small larvae feed for several weeks before hibernating, usually in their second or third instars, in silken retreats spun on the twigs. Activity is resumed in spring, larvae feeding in May and June within rolled leaves on the young shoots or under webs spun on the underside of the leaves. Pupation occurs in spun leaves near the tips of infested shoots or in the larval habitation. In warmer regions there are two generations annually, with young second-brood larvae overwintering.

DAMAGE

Attacks on the foliage are usually unimportant. However, in spring, the larvae sometimes feed on the flowers and fruitlets, causing loss, malformation or blemishing of fruits. In addition, young summer larvae may graze on the surface of ripening apples, pears or plums.

Ptycholoma lecheana (Linnaeus) (696–698)

Leche's twist moth

An often common, polyphagous pest of various trees and shrubs, including apple, pear, apricot, cherry, plum and raspberry. Eurasiatic. Widely distributed in Europe.

DESCRIPTION

Adult: 16–22 mm wingspan; forewings blackish brown, suffused with golden yellow, especially basally, and with silvery-metallic markings; hindwings blackish brown, with pale cilia. **Larva:** up to 20 mm long; body greyish green dorsally, pale yellowish or yellowish green laterally and ventrally; pinacula yellowish or pale yellowish green; head yellowish

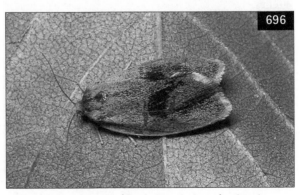

696 Leche's twist moth (*Ptycholoma lecheana*) – female.

brown, partly edged with black posteriorly; prothoracic plate yellowish brown, with a pale leading edge and black lateral patches posteriorly; anal comb small; thoracic legs black. **Pupa:** 9–11 mm long; brownish black to black; cremaster forming a strong projection, armed with four terminal and four subterminal hooked bristles.

LIFE HISTORY

Adults occur in June and July, sometimes earlier. The larvae feed on foliage during the summer and then, while still small, hibernate in silken cocoons spun on the bark of twigs and spurs. They reappear in the early spring to feed on the opening buds and young foliage. Later, they feed in rolled leaves, becoming fully grown in April, May or early June. Pupation occurs in the larval habitation or in a freshly rolled leaf.

DAMAGE

Larvae contribute to damage caused by those of various other lepidopteran pests, but are rarely numerous on fruit crops.

Rhopobota naevana (Hübner) (**699–701**)

Holly tortrix moth
larva = holly leaf tier

Associated mainly with *Ilex aquifolium*, but also a generally common pest of apple and, less frequently, pear and plum. Eurasiatic. Widely distributed in Europe. Introduced into North America, where it is a pest of cranberry.

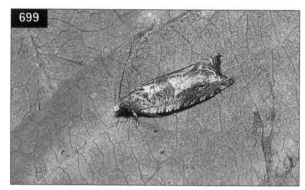

699 Holly tortrix moth (*Rhopobota naevana*).

697 Leche's twist moth (*Ptycholoma lecheana*) – male.

700 Holly tortrix moth (*Rhopobota naevana*) – larva.

698 Leche's twist moth (*Ptycholoma lecheana*) – larva.

701 Holly tortrix moth (*Rhopobota naevana*) – early larval habitation on pear.

DESCRIPTION

Adult: 12–15 mm wingspan; forewings subfalcate, dark grey, marked with blackish and dark rusty brown, and with several whitish strigulae on the costal margin; hindwings grey. **Egg:** 0.7 × 0.5 mm; translucent whitish when laid, but soon becoming yellowish to reddish. **Larva:** up to 12 mm long; body shiny yellowish green to greyish green to yellowish brown; head and prothoracic plate black; anal plate dark brown to green, mottled with black; anal comb usually with two dark prongs; thoracic legs brown. **Pupa:** 5–7 mm long; yellowish brown; tip with four thorn-like spines and with a small hump behind the anal slit.

LIFE HISTORY

In northern Europe, this species overwinters as eggs laid singly during the summer on the smooth bark of the trunks and branches of host trees. The eggs hatch in spring, usually during the blossom period. Larvae then feed in a webbed shelter of young leaves; the larvae also attack unopened and opened flowers. Feeding is completed in June, each larva pupating in a white cocoon spun in a folded leaf or amongst dead leaves or debris on the ground. Adults emerge about 3 weeks later and then deposit the overwintering eggs. In warmer regions, the life cycle is far less protracted; in parts of southern Europe, for example, there may be up to five generations annually.

DAMAGE

Larvae destroy young leaves, flowers and, occasionally, newly set fruitlets; they also kill young lateral shoots.

Sparganothis pilleriana (Denis & Schiffermüller)

An important pest of grapevine in mainland Europe; other hosts include cultivated strawberry, and also various wild trees, shrubs and herbaceous plants. Palaearctic. Widely distributed in Europe; also present in North America.

DESCRIPTION

Adult female: 18–24 mm wingspan; forewings usually reddish brown or ochreous brown and unpatterned; hindwings brownish grey; labial palps prominent and directed anteriorly. **Adult male:** 15–20 mm wingspan; forewings mainly ochreous to greyish ochreous, with a brassy sheen, and with more or less distinct darker markings, including a median fascia, but extremely variable; hindwings brownish grey; labial palps prominent and directed anteriorly. **Egg:** 1.25 × 0.85 mm; emerald green to greenish yellow when laid, later becoming yellow. **Larva:** up to 30 mm long; body yellowish green to greyish green, with a dark dorsal line

and with small, pale pinacula; head and prothoracic plate black; anal plate yellowish brown; thoracic legs black. **Pupa:** 12–14 mm long; dark chestnut-brown; cremaster with eight hook-tipped bristles.

LIFE HISTORY

Adults of this single-brooded species appear in July. Eggs are then laid in batches of 40–60 on the upper side of leaves of the foodplant; on strawberry, however, they are usually laid on leaves close to or touching the soil. The eggs hatch about 1–3 weeks later. Larvae feed in late summer and early autumn on the foliage and developing fruits, each inhabiting a silken retreat. The larvae hibernate during the winter months and resume feeding in the early spring, on strawberry typically coinciding with the commencement of flowering. They then attack the buds and developing leaves, inhabiting folded leaves, webbed shoot tips or other shelters. When fully fed, usually in late June, each larva pupates in a white cocoon formed in a rolled leaf or within a bunch of withered leaves spun together with silk. Adults emerge within 2–3 weeks.

DAMAGE

Grapevine: infestations within developing bunches of grapes in late summer and autumn can result in significant fruit losses, and attacks on leaves and buds in spring can have an adverse effect on growth and may also reduce cropping potential. Webbing within bunches of grapes is also detrimental, and particularly significant on table grapes. **Strawberry:** damage to blossoms may reduce cropping. Larvae also damage the developing fruitlets at any stage of development, resulting in significant crop losses.

Spilonota ocellana (Denis & Schiffermüller) (702–703)

Bud moth

larva = brown apple budworm

A generally common pest of apple; also occurs on pear, quince and, less frequently, cherry, plum, blackberry and raspberry. Wild hosts include *Alnus*, *Corylus avellana*, *Crataegus monogyna*, *Prunus spinosa*, *Quercus*, *Salix* and *Sorbus*. Eurasiatic. Widely distributed in Europe. An introduced pest in North America.

DESCRIPTION

Adult: 12–16 mm wingspan; forewings rectangular, whitish, more or less suffused with grey, marked towards the apex with metallic bluish grey and black, and each with a dark, triangular dorsal spot and a blackish, angular basal patch; hindwings dark grey. **Larva:** up to 12 mm long; body dark purplish brown, with lighter pinacula; head, prothoracic plate and anal plate shiny black or blackish brown. **Pupa:** 6–7 mm long; brown, with outline of wing cases distinctly darker than abdomen; tip blunt.

LIFE HISTORY
Moths occur from mid-June to mid-August. Eggs are laid singly or in small batches on leaves of the foodplant, each female depositing about 50–100. The eggs hatch 1–2 weeks later. The tiny larvae then feed beneath the leaves in August and September, each constructing a small tube-like shelter of silk. Individuals feed mainly on leaf tissue. However, on apple and pear, a leaf may be spun to the surface of a fruit, upon which the larva will also graze. During October, whilst still small, larvae spin

702a, b Bud moth (*Spilonota ocellana*).

703 Bud moth (*Spilonota ocellana*) – larva.

hibernacula in bark crevices. The larvae reappear in April and then invade the swelling leaf and fruit buds. Later, they attack flowers and foliage, each larva sheltering in a web of silk that usually incorporates dead leaves or flower fragments. Pupation takes place in late May and June, within a cocoon spun in the larval habitation or amongst dead leaves. Adults emerge 3–4 weeks later.

DAMAGE
The young larvae cause little or no harm during the summer, although the skin of grazed apples or pears may be sufficiently damaged to result in the fruit being unmarketable. Skin blemishes, however, are usually superficial. Attacks in the spring are more significant, infested buds being hollowed out and killed. Blossoms are also destroyed, but damage to leaves and young shoots is relatively unimportant.

Syndemis musculana (Hübner) (**704–707**)
Autumn apple tortrix moth
larva = autumn leaf roller
A pest of apple. Also associated with *Rubus fruticosus* and forest trees such as *Betula*, *Larix*, *Picea*, *Quercus* and *Tilia*. Eurasiatic. Widely distributed in Europe.

DESCRIPTION
Adult: 15–22 mm wingspan; forewings whitish, suffused with grey, and with dark brown to blackish markings, including a broad median facia. **Larva:** up to 22 mm long; body greyish green or olive green to blackish brown dorsally, paler laterally and ventrally, with pale pinacula; head mainly yellowish brown to orange brown; prothoracic plate greyish brown to yellowish brown, marked with black; anal plate usually yellowish brown or greenish.

LIFE HISTORY
Larvae of this single-brooded species feed from late June or July onwards, completing their development in

704 Autumn apple tortrix moth (*Syndemis musculana*).

the autumn. They then overwinter amongst debris on the ground and eventually pupate in the spring. Adults occur mainly in May and June.

DAMAGE
In addition to feeding on leaves, the larvae graze on the surface of maturing apples, removing extensive areas of tissue. Damaged fruits are unmarketable.

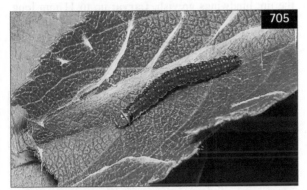

705 Autumn apple tortrix moth (*Syndemis musculana*) – half-grown larva.

706 Autumn apple tortrix moth (*Syndemis musculana*) – final-instar larva.

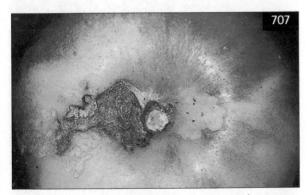

707 Autumn apple tortrix moth (*Syndemis musculana*) – damage to apple.

Family **PYRALIDAE**

A very large and varied family of small to medium-sized moths, with moderately long, narrow bodies, prominent labial palps, usually relatively narrow forewings and broad hindwings. Larvae have five pairs of abdominal prolegs and usually few body hairs; crochets on the abdominal prolegs are biordinal or triordinal, and form a complete circle or a mesal penellipse; the prespiracular plate on thoracic segment 1 bears two setae (three setae present in most other Lepidoptera). Pyralid larvae are often extremely active and many wriggle backwards violently when disturbed.

Cryptoblabes gnidiella (Millière)
A minor, often secondary pest of peach, plum, highbush blueberry, grapevine and various other fruit crops, including citrus, fig and pomegranate. Wild hosts include *Daphne gnidium*, *D. mezereum* and *Tamarix*. Widely distributed in the Mediterranean basin. Also introduced into New Zealand.

DESCRIPTION
Adult: 12–18 mm wingspan; forewings mainly grey or brownish; hindwings mainly whitish and translucent, with dark veins and cilia, and a narrow submarginal band. **Larva:** up to 12 mm long; body varying from greenish to brownish. **Pupa:** 5–7 mm long; brown; tip with a pair of close-set, elongate, angularly hooked spines (cf. *Ectomyelois ceratoniae*).

LIFE HISTORY
In Europe there are up to three generations annually. Infestations often occur on plants infested by pests such as aphids, mealybugs and scale insects, in addition to feeding on plant material the moth larvae then also imbibing the excreted honeydew. Larvae also invade developing fruits. For example, bunches of grapes infested by larvae of European vine moth (*Lobesia botrana*) (family Tortricidae) and fruits with surface blemishes caused by primary pests such as Mediterranean fruit fly (*Ceratitis capitata*) (Chapter 5) and locust bean moth (*Ectomyelois ceratoniae*) are often invaded. Pupation takes place in a silken cocoon spun amongst vegetation or amongst debris on the ground.

DAMAGE
Although often a secondary problem (the larvae invading fruits previously attacked by other pests) the pest can also cause primary damage; on citrus, for example, the larvae often invade the fruits of navel oranges, entering via the sunken remains of the calyx. Fruits invaded by larvae are unmarketable, with rotting areas developing on the skin and in the underlying

tissues, owing to subsequent invasion by pathogenic organisms. However, fruit infested by young larvae may escape detection and result in the pest being discovered only after harvest and perhaps only after crops have been sold and exported.

Ectomyelois ceratoniae (Zeller) (**708–709**)
Locust bean moth

This polyphagous species feeds within dried fruits, nuts and seeds, and is often a pest in stores and warehouses. It occurs outdoors in Mediterranean areas on a wide range of host plants, but especially *Ceratonia siliqua*; other outdoor hosts include *Acacia fanesiana* and *Robinia pseudoacacia*, the larvae feeding on seeds within the pods. Minor infestations may occur prior to harvest on maturing citrus fruits, including oranges. Larvae also occur in the field on mummified almond fruits, the larvae boring within the kernels. Widely distributed in the Mediterranean basin, northwards into parts of central Europe; also now established in South Africa and America.

DESCRIPTION
Adult: 18–24 mm wingspan; forewings mainly grey, with black markings; hindwings whitish, with a greyish-brown border; the moth bears a superficial resemblance to members of the genus *Ephestia*, which includes several notorious stored-product pests, but may be distinguished by details of the hindwing venation (vein 5 present in *Ectomyelois*). **Larva:** up to 15 mm long; body pinkish to dull reddish ochreous; head and prothoracic plate brown, partly bordered with black. **Pupa:** 7 mm long; yellowish brown; tip with a pair of robust, pointed spines (cf. *Cryptoblabes gnidiella*).

LIFE HISTORY
This species breeds continuously under favourable conditions, but in the more northern parts of its range it is single-brooded. Outdoors in southern France, there are two generations annually, with adults occurring in small numbers from April to June and in larger numbers from July or August to September.

DAMAGE
On citrus fruits, the larvae typically bore within the rind close to the base, where they feed amongst accumulated frass and webbing.

Ephestia parasitella Staudinger
A potentially invasive, polyphagous species; usually associated with dry vegetable material, but reported locally (e.g. in southern France and Italy) as a pest of grapevine; also present in, for example, the British Isles, Spain and Switzerland. Precise details of the biology

and distribution of this species are wanting, owing to frequent confusion with *Ephestia woodiella* Richards & Thomson; there are also two subspecies: *parasitella* Staudinger and *unicolorella* Staudinger.

DESCRIPTION
Adult: 14–20 mm wingspan; forewings grey to brownish grey, more or less tinged with purplish red, and each with a pale (often indistinct) median crossband and a blackish-edged (post-median) crossband towards the tip; hindwings greyish. **Larva:** up to 11 mm long; body pinkish, with small, black pinacula; head brown. **Pupa:** 7–8 mm long; brownish orange.

708 Locust bean moth (*Ectomyelois ceratoniae*).

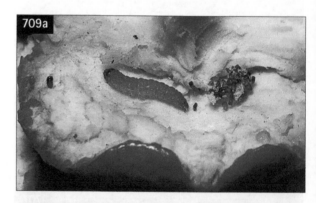

709a, b Locust bean moth (*Ectomyelois ceratoniae*) – larva.

LIFE HISTORY

Adults occur from June onwards, but whether there are one or more generations is uncertain. In vineyards, larvae are reported feeding in the autumn amongst mature bunches of grapes. Fully fed larvae overwinter in sheltered situations (including within shrivelled, dried-up grapes left on vines after harvest) and pupate in the spring, each in a silken cocoon.

DAMAGE

In vineyards, young larvae graze on the fruit stalks and older ones bore into the pulp of the matured fruits. Post-harvest damage to bunches of grapes remaining on the vines is of no immediate significance; however, if populations in vineyards subsequently increase (and, as a result, pre-harvest attacks then escalate), the pest may pose a serious threat to cropping in future years.

Euzophera pinguis (Haworth) (**710**)

Olive pyralid moth

A locally important pest of olive. Widely distributed in Europe on various other Oleaceae.

DESCRIPTION

Adult: 20–25 mm wingspan; forewings mainly pale beige, with blackish markings; hindwings mainly whitish, with a pale, brownish-black border. **Larva:** up to 20 mm long; body light green to pinkish or pale brownish yellow, with a darker, interrupted dorsal line; head and prothoracic plate brownish black. **Pupa:** 10–15 mm long; pale reddish brown to brownish yellow and shiny.

LIFE HISTORY

In areas where olives are grown, adults typically occur in April and May and in those of a summer generation in August and September; when conditions are particularly favourable there may be three generations annually. In northern Europe, however, where the pest is associated with *Fraxinus excelsior*, there is just one generation annually, and adults occur in July and August. Eggs are laid on the bark, most often at wound sites. Following egg hatch the larvae bore into the host tree, where they tunnel between the bark and phloem, each forming a chamber-like gallery about 7 cm long. Pellets of frass, intermingled with sawdust and strands of silk, are expelled from the feeding galleries and these often betray the presence of the pest. Pupation takes place in a silken cocoon formed in the larval habitation. Larvae of the overwintering generation feed slowly throughout the winter and complete their development in the following spring.

DAMAGE

Attacks often result in desiccation of leaves and premature leaf fall. Heavy infestations may result in the flow of sap to branches being entirely cut off, in which case death of branches will occur.

Ostrinia nubilalis (Hübner) (**711**)

European corn borer moth

larva = European corn borer

This polyphagous species, of which there is more than one biological race, is a notorious pest of maize and sweet corn. It has also been reported attacking apple, hop, peach, pear and strawberry. Present throughout the warmer parts of Europe and also recorded, usually as a non-established migrant, in various parts of northern Europe. The Z race of this species is associated with maize and sweet corn, whereas the E race (that sometimes established in northern Europe, as in southern England) is associated primarily with *Artemisia vulgaris*. Eurasiatic. An introduced pest in North America.

DESCRIPTION

Adult: 22–32 mm wingspan; forewings pale yellowish to olive brown, with irregular purplish-grey markings; hindwings mainly light grey, with darker markings. **Larva:** up to 25 mm long; body pale purplish brown, with brownish, pale-centred pinacula; head brown to brownish black; prothoracic plate light brown, with brownish black markings.

710 Olive pyralid moth (*Euzophera pinguis*).

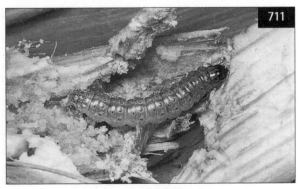

711 European cornborer moth (*Ostrinia nubilalis*) – larva.

LIFE HISTORY

Adults of this migratory species emerge in June. Eggs are then laid on the underside of leaves of host plants, typically in batches of about 20. The eggs hatch about 10 days later. Initially, larvae feed externally. Later, they bore into hosts to continue their development, piles of frass sometimes being visible externally at the points of entry. Fully fed larvae pupate in situ or in the soil, each in a silken cocoon. In northerly parts of its range, this species has just one generation annually; in warmer districts, however, it is capable of completing two or more generations annually. The overwintering stage, although capable of surviving low temperatures, is very susceptible to excessive dampness.

DAMAGE

Larvae burrowing in the shoots or crowns of host plants cause wilting, distortion and death. On apple, the larvae may also burrow into maturing or mature fruits, which then become unmarketable.

Palpita unionalis (Hübner) (**712–713**)
Jasmine moth

A frequent pest of olive in the Mediterranean area, particularly in southern France, Greece, Italy and North Africa; other hosts include *Arbutus unedo*, *Fraxinus excelsior*, *Jasminum* and *Ligustrum vulgare*. Also present in parts of Asia. In northern Europe, including the British Isles, the moth occurs as a rare, non-resident migrant and is not a pest.

DESCRIPTION

Adult: 28–30 mm wingspan; wings mainly white and semitransparent, with a brown leading edge to each forewing; general configuration when at rest reminiscent of the aircraft Concorde. **Egg:** 1.0 × 0.5 mm; creamy white, flat and oval. **Larva:** up to 22 mm long; light green to yellowish green. **Pupa:** 12–16 mm long; brown.

LIFE HISTORY

Adults of the first generation are active in spring, females depositing several hundred eggs on the leaves of host plants. The eggs hatch within a few days under hot conditions, but after a much longer period at lower temperatures. Larvae feed for 3 or 4 weeks, attacking the buds and young terminal leaves; those feeding later in the year also attack the developing fruits. Pupation occurs amongst spun leaves at the shoot tips, adults emerging shortly afterwards. In southern Europe there are usually two or three generations annually; larvae form the overwintering stage.

DAMAGE

Larvae cause extensive damage to young shoots and also destroy the fruitlets, but attacks in mature olive groves are usually unimportant; partially damaged fruits are also unmarketable. In nurseries, however, infestations may result in the development of deformed young trees.

Udea ferrugalis (Hübner) (**714–716**)
Rusty dot moth

An occasional, minor pest of plum and gooseberry, but associated mainly with a wide range of wild herbaceous plants (especially *Arctium*, *Eupatorium cannabinum* and *Stachys*). Also reported in mainland Europe as a minor pest of field crops such as artichoke, red beet and sugar beet. Widely distributed and a well-known migrant in Europe. An introduced pest in the USA.

DESCRIPTION

Adult: 22 mm wingspan; forewings light yellowish brown to dark rusty brown, with indistinct darker markings; hindwings brownish grey. **Larva:** up to 15 mm long; body yellowish to dark green, with white subdorsal lines; head pale ochreous brown, with darker markings; prothoracic plate shiny, with a pair of black marks on either side, the anterior pair larger and elongate, the posterior pair smaller and rounded.

712 Jasmine moth (*Palpita unionalis*).

713 Jasmine moth (*Palpita unionalis*) – larva.

LIFE HISTORY
Adults occur throughout much of the year, but are usually most numerous in summer and autumn. Larvae usually feed beneath a slight silken web on the underside of leaves, with a first brood in June to July and a second, usually larger, brood in the summer or autumn. Fully fed larvae pupate in silk-lined, pod-like cocoons, each formed from a partly excised portion of leaf. Adults appear about 3 weeks later, but sometimes not for several months. Larvae resulting from immigrant moths that have reached northern Europe may occur at any time from July to October.

DAMAGE
Larvae browse on the leaves, typically removing tissue from the underside and leaving the upper epidermis intact. Damage caused is usually of little or no importance.

Udea olivalis (Denis & Schiffermüller) (**717–718**)
A minor pest of strawberry, but associated mainly with wild herbaceous plants such as *Glechoma hederacea*, *Mercurialis perennis*, *Stachys* and *Urtica*. Present mainly in northern Europe.

DESCRIPTION
Adult: 21–24 mm wingspan; forewings grey to pale reddish brown, with white markings; hindwings mainly white, with a broad greyish-brown border. **Larva:** up to 22 mm long; body light green or pale yellow and shiny, with a darker line along the back and very large, black pinacula; head light brown.

LIFE HISTORY
Similar to that of *Udea prunalis*, but larvae develop slightly earlier in the season.

DAMAGE
Unimportant.

714 Rusty dot moth (*Udea ferrugalis*).

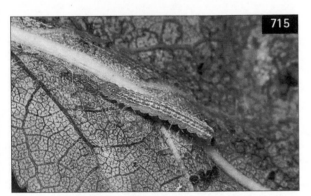
715 Rusty dot moth (*Udea ferrugalis*) – larva.

717 *Udea olivalis* – adult.

716 Rusty dot moth (*Udea ferrugalis*) – pupal cocoon.

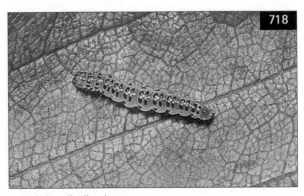
718 *Udea olivalis* – larva.

Udea prunalis (Denis & Schiffermüller) (**719–722**)

A generally common, but minor, pest of damson and plum; in addition to various wild hosts the polyphagous larvae are also found, occasionally, on cultivated pear, cherry, black currant, blackberry and strawberry. Widely distributed in north-western Europe.

DESCRIPTION

Adult: 22–26 mm wingspan; wings brownish grey, with often indistinct blackish and ash-grey markings. **Larva:** up to 25 mm long; body bright green and glossy, with two brilliant white subdorsal lines; head whitish grey; young larva translucent and greenish, with a shiny black head. **Pupa:** 10 mm long; reddish brown; cremaster bluntly elongate, with eight long terminal spines.

LIFE HISTORY

Adults occur in June and July, eggs being laid on leaves of a wide variety of plants. Larvae feed on the underside of leaves during October and then hibernate in silken cocoons spun at the leaf edges. Feeding is resumed in the spring, each larva folding and spinning leaves together with silk. The larvae wriggle rapidly out of their habitation and drop to the ground if disturbed. Pupation takes place in May or June in a white, silken cocoon, shielded by a folded leaf edge.

DAMAGE

In autumn, larvae remove tissue from the lower leaf surface. However, in spring, when somewhat older, they bite right through the leaf tissue. Larvae are rarely sufficiently numerous to cause serious damage to fruit crops, although minor outbreaks have occurred locally on black currant.

Zophodia convolutella (Hübner) (**723–724**)

Gooseberry moth

A pest of black currant and gooseberry in mainland Europe. Widely distributed from Scandinavia southwards, but not present in the British Isles. Also introduced into North America.

DESCRIPTION

Adult: 25–30 mm wingspan; forewings brownish grey to blackish, and elongate; hindwings light brownish grey, with a darker border. **Egg:** 0.5 mm across;

719 *Udea prunalis* – adult.

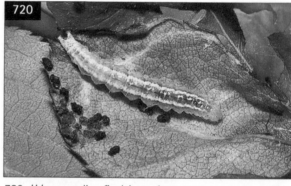

720 *Udea prunalis* – final-instar larva.

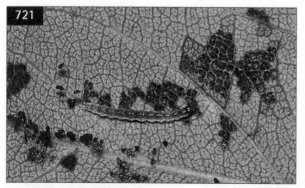

721 *Udea prunalis* – early-instar larva.

722 *Udea prunalis* – pupa.

spherical and white. **Larva:** up to 20 mm long; body light green; head and prothoracic plate black. **Pupa:** 10 mm long; reddish brown.

LIFE HISTORY

Adults appear from late April to mid-May. Eggs are then laid singly or in small batches on the young developing fruits, or on the calyx of open flowers. The eggs hatch about 2 weeks later. The larvae then bore into the young fruitlets to commence feeding. On currant, each larva attacks up to three or four fruits, which they web together with strands of silk. On gooseberry, however, a single fruit may provide adequate nourishment for a larva to complete its development. When fully grown, usually in July, the larvae drop to the ground where they form silken cocoons in which they pupate and then overwinter.

DAMAGE

Infested berries change colour (i.e. appear to ripen) prematurely, and are hollowed out, gooseberries often turning dark red. Also, harvested crops are contaminated by silk and masses of frass.

Family **PAPILIONIDAE**

Large to very large butterflies, often with vivid wing markings. Butterflies (see also the families Pieridae, Lycaenidae and Nymphalidae) have clubbed antennae, but the tips are never hooked (cf. family Zygaenidae, which includes moths with clubbed antennae). The hindwings of swallowtail butterflies (subfamily Papilioninae) each possess a long, tail-like projection. Larvae are stout-bodied and at maturity largely glabrous, apart from the presence of minute secondary hairs; an eversible gland ('osmeterium') is present dorsally on thoracic segment 1.

Iphiclides podalirius (Linnaeus) (**725–726**)

Associated with wild *Prunus*, especially *P. padus*; particularly in southern Europe, also a pest of cultivated *Prunus*, especially almond, cherry and peach. Palaearctic. Widely distributed in central and southern Europe.

DESCRIPTION

Adult: 60–80 mm wingspan; forewings pale creamy yellow, with several black stripes; hindwings pale creamy yellow and black, each with five blue and one

723 Gooseberry moth (*Zophodia convolutella*).

725 *Iphiclides podalirius* – larva.

724 Gooseberry moth (*Zophodia convolutella*) – larva.

726 *Iphiclides podalirius* – pupa.

orange patch, and a very long, mainly black and yellow tail; underside of wings similar, but paler. **Larva:** up to 45 mm long; body green, with a pale dorsal line and pale, chevron-like, segmentally arranged, markings down the back; subspiracular line white. **Pupa:** 23–25 mm long; apple-green, with pale speckles and leaf-vein-like markings.

LIFE HISTORY
Adults appear in the early spring. Eggs are then laid on the leaves of host plants. Larvae feed for several weeks and then pupate. There are two or three generations annually.

DAMAGE
Larvae cause defoliation, but are usually present on fruit trees in only small numbers.

Family **PIERIDAE**

Medium-sized, predominantly white or yellow butterflies. Larvae are elongate and lack spines, but have an even coating of secondary hairs on the body, and setae on the head that arise from raised tubercles (cf. family Lasiocampidae); crochets on the abdominal prolegs are biordinal or triordinal, arranged in a mesoseries.

727 Black-veined white butterfly (*Aporia crataegi*).

728 Black-veined white butterfly (*Aporia crataegi*) – egg batch.

729 Black-veined white butterfly (*Aporia crataegi*) – final-instar larva.

Aporia crataegi (Linnaeus) (727–735)
Black-veined white butterfly

A polyphagous pest of broadleaved (especially rosaceous) trees and shrubs, including apple, pear, medlar, quince, almond, apricot, peach and plum. Palaearctic. Widely distributed in mainland Europe.

DESCRIPTION

Adult: 45–55 mm wingspan; wings mainly white, with black veins; antennae black, except for last (= male) or last four (= female) segments. **Egg:** 1.0 × 0.5 mm; yellow, with distinct longitudinal ribs that terminate in small spheres to form a crown-like rim. **Larva:** up to 40 mm

730 Black-veined white butterfly (*Aporia crataegi*) – young larvae before hibernation.

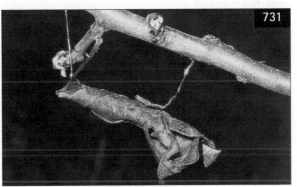

731 Black-veined white butterfly (*Aporia crataegi*) – hibernaculum.

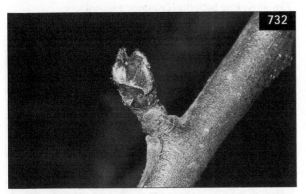

732 Black-veined white butterfly (*Aporia crataegi*) – bud damage in early spring.

733 Black-veined white butterfly (*Aporia crataegi*) – half-grown larvae.

734 Black-veined white butterfly (*Aporia crataegi*) – penultimate-instar larva.

735 Black-veined white butterfly (*Aporia crataegi*) – pupa.

long; body mainly black above, with a sparse coat of fine hairs; body hairs mainly whitish, with yellowish hairs arising from thoracic segment and small, segmentally arranged clusters of short, orange hairs arising from black-centred orange speckles that, together, form a pair of interrupted subdorsal stripes; sides and underside of body mainly purplish grey, with whitish, often black-centred speckles; head black, with short, black (and a few longer, white) hairs; spiracles black. **Pupa:** 22–26 mm long; mainly greenish yellow, speckled with black, and with black stripes on the head, thorax and wing pads; underside black.

LIFE HISTORY
Adults occur in June and July. Eggs are laid in batches on the leaves of host plants and hatch 2–3 weeks later. After devouring their egg shells, the larvae feed gregariously in a flimsy communal web. Eventually, they overwinter in their third instar, when about 4.0–7.5 mm long, typically in small groups within silken cocoons formed in the shelter of a crumpled leaf; this structure forms a hibernaculum and is suspended from a shoot by silken threads. Activity is resumed in the early spring. The larvae then feed gregariously on the buds and developing leaves, returning to the shelter of the hibernaculum during periods of inactivity. Later, the larvae become solitary and disperse to feed elsewhere on the foodplant. When fully grown, usually from mid-May onwards, they pupate on the shoots or stems, and adults emerge about 3 weeks later.

DAMAGE
Young larvae scarify leaves, but such damage is of little or no importance. In spring, however, the overwintered larvae can cause significant damage to developing buds and, later, extensive defoliation; this can have an adverse effect on the cropping and vigour of host trees.

Family **LYCAENIDAE**

Small, often brilliantly coloured butterflies, including 'blues', 'coppers' and 'hairstreaks'. Larvae are stout-bodied and woodlouse-like, and sometimes attended by ants.

Celastrina argiolus (Linnaeus) (**736–738**)
Holly blue butterfly
A minor pest of black currant in parts of mainland Europe; in addition to *Hedera helix* and *Ilex aquifolium*,

736 Holly blue butterfly (*Celastrina argiolus*).

737 Holly blue butterfly (*Celastrina argiolus*) – larva.

738 Holly blue butterfly (*Celastrina argiolus*) – pupa.

wild foodplants include *Rhamnus cathartica* and *Ulex europaeus*. Holarctic. Widely distributed in Europe.

DESCRIPTION
Adult female: 28–34 mm wingspan; forewings silvery blue, with a black marginal band and discal spot; hindwings silvery blue, with six submarginal spots; underside of wings bluish white, with numerous black spots. **Adult male:** similar in appearance to female, but black markings on the forewings less extensive. **Larva:** up to 15 mm long; body light velvet-green, coated with short white hairs; head black and shiny. **Pupa:** 8–9 mm long; brown to yellowish brown, short and stout.

LIFE HISTORY
This species overwinters in the pupal stage. Adults appear from mid-April onwards. Eggs are eventually laid, usually at the base of a flower bud, and hatch about 2 weeks later. Larvae then feed on the developing buds and often burrow into the developing young fruits of host plants. They also feed on young, tender leaves. In regions where this species is double-brooded, larvae occur mainly in late May and June, with those of the second generation feeding from about mid-August to late September. In less favourable, more northerly areas, this species is single-brooded.

DAMAGE
Although larvae damage buds and may destroy berries on black currant, infestations are not of economic importance.

Satyrium pruni (Linnaeus) (**739–741**)
Black hairstreak butterfly

In parts of mainland Europe, larvae of this generally local and rare species may occur in the spring on cultivated damson or plum; the more usual host plant, however, is *Prunus spinosa*. Larvae (up to 17 mm long) are apple-green, with paired, purplish-tipped, wart-like projections dorsally on abdominal segments 2–6. Pupation occurs in May on a leaf or nearby twig, the pupa closely resembling a bird's dropping. The mainly brown adults (35–40 mm wingspan) emerge about 3 weeks later. Unlike the far more numerous brown hairstreak butterfly (*Thecla betulae*), this species has not been reported to cause economic damage to cultivated fruit trees.

739 Black hairstreak butterfly (*Satyrium pruni*).

740 Black hairstreak butterfly (*Satyrium pruni*) – larva.

741 Black hairstreak butterfly (*Satyrium pruni*) – pupa.

Thecla betulae (Linnaeus) (742–745)
Brown hairstreak butterfly

Although most often associated with wild *Prunus*, especially *P. spinosa*, in mainland Europe the larvae frequently also occur on cultivated plum. Palaearctic. Widely distributed in Europe. In some countries (e.g. England & Wales) it is very local and on the decline, owing to modern land-management practices, and is certainly not a pest.

DESCRIPTION
Adult female: 38–45 mm wingspan; forewings bronzy brown, with a broad, orange subcentral band – underside mainly reddish yellow, with a white, partly black-edged, line and a brownish discal bar; hindwings bronzy brown, partly marked with orange at tip – underside mainly reddish yellow to orange, with two irregular white, partly black-edged, lines. **Adult male:** 33–40 mm wingspan; similar in appearance to female, but orange markings on upper surface of forewings replaced by a small patch of ochreous scales and those on hindwings reduced; underside of wings also browner. **Larva:** up to 18 mm long; body light green, with a pair of closely set, creamy-white lines along the back that diverge over the thorax, and diagonal lines down the sides; head black. **Pupa:** 10–12 mm long; brown, smooth and shiny.

LIFE HISTORY
Adults occur from July or August onwards. Eggs are laid in August and September, usually low down on the bark of host trees upon which they overwinter. Larvae feed mainly in May and June, their rate of development fluctuating considerably depending on temperature. Pupation takes place low down on the stems of host plants, but sometimes on or amongst dead leaves or in other sheltered situations. Adults emerge 4–6 weeks later.

DAMAGE
Larvae cause defoliation, but are rarely present in sufficient numbers on plum trees to be of concern.

742 Brown hairstreak butterfly (*Thecla betulae*) – male.

743 Brown hairstreak butterfly (*Thecla betulae*) – larva.

744 Brown hairstreak butterfly (*Thecla betulae*) – prepupa.

745 Brown hairstreak butterfly (*Thecla betulae*) – pupa.

Family **NYMPHALIDAE**

Butterflies with just two pairs of functional legs, the forelegs being reduced and unusable for walking. Larvae are cylindrical and often armed with prominent, branched spines.

Charaxes jasius (Linnaeus) (**746–749**)
Two-tailed pasha butterfly

At least in southern Spain, a minor pest of highbush blueberry. The normal larval foodplant, however, is the closely related *Arbutus unedo*. Locally common in the coastal fringes of southern Europe, from southern Portugal to Greece; also present in North Africa.

DESCRIPTION
Adult: 75–90 mm wingspan; forewings mainly chocolate-brown, with a broad, orange marginal band – underside marbled with black, brown, grey, orange and white; hindwings mainly chocolate-brown to black, with a pale-orange submarginal band and small white to blue eye patches, each with a pair of distinctive, tail-like projections – underside similarly patterned to underside of forewings, but markings more linear and including small purplish-blue submarginal patches. **Larva:** up to 50 mm long; body plump and green, with a yellow or yellowish-white subspiracular line and numerous pale speckles; abdominal segments 3 and 5 each with a more or less distinct, pale, eye-like patch (absent in early instars); head capsule green to purplish red, and crowned by distinctive horn-like, mainly purplish-red processes; anal segment with a pair of divergent, purplish-tipped spines. **Pupa:** 22–25 mm long; light green, plump.

LIFE HISTORY
This species has two generations annually. Eggs from first-generation adults are deposited in May or June, and the resulting larvae feed throughout June and July to produce a second generation of adults from mid-August onwards. Second-brood larvae feed during the autumn and then overwinter, sheltering on the foodplant. They complete their development in the spring.

DAMAGE
Larvae cause noticeable defoliation, particularly in their later instars.

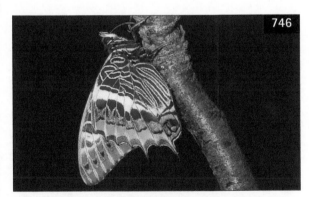

746 Two-tailed pasha butterfly (*Charaxes jasius*).

747 Two-tailed pasha butterfly (*Charaxes jasius*) – larva.

748 Two-tailed pasha butterfly (*Charaxes jasius*) – prepupa.

749 Two-tailed pasha butterfly (*Charaxes jasius*) – pupa.

Polygonia c-album Linnaeus (750–751)

Comma butterfly

Although feeding mainly on wild *Humulus lupulus* and *Urtica dioica*, larvae of this species are sometimes found on cultivated hop; they also occur occasionally on currant and, less frequently, gooseberry bushes. Palaearctic. Widely distributed in Europe.

DESCRIPTION

Adult: 50–60 mm wingspan; wings mainly brownish orange, with darker markings and a characteristic jagged outline – underside mainly brownish grey to yellowish brown, with a shiny white, comma-shaped mark centrally on each hindwing. **Egg:** 0.80 × 0.65 mm; green, elongate and strongly ribbed longitudinally. **Larva:** up to 35 mm long; body mainly black, reticulated with lilac-grey, streaked with orange, and with the dorsal surface of abdominal segments 3–7 suffused with white; body bearing large, branched spines, those on thoracic segment 1 black, those on thoracic segments 2 & 3 and abdominal segments 1 & 2 orange, and those from abdominal segment 3 onwards white; spiracles black, ringed with white; head black, streaked with orange, and with several short, orange, seta-bearing spines.

LIFE HISTORY

Adults overwinter on the branches and trunks of trees upon which, with their cryptic coloration and wing shape, they closely resemble withered leaves. Activity is resumed in the early spring. After mating, eggs are laid singly on the upper side of leaves and hatch 2–3 weeks later. First-brood larvae occur from about May onwards. Some feed rapidly and pupate in early June, but others develop more slowly and do not pupate until the end of June or early July. Pupation usually takes place amongst vegetation, and adults emerge about 2 weeks later. Adults arising from the rapidly developing larvae mate and eventually deposit eggs, their larval progeny then feeding in July and August. Adults produced by the slowly developing larvae, however, do not mate. Instead (along with adults produced by the second-brood larvae), they enter hibernation.

DAMAGE

Although sometimes attracting attention, usually when more or less fully grown, the colourful larvae are rarely numerous and cause no significant damage to crops.

750a, b Comma butterfly (*Polygonia c-album*).

751 Comma butterfly (*Polygonia c-album*) – larva.

Family **LASIOCAMPIDAE**

Medium-sized to very large, stout-bodied moths, with bipectinate antennae; males usually noticeably smaller than females of the same species. Larvae have scattered body hairs, some of which may arise from verrucae; crochets on the abdominal prolegs are biordinal and form a mesoseries; setae on the head do not arise from raised tubercles (cf. family Pieridae).

Eriogaster lanestris (Linnaeus) (**752–757**)
Small eggar moth
Polyphagous on various trees and, in mainland Europe, occasionally present on fruit trees, particularly apple, pear, damson and plum, but occurring mainly on *Crataegus monogyna* and *Prunus spinosa*. Widely distributed in central and northern Europe, but nowadays less common than formerly.

DESCRIPTION
Adult female: 42 mm wingspan; wings thinly scaled, greyish brown to pale reddish brown, with a white crossline and, on each forewing, a white subcentral and a white basal spot; abdomen with a greyish anal hair tuft. **Adult male:** 32 mm wingspan; similar to female, but darker and without the anal hair tuft; antennae strongly bipectinate. **Larva:** up to 50 mm long; body black or greyish black, with reddish to whitish hairs and a series of brown, yellowish-edged patches along the back; head black; early instars darker, and recently moulted final-instar larvae brightly coloured and with a gingery-brown head.

752 Small eggar moth (*Eriogaster lanestris*) – female.

754a, b Small eggar moth (*Eriogaster lanestris*) – late-instar larva.

753 Small eggar moth (*Eriogaster lanestris*) – male.

755 Small eggar moth (*Eriogaster lanestris*) – larva recently moulted to final-instar.

LIFE HISTORY

Adults appear in late winter or early spring. Eggs are deposited on the twigs of host plants in batches about 4 cm long. The batch is then covered by a secreted layer of liquid, into which hairs from the female's anal tuft are incorporated; the liquid soon hardens to form a protective coating. The eggs hatch several weeks later. Larvae feed from May to June or July, living gregariously within dense, silken webs. Pupation occurs in large, yellowish-white to reddish-brown cocoons. Most adults emerge in the following year, but some individuals remain in the pupal stage for two or more winters.

DAMAGE

Larvae cause considerable defoliation, but infestations on fruit trees rarely affect plant growth or cropping.

Gastropacha quercifolia (Linnaeus) (**758–760**)

Lappet moth

A minor pest of fruit trees, particularly apple and plum; wild hosts include *Crataegus monogyna*, *Prunus spinosa* and broadleaved *Salix*. Eurasiatic. Widely distributed in Europe.

DESCRIPTION

Adult: 60–80 mm wingspan; body and wings mainly reddish brown, the wings also with blackish crosslines and partly suffused with purplish grey. **Larva:** up to 100 mm long; body robust, ash-grey to reddish brown, with a coating of downy hairs, and with hairy lateral prominences (especially well developed on the thoracic

758a, b Lappet moth (*Gastropacha quercifolia*).

756 Small eggar moth (*Eriogaster lanestris*) – young larvae on communal web.

759 Lappet moth (*Gastropacha quercifolia*) – late-instar larva.

757 Small eggar moth (*Eriogaster lanestris*) – pupal cocoon.

760 Lappet moth (*Gastropacha quercifolia*) – final-instar larva.

761 Lackey moth (*Malacosoma neustria*) – male.

762 Lackey moth (*Malacosoma neustria*) – egg batch.

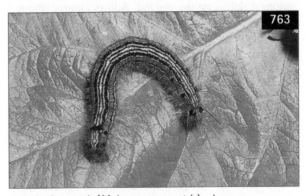

763 Lackey moth (*Malacosoma neustria*) – larva.

764 Lackey moth (*Malacosoma neustria*) – young larvae on communal web.

segments); thoracic segments 2 and 3 each with a velvet-black, often partly white-bordered crossband dorsally; abdominal segment 8 with a hairy dorsal prominence; head grey, marked with black. **Pupa:** 25–30 mm long; dark brown to blackish, coated with a white bloom.

LIFE HISTORY

Adults occur mainly in July. When at rest, they adopt a very characteristic posture, closely resembling a bunch of withered leaves. Eggs are laid in small batches on the leaves or twigs of host trees, each female depositing several hundred during her lifetime. The eggs hatch in about 2 weeks. Larvae then feed for a few weeks before, whilst still relatively small, hibernating in cocoons formed low down on the foodplant. Activity is resumed in the following spring, larvae usually completing their development in June. Each then spins a large, yellowish to greyish-brown cocoon amongst a cluster of leaves and pupates. Adults emerge about 2 weeks later.

DAMAGE

In spring and early summer larvae are capable of causing noticeable defoliation on small fruit trees; however, larvae are usually present in only small numbers.

Malacosoma neustria (Linnaeus) (**761–765**)

Lackey moth

An often common, polyphagous pest of various broadleaved trees and shrubs, including apple, pear, apricot, cherry, peach, plum and raspberry; infestations on fruit crops are usually restricted to gardens and unsprayed orchards. Eurasiatic. Present throughout much of Europe, except for the extreme north.

DESCRIPTION

Adult female: 35–40 mm wingspan; body and wings pale ochreous yellow to dark brown; each forewing with a pair of pale crosslines, usually enclosing a darker crossband. **Adult male:** 30–35 mm wingspan; similar to female, but crossband often less well defined. **Larva:** up to 50 mm long; body greyish blue, clothed in reddish-brown hairs, with a white dorsal stripe, and with orange-red, black-edged stripes running along the back and sides; head blue, with two black spots. **Pupa:** 18 mm long; brownish black and hairy.

LIFE HISTORY

Moths occur from late July to September. About 100–200 eggs are laid by each female in a characteristic band about 6–14 mm wide around a twig or spur of the foodplant. Each egg mass, protected by a clear, varnish-like coating secreted by the egg-laying female, remains in situ throughout the winter, the eggs hatching in late April and May. The larvae feed gregariously in a communal silken web or 'tent'. They often bask in sunshine, stretching out in groups on the outside of their web or on nearby shoots and branches. When all the foliage in or near the tent is consumed, the caterpillars migrate and form a new web elsewhere on the same or on an adjacent tree. The webs may exceed 30 cm in length and are very conspicuous. Larvae are fully grown by late June or early July. They then pupate in white or yellowish, double-walled cocoons spun between leaves, in bark fissures or amongst debris on the ground. Adults emerge about 3 weeks later.

DAMAGE

Defoliation is often severe, with infested branches or even whole trees stripped of leaves and covered in webbing. Growth and fruit production may be seriously affected.

Odonestis pruni (Linnaeus)

Plum lappet moth

A pest of fruit trees, particularly damson and plum, but more frequently associated with *Prunus spinosa*; other hosts include *Alnus*, *Betula*, *Fagus*, *Quercus* and *Tilia*. Widely distributed over much of mainland Europe.

DESCRIPTION

Adult female: 45–65 mm wingspan; body and wings mainly brownish orange; forewings each with a large, white, central spot, a pair of curved, purplish crosslines and the veins partly dusted with purplish. **Adult male:** similar to female, but paler. **Larva:** ash-grey to reddish brown, with pale yellowish markings and a coating of downy hairs.

LIFE HISTORY

Adults occur during the summer, eggs then being laid on the leaves of host plants. Larvae feed on the foliage during the late summer and autumn. They then overwinter and recommence feeding in the spring. Individuals typically lie stretched out along a shoot when in repose, blending in closely with their surroundings. Fully fed larvae pupate in yellowish cocoons, and adults emerge shortly afterwards.

DAMAGE

Larvae are capable of causing noticeable defoliation, especially in spring and early summer, but are usually present in only small numbers.

Poecilocampa populi (Linnaeus) (**766–767**)

December moth

A minor pest of fruit trees, particularly apple and plum; wild hosts include *Betula*, *Crataegus*, *Populus*, *Quercus* and *Tilia*. Widely distributed and locally common in central and northern Europe.

DESCRIPTION

Adult: 35–45 mm wingspan; wings dark brown or blackish, marked with grey and pale ochreous yellow;

765 Lackey moth (*Malacosoma neustria*) – pupal cocoon.

766 December moth (*Poecilocampa populi*) – male.

body rather hairy. **Larva:** up to 45 mm long; body downy-coated, mainly pale yellow to greyish, mottled with black (but underside ochreous yellow and black), and with a yellowish-red crossbar on thoracic segment 1.

LIFE HISTORY
Adults occur in November and December, but sometimes earlier. Eggs hatch in the following spring. Larvae then feed from late April onwards, and usually complete their development in June. They then pupate, each in a greyish cocoon formed in the ground.

DAMAGE
Larvae cause defoliation, but are not important pests as they tend to occur on fruit trees only in small numbers.

Trichiura crataegi (Linnaeus) (**768–769**)
Pale eggar moth
A polyphagous species on various broadleaved trees and shrubs (including *Alnus*, *Betula*, *Crataegus*, *Populus*, *Prunus*, *Sorbus* and *Ulmus*), and occasionally also a minor pest in mainland Europe on rosaceous fruit trees, particularly apple and plum. Eurasiatic. Widely distributed in Europe.

DESCRIPTION
Adult female: 30 mm wingspan; body and legs distinctly hairy; forewings mainly grey, with a darker central crossband; hindwings grey. **Adult male:** similar to female, but antennae strongly bipectinate and coloration usually paler and hindwings greyish brown. **Larva:** up to 45 mm long; body mainly black and hairy, variably marked with orange, white and yellow; spiracles white, with black rims; head black.

LIFE HISTORY
Eggs overwinter in batches on the shoots of host plants and hatch in the spring. Larvae then feed over a 2- to 3-month period, to become fully grown in June. They then pupate, each in a tough, semitransparent, parchment-like cocoon. Adults occur in August and September. The pupal stage can also be protracted, with adults emerging a year or more later. Also, in cooler regions, larvae may not complete their development until the following year.

DAMAGE
As for December moth (*Poecilocampa populi*).

767 December moth (*Poecilocampa populi*) – larva.

768 Pale eggar moth (*Trichiura crataegi*) – female.

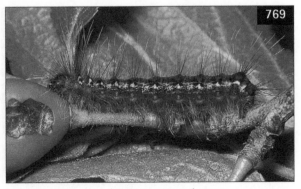

769 Pale eggar moth (*Trichiura crataegi*) – larva.

Family **SATURNIIDAE**

A mainly tropical family of often very large, highly colourful, hairy-bodied moths; antennae of males well developed. The family includes giant silk moths, some of which have a wingspan of 250 mm and are amongst the largest-known of all insects. Larvae are typically stout-bodied, and often armed with spiny projections ('scoli').

Saturnia pavonia (Linnaeus) (**770–773**)
Emperor moth
In mainland Europe a pest of raspberry. Wild hosts include *Calluna vulgaris*, *Prunus spinosa*, *Rubus fruticosus* and *Salix atrocinerea*. Widely distributed in Europe.

DESCRIPTION
Adult female: 70–75 mm wingspan; wing pattern similar to that of male, but mainly whitish, greyish and brownish grey. **Adult male:** 55–60 mm wingspan; forewings whitish, suffused with brownish grey, black, blue and purple, and with a prominent eye-like central patch; hindwings similarly patterned to forewings, but suffused extensively with orange; antennae strongly bipectinate. **Larva:** up to 60 mm long; green, with conspicuous black crossbands, each with ochreous verrucae bearing prominent black setae; early instars mainly black, with an ochreous to reddish-ochreous stripe along each side; penultimate instar mainly black, with light green markings.

LIFE HISTORY
Adults appear in the early spring. The males fly strongly in sunny weather and often gather in numbers around freshly emerged females. The females, which fly only at night, deposit eggs in batches on the stems of host plants. Larvae feed from May onwards, becoming fully grown in June. They then pupate, each forming a strong, dense reddish-brown to ginger-coloured cocoon, surrounded by a finer network of silken threads. The cocoon has a rounded door-like operculum, capable of being opened only from within and which closes again following emergence of the adult.

DAMAGE
Larvae are capable of causing significant defoliation to raspberry foliage, but infestations on cultivated canes are rarely of economic importance.

770 Emperor moth (*Saturnia pavonia*) – male.

771 Emperor moth (*Saturnia pavonia*) – final-instar larva.

772 Emperor moth (*Saturnia pavonia*) – early-instar larva.

773 Emperor moth (*Saturnia pavonia*) – late-instar larva.

Family **THYATIRIDAE**

A small family of small to medium-sized, stout-bodied moths, similar in general appearance and structure to noctuids (family Noctuidae), but with tympanal organs located on the abdomen.

Thyatira batis (Linnaeus) (**774**)
Peach-blossom moth

Larvae of this species occur mainly on *Rubus fruticosus*, but are sometimes also found on cultivated *Rubus*, including raspberry. Eurasiatic. Widely distributed in central and northern Europe.

DESCRIPTION
Adult: wingspan 34–36 mm; forewings olive brown, each with five whitish more or less peach-tinged blotches; hindwings light brown. **Larva:** up to 35 mm long; body light reddish brown, marked with dark brown and speckled with white, and with pale triangular markings on the back; thoracic segments 1 and 2, and abdominal segments 2–6 and 8, each with a divided ridge or swelling; early instars with thoracic segments 2 and 3 whitish dorsally.

LIFE HISTORY
Adults occur mainly in May and June, but sometimes later. Larvae feed from July onwards, and usually complete their development in August or early September.

DAMAGE
Larvae cause defoliation, but are usually of only minor importance.

Family **GEOMETRIDAE**
(geometer moths)

A large family of mainly small to medium-sized, often slender-bodied moths, with relatively large, broad wings, and with tympanal organs located at the base of the abdomen; most are rather weak fliers. Wing pattern of forewings and hindwings often continuous. Some species are superficially butterfly-like, but the antennae are either filiform or bipectinate, never clubbed. Larvae, commonly called 'loopers' (in America, 'inchworms'), usually have just two pairs of functional abdominal prolegs (including the pair of anal claspers) and progress with a characteristic, looping gait. The larvae of some species are brightly coloured, but many are cryptic and, when at rest, bear a close resemblance to shoots or twigs.

Abraxas grossulariata (Linnaeus) (**775–777**)
Magpie moth

A frequent pest of currant and gooseberry, but usually only in gardens; occasionally, damson and plum are also attacked. Natural hosts include *Corylus avellana*, *Euonymus europaeus* and *Prunus spinosa*. Generally distributed in Europe and in temperate parts of Asia.

775 Magpie moth (*Abraxas grossulariata*) – adult.

776 Magpie moth (*Abraxas grossulariata*) – larva.

774 Peach-blossom moth (*Thyatira batis*) – adult (top), early-instar larva (bottom).

DESCRIPTION

Adult: 35–40 mm wingspan; body yellowish orange, marked with black; wings white, with variable black markings; each forewing basally partly yellowish orange and with a yellowish-orange crossband. **Egg:** oval, smooth and creamy. **Larva:** up to 40 mm long; body creamy white, somewhat orange anteriorly, with a yellowish-orange stripe along each side; head black and shiny. **Pupa:** 15 mm long; black, ringed with yellow; cremaster with three hook-like spines.

LIFE HISTORY

The adults fly in July and August. Eggs are then laid singly or in small batches on the underside of leaves, hatching in about a fortnight. The larvae feed from August to May or June, hibernating in any convenient shelter during the winter months. They recommence feeding at bud burst, earlier than larvae of common gooseberry sawfly, *Nematus ribesii* (Chapter 7), with which they are sometimes confused. When disturbed, the larvae drop to the ground or remain suspended in mid-air on a silken thread, adopting a tight U-shaped posture. Fully fed larvae pupate beneath the leaves or elsewhere on the foodplant, each pupa being suspended in a flimsy, transparent cocoon. There is normally a single generation in a season. However, under favourable conditions, there may be at least a partial second.

DAMAGE

Larvae cause some defoliation, but are rarely numerous or of great significance; on gooseberry, for example, they are far less damaging than sawfly larvae.

Agriopis aurantiaria (Hübner) (**778–780**)
Scarce umber moth

A minor pest of apple, cherry, damson, plum and hazelnut; larvae also occur on *Betula*, *Carpinus betulus*, *Fagus sylvatica* and *Quercus*, and on various other forest trees and shrubs. Present and locally common throughout Europe; also found in Asia Minor.

DESCRIPTION

Adult female: wings reduced to stubs 2–4 mm long; body 8 mm long, yellowish brown to black, speckled with pale yellow. **Adult male:** 35–40 mm wingspan; forewings pale golden yellow, with purplish speckles and crosslines; hindwings pale. **Larva:** up to 35 mm long; body greyish to yellowish or brownish, marked with purplish and black lines and patches; head yellowish, mottled with black.

778 Scarce umber moth (*Agriopis aurantiaria*) – female.

779 Scarce umber moth (*Agriopis aurantiaria*) – male.

777 Magpie moth (*Abraxas grossulariata*) – pupa.

780 Scarce umber moth (*Agriopis aurantiaria*) – larva.

LIFE HISTORY

Adults occur in October and November. Larvae feed from April to late May or early June, eventually pupating amongst debris on the ground.

DAMAGE

Larvae contribute to defoliation caused by other spring-feeding geometrid species, e.g. winter moth (*Operophtera brumata*), but numbers on fruit trees are rarely large.

Agriopis leucophaearia (Denis & Schiffermüller) (**781**)

Spring usher moth

Although associated mainly with *Quercus* and often abundant in oak-dominated woodlands, larvae of this widely distributed Eurasiatic species sometimes occur (at least in mainland Europe) on fruit trees. Adults appear in the early spring, eggs being deposited by the females on the bark of host trees. Larvae feed from April to May and then pupate in the soil, adults emerging in the following year. Females (body 5 mm long) are dark grey and wingless; males (28–32 mm wingspan) are extremely variable in appearance, the relatively angular forewings varying from brownish black to whitish, tinged with grey or olive green. Larvae (up to 28 mm long) vary from greenish yellow to green, often with yellowish lines down the back and sometimes with darker markings along each side.

Agriopis marginaria (Fabricius) (**782–784**)

Dotted border moth

A minor pest of fruit trees, particularly apple, cherry, damson and plum; larvae also occur on a wide range of other trees and shrubs, including *Alnus*, *Betula*,

783 Dotted border moth (*Agriopis marginaria*) – male.

781 Spring usher moth (*Agriopis leucophaearia*) – male.

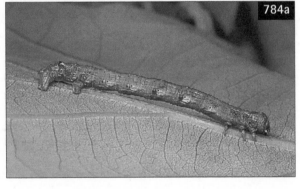

782 Dotted border moth (*Agriopis marginaria*) – female.

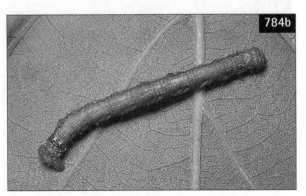

784a, b Dotted border moth (*Agriopis marginaria*) – larva.

Carpinus betulus, Crataegus monogyna, Prunus spinosa, Quercus and broadleaved *Salix*. Generally common in Europe; also present in the Near East.

DESCRIPTION
Adult female: wings reduced to stubs a few millimetres long, hindwings the longer pair; body 7–10 mm long; ochreous to greyish brown, with black and chocolate-brown markings. **Adult male:** 30–35 mm wingspan; forewings yellowish brown, with dark crosslines and a row of black dots along the margin; hindwings greyish white. **Egg:** oval and lime-green. **Larva:** up to 30 mm long; body slender, olive green or reddish brown, with pale, often yellowish, patches along the sides; sometimes also with blackish, X-shaped markings on the back; spiracles black; head prominent. **Pupa:** 10–13 mm long; brown and relatively slender; cremaster terminating in a pair of divergent spines.

LIFE HISTORY
Adults occur in March and April. Eggs are then laid singly on the bark of host plants. Larvae feed on foliage from April to late May or early June. When at rest, they remain motionless, with the body held out at an angle of about 45° and mimicking a small twig. Pupation takes place in the soil a few centimetres beneath the surface.

DAMAGE
As for scarce umber moth (*Agriopis aurantiaria*).

Alcis repandata (Linnaeus) (**785–786**)
Mottled beauty moth
A minor pest of apple, but more frequently associated with *Betula, Corylus avellana, Crataegus monogyna* and *Ulmus*, and various other wild trees and shrubs; also present on coniferous and broadleaved forest trees. Widely distributed in Europe; also present in central Asia.

DESCRIPTION
Adult: 38–44 mm wingspan; greyish white, mottled with yellowish grey and black. **Larva:** up to 40 mm long; body brownish to ochreous, with a dark brown dorsal line and diamond-shaped marks along the back; underside pale, with dark longitudinal stripes; spiracles whitish and black-rimmed; head brownish grey, with darker markings.

LIFE HISTORY
Adults occur mainly in June and July. Under favourable conditions, there is also at least a partial second generation in the autumn. Larvae feed from July onwards, most overwintering and completing their development in the following spring.

DAMAGE
Larvae destroy the foliage, but are of little significance as they are usually present on fruit trees in only small numbers.

Alsophila aescularia (Denis & Schiffermüller) (**787–789**)
March moth
Often common on a wide range of trees and shrubs, including apple, pear, cherry, plum and, sometimes, currant. Eurasiatic. Widely distributed in Europe.

DESCRIPTION
Adult female: wingless; body 8 mm long, shiny greyish brown, with a large anal hair tuft. **Adult male:** 25–30 mm wingspan; forewings rather angular, light grey to brownish grey, with lighter irregular crosslines; hindwings greyish white. **Egg:** dark brown and more or less spherical. **Larva:** up to 30 mm long; body light green, with a darker green dorsal line and yellowish lines along the sides, including one below the spiracles; a vestigial pair of prolegs on abdominal

785 Mottled beauty moth (*Alcis repandata*).

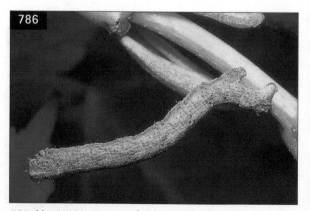

786 Mottled beauty moth (*Alcis repandata*) – larva.

segment 5; head green. **Pupa:** 8–10 mm long; brown and stumpy; cremaster with two curved, divergent spines.

LIFE HISTORY
Adults occur from mid-February to mid-April. The female lays her eggs in a large band around a twig. The eggs hatch in April, the larvae then feeding on foliage and blossom trusses, often in company with those of winter moth (*Operophtera brumata*). Larvae of the latter, however, are usually more numerous. When fully grown, in late May or June, larvae enter the soil and each pupates in a silken cocoon.

DAMAGE
As for winter moth (*Operophtera brumata*).

Apocheima hispidaria (Denis & Schiffermüller) (**790**)
Small brindled beauty moth

A minor pest of fruit trees, particularly apple, pear, cherry and plum, but associated mainly with *Quercus*. Widely distributed in Europe, but local and favouring lower-lying habitats.

DESCRIPTION
Adult female: 10–11 mm long; wingless; body brown to black; legs hairy. **Adult male:** 30–35 mm wingspan; forewings grey to yellowish brown, more or less marked with dark brown; hindwings light grey, with a dark central band; antennae bipectinate. **Larva:** up to 40 mm long; body brown to brownish black or purplish, with small, black, seta-bearing verrucae and the sides sometimes marked with orange; dorsal verrucae on abdominal segment 8 larger than the rest (the largest subdorsal pair occurring on abdominal segment 2).

LIFE HISTORY
Adults occur in February and March, and the larvae feed from April or May onwards. Larvae are fully grown in June. They then pupate in the soil, and adults emerge in the following year.

DAMAGE
Larvae contribute to defoliation caused in the spring by various other related species.

787 March moth (*Alsophila aescularia*) – male.

788 March moth (*Alsophila aescularia*) – egg batch.

789 March moth (*Alsophila aescularia*) – larva.

790 Small brindled beauty moth (*Apocheima hispidaria*) – male.

Biston betularia (Linnaeus) (**791–794**)
Peppered moth
larva = hop-cat
A generally common species, attacking a wide variety of trees and shrubs, including apple, cherry, plum, currant and hop. Widely distributed in Europe; also present in eastern Asia, North Africa and Siberia.

DESCRIPTION
Adult: 42–55 mm wingspan; body and wings white, peppered with black; entirely black (ab. *carbonaria*) and intermediate (ab. *insularia*) forms also occur; male antennae pectinate. **Egg:** 0.7 × 0.5 mm; whitish green. **Larva:** up to 50 mm long; body brown or green, with pinkish markings and reddish spiracles; stick-like, with a pair of dark purplish prominences on abdominal segment 5; head purplish brown, with a distinct central notch. **Pupa:** 20–22 mm long; blackish brown; cremaster terminating in a spike.

LIFE HISTORY
Adults occur in May, June and July, depositing eggs on various plants. The eggs hatch 2–3 weeks later. Larvae feed from July to September or October. At rest, they cling to a twig or shoot with the body held straight out at an angle of about 45°, mimicking a broken twig. When fully grown, larvae enter the soil to pupate, adults emerging in the following year.

DAMAGE
Owing to the late appearance of the larvae, defoliation caused is of little importance; fruits are not attacked. However, larvae sometimes occur as contaminants in trays of mechanically harvested black currants.

Biston strataria (Hufnagel) (**795**)
Oak beauty moth
Although associated mainly with *Quercus*, and also with other broadleaved trees such as *Betula*, *Populus*, *Tilia* and *Ulmus*, larvae of this widely distributed species are, occasionally, found in mainland Europe feeding on the leaves of fruit trees, especially in warmer regions. Adults (48–56 mm wingspan) are mainly white and chocolate-brown. They occur in the early spring. The dark brown, twig-like larvae (up to 55 mm long), feed from May to July.

791 Peppered moth (*Biston betularia*) – female.

792 Peppered moth (*Biston betularia*) – male.

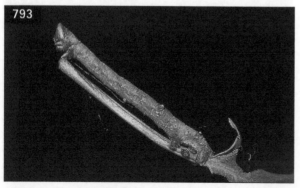

793 Peppered moth (*Biston betularia*) – larva (brown form).

794 Peppered moth (*Biston betularia*) – larva (green form).

Campaea margaritata (Linnaeus) (**796–797**)

Light emerald moth

A minor pest of fruit trees, particularly apple, but more frequently associated with forest trees such as *Betula*, *Quercus* and *Ulmus*. Widely distributed in Europe; also present in Asia Minor.

DESCRIPTION

Adult: 32–38 mm wingspan; whitish green, fading to dirty yellowish white; forewings each with a small, reddish apical mark and two crosslines; hindwings each with one crossline. **Larva:** up to 35 mm long; body mainly greyish, sometimes greenish brown to purplish brown, with a ventral fringe of bristles along each side; head yellowish grey mottled with black; unlike most geometrids, with three pairs of functional abdominal prolegs.

LIFE HISTORY

Adults occur in June and July. Eggs are laid in batches on the underside of leaves; they hatch in September. Larvae feed throughout the autumn and winter, grazing on the bark of the young shoots; when at rest, they lie stretched out flat against the shoots and are difficult to detect. In early spring they attack the young buds and, later, after bud burst, devour the foliage. In warmer regions there are sometimes two generations annually.

DAMAGE

When overwintering larvae remove bark from the young shoots, they expose the pale wood; potentially, this can lead to subsequent infection by secondary pathogens. On fruit trees, damage to buds is most significant.

Chloroclysta siterata (Hufnagel) (**798**)

Red-green carpet moth

A minor pest of apple and other fruit trees in mainland Europe, but more frequently associated with broadleaved trees such as *Betula*, *Fagus*, *Fraxinus*, *Quercus* and *Tilia*. Widely distributed in Europe.

DESCRIPTION

Adult: 24–26 mm wingspan; forewings greyish green to dark green, often with a pinkish tinge; hindwings greyish green. **Larva:** body elongate and yellowish green, with an incomplete red dorsal line; final abdominal segment with a pair of small, pale-tipped projections.

LIFE HISTORY

Adults are active in the late summer or early autumn and again, after hibernation, in the spring. Larvae feed on the leaves of host plants from June or July onwards, completing their development in August. They then pupate in the soil.

795 Oak beauty moth (*Biston strataria*) – male.

796 Light emerald moth (*Campaea margaritata*).

797 Light emerald moth (*Campaea margaritata*) – larva.

798 Red-green carpet moth (*Chloroclysta siterata*).

DAMAGE

Owing to the late appearance of the larvae, defoliation caused is of little or no importance; fruits are not attacked.

Chloroclysta truncata (Hufnagel) (**799–801**)

Common marbled carpet moth

A generally common species, the larvae feeding on, for example, *Betula*, *Crataegus*, *Salix* and various herbaceous plants. Also minor pest on cultivated currant and raspberry, and on both field-grown and greenhouse-grown strawberry. Eurasiatic. Widely distributed in Europe.

DESCRIPTION

Adult: 30–35 mm wingspan; forewings extremely variable (with several colour forms), blackish grey to brownish grey, suffused with ochreous to whitish, and with a dark discal spot and wavy, pale crosslines; hindwings light brownish grey. **Larva:** up to 32 mm long; body slender and greenish, with a darker green dorsal line and a pair of subdorsal lines along the back, and sometimes also with a complete or interrupted reddish stripe along the somewhat warty sides; final body segment with a pair of pointed projections.

799 Common marbled carpet moth (*Chloroclysta truncata*).

800 Common marbled carpet moth (*Chloroclysta truncata*) – ab. *rufescens*.

LIFE HISTORY

Adults appear in May and June, and in the late summer. Larvae occur from late June to late July or early August, and from September to April. Under favourable conditions there may be a third generation and larvae may remain active throughout the winter.

DAMAGE

Larvae cause noticeable defoliation, but the extent of damage is usually insignificant.

Chloroclystis rectangulata (Linnaeus) (**802–804**)

Green pug moth

A common and sometimes persistent pest of apple and pear; occasionally, larvae also feed on apricot, cherry and other fruit crops. Widely distributed in Europe.

DESCRIPTION

Adult: 18–20 mm wingspan; dark green to blackish, with numerous wavy crosslines on the wings. **Larva:** up to 20 mm long; body stumpy, pale yellowish green, with a dark green or red band along the back and, sometimes, with reddish intersegmental rings; head blackish. **Pupa:** 6–7 mm long; stout; yellowish, with an olive tinge; cremaster reddish brown, with several hook-tipped bristles.

LIFE HISTORY

Moths are active in June and July. Each female deposits up to 100 eggs on the dormant buds, where they remain until the following spring. Eggs hatch in late March or April and the larvae then attack the buds. Later, the blossoms and expanded foliage are attacked. Larvae feed until early June; they then enter the soil to pupate. Adults emerge about 2 weeks later.

801 Common marbled carpet moth (*Chloroclysta truncata*) – larva.

DAMAGE

Larvae destroy the stamens and other parts of the flowers, causing considerable deformation. Each larva may attack several blossoms and, if the pest is numerous, fruit set may be significantly reduced. Normally, however, the pest is present in only small numbers and is often overlooked. Attacked leaves tend to be grazed on just one surface, so the tissue is not bitten right through.

Chloroclystis v-ata (Haworth) (**805–806**)

V-pug moth

In addition to various wild hosts (such as *Clematis vitalba*, *Eupatorium cannabinum*, *Lythrum salicaria* and *Solidago virgaurea*), larvae of this generally distributed species are sometimes found on cultivated blackberry, attacking the blossoms and foliage. Although direct damage is slight, the larvae are of some importance, often being accidental contaminants in harvested fruit sent for processing. Adults (18 mm wingspan) occur from May to August, and are most numerous in May; the forewings are green, each bearing a characteristic V-shaped mark. The larvae feed in June and July, with a second brood developing during the autumn. Individuals (up to 20 mm long) are yellowish green or greyish, with three reddish or brownish lines (and often also a series of broad, triangular or lozenge-shaped markings) along the back.

802 Green pug moth (*Chloroclystis rectangulata*).

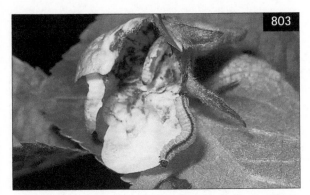

803 Green pug moth (*Chloroclystis rectangulata*) – larva.

805 V-pug moth (*Chloroclystis v-ata*).

804 Green pug moth (*Chloroclystis rectangulata*) – pupa.

806 V-pug moth (*Chloroclystis v-ata*) – larva.

Colotois pennaria (Linnaeus) (**807–810**)

Feathered thorn moth

A polyphagous, generally common woodland species whose larvae sometimes feed on fruit trees. Eurasiatic. Widely distributed in Europe.

DESCRIPTION

Adult: 40 mm wingspan; pale orange to brownish white, marked with brown; male with strongly pectinate antennae. **Larva:** up to 45 mm long; body moderately stout and twig-like, purplish slate-grey or yellowish brown, marked with indistinct, yellowish diamonds down the back and similarly coloured spots along the sides; abdominal segment 8 with two rounded, reddish or brownish-red projections.

LIFE HISTORY

Oblong egg batches, each containing 100–200 olive-green, smooth-shelled eggs, are placed along shoots of trees in October or November, where they remain until the spring. Larvae then feed from April to June before eventually pupating in the soil, each in an earthen cell. Adults emerge in the autumn.

DAMAGE

Larvae destroy leaves, but attacks on fruit trees are of little significance.

Crocallis elinguaria (Linnaeus) (**811–812**)

Scalloped oak moth

Polyphagous on various trees and shrubs, and a minor pest of fruit trees, particularly apple, cherry and plum. Widely distributed in Europe.

DESCRIPTION

Adult: 34–42 mm wingspan; wings pale yellow, with a light brown crossband and a blackish discal spot on each forewing; head and thorax clothed in yellow or brownish-yellow hairs. **Larva:** up to 45 mm long; body twig-like, greyish yellow to greyish black, tinged with purple; abdominal segment 8 with a pair of small, pale-tipped projections.

LIFE HISTORY

Adults occur from June to August, eggs being laid in small batches on the twigs of host plants. The larvae feed during the spring, eventually pupating in June.

807 Feathered thorn moth (*Colotois pennaria*) – female.

808 Feathered thorn moth (*Colotois pennaria*) – male.

809 Feathered thorn moth (*Colotois pennaria*) – larva (light form).

810 Feathered thorn moth (*Colotois pennaria*) – larva (dark form).

811 Scalloped oak moth (*Crocallis elinguaria*).

812 Scalloped oak moth (*Crocallis elinguaria*) – larva.

DAMAGE

Although capable of causing noticeable defoliation, numbers of larvae on fruit trees are usually small.

Ectropis bistortata (Goeze) (**813–814**)
Engrailed moth

A minor pest of apple and other fruit trees; also present on a wide range of coniferous and broadleaved forest trees. Eurasiatic. Widely distributed in Europe.

DESCRIPTION

Adult: 32–44 mm wingspan; whitish to yellowish grey, with blackish markings. **Larva:** up to 45 mm long; body twig-like, varying from brownish black to greyish brown. **Pupa:** 12–14 mm long; brown, cremaster elongate.

813 Engrailed moth (*Ectropis bistortata*).

LIFE HISTORY

Adults of the first generation occur in March and April and those of the second in July and August. In some situations, however, there is just one generation annually. Eggs are laid in small batches on the bark of host plants, each female depositing on average about 200–300 in her lifetime. Larvae feed on the leaves and may also graze upon the rind of the young shoots. They are most numerous in June and July. Pupation takes place on the ground, beneath leaves, moss or other shelter.

814 Engrailed moth (*Ectropis bistortata*) – larva.

DAMAGE

Larvae are rarely sufficiently numerous on fruit trees to cause economic damage to either the foliage or the bark.

Epirrita dilutata (Dennis & Schiffermüller) (**815–816**)
November moth

A minor pest of apple, pear, plum and other fruit trees. Wild hosts include *Alnus*, *Betula*, *Crataegus*, *Prunus*, *Quercus* and *Ulmus*. Widely distributed in Europe and generally common in woodland areas.

815 November moth (*Epirrita dilutata*).

816 November moth (*Epirrita dilutata*) – larva.

817 Mottled umber moth (*Erannis defoliaria*) – female.

DESCRIPTION

Adult: 32–38 mm wingspan; dull greyish to black, the wings also marked with feint wavy crosslines. **Larva:** up to 30 mm long; body bright velvet-green (somewhat whitish below), flecked with pale yellow and with a more or less distinct, often incomplete, yellowish-white stripe along each side; often patterned with purplish-red, crosshatched and diamond-shaped markings and with a purplish-red central line down the back; purplish-red markings may also occur on the sides; spiracles reddish-ringed.

LIFE HISTORY

Eggs overwinter and hatch in the early spring. Larvae then feed on the young foliage of host trees, unlike larvae of many other related species often sitting openly on the upper surface of leaves during the daytime. Larvae are fully grown by late May or early June. They then pupate in the soil. Adults appear in October and November.

818 Mottled umber moth (*Erannis defoliaria*) – male.

DAMAGE

The larvae cause indiscriminate damage to the leaves of host trees, contributing to spring defoliation.

Erannis defoliaria (Clerck) (**817–819**)

Mottled umber moth

Generally common on broadleaved forest trees, particularly *Quercus*; fruit trees, currant and gooseberry are also attacked. Widely distributed in Europe.

DESCRIPTION

Adult female: wingless; body 10–15 mm long, mottled with black, yellow and, sometimes, white scales. **Adult male:** 35–38 mm wingspan; forewings pale yellow to reddish brown, more or less finely peppered with black and often variably decorated with darker cross-markings. **Egg:** 0.9 × 0.5 mm; oval and almost smooth; yellowish to greyish. **Larva:** up to 35 mm long; body reddish brown, with yellow or creamy-white patches on the sides of abdominal segments 1–7.

819 Mottled umber moth (*Erannis defoliaria*) – larva.

Pupa: 12–14 mm long; brown; cremaster elongate, with a bifid tip.

LIFE HISTORY

Adults occur from mid-October to mid-January, and sometimes later. Females deposit an average of 300–400 eggs in bark crevices. The eggs hatch in early April. Larvae then feed on the buds and leaves of host plants. They are easily dislodged from the foodplant, remaining temporarily suspended by a silken thread, with the head and thorax bent back at an angle to the abdomen. The larvae are fully grown in June. They then descend to the ground to pupate in the soil a few centimetres beneath the surface.

DAMAGE
Larvae are capable of causing considerable defoliation and may also damage unopened or opening buds.

Eulithis prunata (Linnaeus) (820)
Phoenix moth
Larvae of this locally distributed species occur in gardens on black currant, red currant and gooseberry, but they are not an important pest. Adults fly in July and August, eggs being laid on the stems and branches of the bushes, where they remain until the following spring. The larvae (up to 40 mm long) are stout-bodied, green to brownish, each with a series of whitish, purplish-edged triangular marks down the back; the thoracic segment 3 is swollen and bears a dark crossband. They feed on foliage from late March or early April to mid-June; each then pupates between two leaves.

Eupithecia assimilata Doubleday (821–822)
Currant pug moth
A minor pest of currant bushes, particularly in gardens and allotments; gooseberry and hop are also attacked. Widespread in the northern half of Europe.

DESCRIPTION
Adult: 18–20 mm wingspan; body and wings olive brown, with blackish markings; forewings tinged with reddish, and each with a prominent black discal mark and a series of white, submarginal spots. **Larva:** up to 20 mm long; body slender and slightly roughened, whitish green, marked with yellow intersegmentally, and often tinged dorsally with purplish red; head light greyish green. **Pupa:** 7–8 mm long; golden brown.

LIFE HISTORY
This species is usually double-brooded, moths of the first generation occurring in May and June. Eggs are laid on leaves of the foodplant, the larvae then feeding on the foliage in June and July. Pupation takes place in a slight web under a leaf or amongst debris on the ground. Second-generation adults appear in August, their larvae feeding in the autumn.

DAMAGE
The larvae feed singly, biting out large holes in the leaves. Leaf symptoms are very obvious, and sometimes extensive, but rarely of economic importance.

820 Phoenix moth (*Eulithis prunata*).

821 Currant pug moth (*Eupithecia assimilata*) – female.

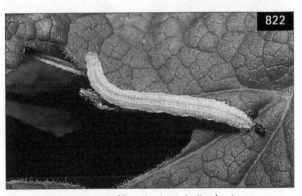

822 Currant pug moth (*Eupithecia assimilata*) – larva.

Eupithecia exiguata (Hübner) (**823–824**)
Mottled pug moth

Although associated mainly with wild hosts such as *Crataegus*, *Prunus spinosa* and *Salix*, larvae of this

823 Mottled pug moth (*Eupithecia exiguata*) – female.

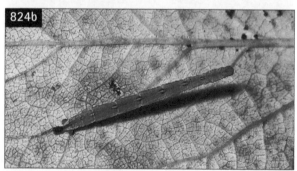

824a, b Mottled pug moth (*Eupithecia exiguata*) – larva.

825 Olive pug moth (*Gymnoscelis rufifasciata*) – larva.

widely distributed and generally common species are found, occasionally, on cultivated *Prunus* (especially bullace and damson) and currant. Adults (18–22 mm wingspan) are mainly greyish, with irregular black markings. They occur in May and June. The larvae (up to 20 mm long) are dark green, long and slender, with a pale, discontinuous, partly dark-red-marked, dorsal stripe and a dark-red, yellow-edged stripe along the sides. They feed on foliage during the autumn. The winter is passed in the soil in the pupal stage. Owing to the late appearance of larvae, damage caused is of no importance.

Gymnoscelis rufifasciata (Haworth) (**825**)
Olive pug moth

A minor, polyphagous pest of citrus and olive; other hosts include *Buxus*, *Clematis*, *Crataegus* and *Ilex aquifolium*. Widely distributed in Europe.

DESCRIPTION

Adult: 17–19 mm wingspan; forewings and hindwings mainly grey to blackish, with two dark crossbands and pale, wavy crosslines; forewings rather angular. **Larva:** up to 17 mm long; body extremely variable, ranging from whitish to dark red, and often with a series of dark markings along the back; head dark brown to blackish.

LIFE HISTORY

First-generation adults appear in the spring. Eggs are then laid on host plants and hatch within 1–2 weeks. Larvae feed for 4–6 weeks, those of the first generation typically boring into flower buds and attacking the developing stamens and other floral parts. Larvae also attack the leaves. Pupation occurs in strong cocoons, adults emerging 2–3 weeks later. There are several generations annually. The winter is passed in the pupal stage.

DAMAGE

Larvae destroy flower buds and floral parts, resulting in a reduction in fruit set. Damage to leaves is of little or no importance.

Lycia hirtaria (Clerck) (**826–827**)
Brindled beauty moth

A minor pest of fruit trees, particularly pear, cherry and plum, but most frequently associated with *Salix*, *Tilia*, *Ulmus* and other broadleaved forest trees. Eurasiatic. Widely distributed in Europe.

DESCRIPTION

Adult: 40–45 mm wingspan; body rather hairy; both sexes fully winged; wings greyish to blackish, marked with ochreous; antennae of male strongly bipectinate. **Larva:** up to 55 mm long; body robust, purplish grey to reddish brown, marked with dark speckles and with a wavy pattern, prominent yellow spots on abdominal segments 2–5 and a yellow crossline just behind the head; head small, marked with black and, near the mandibles, with yellow. **Pupa:** 20–25 mm long; brown; cremaster with a pair of short, thorn-like spines basally and terminating in a pair of long, divergent spines.

LIFE HISTORY
Adults occur in March and April. The larvae feed from May to July and then pupate in the soil a few centimetres below the surface. There is just one generation annually.

DAMAGE
Larvae are capable of causing extensive damage to foliage, but numbers on fruit trees are usually small.

Operophtera brumata (Linnaeus) (**828–831**)
Winter moth
A major pest of fruit trees, particularly apple, pear, cherry, damson and plum; other hosts include currant, gooseberry, highbush blueberry and hazelnut, and a wide range of broadleaved forest trees and shrubs. Eurasiatic. Widely distributed in Europe, particularly in central and northern regions. An introduced pest in Canada.

DESCRIPTION
Adult female: wings reduced to blackish stubs; body 5–6 mm long, dark brown, mottled with greyish yellow. **Adult male:** 22–28 mm wingspan; forewings rounded, greyish brown, with darker wavy crosslines; hindwings brownish white. **Egg:** 0.5 × 0.4 mm; oval, with a pitted surface; pale yellowish green when laid, soon becoming orange red. **Larva:** up to 25 mm long; body light green, with a dark green dorsal stripe and several whitish or creamy-yellow stripes along the back and sides, including a pale yellow line passing through the spiracles; head light green. **Pupa:** 7–8 mm long; brown and stumpy; cremaster bearing a pair of laterally directed spines.

826 Brindled beauty moth (*Lycia hirtaria*) – male.

827 Brindled beauty moth (*Lycia hirtaria*) – larva.

828 Winter moth (*Operophtera brumata*) – female.

829 Winter moth (*Operophtera brumata*) – male.

830 Winter moth (*Operophtera brumata*) – larva.

831a, b Winter moth (*Operophtera brumata*) – damage to apples.

832 Northern winter moth (*Operophtera fagata*) – female.

833 Northern winter moth (*Operophtera fagata*) – larva.

LIFE HISTORY

Adults occur from October to January, but are most abundant in November and December. On emerging from the pupa, the spider-like female crawls up the trunk of a tree and, after mating, lays 100–200 eggs singly in crevices in the bark. Eggs hatch in the spring, from about bud burst to green cluster, but often later. Newly emerged larvae then attack the developing leaves. At this stage, the minute larvae are often blown from tree to tree on strands of silk, and infestations may then spread from adjacent woodlands or hedgerows to previously 'clean' fruit trees. Feeding continues until late May or early June, the rather sluggish larvae often spinning two leaves loosely together with silk or 'capping' the blossoms and feeding within the shelter of the overlying petals. Older larvae attack fruitlets in addition to leaf and flower trusses. When fully grown, larvae drop to the ground and enter the soil to pupate in flimsy cocoons at a depth of about 8–10 cm.

DAMAGE

If larvae become active before bud burst, unopened buds may be destroyed. However, most damage occurs later in the spring, the foliage being devoured indiscriminately. Damage caused to the blossoms is often considerable and if infestations are exceptionally severe all flowers and greenery on the foodplant may be destroyed. Older larvae commonly bite holes into fruitlets, which either subsequently drop prematurely or become malformed and develop corky scars or depressions; sometimes, a damaged fruit will have a deep depression, extending down to the core. Damage caused by winter moth larvae is usually compounded by larvae of several other 'looper' species, notably *Alsophila aescularia* and *Erannis defoliaria*.

Operophtera fagata (Scharfenberg) (**832–833**)
Northern winter moth

A minor pest of apple, but usually only in the vicinity of *Betula* and other more usual wild hosts such as *Carpinus betulus*, *Fagus sylvatica* and *Quercus*. Widely distributed in Europe, particularly in more northerly and montane regions.

DESCRIPTION

Adult female: wings reduced to greyish stubs, but longer than those of *Operophtera brumata*; body 7 mm long, greyish to blackish. **Adult male:** 24–30 mm wingspan; forewings greyish white, with indistinct crosslines; hindwings whitish. **Larva:** up to 21 mm long; body yellowish green, with greyish lines along the back and sides; spiracles black; head black.

LIFE HISTORY
Adults emerge in October or November, and eggs are eventually laid on the twigs of host plants. Larvae occur in May and June.

DAMAGE
Larvae contribute to damage caused by various other related species and, if numerous, can cause noticeable defoliation. Infestations on nursery stock are particularly harmful.

Opisthograptis luteolata (Linnaeus) (**834–837**)
Brimstone moth
A minor pest of fruit trees, particularly apple and plum, but associated mainly with *Crataegus monogyna* and, to a lesser extent, *Prunus spinosa* and other shrubs. Widely distributed in Europe.

DESCRIPTION
Adult: 32–35 mm wingspan; wings pale yellow to sulphur-yellow, with indistinct greyish markings, the forewings also with distinct brownish-orange markings and a small, whitish-blue, brownish-rimmed subcentral patch. **Larva:** up to 25 mm long; body greenish to dark grey, stout and twig-like, with a distinct projection on abdominal segment 3; four pairs of abdominal prolegs; head greenish or brownish, mottled with blackish.

LIFE HISTORY
Adults occur from April to August, but are most numerous in May and June. Larvae occur in spring, summer and autumn. There are usually two generations annually.

DAMAGE
Larvae contribute to defoliation, but are themselves of only minor significance as they usually occur on fruit crops in only small numbers.

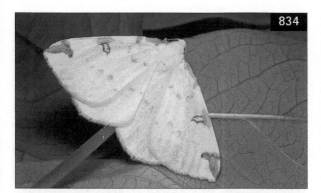

834 Brimstone moth (*Opisthograptis luteolata*).

835 Brimstone moth (*Opisthograptis luteolata*) – larva (green form).

836 Brimstone moth (*Opisthograptis luteolata*) – larva (dark form).

837 Brimstone moth (*Opisthograptis luteolata*) – pupal cocoon.

Ourapteryx sambucaria (Linnaeus) (**838–839**)
Swallow-tailed moth

A minor pest of plum and other cultivated species of *Prunus*. Wild hosts include *Crataegus monogyna*, *Hedera helix*, *Prunus spinosa* and *Sambucus nigra*. Eurasiatic. Generally common and widely distributed in Europe.

DESCRIPTION
Adult: 55–60 mm wingspan; pale yellowish white to greenish white, with greenish-yellow crosslines; hindwings each with a distinctive tail-like projection. **Larva:** up to 60 mm long; body twig-like, slender and distinctly tapered anteriorly, greyish brown, with purplish markings; head very small, greyish.

LIFE HISTORY
Larvae feed from August onwards and again in the following spring, usually completing their development in June. Adults occur in July. Under favourable conditions there may be a partial second generation.

DAMAGE
Larvae cause extensive defoliation, but are rarely sufficiently numerous on cultivated plants to be of significance.

Peribatodes rhomboidaria (Denis & Schiffermüller) (**840–841**)
Willow beauty moth

A locally important pest of grapevine; also a minor pest on fruit trees and generally common on a wide range of other trees and shrubs. Widely distributed in Europe.

DESCRIPTION
Adult: 40 mm wingspan; wings and body greyish to brownish grey, with pale and blackish wavy markings; antennae of male strongly bipectinate. **Larva:** up to 38 mm long; body relatively slender, reddish brown, mottled with ochreous, and with faint diamond-shaped markings along the back; head brownish.

838 Swallow-tailed moth (*Ourapteryx sambucaria*).

840 Willow beauty moth (*Peribatodes rhomboidaria*) – male.

839 Swallow-tailed moth (*Ourapteryx sambucaria*) – larva.

841 Willow beauty moth (*Peribatodes rhomboidaria*) – larva.

LIFE HISTORY

Adults are active in July and August, mated females laying anything up to 500 eggs. The larvae feed briefly during the late summer and autumn. They then overwinter in bark crevices or in the soil. Feeding is resumed in the spring, each individual eventually pupating in late May or early June in a strong, silken cocoon formed in the soil or on a twig or small branch of the foodplant.

DAMAGE

Autumn feeding by young larvae is of no significance. However, in the following spring, attacks on the buds and young foliage can be extensive, particularly in vineyards.

Phigalia pilosaria (Denis & Schiffermüller) (842–844)

Pale brindled beauty moth

A minor pest of fruit trees, particularly apple, pear, cherry and plum, but associated mainly with forest trees and shrubs, including *Betula*, *Carpinus betulus*, *Populus*, *Prunus*, *Quercus*, *Rosa*, *Salix*, *Tilia* and *Ulmus*. Widely distributed in Europe.

DESCRIPTION

Adult female: 12 mm long; virtually wingless; body grey to black, often with a coppery tinge; abdomen with deeply incised bifid scales dorsally. **Adult male:** 38–42 mm wingspan; forewings mainly grey to greenish grey, with yellowish-brown to dark brown markings; antennae bipectinate. **Larva:** up to 45 mm long; body mainly grey, partly suffused with pink, with prominent black seta-bearing verrucae and, often (but not invariably), a pale inverted V-shaped mark on both abdominal segments 2 and 3; verrucae on abdominal segments 2 and 3 noticeably larger than the rest; spiracles white, with a black rim.

LIFE HISTORY & DAMAGE

As for small brindled beauty moth (*Apocheima hispidaria*), although adults sometimes occur earlier (if not also later) in the year.

842 Pale brindled beauty moth (*Phigalia pilosaria*) – female.

843 Pale brindled beauty moth (*Phigalia pilosaria*) – male.

844 Pale brindled beauty moth (*Phigalia pilosaria*) – larva.

Selenia dentaria (Fabricius) (**845–846**)
Early thorn moth

A minor pest of fruit trees, particularly apple and plum; polyphagous and generally common on a wide range of broadleaved trees and shrubs. Widely distributed in central and northern Europe.

DESCRIPTION
Adult: 32–40 mm wingspan; wings brownish white, with reddish-brown markings; underside darker and partly suffused with lilac and white. **Larva:** up to 40 mm long; body twig-like, orange brown to reddish brown, but extremely variable, marked with ash-grey and black, and with paired projections on abdominal segments 4 and 5. **Pupa:** 10–14 mm long; reddish brown.

LIFE HISTORY
The butterfly-like adults occur in April and early May, with a smaller second generation in July and August. Larvae feed on the leaves of host plants. They are most numerous in May and June, but also occur in August and September. Pupation takes place in the soil. Pupae form the overwintering stage.

DAMAGE
Unimportant, as larvae are rarely numerous on fruit trees.

Semiothisa wauaria (Linnaeus) (**847**)
V-moth

A minor pest of currant and gooseberry. Widely distributed in northern and north eastern Europe.

DESCRIPTION
Adult: 25–27 mm wingspan; forewings light greyish brown, each with darker costal markings and a V-shaped discal mark; hindwings brownish white. **Larva:** up to 25 mm long; body greyish or light green, with four white lines down the back and a broad yellow, interrupted stripe along each side, and with numerous black pinacula and short black bristles.

LIFE HISTORY
The larvae feed from April to June, concentrating particularly on the young shoots. They then pupate amongst the foliage in a slight cocoon, adults appearing in July and August.

DAMAGE
Larvae browse on the leaves and also destroy the young shoots.

Theria primaria Haworth (**848–849**)
Early moth

A minor pest of damson, plum and other fruit trees, but associated mainly with *Crataegus monogyna* and *Prunus spinosa*. Widely distributed in Europe.

845

845 Early thorn moth (*Selenia dentaria*) – male.

846

846 Early thorn moth (*Selenia dentaria*) – larva.

847

847 V-moth (*Semiothisa wauaria*).

DESCRIPTION

Adult female: wings reduced, the wing stubs rather angular, mainly grey, and each with a dark crossband; body 8–9 mm long; grey, speckled with black. **Adult male:** 30–32 mm wingspan; wings mainly light greyish brown (hindwings paler), each with an indistinct crossline and a discal spot. **Larva:** up to 20 mm long; body greenish to greenish brown, with distinct white, longitudinal stripes; head bluish green, mottled with dark brown.

LIFE HISTORY

The moths appear in January and February, but are usually overlooked. Eggs are laid on the twigs of host plants and hatch about 2 months later. Larvae feed throughout April and May. They then pupate in the soil, adults emerging in the following year.

DAMAGE

Unimportant, although larvae on fruit trees contribute to defoliation caused in spring by other geometrids, such as winter moth (*Operophtera brumata*).

Family **SPHINGIDAE**
(hawk moths)

Large to very large, stout-bodied, strong-flying moths, with elongate forewings, large tegulae and a very long proboscis. Larvae are stout-bodied, usually with a prominent dorsal horn (caudal horn) arising from abdominal segment 8. Hawk moth larvae often occur in two or more colour forms, and in many species there are oblique lateral stripes on abdominal segments 1–7.

Acherontia atropos (Linnaeus) (**850a, b**)
Death's head hawk moth

Although associated mainly with Solanaceae, including cultivated potato, the larvae of this minor pest sometimes occur on olive and other Oleaceae. Widely distributed in the southern-most parts of Europe; also occurs elsewhere, including Africa and the Middle East. In central and northern Europe, where it is unable to survive the winter, its presence is sporadic and dependent upon the annual immigration of adults.

DESCRIPTION

Adult: 100–135 mm wingspan; forewings dark brown to reddish brown, with irregular black and yellowish markings; hindwings mainly yellow, with brownish-

848 Early moth (*Theria primaria*) – female.

850a Death's head hawk moth (*Acherontia atropos*).

849 Early moth (*Theria primaria*) – larva.

850b Death's head hawk moth (*Acherontia atropos*) – larva.

black bands, and the venation also partly darkened; thorax brownish black, with pale brownish-yellow markings in the form of a skull and crossbones; abdomen pale brownish yellow, with a black crossband on each segment, and an interrupted bluish-grey band dorsally. **Larva:** up to 125 mm long; body green or yellowish green, speckled with purple dorsally, with seven oblique, purple or purplish brown, yellow-edged stripes on either side; caudal horn yellow and S-shaped; spiracles black; head yellow marked with a pair of black stripes. **Pupa:** 50–60 mm long; reddish brown and shiny.

LIFE HISTORY

Adults occur in the spring, and again in late summer, with larvae feeding in the summer and again, usually in greater numbers, in the autumn. Fully grown larvae pupate in flimsy silken cocoons formed in the soil a few centimetres below the surface.

DAMAGE

Larvae cause noticeable, localized defoliation; however, they are rarely sufficiently numerous for attacks to be significant, except on very young trees.

Agrius convolvuli (Linnaeus) (851–853)

Convolvulus hawk moth

Although associated mainly with *Convolvulus arvensis* and *Ipomoea*, the larvae sometimes feed on other plants, including (at least in weedy vineyards) grapevine. Present throughout much of the southern Mediterranean basin, and also established further north in, for example, Crete, southern Greece and Sicily. A frequent migrant to central and southern Europe.

DESCRIPTION

Adult: 95–120 mm wingspan; forewings light grey to dark grey, with darker markings; hindwings mainly grey to brownish grey; body mainly grey, the abdomen partly barred with white, pink and black. **Larva:** up to 100 mm long; *brown form* – body dark brown, with seven oblique greyish to yellowish-brown stripes on either side; spiracles black; caudal horn black; head black, with lighter stripes; *green form* – body green, with seven oblique, yellowish stripes on either side; spiracles located in prominent black patches; caudal horn yellowish red, with a black tip; head green, with black stripes. **Pupa:** 45–48 mm long; reddish brown to black; tongue sheath prominent and characteristically curved; cremaster rugose and triangular, terminating in a pair of short spines.

LIFE HISTORY

In regions where this species is an established resident, larvae occur from April to November, representing several generations; in more northern regions, however, larvae occur mainly from July to September, most being the progeny of summer immigrant moths that arrived from June onwards. These larvae usually pupate to produce an autumn generation of adults in August or September. Pupation usually takes place in the soil some distance from the foodplant and, when present in vineyards, the fully grown larvae are often accidentally squashed (e.g. by motor vehicles) as they wander across roads and tracks. The winter is usually passed in the pupal stage, but few are able to survive under European conditions.

851 Convolvulus hawk moth (*Agrius convolvuli*).

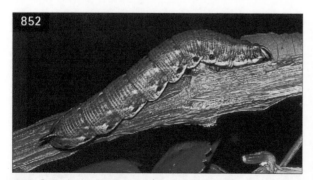

852 Convolvulus hawk moth (*Agrius convolvuli*) – larva (brown form).

853 Convolvulus hawk moth (*Agrius convolvuli*) – larva (green form).

DAMAGE

Larvae strip the foliage from the vines, leaving only the petioles, but infestations are of little or no economic importance.

Daphnis nerii (Linnaeus) (854–859)

Oleander hawk moth

In southern Europe a mainly non-resident, minor pest of grapevine; other hosts include *Nerium oleander* and *Vinca*. A migratory species, reaching southern Europe annually from Africa and southern Asia.

DESCRIPTION

Adult: 90–110 mm wingspan; wings and body deep olive green, with irregular whitish and pinkish-white markings. **Egg:** 1.6 × 1.3 mm; subspherical, smooth and light green. **Larva:** up to 120 mm long; body usually green, with the thoracic segments and last abdominal segments suffused with yellow, and the abdomen with a white dorsal stripe extending to the caudal horn; thoracic segment 3 with a pair of white, black-rimmed eye-spots; spiracles black; caudal horn mainly yellowish orange, short and bulbous; caudal horn of early-instar larva mainly black and very long, with a thread-like tip. **Pupa:** 60–75 mm long; light brown,

854 Oleander hawk moth (*Daphnis nerii*).

855 Oleander hawk moth (*Daphnis nerii*) – final-instar larva.

856 Oleander hawk moth (*Daphnis nerii*) – first-instar larva.

857 Oleander hawk moth (*Daphnis nerii*) – early-instar larva.

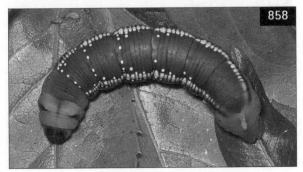

858 Oleander hawk moth (*Daphnis nerii*) – prepupa.

859 Oleander hawk moth (*Daphnis nerii*) – pupa.

speckled with black; spiracles distinct, surrounded by brownish black; cremaster pointed, with a bifid tip.

LIFE HISTORY

Immigrant moths appear in southern Europe in June. These deposit eggs singly on either side of the leaves of host plants. The eggs hatch within 1–2 weeks. Larvae then feed on flowers and young leaves and eventually pupate, each in a flimsy cocoon spun on the ground amongst dried leaves and other debris. A further generation of adults appears in August, and these often migrate further northwards into central and, occasionally, into northern Europe. The autumn brood of larvae eventually pupates; however, the pupae do not survive European winters, except in a few especially favourable areas where the insect is an established resident.

DAMAGE

Larvae, particularly in their later instars, are capable of causing considerable defoliation. However, they are rarely sufficiently numerous on grapevines to be of economic importance.

Deilephila elpenor (Linnaeus) (**860–864**)

Elephant hawk moth

A minor pest of grapevine. More usually, however, associated mainly with *Epilobium*, *Fuchsia*, *Galium* and *Impatiens*. Palaearctic. Widely distributed in temperate Europe.

862 Elephant hawk moth (*Deilephila elpenor*) – final-instar larva (green form).

860 Elephant hawk moth (*Deilephila elpenor*).

863 Elephant hawk moth (*Deilephila elpenor*) – early-instar larva.

861 Elephant hawk moth (*Deilephila elpenor*) – final-instar larva (brown form).

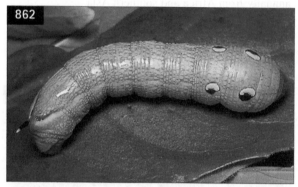

864 Elephant hawk moth (*Deilephila elpenor*) – young larva.

DESCRIPTION

Adult: 60–70 mm wingspan; forewings olive brown, with a pinkish-grey subterminal line and costal margin; hindwings bright pink, basally black; body yellowish brown to olive green, extensively marked with pink. **Larva:** up to 80 mm long; body green or ochreous brown to black; abdominal segments 1 and 2 each marked with a pair of black, yellow- and lilac-marked, eye-like patches; head and thoracic segments retractile; caudal horn relatively short; the earliest instars are green and lack the distinctive eye-like patches. **Pupa:** 40–45 mm long; brown, speckled with darker brown.

LIFE HISTORY

Moths occur from late May to early July and, under favourable conditions sometimes also later in the year, in August. Eggs are laid singly or in pairs on the leaves of host plants and hatch about 10 days later. Larvae feed from June or early July onwards. Individuals often bask in sunshine, if disturbed immediately retracting the head and thoracic segments and dilating the anterior abdominal segments to display the eye-like markings on these segments. Larvae are fully grown by the end of August. They then pupate in fragile, silken cocoons formed on or just beneath the surface of the ground. The pupae usually overwinter, except for those producing second-generation adults.

DAMAGE

As for oleander hawk moth (*Daphnis nerii*).

Deilephila porcellus (Linnaeus) (**865**)

Small elephant hawk moth

A minor pest of grapevine, but associated mainly with *Galium*; other hosts include *Epilobium*, *Impatiens* and *Lythrum salicaria*. Widely distributed in Europe.

DESCRIPTION

Adult: 45–55 mm wingspan; forewings dark yellowish ochreous to golden brown, suffused with brown and pink; hindwings yellowish ochreous to brownish grey, suffused with pink; body brown to greyish brown or deep ochreous, suffused with pink. **Larva:** up to 50 mm long; body usually brown (rarely green), similar in appearance to that of *Deilephila elpenor*, but with the caudal horn reduced to a small double-headed tubercle. **Pupa:** 25–30 mm long; light brown, streaked with darker brown.

LIFE HISTORY

Similar to that of elephant hawk moth (*Deilephila elpenor*), with larvae most numerous from July to late September.

DAMAGE

As for oleander hawk moth (*Daphnis nerii*).

Hippotion celerio (Linnaeus)

Silver-striped hawk moth

In mainland Europe a minor, non-resident pest of grapevine. Other hosts include *Epilobium*, *Fuchsia* and *Galium*. A migratory species, widely distributed in tropical Africa and southern Asia; also established in Australia.

DESCRIPTION

Adult: 70–80 mm wingspan; forewings light brown to ochreous brown, with distinct silvery-white, black-edged streaks; hindwings pinkish red, with black markings; veins black; body mainly ochreous brown, with silvery-white markings. **Larva:** up to 80 mm long; body brown or green, with a pale stripe along each side and with distinctive eye-spots present on abdominal segments 1 and 2; caudal horn straight, brown or brownish black. **Pupa:** 40–50 mm long; elongate, light brown and shiny.

LIFE HISTORY

In southern Europe, immigrant moths arrive from June onwards, with up to three broods of larvae occurring from July to September. Locally produced moths often migrate into central Europe and, rarely, further north. This species cannot survive European winters.

DAMAGE

As for oleander hawk moth (*Daphnis nerii*), but larvae typically devour only about half of the leaf area rather than complete leaves.

865 Small elephant hawk moth (*Deilephila porcellus*).

Hyles livornica (Esper)

Striped hawk moth

A minor pest of grapevine in southern Europe. Wild hosts include *Arbutus unedo*, *Polygonum* and *Rumex*. Although resident in the warmer parts of Europe, especially in the Mediterranean basin, its presence in central and northern Europe, where it cannot survive the winter, is dependent upon the arrival of immigrant moths. Widely distributed in Africa and Asia.

DESCRIPTION

Adult: 65–85 mm wingspan; forewings dark ochreous brown, with whitish veins and a broad, ochreous stripe extending from near the base to the apex; hindwings pinkish red, with the base and border mainly black; head and thorax dark ochreous brown, with whitish stripes; abdomen dark ochreous brown, chequered with white and black. **Larva:** up to 80 mm long; body extremely variable in colour, usually blackish (but underside pale), spotted with yellow, marked with yellow dorsal and subdorsal lines, and often with a white stripe along the sides, interrupted by pinkish patches; caudal horn curved and red, with a black tip; head black and red. **Pupa:** 30–45 mm long; elongate; light brown.

LIFE HISTORY

Adults of this migratory species occur from May onwards, with several overlapping generations annually. Larvae, which occur from June to September, often feed during the daytime and also bask in sunshine. Fully fed individuals usually pupate amongst debris on the ground, each in a flimsy, silken cocoon.

DAMAGE

As for oleander hawk moth (*Daphnis nerii*).

Mimas tiliae (Linnaeus) (**866–867**)

Lime hawk moth

In mainland Europe a minor pest of apple, pear and cherry; larvae also feed on *Betula*, *Fraxinus excelsior*, *Tilia*, *Ulmus* and various other forest trees. At least in the British Isles found mainly on *Tilia* and *Ulmus*, and not a pest of fruit trees. Widely distributed, particularly in central and southern Europe.

DESCRIPTION

Adult: 70–80 mm wingspan; forewings greyish ochreous to pinkish ochreous, suffused with greenish grey, with a darker olive-green median fascia; hindwings ochreous brown, marked with dark, greyish brown; body greyish ochreous to olive green, often with a pinkish tinge. **Larva:** up to 65 mm long; body green, covered with small raised white spots and with seven oblique, whitish stripes on either side; spiracles red-rimmed; caudal horn blue above, yellowish white below.

LIFE HISTORY

Adults occur in May and June. Eggs are then laid singly or in pairs on the underside of leaves of host plants. Larvae feed from July to September, and eventually pupate in flimsy cocoons spun in bark crevices or in the top few centimetres of the soil. The pupae usually overwinter. However, in favourable southern regions, this species produces two generations annually, with second-generation adults emerging in August and their larvae developing in the late summer and autumn.

DAMAGE

Larvae strip the foliage from the branches, leaving only the petioles, but infestations on fruit trees are usually of little or no economic importance, except on nursery stock or small trees.

866 Lime hawk moth (*Mimas tiliae*) – male.

867 Lime hawk moth (*Mimas tiliae*) – larva.

Smerinthus ocellata (Linnaeus) (868–870)
Eyed hawk moth

A minor pest of apple, particularly young trees, but associated mainly with broadleaved *Salix*. Other hosts include *Prunus laurocerasus* and, at least in mainland Europe, trees such as *Alnus*, *Betula*, *Populus* and *Tilia*. Widely distributed in Europe.

DESCRIPTION
Adult: 75–95 mm wingspan; forewings brown, with cloudy, pinkish markings; hindwings brown, suffused with red, and each with a large, blue, grey and black, eye-like mark; thorax chocolate brown, with greyish-brown tegulae; abdomen pinkish brown. **Egg:** 1.6 × 1.4 mm; subspherical and light green. **Larva:** up to 70 mm long; body apple-green or bluish green, covered with small raised white spots, and with seven oblique white stripes on either side; spiracles purple-rimmed; caudal horn bluish green; young larva light green, with a purple caudal horn. **Pupa:** 35–40 mm long; dark brown to blackish brown; rather plump and shiny.

LIFE HISTORY
Adults appear in May, June and July. Each female is capable of laying several hundred eggs, and these are usually deposited singly or in pairs on the underside of leaves or on the petioles and shoots. Larvae feed during June, July and August. When fully grown, they burrow into the soil and pupate in an earthen cell at a depth of a few centimetres, where they eventually overwinter. Occasionally, there may be a partial second generation of adults in August, larvae then also occurring during the late summer and autumn.

DAMAGE
Larvae rapidly defoliate the shoots or branches, also devouring the major veins. Although sometimes causing significant damage to nursery stock and small trees, larvae are not of importance on mature fruit trees and rarely feed upon them.

868 Eyed hawk moth (*Smerinthus ocellata*).

869 Eyed hawk moth (*Smerinthus ocellata*) – final-instar larva.

870 Eyed hawk moth (*Smerinthus ocellata*) – young larva.

Sphinx ligustri Linnaeus (**871–874**)
Privet hawk moth

Although associated mainly with *Fraxinus excelsior*, *Ligustrum vulgare* and *Syringa vulgaris*, larvae are occasionally found on currant bushes (usually in gardens) and olive. Other hosts include *Euonymus*, *Lonicera*, *Sambucus* and *Symphoricarpos rivularis*. Palaearctic. Present throughout most of Europe.

DESCRIPTION
Adult: 100–120 mm wingspan; forewings greyish brown to dark brown, suffused with pinkish basally, and with scattered white scales and the veins partly black; hindwings pinkish white, with distinct black fasciae; abdomen brown, partly banded with pink and black. **Egg:** 2 mm long; rounded, but flattened top and bottom; light green and unsculptured. **Larva:** up to 80 mm long; body green, with seven oblique, white stripes, mainly edged above with purple, on either side; spiracles reddish ochreous; caudal horn black and yellow.

LIFE HISTORY
Adults occur in June. Eggs are then laid singly on the leaves or stems of host plants. Larvae feed in July and August and then pupate in the soil, each in an earthen cell. Adults usually emerge in the following year, but sometimes not until pupae have survived for a further winter. In suitably warm regions, there may be a partial or a full second generation annually, with first-generation adults present from April to May and those of the second in August.

DAMAGE
Larvae cause noticeable defoliation, particularly in their later instars. Numbers of larvae on currant bushes and olive trees, however, are usually restricted to one or two individuals. Thus, unless larvae are present on very small plants, damage is of little or no significance.

871 Privet hawk moth (*Sphinx ligustri*).

872 Privet hawk moth (*Sphinx ligustri*) – final-instar larva.

873 Privet hawk moth (*Sphinx ligustri*) – egg on olive leaf.

874 Privet hawk moth (*Sphinx ligustri*) – young larva alongside egg shell.

Family **NOTODONTIDAE**

A diverse family of medium-sized to large, robust, often downy-looking moths, with relatively long forewings; antennae usually bipectinate in males, and simple or pectinate in females; forewings often with a medium tuft of scales projecting from the dorsum. Larvae often (but not in the following species) have fleshy humps dorsally on one or more segments and, in some species, the anal claspers are modified to form a pair of filamentous, tail-like appendages; crochets on the abdominal prolegs are uniordinal, arranged as a mesoseries; body hairs, when present, never arise in tufts.

Phalera bucephala (Linnaeus) (875–878)
Buff-tip moth

Polyphagous on various broadleaved trees and shrubs, and a sporadic pest of apple, pear, cherry, plum, raspberry, chestnut and hazelnut. Larvae also occur on many other trees and shrubs. Widely distributed in Europe.

DESCRIPTION

Adult: 50–60 mm wingspan; forewings ash-grey to silvery grey, marked with dark brown and reddish brown, and each with a large, pale yellow apical blotch; hindwings pale yellow to whitish. **Egg:** 1 mm across; bluish white above, with a dark central spot; brown below. **Larva:** up to 60 mm long; body downy and yellow, with several incomplete black longitudinal lines and orange intersegmental crosslines; head black. **Pupa:** 25–28 mm long; dark purplish brown; cremaster with two pairs of short spines.

LIFE HISTORY

Adults occur from late May to July, eggs being laid on the underside of leaves in batches of about 50. The larvae feed gregariously during the summer, becoming fully grown by the early autumn. Each then enters the soil and pupates just beneath the surface, but without forming a distinct cocoon.

DAMAGE

Larvae are ravenous feeders and rapidly defoliate the shoots and branches. Large accumulations of blackish frass are often noticed on the ground beneath an infested bush or tree.

875 Buff-tip moth (*Phalera bucephala*).

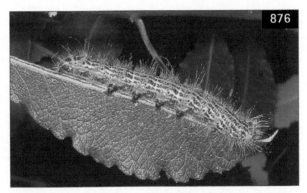

876 Buff-tip moth (*Phalera bucephala*) – final-instar larva.

877 Buff-tip moth (*Phalera bucephala*) – young larvae.

878 Buff-tip moth (*Phalera bucephala*) – early larval damage.

Family **DILOBIDAE**

A small group of medium-sized moths, formerly included in the family Notodontidae. Larvae have dorsal setae arising from raised verrucae.

Diloba caeruleocephala (Linnaeus) (**879–880**)
Figure of eight moth
Polyphagous on various trees and shrubs, including apple, pear, almond, apricot, cherry and plum. Widely distributed in Europe, particularly in temperate regions.

DESCRIPTION
Adult: 30–35 mm wingspan; forewings brownish grey, each with a large, pale greenish-yellow, more or less discrete 'figure of eight' mark; hindwings whitish, with darker markings. **Egg:** hemispherical, ribbed longitudinally and finely reticulate; grey to greenish brown. **Larva:** up to 35 mm long; body plump, bluish white, with numerous black, setose verrucae, an incomplete yellow dorsal stripe and a yellow stripe along each side; head bluish grey, with two black spots.

Pupa: 15 mm long; dull reddish brown; cremaster forming a pair of irregular, flattened lateral projections, each bearing four short spines.

LIFE HISTORY
Adults appear in the autumn, usually during October and November. Eggs are laid singly or in small batches on the spurs, branches or trunks (often at the base of young shoots) and then partially covered with brown hairs from the top of the female's abdomen. Eggs hatch in the spring, at about bud burst. Larvae then feed on the underside of the leaves. Most become fully grown before the end of June, each spinning an oblong-shaped cocoon on the host tree or on nearby wooden supports and then pupating. Pieces of leaf or bark are usually incorporated in the walls of the pupal cocoon. Pupation may also occur in cocoons formed amongst debris on the ground.

DAMAGE
Larvae are rapid, erratic feeders, and are capable of causing significant defoliation. Numbers on fruit crops, however, are usually small.

879 Figure of eight moth (*Diloba caeruleocephala*).

880 Figure of eight moth (*Diloba caeruleocephala*) – larva.

Family **LYMANTRIIDAE**

A group of medium-sized, often hairy moths that lack a proboscis; females often with a distinct hair tuft on the anal segment, used to cover their egg batches; antennae of males strongly bipectinate; females of some species are flightless, with greatly reduced wings (= brachypterous). Larvae are long-haired, with characteristic hair tufts (tussocks) or hair pencils arising from abdominal segments 1–4, or with eversible glands dorsally on abdominal segments 6 and 7; the body hairs are often urticating; the head is often glabrous.

Calliteara pudibunda (Linnaeus) (**881–882**)
Pale tussock moth

A minor pest of hop; also associated with a wide range of broadleaved trees and shrubs including, occasionally, apple, cherry and other fruit trees. Eurasiatic. Widely distributed in central and northern Europe.

DESCRIPTION
Adult female: 45–60 mm wingspan; forewings silvery grey, with darker markings; hindwings mainly white. **Adult male:** 35–45 mm wingspan; forewings darker than in female and less angular; hindwings whitish, suffused with grey. **Larva:** up to 50 mm long; body light green or yellowish, with black markings, especially prominent intersegmentally on the anterior abdominal segments; brushes of whitish-yellow hairs on abdominal segments 1–4 and a long red tuft of hair on abdominal segment 8; underside black. **Pupa:** 15–25 mm long; purplish brown and hairy.

LIFE HISTORY
Adults occur in May and June. Larvae, commonly known as 'hop-dogs', feed from June onwards. They complete their development in late summer or autumn and then spin bulbous silken cocoons amongst the foliage, where they pupate and eventually overwinter.

DAMAGE
The larvae devour considerable amounts of foliage. However, being active relatively late in the season, damage caused is unimportant.

Euproctis chrysorrhoea (Linnaeus) (**883–886**)
Brown-tail moth

A local, polyphagous pest of various broadleaved trees and shrubs, including apple, pear, peach and plum;

881 Pale tussock moth (*Calliteara pudibunda*) – male.

883a, b Brown-tail moth (*Euproctis chrysorrhoea*).

882 Pale tussock moth (*Calliteara pudibunda*) – larva.

883a, b Brown-tail moth (*Euproctis chrysorrhoea*).

occasionally, infestations also occur on strawberry. Palaearctic. Widely distributed in Europe. An introduced pest in the USA.

DESCRIPTION
Adult: 30–40 mm wingspan; head, thorax and legs white and fluffy; much of abdomen dark brown, with a large anal hair tuft; wings white; forewings (especially of male) sometimes with a few black dots. **Larva:** up to 40 mm long; body black to reddish brown, with long, yellowish-brown hair tufts arising from orange-brown to brownish-black verrucae, two often indistinct, interrupted rows of red marks down the back; abdominal segments 1–8 each with a pair of distinct, white subdorsal patches; eversible glands on abdominal segments 6 and 7 orange red; head black; younger larvae darker and less hairy, and the eversible glands very prominent. **Pupa:** 16–20 mm long; brownish black, with tufts of hair.

LIFE HISTORY
Adults occur in July and August. Eggs are laid in elongate batches on leaves or stems and then covered with brown hairs from the female's anal tuft. The eggs hatch from mid-August to early September, and the small larvae soon construct a strong, silken tent in which they shelter during inclement weather. The larvae emerge from their retreat on sunny days to feed gregariously on surrounding foliage. As the season progresses, however, fewer and fewer individuals leave the tent until by the end of October (when in their second or third instar) most if not all activity ceases. By this time the tent has become a tough, opaque and conspicuous greyish mass of silk about 20 cm long. Larvae overwinter in these tents and become active again in April. They then begin to appear on the outside of the webbing, basking in sunshine, but little or no feeding takes place until May. Young foliage is then devoured ravenously. As the larvae grow, they wander further and further from their overwintering tent, spinning additional, less substantial webs and establishing trails of silk along the branches. When about 25 mm long, the larvae wander away to continue feeding elsewhere, often becoming solitary. In late June, when fully grown, each spins a silken cocoon between two or more leaves and then pupates. The pupal cocoons either occur alone or in large gregarious masses. Moths emerge about 2 weeks later.

DAMAGE
Larvae devour considerable quantities of foliage, affecting both plant growth and cropping. In addition, the extensive webbing is often troublesome, as are the larval hairs; the latter, if touched, can cause considerable skin irritation and produce a painful rash (urticaria).

884 Brown-tail moth (*Euproctis chrysorrhoea*) – final-instar larva.

885 Brown-tail moth (*Euproctis chrysorrhoea*) – late-instar larva.

886 Brown-tail moth (*Euproctis chrysorrhoea*) – web.

Euproctis similis (Fuessly) (**887–890**)
Yellow-tail moth

A generally common, but minor, pest of fruit crops such as apple, pear, cherry, plum, loganberry and raspberry. Also associated with a wide range of broadleaved forest trees and shrubs, including *Betula*, *Crataegus monogyna*, *Prunus spinosa* and *Quercus*. Eurasiatic. Present throughout Europe.

DESCRIPTION
Adult: 30–45 mm wingspan, female noticeably larger than male; head, thorax and legs fluffy; wings white, the male with black markings on the trailing edge of each forewing; body with a conspicuous yellow anal tuft. **Larva:** up to 35 mm long; body velvet-black, prominently marked with white, with an interrupted bright red, black-centred dorsal stripe, and a red lateral stripe running longitudinally just above the legs; body hairs blackish to whitish; eversible glands on abdominal segments 6 and 7 orange red; abdominal segment 1 noticeably humped; head black and shiny. **Pupa:** 12–16 mm long; dark brown, plump and slightly hairy.

LIFE HISTORY
Adults occur in July and August. Eventually, eggs are laid in batches on the twigs of host plants and then covered with hairs from the female's anal tuft. The eggs hatch about 7–10 days later. At first, the larvae feed gregariously, grazing away the epidermal tissue. In autumn, each larva then spins a roughly oval, greyish hibernaculum (about 6 mm long) under a bark flake or in other shelter, in which to overwinter. In spring, the larvae emerge and continue their now solitary existence, feeding on the foliage until about June. Each then pupates in an oval, greyish-brown cocoon of silk mixed with body hairs. Adults emerge a few weeks later. Unlike those of *Euproctis chrysorrhoea*, larvae of this species do not form communal webs.

DAMAGE
In autumn, infested leaves turn brown where the upper epidermis has been removed. In spring, larvae cause slight defoliation and may also damage developing fruitlets.

887 Yellow-tail moth (*Euproctis similis*) – female.

888 Yellow-tail moth (*Euproctis similis*) – male.

889 Yellow-tail moth (*Euproctis similis*) – final-instar larva.

890 Yellow-tail moth (*Euproctis similis*) – young larvae and damage to apple foliage.

Lymantria dispar (Linnaeus) (**891–894**)
Gypsy moth

An important forestry pest in mainland Europe, and a sporadic pest of fruit trees (particularly pear and plum); infestations also occur on other cultivated hosts such as hazelnut and walnut. Eurasiatic. Also present in North America.

DESCRIPTION
Adult female: 50–65 mm wingspan; wings and body creamy white, with brownish grey to light brown markings; antennae weakly bipectinate. **Adult male:** 35–45 mm wingspan; wings and body greyish brown, with black markings; antennae strongly bipectinate. **Larva:** up to 55 mm long; body pale bluish grey to pale creamy grey, extensively marked with black, and with whitish sides and a pale dorsal line; verrucae on body segments 1–5 blue, those on segments 6–11 reddish brown (in younger larvae, all verrucae are blue); body hairs, arising in clumps, black dorsally, pale gingery brown laterally; head pale ochreous, marked with black; eversible glands on abdominal segments 6 and 7 reddish brown. **Pupa:** 18–28 mm long; dull reddish brown, stumpy, with small tufts of paler hairs; cremaster dark and prominent.

LIFE HISTORY
Adults occur from July to early September. Males are very active and fly about in search of freshly emerged females. Females, however, do not fly; instead, they either crawl around near their pupal cocoon or may spread their wings and glide to the ground. Eggs are laid in large batches of several hundred on the bark of trees or on other surfaces. Each female then coats the egg clump with hairs from her anal tuft, mixed with a protective liquid that soon hardens once in contact with the air. Although embryonic development is usually

892 Gypsy moth (*Lymantria dispar*) – male.

893a, **b** Gypsy moth (*Lymantria dispar*) – larva.

891 Gypsy moth (*Lymantria dispar*) – female.

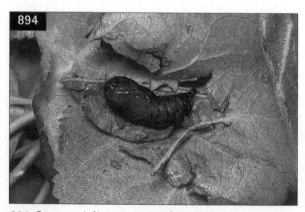

894 Gypsy moth (*Lymantria dispar*) – pupa.

completed in about a month, the larvae do not emerge from the egg shells until the following spring, usually in April or May. The young larvae often disperse by crawling or they may produce strands of silk and then be carried away over considerable distances in wind currents. The larvae feed ravenously on buds, leaves and young shoots; on fruit trees, they may also devour the developing fruitlets. When fully grown, usually within 2–3 months, the larvae pupate, each in a flimsy silken cocoon spun amongst foliage on the foodplant or attached to some artificial support, such as a fence or wall; pupation sometimes also takes place amongst debris on the ground. Adults emerge about 2–3 weeks later, there being just one generation annually.

DAMAGE
Larvae cause extensive defoliation and infested fruit trees can be devastated.

Lymantria monacha (Linnaeus) (895)
Black arches moth
Polyphagous on broadleaved trees and shrubs in mainland Europe, including apple, plum and other rosaceous fruit trees, but associated mainly with conifers (e.g. *Abies*, *Picea* and *Pinus*); in some regions (e.g. England) present mainly on *Quercus*, and not reported as a fruit pest. Eurasiatic. Widespread in Europe.

DESCRIPTION
Adult female: 45–50 mm wingspan; body and forewings mainly white, irregularly marked with black, the abdomen tinged with pink; hindwings mainly pale greyish brown. **Adult male:** 37–42 mm wingspan; similar to female, but antennae strongly

bipectinate. **Larva:** up to 35 mm long; dark grey, with an irregular, brownish-black, black-edged dorsal stripe (the stripe interrupted by a whitish, black-centred, mark on thoracic segment 3 and by whitish, sometimes red-centred, patches on abdominal segments 4–6); pale hair tufts arise from pinacula along the dorsal and spiracular lines; eversible glands on abdominal segments 6 and 7 red.

LIFE HISTORY
Eggs are laid during the summer on the bark of host trees, either singly or in pairs, and they hatch in the following spring. The larvae feed on the foliage until June or early July. They then pupate in silken cocoons spun in bark crevices. Adults usually emerge in late July or August.

DAMAGE
Attacks on fruit trees are usually insignificant, but severe infestations can cause considerable defoliation and affect the vigour of host trees.

Orgyia antiqua (Linnaeus) (896–899)
Vapourer moth
A polyphagous pest of various trees and shrubs, including fruit crops such as apple, pear, cherry, plum and raspberry. Holarctic. Widely distributed in Europe. Often particularly common in urban areas.

DESCRIPTION
Adult female: virtually wingless; body 10–15 mm long, dark yellowish grey, fat and sack-like. **Adult male:** 25–35 mm wingspan; wings ochreous brown or chestnut-brown; each forewing with darker markings and with a large white spot near the hind angle. **Egg:** 0.9 mm across;

895 Black arches moth (*Lymantria monarcha*) – male.

896 Vapourer moth (*Orgyia antiqua*) – female.

rounded, brownish grey to reddish grey, with a central spot and a dark rim-like band. **Larva:** up to 40 mm long; greyish or violet, with various red, black and yellow markings; very hairy, including four brush-like tufts of yellow hairs on abdominal segments 1–4, long pencils of blackish hairs near the head and similar brownish tufts on abdominal segment 8. **Pupa:** 10–15 mm long; shiny brownish black and rather hairy.

LIFE HISTORY
Eggs overwinter en masse on the silken web surrounding the former pupal cocoon of the maternal female. They hatch in the spring and the young larvae wander away to feed on host plants, often also dispersing from plant to plant during the course of their development. When fully grown, each larva spins a tough silken cocoon, incorporating body hairs, on a twig, spur or other suitable support (such as a fence post or nearby shed wall), and then pupates. Adults emerge a few weeks later. Males are active in sunny weather and then fly rapidly in search of freshly emerged females. Females, however, are sedentary and remain in association with their pupal cocoon, where mating takes place and upon which their egg batches are laid. Depending on conditions there are usually one or two, but sometimes three, generations annually.

DAMAGE
Although buds, blossoms and young fruits are liable to be attacked, larval feeding is usually restricted to foliage and is rarely serious on fruit crops.

897 Vapourer moth (*Orgyia antiqua*) – male.

898 Vapourer moth (*Orgyia antiqua*) – egg batch.

899 Vapourer moth (*Orgyia antiqua*) – larva.

Family **ARCTIIDAE**
(e.g. ermine moths and tiger moths)

Adults (subfamily Arctiinae) medium-sized to large, with often brightly coloured hindwings. Larvae (subfamily Arctiinae) are very hairy, with the body hairs arising in tufts from large verrucae; the head is glabrous; crochets on the abdominal prolegs form a mesoseries, with those at each end much reduced in size.

Arctia caja (Linnaeus) (**900–901**)
Garden tiger moth

A minor pest of loganberry, raspberry, strawberry and grapevine, but associated mainly with wild herbaceous plants. Northern Palaearctic. Widely distributed in northern Europe.

DESCRIPTION
Adult: 50–75 mm wingspan; forewings dark brown, with large creamy-white markings; hindwings orange red, with several large, blue-black, yellow-rimmed spots. **Larva:** up to 50 mm long; body mainly blackish, with a thick coat of long, gingery hairs; head black and shiny.

LIFE HISTORY
Adults occur in July and early August. Large batches of eggs are then laid on the underside of leaves of various low-growing plants. Larvae feed in August and September before hibernating. They recommence feeding in the spring and may then often be found wandering about or basking in sunshine. If disturbed, the larvae (commonly known as 'woolly bears') typically roll into a tight ball. Development is completed from late May onwards. Individuals then pupate, each in a large oval cocoon formed amongst debris on the ground.

DAMAGE
Numbers of larvae on fruit crops are usually small, and damage caused to the foliage is of little or no importance.

Hyphantria cunea (Drury) (**902–909**)
American white moth
larva = fall webworm

A polyphagous, North American pest, first noted in Europe in 1940 when infestations were found in Hungary. In spite of quarantine measures, infestations have since spread to several other European countries, including Austria, Bulgaria, the Czech Republic, France, Italy, Poland, Switzerland and the former Yugoslavia. Although associated mainly with forest or amenity broadleaved trees and shrubs, infestations also occur on

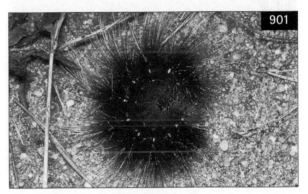

901 Garden tiger moth (*Arctia caja*) – larva.

902 American white moth (*Hyphantria cunea*) – female.

900 Garden tiger moth (*Arctia caja*) – female.

903 American white moth (*Hyphantria cunea*) – male.

fruit trees (mainly apple, pear, cherry, plum and mulberry). Grapevine, walnut and hop are also attacked, but these are not considered particularly favourable hosts.

DESCRIPTION

Adult: 26–30 mm wingspan; mainly white, the forewings sometimes flecked with black; male with noticeably bipectinate antennae. **Egg:** light green. **Larva:** up to 35 mm long; body varying from yellow or yellowish green to deep reddish brown, with tufts of whitish hairs arising from black verrucae; spiracles white, ringed with black; head shiny black. **Pupa:** 10–12 mm long; shiny blackish brown; cremaster with 12 hooks.

LIFE HISTORY

Adults of the first generation occur in the spring from April onwards. The females deposit several hundred eggs in large batches on the spurs or on the underside of

904 American white moth (*Hyphantria cunea*) – egg batch.

905 American white moth (*Hyphantria cunea*) – late-instar larva (light form).

906 American white moth (*Hyphantria cunea*) – late-instar larva (dark form).

907 American white moth (*Hyphantria cunea*) – young larvae on walnut.

908 American white moth (*Hyphantria cunea*) – larval web.

909 American white moth (*Hyphantria cunea*) – pupal cocoon.

expanded leaves; the eggs are then partly covered with whitish hairs. The eggs hatch 7–10 days later. Larvae then feed gregariously on the foliage, usually from May to July, sheltering during the day within a large, but flimsy, communal web. When fully fed, the larvae wander away to pupate on the foodplant, each in a slight, greyish-brown cocoon. Moths of a second generation appear in July and August. These eventually give rise to larvae that complete their development in the autumn. These larvae usually pupate within bark crevices or amongst dead leaves, and then overwinter.

DAMAGE
Host trees are clothed in webbing and defoliation can be extensive, adversely affecting vigour and cropping.

Spilosoma lutea (Hufnagel) (**910–913**)
Buff ermine moth
A minor pest of blackberry, raspberry and strawberry; larvae are also recorded on apple. Palaearctic. Widely distributed in Europe.

DESCRIPTION
Adult: 36–42 mm wingspan; body and wings pale yellow to ochreous, with variable black markings; antennae of male bipectinate. **Egg:** hemispherical, smooth and shiny; whitish green. **Larva:** up to 32 mm long; body greenish brown, with lighter sides and a whitish-yellow to reddish dorsal line; body hairs long, whitish to brownish; head light brown and shiny; early instars paler and less densely haired, the black verrucae clearly visible. **Pupa:** 16–18 mm long; reddish brown and shiny; cremaster with two closely set spines.

LIFE HISTORY
Adults occur in June, laying eggs in large batches on the leaves of various plants, particularly weeds such as *Rumex* and *Taraxacum officinale*. Larvae feed during the summer, pupating in the autumn in relatively large, silken cocoons spun amongst leaves or ground litter. On protected strawberry, there may be a partial second generation; larvae then also feed during the autumn, winter and spring. Two generations annually may occur outdoors in southern Europe.

DAMAGE
Feeding is restricted to leaf tissue and larvae are rarely sufficiently numerous on fruit crops to be harmful.

910 Buff ermine moth (*Spilosoma lutea*) – male.

911 Buff ermine moth (*Spilosoma lutea*) – final-instar larva.

912 Buff ermine moth (*Spilosoma lutea*) – early-instar larva.

913 Buff ermine moth (*Spilosoma lutea*) – pupal cocoon.

Family **NOLIDAE**

A small family of dull-coloured, comparatively small moths, with three scale tufts on each forewing. Larvae are clothed in long hair tufts, arising from verrucae, and possess four pairs of abdominal prolegs (present on abdominal segments 4–6 and 10).

Nola cucullatella (Linnaeus) (**914–915**)
Short-cloaked moth
This moth is locally common in northern Europe, the larvae occurring on *Crataegus monogyna* and *Prunus spinosa*, and on various other trees and shrubs. They also attack the foliage of apple, pear, plum and other fruit trees, but are not important fruit pests. Adults (20 mm wingspan) are mainly grey, with white, brownish and black markings on the forewings. The larvae (up to 14 mm long) are light grey, with somewhat browner sides and reddish pinacula, and are clothed with long grey hairs. They feed in late summer and autumn and, after hibernation, in April and May. Pupation occurs in a strong cocoon, adults appearing in July.

Family **NOCTUIDAE**

One of the largest families of lepidopterans, comprising mostly medium-sized, stout-bodied moths; proboscis well-developed; tympanal organs located in the metathorax. Most species are of drab appearance, but some are brightly coloured. The typical pattern of the forewing includes transverse lines and three more or less distinct markings: a kidney-shaped mark (the reniform stigma), a small circle (the orbicular stigma) and a dart-like mark (the claviform stigma). Eggs are usually more or less hemispherical, with a flat base, and are ribbed vertically; the surface often has a net-like (reticulate) appearance. Larvae of most species have five pairs of abdominal prolegs, but groups (e.g. subfamily Plusiinae) with a reduced number also occur; crochets are of one size, arranged in a half circle. Most larvae are relatively glabrous, but those of the subfamily Acronictinae often have distinct hair tufts.

Acronicta aceris (Linnaeus) (**916–917**)
Sycamore moth
In mainland Europe a minor pest of fruit trees, but associated mainly with forest trees and shrubs such as *Acer*, *Aesculus hippocastanea*, *Betula*, *Corylus*

914 Short-cloaked moth (*Nola cucullatella*).

916 Sycamore moth (*Acronicta aceris*).

915 Short-cloaked moth (*Nola cucullatella*) – larva.

917 Sycamore moth (*Acronicta aceris*) – larva.

avellana, Fagus sylvatica, Tilia and *Ulmus*. Eurasiatic. Widely distributed in Europe.

DESCRIPTION
Adult: 40–45 mm wingspan; forewings whitish to light grey, with darker markings; hindwings whitish, with darker veins. **Larva:** up to 40 mm long, but occasionally much larger; body mainly yellowish brown, with a series of white, black-bordered patches along the back and dense tufts of long, yellow or orange-brown hairs; head brownish black, with a white, inverted, V-shaped mark.

LIFE HISTORY
In northern Europe, adults occur from June to July or early August. Eggs are laid singly on the leaves of host plants and hatch within about a week. Larvae feed from July onwards, becoming fully fed in August or September. Pupation occurs in silken cocoons spun on the ground (e.g. amongst leaf litter) or in bark crevices on host trees. In southern Europe there are two generations annually.

DAMAGE
Larvae cause slight defoliation, but are rarely numerous.

Acronicta alni (Linnaeus) (**918–919**)
Alder moth
In mainland Europe, larvae of this local, but widely distributed, species sometimes feed on the leaves of fruit trees. However, they are more frequently associated with broadleaved forest trees and shrubs. The larvae usually occur singly, often resting fully exposed on the upper surface of a leaf. Older individuals (up to 35 mm long) are black and yellow, with distinctive spatulate body hairs. Their striking and unusual appearance attracts attention, but they cause little or no damage and are of little or no significance as fruit pests. Adults (35–40 mm wingspan) have mainly grey, brownish-grey and blackish-marked forewings, and whitish, dark-bordered hindwings. There are from one to two generations annually, adults occurring from May to early June or July, and those of any second generation from late June to mid-August.

Acronicta auricoma (Denis & Schiffermüller)
Scarce dagger moth
Although associated mainly with non-cultivated plants such as *Betula, Prunus spinosa, Quercus, Rubus fruticosus* and *Vaccinium myrtillus*, larvae of this Eurasiatic species are also recorded in southern Europe attacking cultivated highbush blueberry. Adults (37–40 mm wingspan) are mainly dark grey to brownish grey, with black markings, including

a dagger-like mark basally on each forewing. Larvae (up to 40 mm long) are mainly black, with prominent tufts of orange-red hairs arising from yellowish-red verrucae. There are typically two generations annually.

Acronicta psi (Linnaeus) (**920–921**)
Grey dagger moth
A minor pest of apple, pear, plum, raspberry and, rarely, strawberry. Other foodplants include a wide range of

918 Alder moth (*Acronicta alni*).

919 Alder moth (*Acronicta alni*) – larva.

920 Grey dagger moth (*Acronicta psi*).

921 Grey dagger moth (*Acronicta psi*) – larva.

922 Knotgrass moth (*Acronicta rumicis*).

923 Knotgrass moth (*Acronicta rumicis*) – larva

forest and ornamental trees and shrubs. Eurasiatic. Widely distributed and common in Europe.

DESCRIPTION
Adult: 38–42 mm wingspan; forewings light grey, with black dagger-like markings; hindwings whitish. **Egg:** pale yellowish white when laid, but then darkening. **Larva:** up to 35 mm long; body greyish black, with a broad yellow dorsal stripe, bordered along each side by a blue stripe (interrupted by red, black-edged spots); a prominent black, pointed hump on abdominal segment 1 and a small black hump on segment 8; head black. **Pupa:** 15 mm long; brown and rather slender, tapering towards the tip; cremaster with several strong spines.

LIFE HISTORY
Adults occur from May to September, in warmer regions sometimes earlier. Eggs are laid singly on foliage and hatch about 5 days later. Larvae feed from June or July onwards, often occurring well into the autumn. When fully grown, each pupates in a greyish-brown cocoon constructed in crevices in the bark or elsewhere on the foodplant, often hidden beneath dead leaves. In warmer regions there are two generations annually, but just one or no more than a partial second in northern Europe.

DAMAGE
On apple, pear and plum, defoliation caused is of little or no importance; however, on raspberry, particularly in gardens, damage may be more serious.

Acronicta rumicis (Linnaeus) (922–923)
Knotgrass moth
A minor pest of strawberry and, less frequently, blackberry, raspberry, loganberry and tayberry; sometimes also found on fruit trees, particularly apple, pear, quince and cherry. Eurasiatic. Widely distributed and generally common throughout Europe.

DESCRIPTION
Adult: 35–40 mm wingspan; forewings mainly grey to greyish black, with black markings; hindwings light brown. **Larva:** up to 40 mm long; body dark brownish grey, with a series of red spots (surrounded by black patches) along the back, white spots along each side, and a creamy-white subspiracular line interrupted by raised red spots; abdominal segments 1 and 8 slightly humped; body hairs whitish to black, arising in clumps from distinct verrucae; head black.

LIFE HISTORY
Larvae of this polyphagous species feed mainly on low-growing and herbaceous plants from spring or early summer onwards, and are sometimes noticed on fruit crops. The winter is spent as pupae in cocoons in the soil. Moths emerge in June, but often earlier in warm conditions. The number of generations varies from one in northern Europe to three in southern Europe.

DAMAGE
Larvae cause defoliation, but are rarely sufficiently numerous on fruit crops to cause concern.

Acronicta tridens (Denis & Schiffermüller) (924–925)

Dark dagger moth

A minor pest of rosaceous fruit trees, particularly apple, pear, almond, apricot, cherry and plum. Eurasiatic. Widespread in Europe, but with a more restricted distribution than either grey dagger moth (*Acronicta psi*) or knotgrass moth (*A. rumicis*).

DESCRIPTION

Adult: 33–34 mm wingspan; general appearance similar to that of *A. psi*, but forewings with a slight, violet tint; hindwings white to dusky white. **Larva:** up to 40 mm long; body with a yellow, yellowish or orange-red dorsal stripe, and several very long, pale-tipped hairs; sides marked with black, red and white, the spiracular area and below mainly greyish white to pinkish white; a dorsal hump present on abdominal segments 1 and 8 (black and white, respectively); head black.

LIFE HISTORY & DAMAGE

As for grey dagger moth (*A. psi*).

Agrotis exclamationis (Linnaeus) (926–927)

Heart & dart moth

An often abundant species, attacking a wide range of low-growing plants, including root crops (e.g. carrot, swede and turnip) and, occasionally, strawberry and other fruit crops. Eurasiatic. Widely distributed in Europe.

DESCRIPTION

Adult: 38–40 mm wingspan; forewings whitish brown to dark brown, each with a blackish dart-like mark and prominent, black-outlined stigmata; hindwings whitish (in male) or brownish grey (in female). **Egg:** hemispherical, ribbed and finely reticulate; whitish when laid, later becoming pinkish. **Larva:** up to 40 mm long; body brownish through greenish brown to greyish brown; paler dorsally, but dorsal line usually outlined by, often prominent, darker markings; pinacula blackish; spiracles black and relatively large; head yellowish brown and shiny, marked with a pair of vertical black bars. **Pupa:** 16–18 mm long; reddish brown; cremaster with a pair of divergent spines.

LIFE HISTORY

Adults occur in June and July, and sometimes also (as a partial second generation) in September. Females are capable of laying several hundred eggs, which are usually deposited at random on the soil. The eggs hatch 8–10 days later. Young larvae often browse at night on the leaves of host plants, but usually rest in the soil during the daytime. Eventually the larvae become

924 Dark dagger moth (*Acronicta tridens*).

925 Dark dagger moth (*Acronicta tridens*) – larva.

926 Heart & dart moth (*Agrotis exclamationis*) – male.

927 Heart & dart moth (*Agrotis exclamationis*) – larva.

entirely subterranean, feeding on the roots of nearby plants and, in the case of strawberry, often also tunnelling into the crowns. They feed throughout the summer and autumn, normally overwinter as fully fed larvae and pupate in the spring.

DAMAGE

On strawberry, larvae browse mainly on the roots and burrow into the crowns, causing plants to wilt; attacks are particularly serious under dry conditions. Damage to other fruit crops is restricted mainly to the roots. Owing to their habit of severing plant stems at or about stem level, the larvae of this and various other related species are commonly known as 'cutworms'.

Agrotis ipsilon (Hufnagel) (928)

Dark sword grass moth

A cosmopolitan and polyphagous species of cutworm, attacking a wide range of herbaceous plants (including weeds such as *Galium aparine* and *Taraxacum officinale*). A minor pest of grapevine in mainland Europe, but more notorious worldwide as a pest of cotton, lettuce, tobacco, wheat and various root crops (including mangold, swede and potato). Widely distributed in Europe, with an annual influx of immigrants from North Africa (and a return migration in the autumn).

DESCRIPTION

Adult: 40–55 mm wingspan; forewings dark purplish brown to pale ochreous, with darker markings (including a distinct black 'dart' directed towards the margin); antennae of male bipectinate, those of female simple; hindwings mainly translucent, with dark veins. **Larva:** up to 45 mm long; body purplish brown to ochreous brown dorsally, green or yellow ventrally, head black, with a pair of white spots; pinacula black.

LIFE HISTORY

Adults occur throughout most of the year, but are most numerous from spring to autumn. Eggs are deposited at random on the foliage of low-growing plants, and they hatch 1–3 weeks later. Larvae feed for about a month and then pupate, each in a slight subterranean cocoon. The number of generations, whether one or more annually, depends on temperature.

DAMAGE

As for heart & dart moth (*Agrotis exclamationis*).

Agrotis puta (Hübner) (929–930)

Shuttle-shaped dart moth

An often common, polyphagous species, the larvae feeding on various low-growing plants, including *Polygonum aviculare*, *Taraxacum officinale* and, occasionally, cultivated plants such as strawberry. Eurasiatic. In some regions of local occurrence, although widely distributed in western Europe.

DESCRIPTION

Adult female: 30–32 mm wingspan; forewings light brown, with variable darker markings; hindwings grey, with whitish cilia. **Adult male:** similar to female, but

929 Shuttle-shaped dart moth (*Agrotis puta*) – female.

930 Shuttle-shaped dart moth (*Agrotis puta*) – larva.

928 Dark sword grass moth (*Agrotis ipsilon*) – male.

with pectinate antennae; hindwings mainly whitish, with darker veins. **Egg:** 0.6 mm across; globular and stronly ribbed; yellowish white when laid, later becoming greyish. **Larva:** up to 32 mm long; body plump, olive brown to reddish brown, and distinctly mottled; pinacula blackish brown; spiracles black; head small, light brown and glossy.

LIFE HISTORY
Adults occur, in two or three generations, from May to October. Eggs are laid in batches on a wide range of plants and hatch about a week later. Larvae feed for several weeks before eventually pupating in the soil, those of the final generation overwintering and completing their development in the spring.

DAMAGE
On strawberry, young larvae feed on the foliage, but older individuals tend to browse on the roots. However, larvae are rarely numerous and are usually unimportant pests.

Agrotis segetum (Denis & Schiffermüller) (931–933)
Turnip moth
An often serious pest of vegetable root crops. Also a pest of strawberry and, occasionally, young fruit trees, fruit bushes and grapevines. Eurasiatic. Widely distributed in Europe.

DESCRIPTION
Adult female: 32–42 mm wingspan; forewings brown to blackish brown, with dark-edged markings, including a basal dash and two subcentral stigmata; hindwings pearly white. **Adult male:** similar to female, but characterized by their lighter hindwings and distinctly bipectinate antennae. **Egg:** 0.6 mm across; globular, noticeably ribbed and reticulate; white when laid, soon becoming creamy white. **Larva:** up to 40 mm long; body plump, greyish brown and distinctly glossy, often with a yellowish or pinkish tinge, and with blackish pinacula and indistinct darker lines along the back and sides; spiracles black and relatively small; head yellowish brown. **Pupa:** 15–20 mm long; reddish brown; cremaster with two slightly divergent spines.

LIFE HISTORY
Adults occur in May and June, with in some regions at least a partial second generation in the autumn. Eggs are laid in small batches on a range of low-growing plants and hatch about 2 weeks later. Larvae feed at night, in their early stages ascending plants to browse on the leaves. Later, however, they feed mainly on roots, adopting a typical cutworm habit. The larvae are sluggish and hide during the daytime in the soil near the roots of host plants. Although becoming fully fed in the autumn, most do not pupate until the following spring.

DAMAGE
As for heart & dart moth (*Agrotis exclamationis*).

931 Turnip moth (*Agrotis segetum*) – female.

932 Turnip moth (*Agrotis segetum*) – male.

933 Turnip moth (*Agrotis segetum*) – larva.

Allophyes oxyacanthae (Linnaeus) (**934–936**)
Green-brindled crescent moth

An often common, but minor, pest of apple, cherry and plum; more frequently associated with *Crataegus monogyna* and *Prunus spinosa*. Eurasiatic. Widely distributed in Europe.

DESCRIPTION
Adult: 38–40 mm wingspan; forewings light brown to chocolate-brown, with pale markings, and often dusted with green scales; hindwings greyish. **Larva:** up to 45 mm long; body reddish brown or greyish brown, with whitish, greenish, dark grey and black markings (including three dark lines and a series of diamond-shaped patches along the back); abdominal segment 1 with a pair of reddish-brown, L-shaped markings; abdominal segment 8 slightly humped; head reddish brown, mottled with black. **Pupa:** 15–17 mm long; brown and plump; cremaster with several hook-tipped spines.

LIFE HISTORY
The larvae feed on foliage from April to late May or early June. They then pupate in the soil, each in a flimsy silken cocoon. Adults fly in the autumn. Eggs are then laid singly or in small batches on the bark of trees; they hatch in the following spring.

DAMAGE
Larvae contribute to leaf damage caused by various other species, but they do not occur in sufficiently large numbers to be important.

Amphipyra pyramidea (Linnaeus) (**937–938**)
Copper underwing moth

A minor pest of apple and plum, but associated mainly with broadleaved forest trees and shrubs, particularly *Quercus*; attacks have also been reported on kiwi fruit. Palaearctic. Widely distributed in Europe.

DESCRIPTION
Adult: 45–55 mm wingspan; forewings brown, marked with black and pale yellowish grey; hindwings coppery red. **Larva:** up to 45 mm long; body plump, green to greenish white, dotted with yellow, with three incomplete white (often yellow-tinged) lines along the back and one along each side; abdominal segment 8 humped, with a horn-like apex; head shiny greenish grey; head and prothorax retractile.

LIFE HISTORY
Adults occur from late July to September or October. Larvae feed from April to June, and eventually pupate in cocoons formed in the soil.

DAMAGE
Larvae cause defoliation, but numbers of larvae on fruit trees are usually small.

934 Green-brindled crescent moth (*Allophyes oxyacanthae*).

935 Green-brindled crescent moth (*Allophyes oxyacanthae*) – larva (light form).

936 Green-brindled crescent moth (*Allophyes oxyacanthae*) – larva (dark form).

Ceramica pisi (Linnaeus) (**939**)
Broom moth

A minor pest of raspberry and strawberry; however, associated mainly with non-cultivated plants, including broadleaved trees, herbaceous plants and ferns (especially *Pteridium aquilinum*). Eurasiatic. Present throughout Europe, including northerly regions.

DESCRIPTION
Adult: 33–42 mm wingspan; forewings greyish brown to purplish brown, with a partly broadened, pale yellow subterminal line; hindwings greyish brown. **Larva:** up to 45 mm long; body elongate and slightly tapered anteriorly and posteriorly, green or reddish brown and velvety, with broad yellow (often black-edged) longitudinal stripes; head light brown.

LIFE HISTORY
Adults occur in June and July. Eggs are deposited in batches on the leaves of host plants and hatch about 10 days later. Larvae feed on the leaves from late June or July onwards. Their development is slow and individuals are usually not fully grown until the end of August or September. They then pupate in the soil, each in a flimsy cocoon, adults emerging in the following year.

DAMAGE
Larvae often feed openly on the leaves in full sunshine. However, damage caused is of minor significance, particularly as it occurs rather late in the season.

Conistra vaccinii (Linnaeus) (**940–941**)
Chestnut moth

A woodland species, larvae feeding on a wide range of broadleaved trees and shrubs, but sometimes also a minor pest of fruit trees. Eurasiatic. Widely distributed in Europe.

DESCRIPTION
Adult: 33–36 mm wingspan; forewings reddish brown to brown (but extremely variable), with a distinct blue-black spot; hindwings yellowish brown; abdomen noticeably flattened dorsoventrally. **Larva:** up to 35 mm long; body plump, reddish brown mottled with pale yellow and with three indistinct lines along the back; dorsal pinacula whitish; head brown to greyish brown and inconspicuous; prothoracic plate dark brown, with three whitish-yellow lines; spiracles black.

937 Copper underwing moth (*Amphipyra pyramidea*).

939 Broom moth (*Ceramica pisi*) – larva.

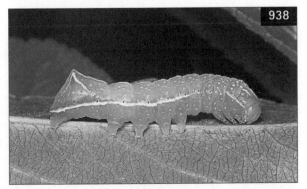

938 Copper underwing moth (*Amphipyra pyramidea*) – larva.

940 Chestnut moth (*Conistra vaccinii*).

LIFE HISTORY

Adults occur in September and October, and then overwinter. Eggs are laid in the spring and hatch within about 2 weeks. Larvae then feed on the foliage of host plants, usually completing their development in June. They then burrow into the soil, to form large, flimsy cocoons, in which they eventually pupate. There is just one generation annually.

DAMAGE
Unimportant.

Cosmia pyralina (Denis & Schiffermüller) (**942**)
Lunar-spotted pinion moth

A minor pest of fruit trees, but associated mainly with broadleaved forest trees and shrubs, including *Betula*, *Populus*, *Prunus*, *Quercus*, *Salix* and *Ulmus*. Eurasiatic. Widely distributed in Europe, particularly in central, southern and south-eastern regions.

DESCRIPTION

Adult: 28–32 mm wingspan; forewings purplish brown, with darker markings and each with a purplish, white-edged blotch towards the apex; hindwings ochreous brown. **Larva:** up to 30 mm long; body stout, yellowish green to bluish green, mottled with yellow, with a distinct creamy-white dorsal line and indistinct subdorsal and subspiracular lines; head pale bluish green, with short black hairs anteriorly.

LIFE HISTORY

Eggs deposited on host plants during the previous summer hatch in the spring. Larvae then feed on the foliage, eventually pupating in late May or early June. Adults occur from mid-June to the end of August.

DAMAGE
Unimportant.

Cosmia trapezina (Linnaeus) (**943–944**)
Dun-bar moth

Polyphagous on broadleaved trees and shrubs, and a minor pest of fruit trees; larvae are both phytophagous and carnivorous. Eurasiatic. Widespread and often common in Europe.

DESCRIPTION

Adult: 25–35 mm wingspan; forewings whitish grey to yellowish grey, with a pinkish tinge and with a more or less darkened central band; hindwings pale to greyish.

941 Chestnut moth (*Conistra vaccinii*) – larva.

943 Dun-bar moth (*Cosmia trapezina*).

942 Lunar-spotted pinion moth (*Cosmia pyralina*).

944 Dun-bar moth (*Cosmia trapezina*) – larva.

Egg: globular and finely ribbed; pearly white when laid, later darkening. **Larva:** up to 40 mm long; body flattened dorsoventrally, green, with three white lines along the back and a yellow line (often edged above with blackish green) along each side; pinacula black, surrounded by white; spiracles black; head light green. **Pupa:** 13 mm long; light brown, with a whitish bloom; cremaster bearing two close-set terminal spines with sharply divergent tips.

LIFE HISTORY
Adults occur in July and August. Eggs form the overwintering stage. Larvae feed on the foliage of various trees and shrubs from April to June, but are of greater significance as predators of other caterpillars, especially those of winter moth (*Operophtera brumata*) (family Geometridae).

DAMAGE
Larvae browse on buds, leaves and blossoms, but damage caused is of little or no importance.

Dysgonia algira (Linnaeus) (945)
Passenger moth
This subtropical, Mediterranean species occurs on several wild plants (including *Genista*, *Ricinus communis* and *Rubus dalmaticus*) and is also found, occasionally, on cultivated pomegranate. Damage caused, however, is unimportant. Adults (44–48 mm wingspan) are deep olive brown, with broad, light brown to greyish-white markings. Larvae (up to 55 mm long) are narrow-bodied, yellowish green to greyish brown, with indistinct longitudinal lines along the body; the prolegs on abdominal segment 3 are non-functional. There are two generations annually, with adults occurring in spring and summer; larvae are present in two broods, from June to October.

Eupsilia transversa (Hufnagel) (946–947)
Satellite moth
Primarily in northern Europe, a minor pest of apple, pear and other fruit trees, but more usually associated with broadleaved forest trees and shrubs. Eurasiatic. Widely distributed in Europe.

DESCRIPTION
Adult: 40–45 mm wingpsan; forewings mainly brown or orange brown to blackish brown, with black markings and a prominent white or yellowish reniform stigma and

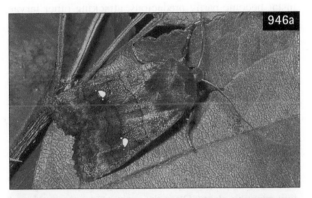

946a, b Satellite moth (*Eupsilia transversa*).

945 Passenger moth (*Dysgonia algira*) – larva.

947 Satellite moth (*Eupsilia transversa*) – larva.

two adjacent spots; hindwings brown. **Egg:** subspherical, stongly ribbed and finely reticulate; pale yellow when laid, later becoming brownish pink. **Larva:** up to 45 mm long; body brownish black, with indistinct bluish-black lines along the back and an often much broken and reduced white stripe along each side; prothoracic and anal plates black; underside of body light green; head dark reddish brown. **Pupa:** 18 mm long; brown; cremaster with a pair of long, stout, more or less parallel spines (with divergent tips) and two long bristles.

LIFE HISTORY

Adults appear in the autumn and survive through the winter to deposit eggs in the early spring. Larvae browse on the leaves of host plants and in their final instar become partly predacious, attacking other larvae and also pests such as aphids. Fully fed larvae enter the soil to form tough cocoons in which they aestivate for 2–3 months before eventually pupating.

DAMAGE

Although larvae often feed voraciously, damage to fruit trees is usually insignificant.

Euxoa tritici (Linnaeus) (**948**)

White-line dart moth

A very polyphagous pest, attacking a wide range of low-growing plants. Often the most numerous species of cutworm to attack grapevine (see also *Agrotis* spp.); sometimes also damaging in nurseries to young fruit trees and bushes. Eurasiatic. Widely distributed in Europe.

DESCRIPTION

Adult: 30–40 mm wingspan; forewings extremely variable, ranging from dark brown through reddish brown to ochreous brown (each with a pale costal streak, pale reniform and pale orbicular stigmata and crosslines), with blackish markings (including the claviform stigma); hindwings mainly whitish to greyish

948 White-line dart moth (*Euxoa tritici*).

white, with darker veins, and suffused with brownish grey towards the margins. **Larva:** up to 40 mm long; body plump and brownish grey, with broadly dark-bordered dorsal and spiracular lines; prothoracic plate speckled or barred with black; spiracles black; head greyish to dark brown. **Pupa:** 15–18 mm long; yellowish brown and shiny; abdomen somewhat truncated; cremaster with two short spines.

LIFE HISTORY

Adults occur in July and August. Eggs are laid singly or in small batches on the soil and hatch 1–3 weeks later. Larvae then feed on the roots of host plants from the end of August or early September onwards, completing their development in the following year. They then pupate, each in an earthen cell, and adults emerge 2–3 weeks later.

DAMAGE

Larvae browsing on the roots cause wilting of small plants, particularly under dry conditions. At night, the larvae may also ascend low-growing plants to attack the young shoots.

Graphiphora augur (Fabricius)

Double dart moth

Reported as a pest of various fruit crops, including currant, gooseberry and raspberry, but more frequently associated with wild hosts such as *Betula*, *Crataegus monogyna*, *Prunus spinosa*, *Salix atrocinerea* and *Rumex*. Holarctic. Widely, but locally, distributed in Europe from northern Italy and south-western France northwards.

DESCRIPTION

Adult: 38–48 mm wingspan; forewings broad, greyish brown, with irregular, blackish crosslines and black-marked stigmata; hindwings greyish brown. **Larva:** up to 47 mm long; body reddish brown, with V-shaped marks along the back and a pale subspiracular line; each segment with two pairs of white spots; spiracles white; head light brown and shiny, with dark markings.

LIFE HISTORY

Adults occur in June, July or early August, eggs being deposited in neat batches on *Rumex*, and possibly on other low-growing plants. The larvae feed briefly during the late summer and early autumn, and then hibernate. They reappear early in the following spring, to feed on the buds, young shoots and expanded leaves of a wide range of broadleaved trees and shrubs. Fully fed larvae pupate in the soil or amongst leaf litter, and adults emerge 3–4 weeks later. There is just one generation annually.

DAMAGE

In early spring, larvae destroy the buds; they will also damage the growing points and developing shoots. On raspberry, larval feeding can have an adverse effect on the production of both vegetative and fruiting canes; crop yields may also be reduced and harvest dates delayed.

Hydraecia micacea (Esper) (**949–950**)

Rosy rustic moth

larva = potato stem borer

An often common pest of crops such as beet, onion, potato and rhubarb; also damaging to hop and, less frequently, raspberry and strawberry. Wild hosts include *Carex*, *Plantago* and, especially, *Rumex*. Eurasiatic. Widely distributed in Europe. Also present in Canada.

DESCRIPTION

Adult: 35–45 mm wingspan; forewings reddish brown, somewhat darker centrally; hindwings pale, each with a dark crossline. **Larva:** up to 45 mm long; body dull pinkish, with a slightly darker dorsal line and several dark pinacula, each bearing a pinkish-brown hair; head yellowish brown; prothoracic plate yellowish brown, with dark margins. **Pupa:** 20 mm long; ochreous; cremaster with a single spike.

LIFE HISTORY

Adults occur from late July to October or November and are most numerous in the autumn. Eggs are laid low down on the stems of various plants, especially grasses. They hatch in the following spring. Larvae then bore into the stems or roots of suitable host plants. They feed from late April to July or August and then pupate in the soil a few centimetres beneath the surface.

DAMAGE

Infestations are usually most severe on or near headlands and in weedy sites. **Raspberry:** larvae bore into the basal parts of the young stems and then tunnel immediately beneath the epidermis, causing canes to wither. **Strawberry:** larvae tunnel into the crowns below soil level; infested plants may wilt and die, particularly under dry conditions. **Hop:** young larvae attack the aerial parts of the bines, usually burrowing into tissue within 30 cm of the ground; they then tunnel within the pith. In June, larvae vacate the bines through a distinct exit hole and enter the soil to attack the crowns. Damaged bines often split open or snap at or near the larval exit holes, and are susceptible to invasion by pathogens; attacks may lead to significant yield reductions. In addition, growth from young 'hills' can be severely checked by the pest and newly planted sets may be killed.

Hypena rostralis (Linnaeus) (**951**)

Buttoned snout moth

A minor pest of hop in various parts of Europe; also associated with *Humulus lupulus* growing in hedgerows and other situations. Eurasiatic. Widely distributed, and often well-established in hop-growing regions, but apparently less numerous than formerly.

DESCRIPTION

Adult: 29–32 mm wingspan; forewings narrow and the outer margins slightly emarginate, light brown to

949 Rosy rustic moth (*Hydraecia micacea*).

950 Rosy rustic moth (*Hydraecia micacea*) – larva.

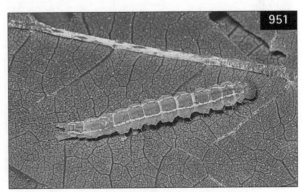

951 Buttoned snout moth (*Hypena rostralis*) – larva.

purplish-brown, with scale tufts arising from the orbicular and reniform stigmata; hindwings broad and mainly light brown; labial palps long; legs long and slender. **Larva:** up to 25 mm long; body slender, bright green, marked with white subdorsal and spiracular lines; pinacula small and dark brown, each bearing a short seta; abdominal prolegs long and flanged, but reduced in number (absent from abdominal segment 3); head green to yellowish brown.

LIFE HISTORY

Adults appear in late summer and eventually hibernate. They reappear in the following spring. Eggs are laid on the young growth of hop plants, usually in May or June. Larvae, which rest on the underside of leaves during the daytime, feed from June onwards, and complete their development by the end of July. Pupation then occurs in flimsy, white cocoons spun on the foodplant or amongst debris on the ground. In warmer parts of Europe there are two generations annually, with second-brood larvae feeding in the autumn.

DAMAGE

Larvae cause noticeable holing of leaves and also graze on the developing strobili.

Lacanobia oleracea (Linnaeus) (952–954)

Tomato moth

A minor, polyphagous pest of fruit crops such as strawberry, raspberry and grapevine, but more frequently associated with outdoor and greenhouse-grown tomato crops and herbaceous ornamentals (e.g. *Chrysanthemum* and *Dianthus*); wild hosts include *Atriplex patula*, *Chenopodium album* and *Taraxacum officinale*. Eurasiatic. Widely distributed in Europe.

DESCRIPTION

Adult: 35–45 mm wingspan; forewings reddish brown to purplish brown, with a small, yellowish stigma and a whitish subterminal line; hindwings light brownish grey. **Egg:** hemispherical, finely ribbed and reticulate; apple-green when laid, becoming yellowish green. **Larva:** up to 40 mm long; body green, yellowish brown or brown, finely speckled with white and, less densely, with black; dorsal and subdorsal lines pale, the

953a, b Tomato moth (*Lacanobia oleracea*) – larva.

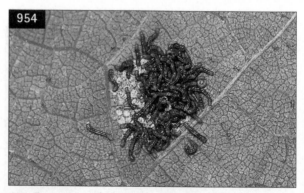

952 Tomato moth (*Lacanobia oleracea*).

954 Tomato moth (*Lacanobia oleracea*) – newly emerged larvae on remains of egg batch.

spiracular line broad and yellow or orange yellow, edged with black above; spiracles white, ringed with black; head light brown. **Pupa:** 16–19 mm long; dark brown to black, and coarsely punctured; cremaster with a pair of blunt-headed spines.

LIFE HISTORY

Adults emerge out-of-doors from late May or early June onwards. Eggs are deposited on the underside of leaves, in batches of 30–200, and hatch 1–2 weeks later. Larvae then graze in groups on the lower surface of the leaves. After the second instar, the larvae disperse and tend to occur singly, each feeding voraciously and becoming fully grown within a few weeks. Larvae are most abundant from June or July to September. Pupation takes place in a flimsy cocoon formed in the soil or amongst debris or attached to a suitable surface. In favourable conditions there is a second generation of adults in the autumn.

DAMAGE

Although young larvae merely graze away the lower surface of leaves, older larvae bite through the complete lamina and, if numerous, can reduce foliage to a skeleton of major veins.

Mamestra brassicae (Linnaeus) (955–956)
Cabbage moth

A generally common, very polyphagous and often serious pest of vegetable crops (e.g. Brussels sprout, cabbage, cauliflower, lettuce, onion, red beet and tomato), and also sometimes present on fruit crops such as apple, currant and strawberry. Eurasiatic if not Holarctic. Widely distributed in Europe.

DESCRIPTION

Adult: 35–45 mm wingspan; forewings dull greyish brown, with irregular black and white markings; hindwings brownish. **Egg:** hemispherical, distinctly ribbed and reticulate; whitish when laid, later becoming purplish brown. **Larva:** up to 45 mm long; body brownish black to greenish, often speckled with yellow and with a series of black subdorsal bars and a broad, creamy-yellow or greenish, subspiracular stripe; abdominal segment 8 slightly humped; head light brown; young larva light green, speckled with black, banded with yellow intersegmentally. **Pupa:** 17–22 mm long; reddish brown; cremaster with two short, hooked spines.

LIFE HISTORY

Adults of the first generation occur from late April or early May to the end of June, and those of a second generation from mid-August to the end of September. Eggs are laid in large batches on the foliage of a wide range of plants and hatch about a week later. Larvae occur from mid-May onwards, those of the second brood completing their development by the end of October. Fully fed larvae pupate in the soil, each in a flimsy silken cocoon. The period of development of the various life stages varies considerably with temperature, the period from egg to adult extending over 45 days at 25°C and increasing to over 65 days at 20°C. Pupae usually form the overwintering stage. In more northerly regions, this species is considered to have just one generation annually. However, the appearance of adults and larvae is extremely variable and suggests that even here there may be a partial second generation, at least when conditions are favourable. The insect frequently occurs in greenhouses.

DAMAGE

Larvae are voracious feeders and capable of causing extensive defoliation. Numbers on fruit crops, however, are usually small.

955 Cabbage moth (*Mamestra brassicae*).

956 Cabbage moth (*Mamestra brassicae*) – larva.

Melanchra persicariae (Linnaeus) (**957–959**)
Dot moth

A minor pest of apple, cherry, plum, currant, gooseberry, raspberry and strawberry; larvae also feed on a wide variety of other hosts, particularly herbaceous plants. Eurasiatic, if not Holarctic. Widely distributed and locally common in western and northern Europe.

DESCRIPTION
Adult: 38–48 mm wingspan; forewings bluish black, each with a prominent, white reniform stigma; hindwings greyish brown, paler basally. **Egg:** hemispherical, ribbed and finely reticulate; greenish white when laid, later becoming pinkish brown. **Larva:** up to 45 mm long; body light green or light brown, with darker chevron-like markings on the back and sides, and a thin, pale, dorsal stripe; abdominal segment 8 with a pointed hump; head light brown. **Pupa:** 17 mm long; dark chestnut-brown; cremaster with two flat-capped spines.

LIFE HISTORY
Moths occur in June, July and August. Eggs are laid in batches on leaves of various plants and hatch in about 8 days. Larvae feed from July to September or October. Pupation then takes place in the soil, adults emerging in the following summer.

DAMAGE
Larvae devour large amounts of foliage and rapidly strip the leaves from the shoots or branches. However, attacks are usually important only on young plants or when larvae are numerous. Occasionally, fruits are damaged directly; on apples, for example, large, irregular areas may be grazed away from the surface.

957 Dot moth (*Melanchra persicariae*).

958 Dot moth (*Melanchra persicariae*) – larva (brown form).

959 Dot moth (*Melanchra persicariae*) – larva (green form).

Naenia typica (Linnaeus) (**960–962**)
Gothic moth

A minor pest of apple and other fruit trees, loganberry and raspberry; also found on a wide range of other hosts, including broadleaved forest trees, shrubs and herbaceous plants. Eurasiatic. Widely distributed in northern and western Europe.

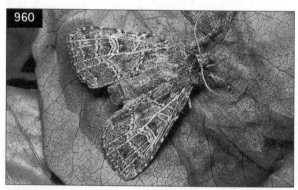

960 Gothic moth (*Naenia typica*).

DESCRIPTION

Adult: 36–46 mm wingspan; forewings whitish brown, suffused with blackish brown, each marked with pale stigmata and crosslines, the reniform stigma enclosing a pale line; hindwings brownish grey. **Larva:** up to 45 mm long; body greyish brown, flecked with darker brown, the sides marked with pale oblique streaks and a pinkish, undulating spiracular line edged above by black; second and third thoracic segments with a pair of creamy-white spots; abdominal segments 7 and 8 each with distinctive, black, oblique markings; pinacula whitish; head light brown, marked with brown; young larva with thoracic segments relatively narrow.

LIFE HISTORY

Adults occur in June and July. Eggs are then laid in batches on the upper surface of leaves and hatch about 2 weeks later. The young larvae feed gregariously on the underside of leaves, from late July or August onwards, grazing away the epidermal tissue. In the following spring, after hibernation, the larvae become solitary. They are fully grown in May and then pupate, each in an earthen cell.

DAMAGE

Young larvae, whilst feeding gregariously in autumn, cause extensive blanching of leaves, but the lateness of attacks lessens their impact.

Noctua comes (Hübner) (**963–964**)

Lesser yellow underwing moth

A polyphagous species, the larvae feeding on a wide range of weeds (e.g. *Taraxacum officinale*), trees and shrubs; attacks sometimes occur on grapevine. Western Palaearctic. Widely distributed in central and southern Europe, and northwards to southern Scandinavia.

DESCRIPTION

Adult: 38–48 mm wingspan; forewings brownish grey to reddish grey, with inconspicuous crosslines and more or less pronounced orbicular and reniform

961 Gothic moth (*Naenia typica*) – young larvae.

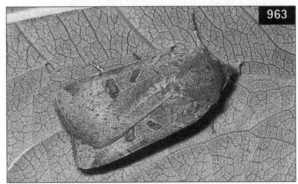

963 Lesser yellow underwing moth (*Noctua comes*).

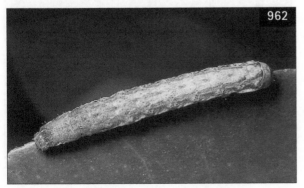

962 Gothic moth (*Naenia typica*) – late-instar larva.

964 Lesser yellow underwing moth (*Noctua comes*) – larva.

stigmata; hindwings orange yellow, with an irregular black border and a black discal spot. **Egg:** 0.65 mm across; hemispherical, strongly ribbed and reticulate; creamy white when laid, later becoming dark brown. **Larva:** up to 50 mm long; body light brown to pale ochreous grey, with a narrow, inconspicuous dorsal line and a broad, subspiracular stripe; abdominal segments 6–8 with dark wedge-shaped marks, segment 8 also with a dark transverse band posteriorly; spiracles small, white and ringed with black.

LIFE HISTORY
Adults are active from July to September, and eggs are laid in batches in a range of situations. The eggs hatch within 10–14 days. Larvae feed at night from late August or September onwards. They eventually overwinter, and resume feeding in the spring. Pupation occurs in the soil, in an earthen chamber.

DAMAGE
Notably in the spring, overwintered larvae may ascend grapevines to browse on the opening buds and young shoots. Also, older larvae are capable of causing noticeable defoliation. Attacks are most likely to occur in weedy sites.

Noctua fimbriata (Schreber) (**965–968**)
Broad-bordered yellow underwing moth
A usually minor pest of grapevine, but associated mainly with a range of forest trees and shrubs such as *Acer pseudoplatanus*, *Betula* and broadleaved *Salix*; larvae will also feed on currant and certain herbaceous plants, particularly *Rumex*. Western Palaearctic. Widely distributed in Europe, northwards to Finland and southern Scandinavia.

DESCRIPTION
Adult: 50–58 mm wingspan; forewings pale ochreous brown (= female) to greyish brown, olive brown or mahogany brown (= male), often tinged with greenish or reddish brown and with a smooth appearance; hindwings deep orange red, with a very broad black border. **Egg:** 0.76 mm across; hemispherical, strongly ribbed and finely reticulate; apple-green when laid, later becoming

966 Broad-bordered yellow underwing moth (*Noctua fimbriata*) – male.

967 Broad-bordered yellow underwing moth (*Noctua fimbriata*) – final-instar larva.

965 Broad-bordered yellow underwing moth (*Noctua fimbriata*) – female.

968 Broad-bordered yellow underwing moth (*Noctua fimbriata*) – penultimate-instar larva.

darker. **Larva:** up to 55 mm long; body ochreous brown, mottled with cream and black, and with a fine dorsal line and a broad subspiracular stripe; spiracles large, white and black-rimmed, and those on abdominal segments with an adjacent black patch posteriorly (the black patches not present in earlier instars); abdominal segments 8 and 9 each with dark wedge-shaped marks dorsally and, posteriorly, a pale transverse band.

LIFE HISTORY

Adults appear in July. They then undergo a brief period of aestivation, reappearing a few weeks later. Eggs are then laid in batches, mainly in August and September. The eggs hatch about 10 days later. Larvae then feed into late autumn before eventually overwintering. Activity is resumed in the spring, the developing larvae usually ascending trees and shrubs at night to feed on the buds and young leaves. Pupation occurs in the soil in May.

DAMAGE

As for lesser yellow underwing moth (*Noctua comes*).

Noctua pronuba (Linnaeus) (969–971)

Large yellow underwing moth

An abundant species, attacking a wide variety of low-growing wild and cultivated plants. Larvae are often damaging to strawberry, particularly in newly planted fields and in allotments and gardens; infestations may also occur on grapevine. Palaearctic. Widely distributed in northern and western Europe.

DESCRIPTION

Adult: 50–60 mm wingspan; forewings yellowish brown, through greyish brown to dark rusty brown or blackish brown, and extremely variable; hindwings orange yellow, with a blackish-brown border. **Egg:** globular, ribbed and reticulate; creamy white when laid, later becoming purplish grey. **Larva:** up to 50 mm long; body stout, tapering anteriorly, greyish brown, dull yellowish or greenish, with pale lines along the back, the outer ones bordered with short blackish bars on abdominal segments 1 and 8; spiracles black; head relatively small, light brown, marked with two black stripes. **Pupa:** 22–25 mm long; plump and reddish brown, smooth and shiny; cremaster with two strong, divergent spines and a pair of small bristles.

LIFE HISTORY

Adults occur from June or July onwards. Although active mainly at night, the moths are readily disturbed during the daytime; they then career wildly through the air before resettling. Eventually, after an obligatory delay of at least a month, eggs are laid in very large batches on the underside of leaves, or at the tips of

grasses and other similar plants. The eggs hatch about 10 days later. Although the larvae often behave as cutworms, inhabiting the soil and, at night, biting into plant roots and crowns at or just below soil level, they will also ascent their foodplants to attack the foliage – hence their colloquial name of 'climbing cutworms';

969 Large yellow underwing moth (*Noctua pronuba*).

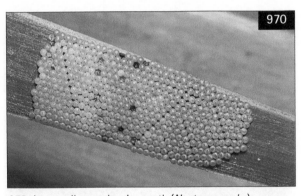

970 Large yellow underwing moth (*Noctua pronuba*) – egg batch.

971 Large yellow underwing moth (*Noctua pronuba*) – larva.

they often also rest during the daytime on the underside of the leaves. Feeding continues throughout the late summer and autumn, and larvae complete their development in the spring. They then pupate in the soil.

DAMAGE

Damage to foliage is of little or no importance, but attacks on roots and crowns may cause strawberry plants to wilt and die. The larvae are most harmful in dry, warm conditions.

Orthosia cruda (Denis & Schiffemüller) (972–973)

Small quaker moth

A minor pest of apple and other fruit trees, but associated mainly with broadleaved forest trees and shrubs such as *Corylus avellana*, *Quercus* and broadleaved *Salix*. Eurasiatic. Widely distributed in Europe.

DESCRIPTION

Adult: 28–32 mm wingspan; forewings ochreous to reddish brown, with paler markings; orbicular and reniform stigmata usually darkened; hindwings light brown to greyish brown. **Egg:** hemispherical, but flattened, finely ribbed and reticulate; whitish when laid, but soon darkening to grey. **Larva:** up to 30 mm long; body light green to olive green, with relatively large black pinacula, whitish dorsal and subdorsal lines, a transverse white crossbar towards the tip of the abdomen, and a broad yellowish-white subspiracular stripe; spiracles white, outlined with black; head and prothoracic plate black.

LIFE HISTORY

Adults occur in the spring, usually from mid- or late March to the end of April. Eggs are laid in sheltered situations on the bark of the foodplant and hatch 7–10 days later. Larvae feed on the foliage in May and early June. They then enter the soil and pupate, each in an earthen cell. There is just one generation annually.

DAMAGE

Larvae cause slight defoliation, but numbers on fruit trees are usually small. In common with other members of the genus *Orthosia*, older larvae are sometimes partly carnivorous.

Orthosia gothica (Linnaeus) (974–975)

Hebrew character moth

A minor pest of apple, pear, plum, black currant and raspberry. Also associated with a wide range of other plants, including many broadleaved trees and shrubs. Eurasiatic. Widely distributed throughout western and northern Europe.

DESCRIPTION

Adult: 32–34 mm wingspan; forewings purplish brown to reddish brown, with distinctive black

972 Small quaker moth (*Orthosia cruda*).

973 Small quaker moth (*Orthosia cruda*) – larva.

974 Hebrew character moth (*Orthosia gothica*) – female.

markings; hindwings light greyish brown. **Egg:** hemispherical, finely ribbed and reticulate; greenish white when laid, but soon becoming darkened. **Larva:** up to 40 mm long; body greenish to yellowish green, paler intersegmentally, with whitish dorsal and subdorsal lines; spiracular line white or greenish white, often bordered above with black; head greenish. **Pupa:** 15 mm long; reddish brown and plump; cremaster with two divergent spines.

LIFE HISTORY

Adults occur in the spring, mainly in March and April. Eggs are laid in loose batches and hatch about 10 days later. Larvae feed from May onwards and usually complete their development by mid-June, but somewhat later in more northerly regions.

DAMAGE

Larvae cause defoliation and may also damage buds, fruitlets and young shoots.

Orthosia gracilis (Denis & Schiffermüller) (976–980)
Powdered quaker moth

A pest of apple, blackberry, raspberry and strawberry. Larvae also feed on many other plants, including *Rubus fruticosus* and *Salix atrocinerea*. Eurasiatic. Widely distributed in western and northern Europe and often common.

DESCRIPTION

Adult: 35–40 mm wingspan; head yellowish brown; forewings whitish brown, more or less tinged with greyish, and marked with a few black dots and a pale yellowish or pinkish submarginal line; hindwings whitish. **Egg:** hemispherical, noticeably ribbed and finely reticulate; yellowish white when laid, soon becoming brownish red. **Larva:** up to 45 mm long; body yellowish green, speckled with white or creamy white, and with three pale lines along the back, and a yellow stripe, broadly edged above with greenish black, along each side; head yellowish brown. **Pupa:** 15–17 mm long; reddish brown, cremaster with two slightly divergent spines.

976 Powdered quaker moth (*Orthosia gracilis*) – male.

975a, b Hebrew character moth (*Orthosia gothica*) – larva.

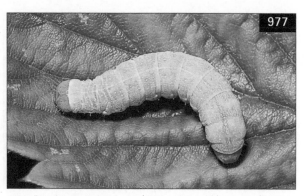

977 Powdered quaker moth (*Orthosia gracilis*) – larva (light form).

LIFE HISTORY

Adults appear in April and May, rather later than other members of the genus. Larvae feed from May to July. They then enter the soil to pupate, but do not form silken cocoons.

DAMAGE

Apple and cane fruit: larvae feed on flowers and foliage, but they are rarely numerous and damage is usually unimportant. **Strawberry:** young larvae sometimes destroy flowers, feeding within 'capped' blossoms, with the petals webbed together; at this stage, the larvae are often mistaken for those of certain tortricids (family Tortricidae); older larvae attack the foliage and, in patches, may cause significant defoliation.

Orthosia incerta (Hufnagel) (**981–982**)

Clouded drab moth

A potentially important pest of apple and pear; other crops, including raspberry and hop, are also attacked. The polyphagous larvae also feed on a wide range of forest trees and shrubs, including *Crataegus*, *Populus*, *Quercus*, *Salix*, *Tilia* and *Ulmus*. Eurasiatic. Widely distributed in western and northern Europe.

DESCRIPTION

Adult: 34–40 mm wingspan; forewings light grey to reddish brown or purplish brown, with darker markings and a pale submarginal line partly edged with brown; hindwings greyish; a large number of named colour forms are produced. **Egg:** 0.7 mm across; hemispherical, finely ribbed and reticulate; dirty creamy white when laid, but then darkening. **Larva:** up to 40 mm long; body blackish green, bluish green or light green, dotted with white; three prominent white, dark-edged stripes down the back, a wide, white stripe along each side, edged above with blackish green, and also sometimes below with yellow; head pale bluish green or yellowish green. **Pupa:** 14 mm long; shiny, dark reddish brown; cremaster with two spines.

978 Powdered quaker moth (*Orthosia gracilis*) – larva (dark form).

979 Powdered quaker moth (*Orthosia gracilis*) – early-instar larva in partly 'capped' strawberry blossom.

980 Powdered quaker moth (*Orthosia gracilis*) – larva and destroyed strawberry blossom.

981 Clouded drab moth (*Orthosia incerta*).

LIFE HISTORY

Adults occur from March to late May or early June, but are most numerous in April and early May. Eggs are laid in batches in cracks and crevices in the bark. They hatch in 10–14 days and the tiny, very mobile, larvae immediately invade the bursting buds or unfurling leaves. First-instar larvae crawl with a looping gait, the anterior-most pair of abdominal prolegs being non-functional. If disturbed, larvae drop from the foodplant and remain temporarily suspended on a strand of silk. Young larvae feed in the shelter of young blossom trusses and partially unfurled leaves, but older individuals feed in exposed situations. Larvae develop through six instars and, in June, when about 6 weeks old, they enter the soil to pupate a few centimetres beneath the surface.

DAMAGE

Larvae are rarely sufficiently numerous on fruit crops for damage to blossom trusses and leaves to be important. However, fifth- and sixth-instar larvae often bite large chunks out of apple and pear fruitlets; such attacks can be serious, resulting in the development of malformed and unmarketable fruits, some with noticeable corky scars or with large holes extending down to the core.

Orthosia munda (Denis & Schiffermüller) (983–984)

Twin-spotted quaker moth

This primarily woodland species is an occasional pest of apple; attacks are also reported on hop. Eurasiatic. Locally distributed in northern Europe.

DESCRIPTION

Adult: 38–44 mm wingspan; forewings light greyish ochreous to reddish ochreous, usually with a pair of prominent black spots towards the margin; hindwings light greyish brown. **Egg:** hemispherical, ribbed and reticulate; creamy white when laid, later becoming grey. **Larva:** up to 45 mm long; body light brown to greyish yellow, speckled with brown and with a wavy, yellowish spiracular line, edged above by black; head light greenish brown.

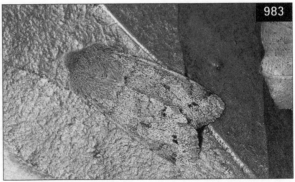

983 Twin-spotted quaker moth (*Orthosia munda*).

982a, b Clouded drab moth (*Orthosia incerta*) – larva.

984 Twin-spotted quaker moth (*Orthosia munda*) – larva.

LIFE HISTORY

Adults occur in March and April, and are often attracted to the catkins of broadleaved *Salix*. Larvae feed from May to mid-June. They then pupate in the soil, adults emerging in the following spring.

DAMAGE

Larvae browse on the leaves of host plants, and have a preference for the younger shoots. Damage on cultivated plants is usually of only minor significance.

Orthosia stabilis (Denis & Schiffermüller) (985–986)

Common quaker moth

This generally common species occasionally attacks apple, cherry and plum, but is more usually associated with forest and hedgerow trees. Eurasiatic. Widespread in Europe.

DESCRIPTION

Adult: 32–35 mm wingspan; forewings reddish ochreous to light greyish ochreous, with paler markings; hindwings greyish brown. **Egg:** 0.7 mm across; hemispherical, finely ribbed and reticulate; whitish green when laid, but soon darkening and becoming grey. **Larva:** up to 40 mm long; body plump, yellowish green and finely dotted with yellow, with three yellow stripes along the back, a broader stripe along each side, and a prominent yellow bar across the first (and another across the last) body segment; head bluish green.

LIFE HISTORY

Similar to that of clouded drab moth (*Orthosia incerta*), but with adults occurring slightly earlier in the spring (mainly in March and April). Larvae feed from April to June, but numbers on fruit trees are usually small.

DAMAGE

As for clouded drab moth (*Orthosia incerta*).

Phlogophora meticulosa (Linnaeus) (987–989)

Angle-shades moth

An often common pest, attacking a wide variety of low-growing and herbaceous plants, including greenhouse crops. Occasionally a minor problem on both protected and outdoor strawberry crops; other hosts include apple, pear, grapevine and hop. Eurasiatic. Widely distributed in Europe, but regarded as an immigrant in the more northern parts of its range (e.g. southern Fennoscandia).

DESCRIPTION

Adult: 40–50 mm wingspan; forewings pale pinkish brown, each with a large inverted triangular olive-green mark; hindwings pale. **Egg:** dome-shaped, flattened and strongly ribbed; pale yellow and darkly mottled. **Larva:** up to 40 mm long; body rather plump, yellowish green or brownish, finely dotted with white, and with a pale, incomplete, dorsal line, a

985 Common quaker moth (*Orthosia stabilis*) – male.

986 Common quaker moth (*Orthosia stabilis*) – larva.

987 Angle-shades moth (*Phlogophora meticulosa*).

series of dusky V-shaped marks down the back and a pale line along each side; head translucent, pale yellowish green or brownish. **Pupa:** 18 mm long; reddish brown and plump; cremaster with a pair of elongate spines.

LIFE HISTORY

Adults may occur throughout the year, but tend to be most numerous in May and June and again in the autumn. Eggs are deposited on leaves, either singly or in small batches. Larvae feed in greater or lesser numbers during most months of the year. However, outdoors, they are most abundant from July to September. Pupation takes place in the soil or amongst debris, within a flimsy cocoon.

DAMAGE

Larvae can cause considerable defoliation; they may also damage flower buds and blossom trusses; infestations tend to be of particular significance on strawberry.

Spodoptera littoralis (Boisduval) (**990–991**)

Mediterranean brocade moth
larva = Mediterranean climbing cutworm

A notorious pest of various subtropical and tropical crops in Africa and the Mediterranean basin. Now widely distributed in Spain, where infestations are reported on, for example, highbush blueberry; also established in southern France, Greece, Italy and southern Portugal. Sometimes introduced accidentally into northern Europe on, for example, *Chrysanthemum* cuttings, and then sometimes a temporary resident in greenhouses.

DESCRIPTION

Adult: 30–40 mm wingspan; forewings blackish brown to olive brown, with lighter markings, and often with a purplish sheen; hindwings mainly white. **Larva:** up to 45 mm long; body usually light brown to blackish brown, finely speckled with white; dorsal and subdorsal lines orange brown, the latter interrupted, at least on abdominal segments 1 and 8, by black patches;

988 Angle-shades moth (*Phlogophora meticulosa*) – larva (brown form).

990 Mediterranean brocade moth (*Spodoptera littoralis*).

989 Angle-shades moth (*Phlogophora meticulosa*) – larva (green form).

991 Mediterranean climbing cutworm (*Spodoptera littoralis*).

subspiracular line broad and pale reddish brown; dark forms are mainly blackish brown. **Pupa:** 15–20 mm long; reddish brown; cremaster with two small spines.

LIFE HISTORY

Eggs are deposited in large batches on the underside of leaves of host plants or on nearby surfaces. They are then coated with scales from the female's abdomen, and usually hatch a few days later. At first, the larvae feed gregariously. However, from the fourth instar onwards they become solitary. The larvae are voracious feeders and produce noticeably wet faecal pellets. Development is completed in 2–4 weeks, but will take considerably longer under cooler conditions. Pupation occurs in flimsy cocoons, and adults emerge about 2 weeks later. Under favourable conditions there are several generations annually; activity outdoors in southern Europe typically extends from May to October.

DAMAGE

Young larvae 'window' the leaves; older individuals bite right through the leaf tissue, causing considerable defoliation; they also attack the buds, flowers, stems and developing fruits.

Xestia c-nigrum (Linnaeus) (**992–993**)

Setaceous hebrew character moth

A polyphagous pest of a wide range of agricultural and horticultural crops, including currant, gooseberry, raspberry, strawberry and grapevine. Wild hosts include low-growing plants such as *Epilobium*, *Lamium*, *Plantago*, *Senecio*, *Stellaria* and *Verbascum*. Holarctic. Widely distributed in western Europe.

DESCRIPTION

Adult: 37–44 mm wingspan; forewings greyish ochreous to reddish ochreous, suffused and marked with black, including a subcentral indented wedge-shaped patch; hindwings whitish, suffused with greyish brown. **Egg:** hemispherical, strongly ribbed and finely reticulate; creamy white when laid, later becoming darkened. **Larva:** up to 35 mm long; body olive brown or greyish brown, with a faint, dark-edged dorsal line, and with black, wedge-shaped subdorsal streaks on the abdominal segments; spiracular line broad, greenish yellowish or yellowish, sometimes tinged with orange; abdominal segment 8 with a pale transverse band; head and prothoracic plate brownish to reddish brown.

LIFE HISTORY

Adults occur in greatest numbers in the autumn, with a smaller period of adult activity in May, June and July. Eggs are deposited singly on the foodplant or on the soil and hatch 8–9 days later. Larvae resulting from autumn-laid eggs overwinter and recommence feeding in the early spring. When fully grown, usually in April or May, they pupate in the soil, adults emerging in the autumn. Larvae found feeding in the summer are probably the progeny of adults that have emerged in the spring and early summer.

DAMAGE

Most damage to fruit crops is caused in early spring, following the emergence of larvae from hibernation; attacks on the young shoots of grapevines are particularly significant.

992 Setaceous hebrew character moth (*Xestia c-nigrum*).

993 Setaceous hebrew character moth (*Xestia c-nigrum*) – larva.

Chapter 7
Sawflies, ants and wasps

Family PAMPHILIIDAE

A small group of primitive, flattened and broad-bodied sawflies, with long, thread-like (filiform), many-segmented antennae (18–24 segments); adults are rapid fliers and sun loving. Larvae lack abdominal prolegs, are web forming and often gregarious; antennae well developed; body with short, segmented anal cerci.

Neurotoma nemoralis Linnaeus (994)
Social peach sawfly
A pest of almond, apricot and peach; also found, occasionally, on cherry, plum and *Prunus spinosa*. Widely distributed in the warmer parts of central and southern Europe.

DESCRIPTION
Adult: 7–8 mm wingspan; head large, black, marked with yellowish white; antennae black; abdomen black, marked with white; legs brown to yellowish; wings transparent, but smoky. **Egg:** 1.6 × 0.6 mm; elliptical, translucent to yellowish. **Larva:** up to 15 mm long; body dark green, with darker longitudinal lines dorsally; head and a pair of rounded, lateral prothoracic plates black; thoracic legs well developed.

LIFE HISTORY
Adults first emerge in early May (earlier or later, depending on local weather conditions). Eggs are laid in batches on the underside of leaves; they hatch 1–2 weeks later. Larvae then feed gregariously for about 4–6 weeks,

sheltered within a brownish communal web. Fully grown larvae enter the soil, where each forms a cavity several centimetres beneath the surface; here they overwinter and, eventually, pupate.

DAMAGE
Larvae cause considerable defoliation and the webs (at least on apricot and peach) may interfere with fruit development; cherries, however, even when surrounded by a mass of webbing, develop and ripen normally. Attacks are of particular significance on young trees.

Neurotoma saltuum (Linnaeus) (995–996)
Social pear sawfly
A local pest of pear; occasionally, cherry, medlar and plum are also attacked. Other hosts include *Cotoneaster*

995 Social pear sawfly (*Neurotoma saltuum*) – 'nest' of young larvae.

994 Social peach sawfly (*Neurotoma nemoralis*) – vacated web on cherry.

996 Social pear sawfly (*Neurotoma saltuum*) – final-instar larva.

and *Crataegus monogyna*. Widely distributed in central and southern Europe.

DESCRIPTION
Adult: 11–14 mm long; head black, with a yellow patch between the antennae; thorax black, with yellow tegulae; abdomen black basally, the apical segments orange, and the sides with yellow triangular markings; legs and basal segments of the antennae yellow. **Egg:** 2.5 × 1.0 mm long; yellow and oblong. **Larva:** up to 25 mm long; body yellowish orange, darker above and with a pale yellowish stripe along each side; head and a pair of rounded, lateral prothoracic plates shiny black; thoracic legs well developed; antennae and anal cerci prominent.

LIFE HISTORY
Adults occur in May and June. Eggs are laid on the underside of leaves, in rectangular batches of up to 50 or more, and then covered by a sticky secretion. After the eggs have hatched, the larvae inhabit communal webs within which they feed and develop. A web may be no longer than 10–20 cm. However, if larvae are numerous, the silk may sometimes envelop a complete branch. The larvae are very active and constantly wriggle about within their protective tents; if disturbed, each larva can exude a protective drop of red liquid from a gland situated behind the head. Larvae feed in June and July, becoming fully grown within a few weeks. They then descend to the ground on silken threads and eventually pupate in chambers formed deeply within the soil. There is a single generation annually.

DAMAGE
Tents most often occur on young trees and the larvae may completely strip the foliage from the webbed portions of the branches. Fortunately, however, attacks on fruit trees tend to be sporadic.

Family **CEPHIDAE**
(stem sawflies)

A small family of relatively small, mainly black- and yellow-banded, slender-bodied sawflies, with long, thread-like (filiform) antennae; abdomen constricted between the first and second segments; cenchri absent. Larvae, although essentially apodous, possess three pairs of small, leg-like thoracic tubercles, and the abdomen terminates in a fleshy protuberance above the anus; ocelli vestigial, metathoracic spiracles well developed; sub-anal appendages present, but vestigial. The larvae bore within shoots or stems; typically, each pupates in a transparent cocoon formed within the feeding gallery.

Janus compressus Linnaeus (**997–999**)
Pear shoot sawfly
A pest of pear; other hosts include apple and *Crataegus monogyna*. Widely distributed in mainland Europe.

DESCRIPTION
Adult: 6–9 mm long; head, antennae and thorax mainly black; abdomen elongate, mainly yellowish brown to

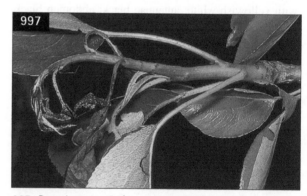

997 Pear shoot sawfly (*Janus compressus*) – shoot damage.

998 Pear shoot sawfly (*Janus compressus*) – larva.

reddish brown; wings transparent, with a distinct stigma and black veins; legs of male yellow, those of female mainly black. **Larva:** up to 15 mm long; body brownish white and translucent; protuberance at tip of abdomen brown.

LIFE HISTORY
Adults occur from mid-May to early June. Eggs are laid singly in the young shoots of host plants, the female then making a spirally arranged series of punctures around the shoot just below the oviposition point. The eggs hatch 2–3 weeks later. Larvae bore within the pith of the shoots throughout the summer, each forming a frass-filled gallery up to 15 cm in length. Individuals are fully fed in September or October. They then overwinter, each in a feeble silken cocoon formed at the end of the feeding gallery. Pupation occurs in April.

DAMAGE
The tips of infested shoots wilt and eventually desiccate and turn black.

Family **CIMBICIDAE**

Medium-sized to large, robust-bodied, fast-flying sawflies, with distinctly clubbed antennae. Larvae eruciform, with eight pairs of abdominal prolegs.

Palaeocimbex quadrimaculata (Müller) **(1000–1002)**
Almond sawfly
A minor pest of almond; other hosts include apricot and peach. Established mainly in Mediterranean regions, but its range extending northwards to parts of south-central Europe.

DESCRIPTION
Adult: 16–23 mm long; body black and yellow, antennae orange red; legs and wings reddish brown. Egg: 2 mm long; light green. **Larva:** up to 45 mm long; body pale bluish grey to greenish grey, with black markings dorsally and subdorsally; spiracles black; head pale bluish grey to greenish grey; eight pairs of abdominal prolegs.

999 Pear shoot sawfly (*Janus compressus*) – punctured shoot.

1000a, b Almond sawfly (*Palaeocimbex quadrimaculata*) – female.

1001a, b Almond sawfly (*Palaeocimbex quadrimaculata*) – larva.

1002 Almond sawfly (*Palaeocimbex quadrimaculata*) – pupal cocoon.

LIFE HISTORY

Adults occur in the spring and, after mating, the females deposit eggs singly on the upper side of the leaves of almond and other host plants. Eggs hatch in 2–3 weeks and larvae feed for about a month, passing through five instars. The larvae remain curled on the foliage during the daytime and, superficially, are similar in appearance to snail shells. When fully fed, usually in early June, they enter the soil to form large (c. 22 × 10 mm), tough, reddish-brown cocoons about 5–10 cm beneath the surface. Pupation takes place in the following spring, and adults emerge about 3 weeks later. Under certain conditions, the period of larval diapause may be extended for a further year.

DAMAGE

Larvae browse on the leaves and cause minor defoliation. Attacks are usually of little significance, but can be harmful to young trees.

Family **TENTHREDINIDAE**

Small to medium-sized sawflies. Adults are typically wasp-like (although without a narrow 'waist'), with (in females) a characteristic, saw-like ovipositor; antennae with 7–15 (but usually nine) segments; cenchri present. Larvae are usually eruciform, with from six to eight pairs of abdominal prolegs. The larvae feed mainly on the leaves of trees, shrubs and herbaceous plants, but some are leaf miners and some bore into fruits.

Allantus cinctus (Linnaeus) (**1003–1007**)
Banded rose sawfly

A minor pest of raspberry and strawberry, but more frequently reported in association with cultivated *Rosa*. Widely distributed in Europe; also present in North America.

DESCRIPTION
Adult: 7–10 mm long; mainly shiny black, the female with a distinct white or creamy band on abdominal segment 5; wings pale yellow, with brown veins; legs mainly black and brown, with an apical patch of white on each tibia. **Larva:** up to 15 mm long; body greyish green above, pale whitish green below, with small (but distinct), white pinacula; head pale yellowish brown; eight pairs of abdominal prolegs.

LIFE HISTORY
Adults emerge in May. They are strong fliers and very active in sunny weather. Eggs are laid in slits made in the underside of leaves of host plants, usually one or two per leaf (but sometimes more). The eggs swell considerably after being laid, causing conspicuous bulges on the upper surface of the leaf. They hatch within about 2 weeks. The larvae feed from June or July onwards, curling into a ball on the underside of a leaf when at rest and dropping to the ground if disturbed. They are fully fed in 3–4 weeks, males passing through six instars and females through seven. Individuals then vacate the leaves and tunnel into

1005 Banded rose sawfly (*Allantus cinctus*) – egg-laying punctures in strawberry leaf.

1003 Banded rose sawfly (*Allantus cinctus*) – female.

1006 Banded rose sawfly (*Allantus cinctus*) – larva feeding.

1004 Banded rose sawfly (*Allantus cinctus*) – male.

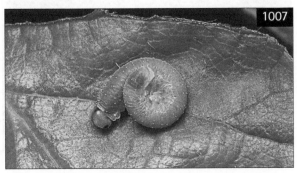

1007 Banded rose sawfly (*Allantus cinctus*) – larva at rest.

decaying wood or other suitable shelter to pupate in flimsy, semitransparent, greenish cocoons. On raspberry, the larvae usually bore into the pith of pruned shoots or cut-back remains of the previous year's fruiting canes. Adults emerge a few weeks later. Larvae of the second brood occur from August onwards and become fully grown in the autumn; although then spinning cocoons, they do not pupate until the following spring.

DAMAGE
At first, the larvae feed on the underside of the foliage; they graze away the tissue, but leave the upper epidermis intact (= 'windowing'). Later, however, they make irregular holes in the leaves, but damage is rarely serious.

Ametastegia glabrata (Fallén) (**1008**)
Dock sawfly
Although larvae of this generally common species feed on herbaceous weeds, they are also minor pests of fruit trees (especially apple and pear) and grapevine. Widely distributed in Europe; also present in North America.

DESCRIPTION
Adult: 5.5–8.0 mm long; body shiny purplish to bronzy black; legs mainly red. **Larva:** up to 18 mm long; body apple-green above, paler on sides and below; head yellowish brown, with a large, dark brown patch on the face and one above each eye; pinacula whitish and minute; eight pairs of abdominal prolegs.

LIFE HISTORY
Adults occur throughout the summer, there being two and sometimes three generations annually. Eggs are laid on the leaves of weeds such as *Chenopodium album*, *Polygonum aviculare* and *Rumex*. The larvae feed on leaves of these plants from June onwards and, when fully grown, bore into hollow or pithy stems of *Rubus fruticosus* and various

other plants in order to pupate. Larvae of the autumn generation hibernate in such situations, and then pupate in the spring. Autumn larvae reared on weeds in orchards may burrow into apples, but finding them unsuitable as hibernation sites usually then migrate to other quarters.

DAMAGE
Apples rarely contain larvae, but damage is characteristic. Invaded fruits possess a round entry hole, about 2 mm in diameter, surrounded by a reddish ring. The tunnel into the flesh is sometimes shallow, but may extend for up to 5 cm; its course is usually straight. Apples damaged by this pest are liable to be attacked by brown rot (*Sclerotinia fructigena*) and other fungal pathogens. Fully grown larvae may also damage young apple, plum or other fruit trees and grapevines by tunnelling into the pith of branches, usually entering through pruning cuts. They can also cause damage to plastic irrigation pipes.

Caliroa cerasi (Linnaeus) (**1009–1012**)
Pear slug sawfly
larva = pear & cherry slugworm
An often common pest of pear and cherry, particularly

1009 Pear slug sawfly (*Caliroa cerasi*).

1008 Dock sawfly (*Ametastegia glabrata*) – damage to mature apple.

1010 Pear slug sawfly (*Caliroa cerasi*) – larva.

in gardens. Various other rosaceous trees and shrubs (including, occasionally, apple and plum) are also attacked, particularly *Sorbus*. Eurasiatic. Present throughout Europe; also introduced accidentally to various other parts of the world, including Africa, North America, South America and Australia.

DESCRIPTION

Adult: 4–6 mm long; black and shiny. **Larva:** up to 10 mm long; body pear-shaped, broadest anteriorly, at first white, but soon becoming greenish yellow to orange yellow and covered with shiny, olive-black slime (except immediately following moulting from one instar to the next); head and legs, including seven pairs of abdominal prolegs, inconspicuous.

LIFE HISTORY

Adults first appear in late May and June. Males are very rare, development being typically parthenogenetic. Eggs are laid in small slits on the underside of leaves, often from two to five in the same leaf. The larvae feed on the upper epidermis, becoming fully grown by about July. Pupation occurs in small, black cocoons formed in the soil about 10 cm below the surface. A second generation of adults occurs in late July and in some seasons or favourable locations there may be at least a partial third during the autumn. The larvae are very sluggish, but feed voraciously. They may be found throughout the summer and early autumn.

DAMAGE

Larval feeding is confined to the upper leaf surface. Heavy attacks can cause considerable leaf damage and discoloration (affected tissue turning brown), and lead to premature leaf fall; tree growth in the following year may also be affected adversely.

Cladius difformis (Jurine in Panzer) (**1013–1016**)

Lesser antler sawfly

A pest of raspberry and strawberry, but more often noted on cultivated *Rosa*. The larvae also feed on *Filipendula ulmaria* and *Potentilla palustris*. Generally common throughout Europe; also present in North America, but probably introduced.

DESCRIPTION

Adult: 5–7 mm long; body black; legs yellowish white; antennae of male with characteristic, long projections

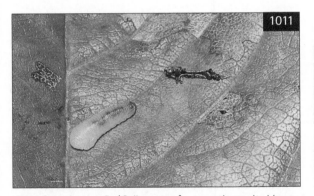

1011 Pear slug sawfly (*Caliroa cerasi*) – recently moulted larva.

1013 Lesser antler sawfly (*Cladius difformis*) – female.

1012 Pear slug sawfly (*Caliroa cerasi*) – damage on almond.

1014 Lesser antler sawfly (*Cladius difformis*) – male.

on the two basal segments and smaller projections on segments 3 and 4; female with slight projections on the two basal-most antennal segments. **Larva:** up to 15 mm long; body somewhat flattened dorsoventrally, distinctly hairy, translucent, yellowish to greenish, with darker subdorsal stripes; body hairs blackish or whitish, arising from pale verrucae; head light brown to reddish brown; seven pairs of prolegs; young larva pale-bodied, with a blackish head. **Pupa:** 5–7 mm long; white.

LIFE HISTORY
Adults of the first generation occur in May. Eggs are then laid singly in leaf stalks. After egg hatch, the larvae browse on the underside of leaves, most often attacking fully expanded leaflets. They mature in 4–5 weeks and then pupate in thin, double-walled, greyish-brown cocoons formed on the leaves or amongst debris on the ground. Adults emerge 2–3 weeks later, usually in late July or early August. Larvae of the autumn generation occur in August and September or October. These eventually overwinter in their cocoons and pupate in the spring.

DAMAGE
At first, the larvae 'window' the leaves (removing the lower epidermis and mesophyll, but leaving the upper

1015 Lesser antler sawfly (*Cladius difformis*) – larva.

1016 Lesser antler sawfly (*Cladius difformis*) – prepupa.

epidermis intact). Later, they bite irregular, elongate holes through the lamina and sometimes also remove tissue from the edges. The major veins, however, are avoided. In strawberry plantations, attacks are often concentrated on plants at the ends of rows. Although infestations sometimes attract attention, particularly in garden plots, larval damage is rarely, if ever, sufficiently extensive to cause significant harm.

Cladius pectinicornis (Geoffroy in Fourcroy)
Antler sawfly
This species is generally common in Europe; it also occurs in Asia. In common with the lesser antler sawfly (*Cladius difformis*), although reported mainly on *Rosa* it will also attack strawberry. Both species have a similar life cycle and cause similar damage. Adults of *C. pectinicornis* are distinguished from those of *C. difformis* (with difficulty in the female) by the increased number of antennal projections. There are no reliable characters to distinguish between the larvae of these two species.

Claremontia confusa (Konow)
Strawberry leaf sawfly
A minor pest of strawberry. Present throughout Europe, but usually of local occurrence.

DESCRIPTION
Adult female: 5–6 mm long; black and shiny; wings brownish, with black veins. **Larva:** up to 15 mm long; body dark green, with numerous 1- or 2-pronged spines; prothorax with 5-pronged spines; head yellow to greenish yellow; eight pairs of abdominal prolegs.

LIFE HISTORY
Adult females of this parthenogenetic species occur in May and June, with eggs being deposited in the leaves of cultivated and wild strawberry plants. Larvae occur from late May to early July. When fully grown they enter the soil and spin cocoons in which to overwinter. Pupation occurs in the spring shortly before the emergence of the adults. There is just one generation annually.

DAMAGE
Larvae eat out irregular holes in the foliage, leaving the major veins, but damage is usually of little or no significance.

Croesus septentrionalis (Linnaeus) (**1017–1018**)
Hazel sawfly
A minor pest in hazelnut plantations. Often common in the wild on various broadleaved trees and shrubs, particularly *Betula* and *Corylus avellana*. Eurasiatic. Widely distributed in Europe.

DESCRIPTION
Adult: 8–10 mm long; head and thorax black; abdomen with the basal two and the apical two or three segments black, the rest reddish brown; hind basitarsus and tip of hind tibia greatly expanded; wings mainly transparent, but apex of each forewing cloudy. **Larva:** up to 22 mm long; body yellowish to bluish green, variably marked with reddish yellow and black; head shiny black; seven pairs of abdominal prolegs.

LIFE HISTORY
The adults fly in May and June with, except in more northerly areas, a second flight in August. Eggs are laid in slits cut into the leaf veins. After egg hatch, the larvae feed gregariously in rows on the leaves, each holding onto the leaf edge with the anterior legs and arching the body over the head. When disturbed, the larvae thrash their bodies violently in the air. When fully fed, they enter the soil to pupate, each spinning an elongate, brown cocoon a short distance below the surface. Larvae of the second generation overwinter in their cocoons and eventually pupate in the spring. Larvae may be found on foliage from mid-June to the end of September.

1017 Hazel sawfly (*Croesus septentrionalis*) – female.

1018 Hazel sawfly (*Croesus septentrionalis*) – larva.

DAMAGE
The larvae feed voraciously and will quickly defoliate a branch. However, infestations are not serious, except on small trees.

Empria tridens (Konow)
Raspberry sawfly
A minor pest of cultivated raspberry. Often common on wild *Rubus idaeus*; also associated with *Geum*. Widely distributed in central and western Europe.

DESCRIPTION
Adult: 5.5–7.0 mm long; body black, the abdomen with several white marks on each side. **Larva:** up to 17 mm long; body greyish green, paler below; anal plate with a triangular black patch; head greenish or yellowish brown; eight pairs of abdominal prolegs.

LIFE HISTORY
Adults occur from late April to June and are very active. Eggs are usually deposited in cuts made in leaf stalks or in the edge of young leaves. They hatch in about 2 weeks. The larvae then feed on the underside of the leaves; if disturbed, each rolls into a tight ball and drops to the ground. Larvae mature in 4–6 weeks, usually in July. Fully grown larvae migrate from the host plant to tunnel into cane stumps, pithy stems of nearby plants, soft wood or dead wood. Here, each forms an oval cavity in which to overwinter and, eventually, pupate.

DAMAGE
Larvae bite out elongate holes between the veins of the young foliage, giving the leaves a fern-like appearance. Plant growth is not affected.

Endelomyia aethiops (Fabricius) (**1019**)
Rose slug sawfly
Although associated mainly with *Rosa* (including cultivated bushes), larvae of this generally common

1019 Rose slug sawfly (*Endelomyia aethiops*) – larva.

species are sometimes reported feeding on the foliage of rosaceous fruit trees (at least in mainland Europe). Present throughout Europe; also found in North America.

DESCRIPTION

Adult: 4–5 mm long; black, with smoky wings; legs mainly black, with the tibiae (and knees of the forelegs and mid-legs) yellowish white. **Larva:** up to 15 mm long; body translucent ochreous yellow, with the greenish gut contents clearly visible; head yellowish brown.

LIFE HISTORY

Adult females of this mainly parthenogenetic species occur from May to June, depositing eggs on the underside of rose leaves in slits cut close to the edges. The eggs hatch in 1–2 weeks and the larvae then feed on the upper surface; they graze away the tissue, but leave the lower surface more or less intact. Larvae are fully fed in 3–4 weeks; they then enter the soil, usually in late June or early July, to construct silken cocoons in which to overwinter. Individuals pupate in the following spring, shortly before the emergence of adults; however, sometimes development may be delayed for a further year.

DAMAGE

Owing to the loss of tissue (i.e. the upper epidermis and mesophyll), infested leaves become extensively blanched and may eventually shrivel up.

Hoplocampa brevis (Klug)

Pear sawfly

A destructive pest of pear. Eurasiatic. Widespread in central Europe.

DESCRIPTION

Adult: 5.0–5.5 mm long; head and thorax black, the former with light brown areas dorsally; abdomen black dorsally, brownish yellow ventrally; legs brownish yellow; wings more or less transparent, with yellow veins. **Egg:** 0.7 × 0.3 mm; white and translucent. **Larva:** up to 10 mm long; body light green; head brown; seven pairs of abdominal prolegs; early instars with a whitish body and brown plates on the last two or three abdominal segments.

LIFE HISTORY

This species is single brooded. Adults occur in April, and are much attracted to pear blossom. Eggs are laid singly in the receptacle of young fruitlets, being inserted just below the sepals. They develop within prominent, blister-covered cavities and hatch within a week. The young larvae then burrow into the flesh of the developing fruitlets. Larval growth is rapid, individuals passing through five instars and becoming fully fed by late May or early June. Larvae may attack several fruitlets during their development. They then enter the soil and form oval, dark brown cocoons about 6 mm long, at a depth of from 5 to 20 cm. The larvae overwinter in their cocoons, pupating in the spring. Males are rare and reproduction is mainly parthenogenetic.

DAMAGE

Attacked areas of a fruitlet turn black and masses of wet, black frass is exuded from a small, black-rimmed hole at the entrance to the feeding burrow. All fruitlets in a truss may be destroyed by just one larva.

Hoplocampa flava (Linnaeus) (**1020–1024**)

Plum sawfly

A locally important pest of damson and plum. Eurasiatic. Widely distributed in Europe.

1020 Plum sawfly (*Hoplocampa flava*).

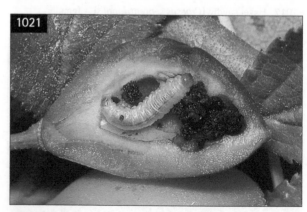

1021 Plum sawfly (*Hoplocampa flava*) – larva.

DESCRIPTION

Adult: 3.5–5.5 mm long; head and thorax dull brownish, the latter with some black markings; abdomen yellowish and shiny; antennae and legs yellowish; forewings basally yellowish, with yellow veins; hindwings transparent. **Egg:** 0.6 × 0.3 mm; elongate when laid, but soon swelling; whitish. **Larva:** up to 11 mm long; body creamy white, with (except in

final instar) yellowish-brown plates on the last two or three body segments; unlike that of *Hoplocampa minuta* the body does not narrow posteriorly; head yellowish brown to pale orange, the mandibles broader and less deeply incised than in *Hoplocampa minuta;* seven pairs of abdominal prolegs.

1022 Plum sawfly (*Hoplocampa flava*) – pupal cocoon.

LIFE HISTORY

Adults appear in April. They are active in sunny weather and much attracted to plum blossom. Males are common and mating occurs soon after emergence. Eggs are then laid singly through elongate, jagged slits made in the receptacle of young fruitlets, usually one per fruit. Occasionally, eggs are deposited directly within the flowers. After egg hatch, the larvae bore into the developing fruits, where they feed for about 9 days. After moulting, each larva then invades another fruitlet, feeding within the flesh for about a week. Later-instar larvae continue to migrate from fruitlet to fruitlet, tunnelling through the flesh to feed on the stones. Larvae are fully fed in 4–5 weeks. Each then drops to the ground and constructs a brown, oval cocoon (c. 6 × 3 mm) in the soil, usually at a depth of 5–25 cm. Larvae overwinter in their cocoons and either pupate in the following spring (adults emerging about 3 weeks later) or after a second winter. Adults live for about 2 weeks.

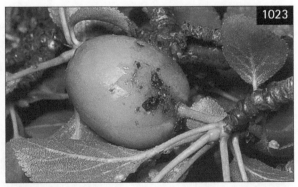

1023 Plum sawfly (*Hoplocampa flava*) – external damage to plum fruitlet.

DAMAGE

The severity of attacks in an orchard may vary considerably from year to year. In extreme cases, 90% or more of a crop may be destroyed, yet this may be followed by an exceptionally light attack in the following season. Infested fruitlets drop prematurely, usually after the larva has escaped. Attacked fruitlets are recognized by the obvious entry hole, and by the associated exudation of gum and masses of wet, black frass. Several, if not all, fruitlets in a cluster may be damaged; some may receive only surface bites and will usually mature more or less normally. Some cultivars (e.g. cvs Czar and Victoria) are more susceptible to attack than others. Earlier-flowering cultivars and later-flowering ones tend to be attacked only lightly, probably because they flower outside the main flight period of the comparatively short-lived adults.

Hoplocampa minuta (Christ)

Black plum sawfly

A pest of damson, myrobalan and plum; other hosts include *Prunus spinosa* and, rarely, apricot. Eurasiatic. Widely distributed in temperate parts of mainland Europe.

DESCRIPTION

Adult: 4–5 mm long; mainly black, with light brown legs. **Larva:** up to 11 mm long; body whitish yellow and tapered posteriorly; head brownish orange, the

1024 Plum sawfly (*Hoplocampa flava*) – internal damage to plum fruitlet.

mandibles relatively long and more deeply incised than in *Hoplocampa flava*; seven pairs of abdominal prolegs.

LIFE HISTORY & DAMAGE
As for plum sawfly (*Hoplocampa flava*).

Hoplocampa testudinea (Klug) **(1025–1030)**
Apple sawfly
An often serious pest of apple. Widely distributed in Europe; also introduced into North America.

DESCRIPTION
Adult: 6–7 mm long; body mainly orange, with (apart from the tegulae and the apical abdominal segments) the thorax and abdomen mainly shiny black above; head with a conspicuous black central patch; wings more or less transparent, with dark brown veins. **Egg:** 0.8 mm long; white, elongate and slightly curved. **Larva:** up to 12 mm long; body yellowish white; head yellowish brown; thorax noticeably stouter than abdomen, the latter tapering apically; seven pairs of abdominal prolegs; earlier instars whitish and translucent, with a blackish head, and with blackish dorsal plates on the last three abdominal segments.

LIFE HISTORY
Adults appear in apple orchards in the spring, flying rapidly on sunny days during the blossom period; they are most active during the morning, particularly around mid-day. After mating, eggs are laid singly in the flowers. Firstly, the female penetrates the receptacle with her saw, making a small slit-like cut just below the ring of sepals. A small cavity is then made at the top of the receptacle, within the circle of stamens, and an egg inserted from

1025a, b Apple sawfly (*Hoplocampa testudinea*) – female.

1027 Apple sawfly (*Hoplocampa testudinea*) – young larva.

1026 Apple sawfly (*Hoplocampa testudinea*) – egg-laying puncture in receptacle of fruitlet.

1028 Apple sawfly (*Hoplocampa testudinea*) – final-instar larva.

below. Occasionally, however, eggs are laid directly into the flowers from above. Eggs swell soon after deposition, hatching in 1–2 weeks; development may take longer in cold weather. Each young larva burrows into the receptacle by tunnelling between two adjacent sepals; occasionally, however, a fruitlet may be entered directly through the eye. In most cases, the larva bores down to the ovary and feeds on one or more of the seeds (pips), thereby preventing further development of the fruit. If the larva fails to reach the ovary it usually dies; the fruit will then continue to grow and mature. After feeding for about 2 weeks, the larva (now in its third instar) vacates the original fruitlet and attacks another, usually entering the basal half; these secondarily invaded fruits frequently contain two or more larvae. Larvae again tunnel deeply into the flesh, causing considerable internal damage. Further migration to other fruitlets may occur, either before or after infested ones have dropped to the ground. Most larvae are fully fed by late June or early July. They then enter the soil and spin soil-coloured, parchment-like cocoons (c. 8 × 4 mm) at depths down to 25 cm or more. They overwinter in the cocoons and pupate in the following spring, adults emerging 3–4 weeks later. A proportion of larvae, however, may spend a second or even a third winter in the ground.

DAMAGE
The egg-laying slit in the side of a receptacle turns brown soon after it is made, and this is a readily visible early sign of an infestation; the eggs may be found in the flowers if the stamens are removed. The larvae, which give off an unpleasant odour, bite out large cavities inside the fruitlets and produce masses of wet, black frass which is exuded through a hole, about 1.5 mm across, in the side of each infested fruit. Fruitlets infested by first- and second-instar larvae remain small, and are malformed, dark in colour and, owing to their arrested development, noticeably pubescent. They usually drop before or soon after the larvae have deserted them for larger fruitlets, but some may remain shrivelled on the trees. Small, infested fruitlets may also bear elongated, open scars which radiate out from the calyx end. These scars are formed by the breakdown of the epidermis over tunnels made just beneath the skin by the young larvae before they penetrated deeply into the flesh; if the larva dies without reaching the ovary, these scars become larger and corky as the apple continues to develop; their characteristic, ribbon-like shape at once distinguishes them from corky scars caused by other pests such as capsid bugs and moth caterpillars. These scars are not formed on secondarily infested fruitlets. Fruit losses can be extremely severe, both in commercial orchards and in gardens. Culinary apples generally suffer only light attacks.

1029a, b Apple sawfly (*Hoplocampa testudinea*) – damage to apple fruitlets.

1030 Apple sawfly (*Hoplocampa testudinea*) – corky scar on mature apple.

Metallus albipes (Cameron)
A raspberry leaf-mining sawfly
This apparently parthenogenetic species is cited as attacking blackberry or raspberry in some parts of Europe (e.g. England, the Netherlands and Scandinavia), but its true pest status is uncertain owing to confusion with *Metallus pumilus*. Adult females (2.5–4.0 mm long) are black and shiny, with whitish legs and (compared with *M. pumilus*), relatively narrow antennae.

Metallus pumilus (Klug) (**1031–1035**)
A raspberry leaf-mining sawfly

A common, but usually minor, pest of blackberry, loganberry and raspberry. Present throughout Europe.

DESCRIPTION

Adult: 3.5–4.5 mm long; body shiny black; wings somewhat smoky; each hind leg yellow, with the femur darker basally; antennae moderately long and noticeably thickened, especially in the male (cf. *Metallus albipes*). **Larva:** up to 10 mm long; body dirty brownish white and dorsoventrally flattened, the thoracic area broad and the abdomen tapered; thorax (dorsally) with a large prothoracic plate, and with small, narrow plates on the front margin of both the mesothoracic and metathoracic segments; a large ventral plate on the prothorax and small ventral plates on the mesothorax, the metathorax and abdominal segment 1 – these plates are lost on moulting to the final instar; abdominal prolegs poorly developed; head light brown and wedge-shaped. **Pupa:** 4 mm long; creamy white. **Leaf mine:** an irregular brown blotch on the upper surface.

LIFE HISTORY

Adults appear in late spring and early summer, with a second flight in late summer. Eggs are laid singly on leaves of cultivated and wild *Rubus*. Larvae of the first brood feed within the leaves from July onwards, each forming a large, irregular, blotch-like mine. Larvae of the second brood may be found feeding in the autumn.

1033 A raspberry leaf-mining sawfly (*Metallus pumilis*) – larva, showing ventral plates.

1031 A raspberry leaf-mining sawfly (*Metallus pumilus*) – female.

1034 A raspberry leaf-mining sawfly (*Metallus pumilus*) – pupa.

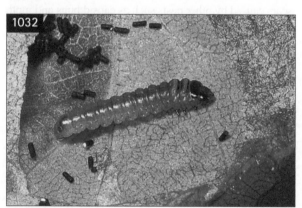

1032 A raspberry leaf-mining sawfly (*Metallus* sp.) – larva, showing dorsal plates.

1035 A raspberry leaf-mining sawfly (*Metallus pumilus*) – mine.

Fully fed larvae overwinter in the soil, and pupate in the spring shortly before the emergence of the adults. Unlike certain other leaf-mining insects (e.g. larvae of the moth *Tischeria marginea* – see Chapter 6), much of the larval frass is expelled at intervals through a small slit in the wall of the feeding gallery.

DAMAGE

A single mine may occupy more than half the total leaf area, and heavy infestations are capable of reducing crop yields. Potentially more harmful to loganberry and raspberry than to blackberry.

Micronematus abbreviatus (Hartig) (**1036–1037**)
Pear leaf sawfly

A locally important pest of pear in various parts of Europe, including Germany, the Netherlands, Scandinavia and Switzerland; present in southern England, but rarely noted on cultivated trees. Also recorded from North America.

DESCRIPTION

Adult: 4–5 mm long; mainly black and shiny, with a partial greyish pubescence; wings transparent and iridescent, with the costa and stigma brownish black. **Larva:** up to 13 mm long; body greyish green, paler below; head light brownish green; six pairs of abdominal prolegs (none on anal segment).

LIFE HISTORY

Adults occur in April and May, eggs then being laid singly in the midribs of expanded leaves. Larvae feed on the leaves of pear in May or early June. At first, the larvae form distinct rounded holes in the laminae; individuals later feed more indiscriminately, often whilst stretched out along the leaf edges. When fully grown, larvae enter the soil to form cocoons, each measuring about 6 × 3 mm, in which pupation eventually takes place. There is a single generation annually.

DAMAGE

When numerous, larvae can cause considerable defoliation.

Monophadnoides geniculatus (Hartig) (**1038**)
Geum sawfly

This generally common species is associated with *Filipendula*, *Geum* and *Rubus*, including cultivated

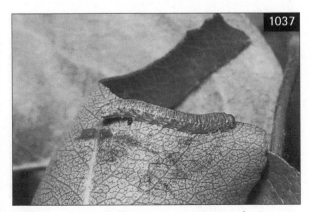

1036 Pear leaf sawfly (*Micronematus abbreviatus*) – female.

1037 Pear leaf sawfly (*Micronematus abbreviatus*) – larva.

1038a, b Geum sawfly (*Monophadnoides geniculatus*) – larva.

raspberry. Reports of larvae infesting strawberry are due, at least in part, to confusion with the closely related strawberry leaf sawfly (*Claremontia confusa*). Eurasiatic. Widely distributed in Europe.

DESCRIPTION
Adult: 5–6 mm long; stout-bodied and mainly black.
Larva: up to 14 mm long; body green to dark green, with numerous whitish, mainly 1- or 2-pronged, thorn-like spines (some spines 3-pronged); head greenish yellow; eight pairs of abdominal prolegs.

LIFE HISTORY
Adults occur in May and early June, females depositing eggs in the underside of leaves of host plants. The larvae feed from late May to mid-July. They then enter the soil and spin cocoons, individuals eventually pupating and adults emerging in the following spring. There is just one generation annually.

DAMAGE
Larvae form large, irregular holes in the foliage, leaving a skeleton of major veins, but such damage is usually of little significance.

Nematus leucotrochus Hartig (**1039–1040**)
Pale-spotted gooseberry sawfly
A minor pest of gooseberry; also, occasionally, found on red currant and white currant. Widespread in central and northern Europe.

DESCRIPTION
Adult: 6–7 mm long; head and thorax mainly black; abdomen mainly yellowish orange, but basally partly black; legs yellowish; antennae blackish above and below; wings transparent and iridescent, with brown veins. **Egg:** 1.25 mm long; pale yellowish white. **Larva:** up to 15 mm long; body mainly green, with the first and the last two segments yellowish; pinacula small, black and seta-bearing (unlike those of *Nematus olfaciens* they do not coalesce); head green, feintly speckled with black; seven pairs of abdominal prolegs. **Prepupa:** body mainly light green, with yellowish markings; head greenish brown.

LIFE HISTORY
Unlike black currant sawfly (*Nematus olfaciens*) and common gooseberry sawfly (*N. ribesii*), this species is single brooded. Adults appear in early May and live for about 2 weeks. After mating, eggs are laid singly on either leaf surface, a female depositing 30 to 50 eggs during her lifetime. Occasionally, there may be two or three eggs on one leaf. The eggs hatch in 10–14 days. Larvae occur in May and June, becoming fully grown after 2 or 3 weeks. They pass through four (if male) or five (if female) instars, and then enter a non-feeding, migratory, prepupal stage. The prepupae enter the soil and spin brown, parchment-like cocoons (8–9 mm long) a few centimetres beneath the surface. Individuals overwinter and then pupate in the spring, shortly before adults emerge.

DAMAGE
The youngest larvae bite holes through the leaf lamina; after a few days, these holes reach the leaf margin and subsequent feeding continues along the edges. Larvae are usually present in only small numbers and damage is rarely, if ever, significant.

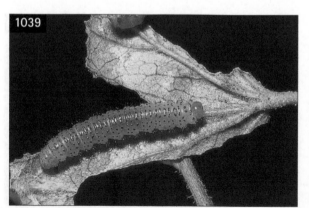

1039 Pale-spotted gooseberry sawfly (*Nematus leucotrochus*) – larva.

1040 Pale-spotted gooseberry sawfly (*Nematus leucotrochus*) – prepupa.

Nematus lucidus (Panzer) (**1041–1042**)

A minor pest of damson and plum, but most frequently associated with wild *Prunus*, particularly *P. spinosa; Crataegus* is also a host. Widely distributed in Europe.

DESCRIPTION

Adult: 8–11 mm long; body cigar-shaped and mainly black above, with a red band on the abdomen. **Larva:** up to 20 mm long; body translucent green, with numerous short black setae; setae most numerous laterally in clusters above the legs; six pairs of abdominal prolegs; a prominent pair of light brown, spatulate processes at hind end; head brownish orange and shiny, with small blackish setae.

LIFE HISTORY

Adults occur in the spring from April onwards, and eggs are eventually laid in the leaves of host plants. Larvae feed along the edge of expanded leaves from May to June. When fully grown they enter the soil and spin cocoons in which to overwinter. There is just one generation annually.

DAMAGE

Larvae are capable of causing extensive defoliation, but numbers on cultivated *Prunus* are usually small.

Nematus olfaciens Benson (**1043–1045**)

Black currant sawfly

A pest of black currant; red currant and gooseberry are also attacked. First recognized as a distinct species in 1952,

1043 Black currant sawfly (*Nematus olfaciens*).

1041 *Nematus lucidus* – young larva.

1044 Black currant sawfly (*Nematus olfaciens*) – larva.

1042 *Nematus lucidus* – late-instar larva.

1045 Black currant sawfly (*Nematus olfaciens*) – prepupa.

having been mistaken previously for common gooseberry sawfly (*Nematus ribesii*) which does not attack black currant. Probably widely distributed in Europe.

DESCRIPTION

Adult: 5.0–6.5 mm long; head and thorax mainly black; abdomen mainly yellowish orange, but basally partly black; antennae blackish; legs yellow; wings transparent, with brown veins. **Larva:** up to 20 mm long; body green, with the first and the last two segments partly yellow; except on the last segment, setae arise from small black pinacula, some of which coalesce to form distinct blotches; head green, speckled with black; seven pairs of abdominal prolegs. **Prepupa:** intermediate in colour between that of *Nematus leucotrochus* and *N. ribesii*.

LIFE HISTORY

Adults appear in late April and May. Eggs are then laid on the leaves of black currant and, occasionally, other species of *Ribes*. The larvae feed in May and June, passing through four (= male) or five (= female) instars. The active prepupal stage then deserts the bush and spins a cocoon in the soil, in which to pupate. The next generation of adults emerges from late June onwards and the second brood of larvae feeds in July and August. Occasionally, there may be a third generation. Prepupae of the final brood overwinter in their cocoons and pupate in the spring.

DAMAGE

Larvae defoliate the bushes and damage can be very serious. In addition, the larvae are often included, accidentally, as unwanted contaminants in consignments of harvested fruit.

Nematus ribesii (Scopoli) **(1046–1049)**

Common gooseberry sawfly

A common and often serious pest of gooseberry. Red currant and white currant (but not black currant) are also attacked; other hosts include jostaberry and Worcesterberry. Present throughout Europe; an introduced pest in North America and in many other temperate parts of the world.

DESCRIPTION

Adult female: 6–7 mm long; head black; body yellow, with black marks on the thorax; antennae dark above, pale below; wings transparent, with brown veins. **Adult male:** 5–6 mm long; head and thorax mainly black; abdomen yellow, but mainly black above; antennae dark above and below. **Egg:** 1.2 mm long; pale greenish white. **Larva:** up to 20 mm long; body green, with the first, part of the second and the last two segments

orange; pinacula large and shiny black, each bearing one or more setae; anal segment with black setae and a prominent black patch; anal cerci black; thoracic legs black; head shiny black; seven pairs of abdominal

1046 Common gooseberry sawfly (*Nematus ribesii*).

1047 Common gooseberry sawfly (*Nematus ribesii*) – eggs.

1048 Common gooseberry sawfly (*Nematus ribesii*) – larva.

prolegs. **Prepupa:** body pale bluish green, but tinged with yellowish orange on the first two and last two segments; head pale.

LIFE HISTORY
Adults first appear in April and May. Eggs are laid in rows on the underside of leaves, each held in an incision made in a major vein. The eggs hatch 8–10 days later. Larvae then feed gregariously, developing rapidly over a period of about 3 weeks. In most cases, eggs are laid in leaves near the ground and in those at the centre of bushes. Larvae initially feed close to the oviposition sites; however, when about half grown, they disperse and feed on foliage throughout the bushes. Individuals pass through four (= male) or five (= female) instars and then moult to the active, but non-feeding, prepupal stage. Prepupae burrow into the soil to a depth of 10–15 cm and spin dark brown, oval, parchment-like cocoons about 8–10 mm long. They pupate after about a week and adults emerge a further week later. Larvae occur at any time from May to October, usually in three distinct broods: e.g. in May and June, in July, and in August and September. Prepupae of the autumn generation overwinter within their cocoons and pupate in the spring.

DAMAGE
Young larvae, although feeding gregariously, at first make small, individual holes in the leaves. Such leaves have a shot-holed appearance, but soon complete leaf laminae (with the exception of the main veins) are devoured. Early infestations occur mainly low down in the centre of bushes. The larvae feed ravenously and infested bushes are rapidly defoliated. Attacks tend to be most severe in gardens and on two- or three-year-old bushes.

Pachynematus pumilio (Konow) (**1050**)
Black currant fruit sawfly
A pest of black currant in Finland, Sweden and the former USSR. In recent years this species has extended its range both southwards and westwards; it is now established, for example, in Poland.

DESCRIPTION
Adult: 4–5 mm long; mainly black. **Larva:** up to 6 mm long; body creamy white; head yellowish brown.

LIFE HISTORY
Eggs are laid singly in the developing fruit. Larvae then feed within the flesh, forming a central cavity within which brownish frass accumulates. In late June or early July the fully grown larva vacates the fruit through a small, rounded hole. It then enters the soil to overwinter and, eventually, pupate. Adults emerge in the spring.

DAMAGE
Infested fruits ripen prematurely and are unmarketable; heavy infestations can result in considerable losses of crop.

1049 Common gooseberry sawfly (*Nematus ribesii*) – prepupa.

1050 Black currant fruit sawfly (*Pachynematus pumilio*) – infested black currant fruit.

Priophorus morio (Lepeletier) (**1051–1054**)
Small raspberry sawfly

A minor, locally common pest of raspberry, particularly in gardens and allotments. Larvae also feed on other species of *Rubus* (including cultivated blackberry and loganberry) and on *Sorbus aucuparia*. Widely distributed in Europe; an introduced pest in New Zealand.

1051 Small raspberry sawfly (*Priophorus morio*) – female.

1052 Small raspberry sawfly (*Priophorus morio*) – young larva.

1053 Small raspberry sawfly (*Priophorus morio*) – final-instar larva.

DESCRIPTION

Adult: 4.5–7.0 mm long: body more or less black, with the centre of abdominal segment 1 lighter; legs, except for bases, mainly pale; antennae tapering noticeably towards the apex; saw sheath, viewed from above, with the tip expanded and terminating in a blunt point.
Larva: up to 12 mm long; body brownish green, greyish green or blackish dorsally, whitish ventrally, with single hairs arising from small, pale pinacula; head black or brown and shiny; seven pairs of abdominal prolegs.

LIFE HISTORY

Adults are active from May to August, there being two or more generations in a season. Males are rare, development usually being parthenogenetic. Eggs are laid on the underside of leaves. Larvae at first graze on the lower leaf surface, usually completing their development on a single leaf, where the cast-off skins of earlier instars may be found. Pupation takes place in a semitransparent, light brown, double-walled cocoon spun on the leaves or amongst debris on the ground.

DAMAGE

Infested leaves have elongate holes bitten through them; however, damage, although obvious, is rarely important.

Priophorus pallipes (Lepeletier)
Plum leaf sawfly

A generally common, but usually unimportant, pest of pear and plum. Wild hosts include *Crataegus monogyna*, *Prunus spinosa* and *Sorbus*. Widely distributed in Europe.

DESCRIPTION

Adult: 5–8 mm long; body black or dark brown; legs usually white, but variable; saw sheath, viewed from above, with the bluntly rounded tip almost parallel-sided.
Larva: up to 10 mm long; body green or grey, paler on the sides and below, with single hairs arising from pinacula;

1054 Small raspberry sawfly (*Priophorus morio*) – early damage to loganberry leaf.

head hairy and usually orange, with a distinct black mark dorsally; seven pairs of abdominal prolegs.

LIFE HISTORY

Adults fly in May and June, and throughout the summer, laying eggs singly on the underside of the leaves of pear, plum and various other rosaceous plants. Larvae feed and shelter on the underside of the leaves, and readily drop to the ground if disturbed. Development from egg to adult takes about 5 weeks, and there are two or more broods annually. Larvae from the final generation overwinter in dark brown cocoons in the soil and pupate in the spring. Unlike the previous species, males are common.

DAMAGE

The youngest larvae graze on the lower surface of leaves; later, larger and larger holes are made in the foliage. Attacks are most evident in September and October, when the autumn brood of larvae is active. On occasions, considerable leaf-holing is caused on both pear and plum, but such damage is usually of little consequence.

Pristiphora maesta (Zaddach)
Apple leaf sawfly

This locally distributed sawfly occurs on apple, but is rarely reported as a pest. Although widely distributed in central and northern Europe, it is usually rare; in many countries it is considered an endangered species.

DESCRIPTION

Adult: 5.5–7.0 mm long; body mainly black, with pronotum, tegulae and parts of some abdominal segments (ventrally and laterally) pale; legs mainly yellowish brown. **Larva:** up to 10 mm long; body yellowish green, with a row of shiny black tubercles along each side; head yellowish orange; seven pairs of abdominal prolegs; early instars with a black head.

LIFE HISTORY

Adults occur in April and May, and eggs are deposited in rows in the expanded leaves of host plants, each inserted into tissue at the leaf margin. The eggs hatch from early May onwards. Larvae then feed gregariously around the leaf margin. In repose, the larvae often adopt an S-shaped posture, with the tip of the abdomen directed away from the leaf. Individuals are fully grown by mid-July. They then enter the soil and overwinter, each in a silken cocoon. Pupation occurs in the spring, and adults emerge about 2 weeks later.

DAMAGE

Larvae feed ravenously and cause noticeable defoliation. Heavily infested shoots and branches are soon stripped of foliage; this weakens hosts and reduces cropping potential.

Pristiphora rufipes (Lepeletier) (**1055–1057**)
Small gooseberry sawfly

A relatively common pest of gooseberry and red currant; also reported on black currant. Present throughout the temperate parts of the northern hemisphere.

1055 Small gooseberry sawfly (*Pristiphora rufipes*).

1056 Small gooseberry sawfly (*Pristiphora rufipes*) – final-instar larva.

1057 Small gooseberry sawfly (*Pristiphora rufipes*) – early-instar larva.

DESCRIPTION

Adult: 4.5–5.5 mm long; body mainly black, with a broad abdomen; legs yellowish white; wings iridescent. **Egg:** 1.1 × 0.4 mm; elliptical, whitish and translucent. **Larva:** up to 10 mm long; body light green or yellowish green, relatively long and slender, and distinctly wrinkled above; head greenish brown, marked with blackish brown; seven pairs of abdominal prolegs.

LIFE HISTORY

Adults occur from April onwards, with up to four generations in a season. Eggs are laid mainly in the edges of leaves, usually one per leaf. They hatch after about 10 days and the larvae then feed whilst stretched out along the leaf edge, with the tip of the abdomen curved slightly downwards. The larvae feed for 2–3 weeks and then pupate in silken cocoons spun in folded leaves or in other shelter on host bushes, or amongst debris on the ground. Adults emerge 2–3 weeks later. Larvae of the autumn generation enter the soil and spin their cocoons 5–10 cm beneath the surface. Here they overwinter, eventually pupating in the spring. Unlike other gooseberry-feeding sawflies (i.e. those in the genus *Nematus*) this species is mainly parthenogenetic, males being very rare; also, there is no active prepupal stage in the life cycle.

DAMAGE

Larvae frequently occur on bushes infested by common gooseberry sawfly (*Nematus ribesii*), but are usually present in only small numbers. When numerous, the larvae can be destructive by devouring significant amounts of foliage, but serious attacks are uncommon. Gooseberry cultivars that have very hairy leaves tend not to be attacked, as egg-laying upon them is difficult; however, smooth-leaved cultivars (e.g. cv. Careless) are particularly vulnerable.

Family **CYNIPIDAE**
(gall wasps)

Small or minute, mostly shiny-black or brown, winged or wingless, gall-forming insects, with a laterally compressed gaster and, in winged individuals, a characteristic, reduced venation. Most species are associated with *Quercus*, their larvae developing in characteristic galls.

Diastrophus rubi (Bouché) (**1058**)
Rubus gall wasp

In the wild, this species infests various kinds of *Rubus*, including *R. caesius*, *R. fruticosus* and *R. idaeus*; only very rarely is it likely to occur on cultivated blackberry and raspberry, and then most frequently in unkempt rural allotments and gardens. Locally distributed in several parts of Europe, including Austria, Germany, Great Britain and Sweden.

DESCRIPTION

Adult: 2.0–2.5 mm long; mainly back, sometimes marked with red; legs reddish yellow; antennae long and 14-segmented; wings transparent, with brownish or blackish veins. **Larva:** up to 3 mm long; translucent, with a distinct head capsule. **Gall:** an elongate, multilocular, spindle-shaped swelling on the stem, usually several centimetres long.

LIFE HISTORY

Adults occur in the spring and deposit eggs in the stems of host plants. Infested stems develop distinct swellings around the oviposition sites, with larvae feeding and developing in small individual cells (each c. 2.5 mm in diameter) formed in the outer (medullary) sheath – cf. galls formed on *Rubus* by blackberry stem

1058 Rubus gall wasp (*Diastrophus rubi*) – gall on raspberry.

gall midge (*Lasioptera rubi*) (Chapter 5). The galls are at first greenish in colour, but eventually become brownish; typically, the walls surrounding the larval cells tend to be somewhat yellower than the rest of the inner tissue. Larvae are fully grown by the following April or May; they then pupate and adult midges emerge a few weeks later.

DAMAGE
Although galls stunt the growth of infested plants, and could have an adverse effect on both leaf and fruit production, infestations are not of economic importance.

Family **EURYTOMIDAE**
(seed wasps)

Minute insects, with a much reduced wing venation and a strongly developed pronotum; body usually black and, unlike members of many other closely related families, non-metallic; abdomen with a distinct petiole; gaster of female compressed laterally, and usually oval or rounded.

Eurytoma amygdali Enderlein
Almond seed wasp
A North American pest of almond, now established as an introduced species in parts of central, eastern and southern Europe. Other reported hosts include apricot, cherry, plum and various wild species of *Prunus*.

DESCRIPTION
Adult: 2 mm long; mainly black. **Larva:** up to 2.5 mm long; creamy white.

LIFE HISTORY
Adults emerge in the spring. After mating, eggs are laid in developing almonds. Larvae feed during the summer on the developing kernels. Fully grown larvae overwinter within the mummified fruits, and may continue in diapause for up to 3 years. There is one generation annually.

DAMAGE
Attacked fruits, which usually remain attached to the host tree, fail to develop and become mummified, with the epicarp adhering closely to the shell (endocarp). Heavy infestations can lead to considerable yield losses.

Family **TORYMIDAE**
(e.g. seed wasps)

Minute insects, with an elongate, often brilliantly metallic-green, body and a very long ovipositor; hind legs with enlarged coxae; tibiae 5-segmented. Although mainly parasitoids of gall-forming insects, some species are phytophagous.

Torymus varians (Walker)
Apple seed wasp
A minor pest of apple; also associated with *Crataegus*, *Pyrus* and *Sorbus*. Holarctic. Also now established in other parts of the world, including New Zealand and Tasmania. Widely distributed in Europe.

DESCRIPTION
Adult female: 3 mm long; metallic green, with mainly pale, reddish-brown legs; ovipositor long and prominent. **Larva:** up to 4.5 mm long; whitish and translucent.

LIFE HISTORY
Adults occur in May and June. Eggs are inserted into young apple fruitlets, and they hatch about a month later. Larvae then feed on the endosperm of the developing seeds (pips), and usually complete their development by the end of September. Fully grown larvae overwinter inside fallen fruits. They then pupate, and adult wasps emerge about 2 weeks later.

DAMAGE
Egg-laying punctures subsequently result in the appearance of elongated corky scars on the skin of maturing fruits, each located in a small depression. Affected fruits may be discarded or downgraded.

Family **FORMICIDAE**
(ants) (**1059**)

Ants frequently climb fruit trees or bushes to collect honeydew excreted by aphids, psyllids, scale insects and other honeydew-producing insects. Indeed, their presence is often a useful guide to the presence of such pests. Less often, ants will invade apple trees to bite the young foliage and buds; such damage is of little or no importance and is most often reported in dry conditions. Ants will also damage apple blossom by biting off some of the stamens in order to gain access to the nectaries and steal nectar; these attacks have been noted most frequently on cv. Bramley's Seedling and, again, usually take place in periods of dry weather. Workers of the common black ant (*Lasius niger* (Linnaeus)) are often the culprits, but other species are probably also involved.

1059 Ant attacking apple flower bud.

Family **VESPIDAE**
(true wasps)

A family of social and solitary (usually mainly black and yellow) wasps, with deeply notched compound eyes and the pronotum extending back to the tegulae; in addition, unlike other groups of wasps, the wings are folded longitudinally when in repose. Members of the subfamily Vespinae include social species; these form colonies, and have male, queen and worker castes.

Vespa crabro Linnaeus (**1060**)
Hornet
In late summer, hornets are sometimes troublesome in orchards and vineyards, where they may feed on, and cause damage to, ripening, ripe or overripe fruits. In spring, adults are also sometimes damaging to forest trees, as they often remove sections of bark from young shoots; this bark, masticated with saliva, is then used to construct their nests. The life cycle of hornets is similar to that of social wasps (*Vespula* spp.), with just the young, fertilized queens overwintering. Adult hornets are usually readily distinguished from those of *Vespula* by their mainly reddish-brown thorax, larger size (queens up to 35 mm long; workers up to 24 mm long), and by the relative position of the ocelli (all three of which lie anterior to the hind margin of the compound eyes and, owing to the extended vertex, at some distance from the back of the head).

Vespula germanica (Fabricius) (**1061–1064**)
German wasp
An abundant and well-known European insect, the workers often causing damage to ripening, ripe and overripe apples, pears, peaches, plums, grapes and

1061 A social wasp (*Vespula* sp.).

1062 A social wasps (*Vespula* sp.) attacking ripe apple.

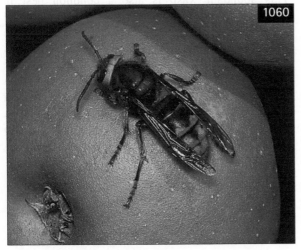

1060 Hornet (*Vespa crabro*).

1063 A social wasp (*Vespula* sp.) attacking ripe plum.

1064 A social wasp (*Vespula* sp.) – abandoned nest.

cells and pupate. Adult worker wasps soon emerge and they then assist the maternal queen in performing various tasks within the colony. The number of brood cells in the colony gradually increases and the outer canopy is also expanded, the whole nest eventually assuming the proportions of a football. In a successful colony, several thousand workers are reared; also, later in the season, males and new queens are produced. Unlike larvae, adult wasps feed mainly on carbohydrates, including nectar and various other sugary substances such as honeydew and fruit juices. They are particularly numerous at such food sources during the summer and early autumn, when adult populations are at their height. Wasp numbers gradually decline in the autumn as each colony, including the original queen, the workers and the males, eventually dies out, leaving only young, mated queens to survive the winter.

DAMAGE

Most problems are caused in late summer and autumn, when large numbers of mainly worker wasps feed voraciously on maturing or overripe fruit. Feeding is concentrated on fruits previously damaged by birds or other agents, and on those with growth splits. However, once an attack is established, neighbouring sound fruit (particularly soft-textured apples, grapes and autumn-fruiting strawberries) may be damaged. Wasps are sometimes attracted into orchards to feed on honeydew excreted by aphids and other pests. They may also occur in vast numbers on aphid-infested windbreak trees; wasp damage to fruit on trees adjacent to such windbreaks is often severe. The presence of wasps in an orchard at harvest-time is also a potential hazard to fruit pickers.

Vespula vulgaris (Linnaeus)

Common wasp

This abundant wasp is also a common pest in European orchards and vineyards; its life cycle and habits are similar to those of *Vespula germanica*. The two species are distinguished by slight, but not always distinct, differences in the black pattern of the abdomen and by reference to the black markings on the head; in *V. vulgaris* the latter typically form a broad, axe-like patch on the clypeus. More reliably, the two species may be distinguished by reference to the colour of the nest canopy which, in *V. vulgaris*, is light brown, rather than grey.

many other fruits. During the spring and summer, however, when they prey upon harmful caterpillars and other pests, their activities are often beneficial to the fruit grower and viticulturist. Widely distributed in Europe. Also now present in various other parts of the world, including South Africa, North America, Australia and New Zealand.

DESCRIPTION

Adult queen: 20 mm long; brightly coloured black and yellow; narrow-waisted, with a pointed abdomen and a sting; wings long, narrow and rather dusky; clypeus typically with three black spots; antennae 12-segmented. **Adult worker:** similar to queen, but smaller. **Adult male:** 15–18 mm long; similar to female (i.e. queen and worker), but abdomen less pointed and without a sting; antennae 13-segmented and noticeably longer than those of workers.

LIFE HISTORY

Young, fertilized queen wasps overwinter in various sheltered situations. They emerge in the spring and having found a suitable nesting site, such as an underground cavity, each begins to build a nest of masticated wood mixed with saliva. The new nest usually consists of four downwardly directed paper cells suspended within an opaque, grey canopy or envelope. An egg is laid in each cell, the resulting larvae being fed mainly on proteinaceous food, such as dead insects. When fully grown, the larvae seal themselves in their

Chapter 8
Mites

Order PROSTIGMATA

Family **PHYTOPTIDAE**
(gall mites)

Minute, elongated mites with two pairs of legs, each leg terminating in a simple (i.e. unbranched) feather-claw; body more or less annulated and narrowly cylindrical (vermiform), with a distinct prodorsal shield usually bearing four setae. Eggs are minute, oval and translucent. Nymphs are usually similar in appearance to adults, but smaller.

Phytoptus avellanae Nalepa (**1065–1066**)
Filbert bud mite
Generally common on *Corylus avellana* and a minor pest of cultivated hazelnut. Widely distributed in Europe; also present in North America, Asia and Australia.

DESCRIPTION
Adult female: 0.22–0.30 mm long; whitish and vermiform; prodorsal shield with a pair of frontal and a pair of dorsal setae; hysterosoma finely cross-striated, with c. 70 narrow tergites and sternites; feather-claws either 4- or 5-rayed. **Protonymph and 'winter' deutonymph:** similar to adult, but smaller. **'Summer' deutonymph:** body flat, the hysterosoma with few tergites; each tergite very broad and with lateral projections.

LIFE HISTORY
During the winter months, all stages of the mite (adult females, eggs, protonymphs and deutonymphs, other than 'summer' deutonymphs) may be found within enlarged buds ('big buds'). The mites also occur in the female flowers and male catkins. In March and April, adult females disperse to the underside of leaves, where eggs are laid. These eggs hatch into active protonymphs that feed and eventually moult into more or less sedentary 'summer' deutonymphs, normally located alongside the leaf veins. By July, these unusual 'summer' deutonymphs (see discription above) moult into 'normal' adults, which then invade new terminal buds. Infested buds soon begin to swell, the mite-induced 'big buds' becoming very noticeable from September onwards. Breeding continues within the buds throughout the autumn and winter.

DAMAGE
Infested buds swell and fail to open. They are fleshy and contain warted inner cells and thick masses of hair. Attacked female flowers may die and invaded male catkins become misshapen and brittle, producing little pollen. Infestations cause blind shoots, but have little or no effect on yield.

1065 Filbert bud mite (*Phytoptus avellanae*) – galled buds ('big buds').

1066 Filbert bud mite (*Phytoptus avellanae*) – old galled bud ('big bud').

Family **ERIOPHYIDAE**
(gall mites and rust mites)

Members of the family Eriophyidae are similar to those of the family Phytoptidae, but the dorsal shield bears two or no setae; the feather-claws are either simple or divided, in the latter case typically 4- or 5-rayed. The annulated body may be narrowly cylindrical (vermiform), broadly elongate (fusiform) or pear-shaped (pyriform), in the last-mentioned case narrowing gradually towards the hind end.

Several eriophyid species are deuterogenous. These have two adult female forms, protogynes and deutogynes, which differ both structurally and physiologically. The female protogyne and the male are the primary (summer) forms; they breed normally. Deutogynes, for which there is no equivalent male stage, overwinter and deposit eggs only after a period of winter cold.

Acalitus essigi (Hassan) (**1067**)
Blackberry mite
Generally common on wild *Rubus fruticosus* and sometimes a pest of cultivated blackberry bushes; also occurs on other kinds of *Rubus*. Widely distributed in Europe; also present in Australia, New Zealand and the USA.

DESCRIPTION
Adult female: 0.12–0.18 mm long; body whitish and vermiform; prodorsal shield with a pair of long, posteriorly directed setae arising from tubercles located on the hind margin; feather-claws 4-rayed.

LIFE HISTORY
Mites overwinter in small numbers beneath bud scales and within mummified fruits still surviving on the canes from the previous summer. They reappear in the early spring, and then invade the new growth to live and breed amongst hairs on the petioles and underside of the leaves. Numbers of the mite increase rapidly as temperatures rise, there being a series of overlapping generations throughout the season. At blossom time, mites enter the flowers and then feed on the developing drupelets, especially those sheltered by the calyx at the base of the fruits. Mite populations diminish considerably from autumn onwards, but there is little apparent dispersal to the buds, natural mortality at this time being very high.

DAMAGE
Mites do no harm to the foliage or flowers, but uneven ripening of fruits occurs following the injection of toxic saliva into developing drupelets. Attacked drupelets remain red, or greenish to reddish, and hard. These unripened drupelets most frequently occur at the base of the fruits, with only a proportion of drupelets on infested fruits attaining the normal blackish coloration at maturity. The condition is commonly known as 'red-berry' disease. The incidence of fruit damage in a plantation tends to increase throughout the picking period and late-maturing fruits tend to be most severely affected.

Acalitus phloeocoptes (Nalepa)
Plum spur mite
This species is sometimes found in Europe on cultivated plum trees, but is not an important pest; infestations also occur on almond. Eurasiatic, but of local occurrence. Also present in the USA.

DESCRIPTION
Adult female: 0.13–0.15 mm long; body whitish and vermiform; prodorsal shield smooth, with a pair of long, posteriorly directed setae arising from tubercles located on the hind margin; feather-claws 5-rayed.

LIFE HISTORY
In spring, adult females invade new, unopened buds to feed beneath the basal scales. Here, swellings are produced which become obvious from May onwards. Females then begin to lay eggs in small cavities within this galligenous tissue; each chamber may become packed with several thousand mites and masses of eggs. Breeding continues throughout the summer, development from egg to adult taking 3–4 weeks, depending on temperature. The galls gradually harden as they age and the surviving female mites remain trapped within them throughout the winter, eventually escaping in the spring as the dead, dry outer tissue finally cracks open.

1067 Blackberry mite (*Acalitus essigi*) – damage to fruit.

DAMAGE

Galls formed on the spurs are brownish, more or less rounded and about 1–2 mm in diameter and are usually clustered in groups of up to 30 around the base of the spurs; later, they become wart-like. Each gall consists of several separate chambers (formed within a fleshy mass of often purplish or reddish cells) surrounded by a protective outer wall that eventually becomes dry and corky. Infestations on plum are relatively harmless, although heavily infested spurs may become distorted; adverse effects on almond tend to be more persistent.

Aceria erineus (Nalepa) (**1068–1070**)

Walnut leaf erineum mite

Very common on walnut, but of little or no importance. Eurasiatic. Also present in North America, Australia and New Zealand.

DESCRIPTION

Protogyne (summer female): 0.20–0.24 mm long; whitish and vermiform; prodorsal shield with a pair of moderately long, posteriorly directed setae arising from tubercles located on the hind margin; hysterosoma with numerous narrow tergites and sternites; feather-claws 3-rayed. **Gall:** a large, conspicuous, often reddish-tinged, blister-like swelling (up to 15 mm long, or even more) on the upper surface of an expanded leaf; the gall contains a felt-like mass of whitish hairs (forming an erineum), visible from below.

LIFE HISTORY

This species is deuterogenous. Deutogynes hibernate under bud scales. In spring, they migrate to the underside of leaves to feed and lay eggs amongst the leaf hairs. Here, conspicuous galls are produced within which all stages (eggs, nymphs, female protogynes and males) occur together. There are several overlapping generations of protogynes and males throughout the summer, new foliage being invaded as and when it is produced. Towards the end of the season, deutogynes are produced. These young deutogynes do not breed; instead, they disperse to the buds, where they eventually overwinter.

DAMAGE

There are often several galls on an infested leaf, but they appear to have no effect on tree growth or on cropping, even when numerous. Heavily infested nursery plants, however, are unsightly and their growth may be affected adversely.

1068 Walnut leaf erineum mite (*Aceria erineus*) – galls, viewed from above.

1069 Walnut leaf erineum mite (*Aceria erineus*) – galls, viewed from below.

1070 Walnut leaf erineum mite (*Aceria erineus*) – infested leaf showing necrosis.

Aceria ficus (Cotte) (**1071–1072**)
Fig bud mite

A widely distributed pest of fig, occurring wherever the host plant is grown. Generally common in Europe. Also present in South Africa, North America, India and Japan.

DESCRIPTION
Adult female: 0.16–0.17 mm long; body brownish white and vermiform; prodorsal shield with a pair of long, posteriorly directed, divergent setae arising from tubercles located on the hind margin; hysterosoma with a large number of very narrow tergites and sternites; feather-claws 5-rayed.

LIFE HISTORY
Mites overwinter in the shelter of buds. Activity is resumed in the spring, the mites then invading the leaves and stems. Eggs are laid mainly in the shelter of the hairs on the underside of young leaves. Development from egg to adult is accomplished in about a week, and there are many overlapping generations annually. Mites tend to be particularly numerous on well-fertilized, non-viruliferous plants.

DAMAGE
The mites have an adverse effect on plant growth and cause a wide range of symptoms, including distortion of buds, chlorosis, bronzing or flecking of foliage and fruits; especially severe infestations may result in defoliation. The mites are also vectors of fig mosaic virus.

Aceria oleae (Nalepa) (**1073**)
Olive bud mite

A pest of olive in the Mediterranean basin, particularly in warmer regions.

DESCRIPTION
Adult female: 0.1 mm long; body brownish white and vermiform; prodorsal shield with a pair of long, posteriorly directed, divergent setae arising from tubercles located on the hind margin; hysterosoma with a large number of very narrow tergites and sternites; feather-claws 4-rayed.

LIFE HISTORY
Mites are active from spring to autumn, and occur on the buds and underside of leaves. They also congregate at the base of developing fruits, often sheltering in large numbers beneath the remnants of the sepals. There are several generations annually.

DAMAGE
Young olive trees are especially susceptible, the mites causing discoloration, deformation and stunting of foliage and fruits, and reductions in yield; in severe cases, shoots may be killed. Death of hairs on leaves results in the appearance of green patches; these later become necrotic and turn brown. When fruits are damaged, silvery areas are produced, upon which minute brown droplets of exuded gum accumulate.

1071 Fig bud mite (*Aceria ficus*) – mites swarming over young leaf.

1072 Fig bud mite (*Aceria ficus*) – leaf damage.

1073 Olive bud mite (*Aceria oleae*) – leaf damage.

Aceria sheldoni (Ewing)
Citrus bud mite

This species occurs only on citrus, particularly grapefruit and lemon. It is found in the Mediterranean basin (including Italy, Sicily and Spain) and is also present in Africa, North America, South America and eastern Australia.

DESCRIPTION
Adult female: 0.16–0.18 mm long; body yellowish or pinkish and vermiform; prodorsal shield with a pair of moderately long, posteriorly directed setae arising from tubercles located close to the hind margin; hysterosoma with a large number of very narrow tergites and sternites (cf. *Phyllocoptruta oleivora*); feather-claws 5-rayed.

LIFE HISTORY
This sporadically important pest breeds continuously throughout the year. Eggs hatch within a week of being laid, often in no more than a couple of days during the summer, development from egg to adult taking from 15 to 30 days, depending on temperature. The mites occur mainly in sheltered situations, within the buds or developing blossoms, under the bud scales, and under the sepals at the bases of developing fruits.

DAMAGE
Lemon: the scales of buds invaded by the mites turn black and the buds often die. The mites also cause significant distortion of blossoms and leaves; further, they cause abnormal development of shoots, so that internodes are shortened and twigs become twisted. Multiple budding and hideous deformation of fruits is also a characteristic symptom. **Orange:** fruits may become somewhat flattened and the skin (rind) ridged and furrowed. Otherwise, symptoms on orange are similar to those on lemon although less severe.

Aceria tristriatus (Nalepa) (**1074–1076**)
Walnut leaf gall mite

A locally common pest of walnut. Eurasiatic. Also present in the USA.

DESCRIPTION
Adult female (protogyne): 0.20–0.24 mm long; pinkish to reddish and vermiform; prodorsal shield with a pair of moderately long, posteriorly directed setae arising from tubercles located on the hind margin; hysterosoma with numerous narrow tergites and sternites; feather-claws 3-rayed. **Gall:** a small (1–2 mm diameter), hard, dark red, pimple-like pustule on the upper surface of the leaf, with an opening through the lower surface.

LIFE HISTORY
Similar to that of walnut leaf erineum mite (*Aceria erineus*).

DAMAGE
Infested leaves often contain large numbers of galls, particularly along the midrib and other major veins; heavily infested leaves can become distorted and misshapen, but there appears to be little or no effect on growth or on cropping.

NOTE
In addition to *Aceria tristriatus* (and *A. erineus*), two other eriophyid species of *Aceria* (namely, *A. brachytarsus* (Keifer) and *A. microcarpae* (Keifer)) also initiate galls on walnut leaves. One or other of these species may be responsible for the relatively large (up to 4 mm in diameter), pale and blister-like galls that cause considerable distortion of infested foliage. Such galls (that are quite unlike those inhabited by *A. tristriatus*) have an opening through the underside at the tip of a short (c. 3 mm long), more or less cylindrical outgrowth; there may also be an opening through the upper surface. The galls develop on the leaves from May onwards.

1074 Walnut leaf gall mite (*Aceria tristriatus*) – galls.

1075 *Aceria* sp. – galls on walnut leaf, viewed from above.

1076 *Aceria* sp. – galls on walnut leaf, viewed from below.

1077 Plum rust mite (*Aculus fockeui*) – mites swarming on bronzed leaf.

Aculops pelekassi (Keifer)
Pink citrus rust mite

A widely distributed pest of citrus, particularly loose-skinned cultivars of mandarin and orange. In Europe, recorded from Italy and Sicily; also present in North America, South America and southern Asia.

DESCRIPTION
Adult female: 0.14–0.15 mm long; body pink and fusiform; prodorsal shield with a pair of posteriorly directed setae arising from tubercles located on the hind margin; hysterosoma with c. 35 tergites and 50 sternites; feather-claws 4-rayed.

LIFE HISTORY
This species occurs mainly on the new growth, the mites often congregating around the leaf margins. The translucent-white eggs are laid at random on the leaves and fruits. They hatch shortly afterwards. Development of the mites is rapid and breeding is continuous throughout the year, with a large number of generations annually.

DAMAGE
Mites cause russeting of leaves and fruits. They also distort the new growth. Other symptoms include chlorosis and the development of brown, necrotic lesions on leaves, which may lead to premature leaf fall.

Aculus cornutus (Banks)
Peach silver mite

A worldwide pest of almond, nectarine and peach. Widely distributed in the warmer parts of Europe and North America.

DESCRIPTION
Protogyne (summer female): 0.19–0.20 mm long; brownish white, fusiform and dorsoventrally flattened; prodorsal shield with a pair of moderately long setae arising from tubercles located just in front of the hind margin; hysterosoma with obvious microtubercles and with numerous, narrow tergites and sternites; feather-claws 4-rayed. **Deutogyne (winter female):** similar to protogyne, but with broader tergites and with microtubercles less distinct.

LIFE HISTORY
Deutogynes of this deuterogenous species overwinter at the base of lateral buds on the shoot tips. They become active in the spring and then migrate to the developing leaves, where eggs are laid. There are several generations of protogynes and males throughout the summer, the mites occurring on both sides of leaves. New deutogynes are produced in the autumn.

DAMAGE
Infested young leaves become speckled with yellow and their margins turn upwards. Also, as the season progresses, mature leaves become silvery; heavy infestations result in general debilitation of host trees, reduced cropping and premature leaf fall.

Aculus fockeui (Nalepa & Trouessart) (**1077–1079**)
Plum rust mite

A pest of damson, myrobalan and plum; also attacks apricot, cherry, peach and various other kinds of *Prunus*. Widely distributed and common in Europe, but with a more northerly distribution than *Aculus cornutus*; also present in North America.

DESCRIPTION
Protogyne (summer female): 0.15–0.23 mm long; brownish white and relatively broad bodied; prodorsal shield with an anterior lobe, two short projecting anterior spines and a pair of posteriorly directed setae arising from tubercles located on the hind margin; hysterosoma with c. 55 tergites and 60 sternites, and with distinct microtubercles; feather-claws 4-rayed. **Deutogyne (winter female):** slightly smaller than

protogyne; prodorsal shield without a pair of projecting anterior spines; tergites and sternites slightly narrower and microtubercles less distinct.

LIFE HISTORY

Deutogynes of this deuterogenous species overwinter under bud scales, mainly on one-year-old wood. During the spring and summer, the mites are free-living on the underside of leaves. There are several overlapping generations of protogynes and males, with new deutogynes being produced in increasing numbers in the later generations.

DAMAGE

Attacked leaves become inwardly rolled and brittle, their lower surface turning brown and the upper surface developing a silvery coloration. Terminal buds may be killed by heavy infestations, and shoot growth may become distorted and stunted; the rind of such shoots may also be discoloured. Mites may also cause yellow flecking of leaves. Serious attacks have occurred on fruiting plum trees (cv. Victoria and, to a lesser extent, cv. Czar), but infestations are usually important only on nursery stock.

1078 Plum rust mite (*Aculus fockeui*) – damage to young shoots.

1079 Plum rust mite (*Aculus fockeui*) – leaf-flecking damage.

Aculus schlechtendali (Nalepa) (**1080**)

Apple rust mite

A pest of apple, but rarely important; at least in some regions (e.g. in southern England) it also occurs, occasionally, on pear. Eurasiatic. Widely distributed and common in Europe; also present in North America.

DESCRIPTION

Protogyne (summer female): 0.16–0.18 mm long; brownish orange and fusiform; prodorsal shield with two short spines projecting beyond the anterior lobe, with a pair of posteriorly directed setae arising from tubercles located on the hind margin, and with a granulated pattern; hysterosoma with numerous narrow tergites and sternites, the former each overlapping one or two sternites; feather-claws 4-rayed. **Deutogyne (winter female):** smaller than protogyne; prodorsal shield without granules.

LIFE HISTORY

Deutogynes of this deuterogenous species overwinter in groups under loose bark, in cracks close to buds on the twigs, and beneath bud scales (particularly those of permanently dormant buds); they also occur beneath bark cover such as algae and dead mussel scales (*Lepidosaphes ulmi*) (Chapter 3). Mites emerge in early spring, and then invade the swelling or opening buds to feed. Eggs are deposited on the green tissue of both fruit and vegetative buds, and these give rise to an initial generation of males and protogynes. Breeding continues throughout the spring and summer, there being several overlapping generations of primary forms (i.e. males and protogynes), with new deutogynes appearing in increasing numbers from late June to early July onwards. Population growth is rapid, individuals developing from egg to adult in 1–2 weeks at normal summer temperatures. Most mites occur on the underside of leaves, with greatest numbers (often several hundred per leaf) present in August. Populations on the foliage then gradually decline, as the new deutogynes enter hibernation and breeding eventually ceases.

1080 Apple rust mite (*Aculus schlechtendali*) – damaged leaf.

DAMAGE

Mites often cause a patchy, felt-like malformation and a yellowing of hairs on the underside of the leaves; also, the upper surface of infested foliage becomes speckled, dull and faded. More heavily infested leaves become silvery and may eventually turn brown; severely damaged foliage shrivels and growth of the shoots may be affected. Infestations are of particular significance on young, unsprayed trees. Under certain conditions, the mites cause fruit russeting, which can result in harvested crops being downgraded.

Anthocoptes ribis Massee
Currant rust mite

Although originally described as an inquiline in big-bud galls formed on black currant by the black currant gall mite (*Cecidophyopsis ribis*), this mite can cause leaf bronzing; damage occurs mainly on red currant, where infestations on developing leaves may also result in bronzing and a reduction in leaf size. Adult females are 0.19 mm long, brownish white and fusiform, with a pair of posteriorly directed prodorsal shield setae arising from tubercles located on the hind margin; characteristically, the hysterosoma has very few (mainly very broad) tergites yet a large number of narrow sternites. The mite is widely distributed in Europe, and is also reported from Canada.

Calepitrimerus vitis (Nalepa)
Vine rust mite

A pest of grapevine. Widely distributed in central Europe; important attacks have been reported recently in parts of southern Germany and northern Italy. Also found in North America and New Zealand.

DESCRIPTION

Protogyne (summer female): 0.15 mm long; brownish white to yellowish, and fusiform; prodorsal shield with a pair of moderately long, upwardly and inwardly directed setae arising from tubercles located in front of the hind margin; hysterosoma with c. 50 narrow tergites and c. 60 narrow sternites, the latter with numerous microtubercles; feather-claws 4-rayed. **Deutogyne (winter female):** body without microtubercles and feather-claws 5-rayed; once thought to be a different species, namely *Phyllocoptes vitis* Nalepa.

LIFE HISTORY

Deutogynes of this deuterogenous species overwinter in the shelter of buds and under bark at the base of the vines. In spring, at bud burst, they invade the young growth where eggs are laid. The first protogynes appear from about mid-May onwards and, by the beginning of June, they will usually have replaced all of the overwintered deutogynes. As further new growth is produced by the vines, this is constantly invaded by protogynes and males. However, once shoots have produced about 10 leaves, the mites migrate to the underside of the leaves, where breeding continues through to late summer. New deutogynes are produced in increasing numbers from August onwards, entirely replacing protogynes by the autumn. The deutogynes, which do not lay eggs until the following spring, gradually disappear from the foliage as they seek hibernation sites.

DAMAGE

Attacks in the spring restrict growth, so that internodes are short and the foliage tightly clustered together; following death of the terminal bud, shoot proliferation occurs, resulting in a symptom similar to that of a typical mite-induced witches' broom. Infestations are particularly important on young vines. Later in the season, feeding by the mites results in discoloration, russeting and distortion of leaves and premature leaf fall. Attacks may also have an adverse effect on cropping.

Cecidophyopsis grossulariae Collinge
Gooseberry bud mite

This species is essentially similar to black currant gall mite (*Cecidophyopsis ribis*), but occurs on gooseberry. Infested buds usually do not swell, but simply shrivel up and die.

Cecidophyopsis ribis (Westwood) (**1081–1082**)
Black currant gall mite

An important pest of black currant. Widely distributed in Europe; also present in North America and New Zealand.

DESCRIPTION

Adult female: 0.21–0.25 mm long; creamy white and vermiform; prodorsal shield without setae; hysterosoma finely cross-striated, with equal numbers of tergites and sternites; feather-claws 5- or 6-rayed. **Gall:** a large swelling of a bud on the host plant.

1081 Black currant gall mite (*Cecidophyopsis ribis*) – galled buds ('big buds').

LIFE HISTORY

Mites live and breed within black currant buds, causing a condition commonly known as 'big bud'. At the grape stage, the mites begin to appear outside the galled buds in vast numbers. They then crawl or leap from leaf to leaf and from branch to branch, and are also carried about by rain, wind and insects. This dispersal period may continue into June or July, but is at its height from the beginning of flowering until early fruit swelling. In June or early July, the mites invade new buds and egg-laying begins. Eggs and nymphal development is very rapid, the mites feeding on tissue within the shelter of the enlarging buds. Breeding reaches a peak in September and, after a temporary halt in early winter, begins again in January to reach a second peak in the early spring. By this time there may be several thousand mites inside a single bud. In spring, mites may also infest the growing points of young apical shoots.

DAMAGE

Attacked black currant buds begin to swell during the summer, soon becoming noticeably rounded and distorted; they are very conspicuous after leaf fall. In the following spring, they swell further and although opening they do not usually produce leaves or flowers. 'Big buds' measure up to 15 mm or more across; they remain on the bushes until June or July, eventually drying out and dying. Leaves developing from infested apical shoots are often badly deformed, the main lobes developing rounded outlines. Heavy infestations of buds and shoots affect bush development and also reduce cropping. However, the role of the mites as vectors of the virus causing reversion disease is of greater significance. The introduction of this virus can limit the economic life of a plantation. Reverted bushes tend to grow vigorously, but fruit production and cropping is seriously impaired. Some black currant cultivars (e.g. cv. Foxendown) show resistance to attack.

Cecidophyopsis selachodon van Eyndhoven
Red currant gall mite

This species is essentially similar to black currant gall mite (*Cecidophyopsis ribis*), but forms 'big buds' on red currant. Reported microscopical differences in the structure of *C. selachodon* and *C. ribis* are not always consistent.

Cecidophyopsis vermiformis (Nalepa)

This species occurs during the winter as an inquiline in 'big bud' galls formed on *Corylus avellana* and cultivated hazelnut by *Phytoptus avellanae* (family Phytoptidae), from which it may be distinguished by the absence of setae on the prodorsal shield. Adults are 0.17–0.25 mm long and vermiform. They feed on the underside of leaves during the summer, where they cause malformation of leaf hairs. Infestations, however, are usually of no significance.

Colomerus vitis (Pagenstecher) (**1083–1086**)
Vine leaf blister mite

An often abundant pest of grapevine, both in nurseries and in established vineyards. Of worldwide distribution in grape-growing areas.

1083 Vine leaf blister mite (*Colomerus vitis*) – galled leaf, viewed from above.

1082 Black currant gall mite (*Cecidophyopsis ribis*) – heavily infested bushes.

1084 Vine leaf blister mite (*Colomerus vitis*) – galled leaf, viewed from below.

DESCRIPTION

Adult female: 0.16–0.20 mm long; body whitish to creamy white and vermiform; prodorsal shield with a pair of moderately long, anteriorly directed setae arising from tubercles located in front of the hind margin; hysterosoma with numerous narrow tergites and sternites, and with distinct microtubercles; feather-claws 5-rayed.

LIFE HISTORY

Three physiologically distinct forms of this mite are recognized: a bud, a leaf-curling and an erineum (gall-forming) strain. The last-mentioned is most often reported. Mites of the erineum strain overwinter beneath bud scales. In spring, they become active and eventually migrate to the leaves to live amongst felt-like masses of hairs developed in galls on the lower surface. There are several, usually about seven, generations during the summer, mites eventually dispersing from the foliage to take up their winter quarters.

DAMAGE

Bud strain: this often causes death of overwintering buds; also, during the growing season, leaves arising from infested buds are distorted and growing points stunted; damage is particularly severe on the more basally located buds and, in the case of heavy infestations, results in noticeable damage to the new shoots. **Leaf-curling strain:** this causes leaves to curl and also induces leaf hairs at sites of infestation to develop abnormally. **Erineum strain:** this produces felt-galls (erinea) on the underside of the leaves, each several millimetres across. The galls are at first whitish, but then turn yellowish; later, they darken through reddish to dark brown. From above, their position is indicated by the development of large blisters, the surface of which may eventually become reddish brown; unlike galls formed on leaves by vine leaf gall midge (*Janetiella oenophila*) (see Chapter 5), the mite-induced blisters are soft to the touch. Severely infested leaves are distorted, but any effect on the vines is slight. Some cultivars are resistant to attack. When mites feed within developing flower clusters they cause a polyp-like enlargement of cells, especially on the bracts and peduncles; later, complete flower clusters may become desiccated and discoloured and will then drop off prior to setting fruit. Erinea are sometimes located along the major leaf veins.

Diptacus gigantorhynchus (Nalepa)
Big-beaked plum mite

This relatively large, cosmopolitan mite is free-living on cultivated plum, causing similar leaf symptoms to plum rust mite (*Aculus fockeui*). Adults are 0.20–0.28 mm long; the body is slightly furrowed longitudinally, with (except posteriorly) noticeably fewer abdominal tergites than sternites (c. 50 and 80 respectively); the prodorsal shield bears a pair of very short setae arising from tubercles located well in front of the hind margin. The mite is not of economic importance on field-grown trees, but may be of significance in nurseries.

Epitrimerus piri (Nalepa) (**1087–1088**)
Pear rust mite

An increasingly common pest of pear, probably as a result of the reduced use of lime sulphur for control

1085 Vine leaf blister mite (*Colomerus vitis*) – galled leaf veins, viewed from below.

1086 Vine leaf blister mite (*Colomerus vitis*) – galled inflorescences.

1087 Pear rust mite (*Epitrimerus piri*) – damage to pear leaves.

of pear scab (*Venturia pirina*) and the introduction of non-acaricidal fungicides. Widely distributed in Europe; also present in North America, Japan and New Zealand.

DESCRIPTION

Protogyne (summer female): 0.14–0.16 mm long; body yellowish, fusiform and dorsoventrally flattened, with a mid-dorsal ridge; prodorsal shield with a pair of short, inwardly and upwardly directed setae arising from tubercles located in front of the hind margin; prodorsal shield with a small, but distinct, anterior lobe; hysterosoma covered with microtubercles; feather-claws 4-rayed. **Deutogyne (winter female):** similar in appearance to protogyne, but smaller and less flattened; prodorsal shield with a small anterior lobe; hysterosoma without microtubercles.

LIFE HISTORY

This species is deuterogenous. Deutogynes overwinter under bud scales (particularly those of the small, permanently dormant buds on the spurs of main branches), under loose bark and in vegetative or fruit buds. The mites emerge in spring and begin to feed and to lay eggs on the freshly formed scale scars around the base of the bursting fruit buds. Mites also wander to the flowers and, later, they invade the foliage. The mites are free-living (i.e. they do not form galls) and are usually most numerous on the underside of the leaves. However, they also occur on the fruits, particularly around the 'eye', beneath the remains of the calyx. Eggs laid by the deutogynes produce males and protogynes. There are several overlapping generations annually, development from egg to adult taking from 1 to 2 weeks, or even less, depending on temperature. Breeding of the primary forms continues into September, but the overwintering deutogynes are produced in increasing numbers throughout the season. These overwintering forms begin to enter buds from July onwards, as the leaves harden.

1088 Pear rust mite (*Epitrimerus piri*) – damage to pear fruitlet.

DAMAGE

Heavy infestations cause a browning of the underside of leaves; ripening fruits also become russeted, damage being concentrated around the calyx.

Eriophyes padi (Nalepa) (**1089–1091**)
Plum leaf gall mite

A minor pest of damson and plum; also occurs on other species of *Prunus*, including *P. spinosa*. Widely

1089 Plum leaf gall mite (*Eriophyes padi*) – galls developing on young foliage.

1090 Plum leaf gall mite (*Eriophyes padi*) – young galls.

1091 Plum leaf gall mite (*Eriophyes padi*) – 'mature' galls.

distributed in Europe, but possibly with a more southerly distribution than plum pouch-gall mite (*Eriophyes similis*).

DESCRIPTION

Adult female: 0.20–0.22 mm long; body whitish and vermiform; prodorsal shield with a pair of short, inwardly directed setae arising from tubercles located in front of the hind margin; hysterosoma with numerous narrow tergites and sternites. **Gall:** finger-shaped (up to 6 mm long), red or dark red, and projecting upwards from the upper surface of the foliage (cf. galls formed by *E. similis*).

LIFE HISTORY

Female mites overwinter under bud scales or in bark crevices. In spring, they feed on the young leaves; flowers or young fruitlets may also be attacked. Characteristic galls are formed on the foliage and within these the mites begin to breed. There are several overlapping generations during the summer, development from egg to adult taking about 2–5 weeks, depending on temperature. All stages and both sexes of the mite occur together in the galls, but the young females disperse before leaf fall to seek shelter for the winter.

DAMAGE

The galls are often clustered closely together; however, they cause little or no distortion of leaves, unless very numerous.

Eriophyes pyri (Pagenstecher) (**1092–1094**)

Pear leaf blister mite

A pest of pear and, occasionally, apple. Eurasiatic. Widespread and often common in Europe; also present in South Africa, North American and New Zealand.

DESCRIPTION

Adult female: 0.20–0.24 mm long; body whitish to light brown, and vermiform; prodorsal shield with a pair of long, divergent, anteriorly directed setae arising from tubercles located in front of the hind margin; hysterosoma with numerous narrow tergites and sternites; feather-claws 4-rayed. **Egg:** oval, whitish and translucent. **Gall:** a small, pimple-like or blister-like pustule on a leaf, fruitlet or pedicel.

LIFE HISTORY

Mites, which overwinter beneath the outer bud scales, become active in the early spring as buds begin to swell. They then penetrate deeper within the buds, to begin feeding and to deposit eggs at the base of the inner scales. Later, young leaves and developing flower buds are attacked. At first, mites occur mainly on the

1092 Pear leaf blister mite (*Eriophyes pyri*) – galled pear leaves.

1093 Pear leaf blister mite (*Eriophyes pyri*) – young galls on apple leaf.

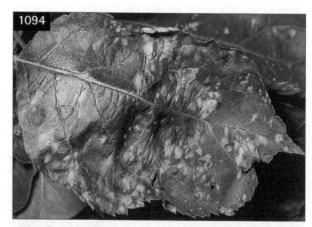

1094 Pear leaf blister mite (*Eriophyes pyri*) – galled apple leaf.

underside of the developing leaves, causing blistering of the tissue. However, following death and collapse of the centre-most surface cells of these blisters, they gain access to the inside of the leaf. Here, distinctive pocket-like galls are formed, each with a small hole remaining at its base. Young adult mites frequently disperse to the young shoots and then initiate new blisters that will also develop into galls; mites may also invade the fruitlets. Mites may be found in the galls throughout the spring and early summer, although older galls are abandoned as necrosis within them eventually makes them uninhabitable. There are two, sometimes three, generations annually; young females of the final generation seek overwintering sites from mid-summer onwards.

DAMAGE

In spring, galling is evident as soon as infested leaves unfurl. At first the galls appear as greenish pimples which soon become yellow; later, reddish blisters develop on the upper surface, each about 2–4 mm in diameter. They are particularly numerous along either side of the midrib, but as attacks develop they become more widely distributed and may cover much of the leaf surface. These blisters eventually turn brown and finally black, badly infested leaves dying and falling off. Infestations may spread to the fruitlets and fruit stalks (pedicels), which then become marked with reddish or brownish to blackish pustules. Damaged fruitlets are often distorted and they may drop prematurely.

Eriophyes similis (Nalepa) (**1095–1096**)
Plum pouch-gall mite
A minor pest of damson and plum; also occurs on *Prunus padus* and *P. spinosa*. Widely distributed in Europe, but possibly with a more northerly distribution than plum leaf gall mite (*Eriophyes padi*).

DESCRIPTION
Adult female: similar in appearance to *E. padi*. **Gall:** a pouch-shaped swelling (c. 5 mm long and 3 mm wide); yellowish, whitish or pinkish above, but greenish, tinged with red, and somewhat hairy, below.

LIFE HISTORY
Essentially similar to that of plum leaf gall mite (*E. padi*).

DAMAGE
The galls are often clustered along the main veins and at the edges of infested leaves. When numerous, they cause some leaf distortion and young leaves may become dwarfed and malformed. Direct mite damage to fruits may be severe, their surface becoming very irregular, with raised and sunken areas developing. Some cultivars (e.g. cvs Monarch and Victoria) are resistant to attack, but others (e.g. cvs Rivers and Yellow Egg) are very susceptible.

1095a, b Plum pouch-gall mite (*Eriophyes similis*) – galled leaves.

1096 Plum pouch-gall mite (*Eriophyes similis*) – damaged fruitlets (cv. Yellow Egg).

418

Phyllocoptes gracilis (Nalepa) (1097)

Raspberry leaf & bud mite

A pest of raspberry; other hosts include blackberry, boysenberry and loganberry. Widely distributed in Europe. Also established in North America.

DESCRIPTION

Adult female: 0.11–0.13 mm long; body brownish and fusiform; prodorsal shield with a pair of setae of medium length arising from tubercles located in front of the hind margin; hysterosoma with numerous narrow tergites and sternites; feather-claws 5-rayed.

LIFE HISTORY

Female mites overwinter beneath bud scales and, less frequently, in cracks in the cane bark. They emerge at bud burst and migrate to the underside of leaves, where eggs are laid amongst the leaf hairs. Breeding continues throughout the summer, there being several overlapping generations annually. Mite populations on the fruiting canes usually reach their peak during mid-summer. However, on the new vegetative growth, which becomes infested during the early summer as foliage on the fruiting canes hardens, mites are most numerous in late summer or early autumn. Particularly when populations are large, mites also invade the flowers and developing fruits. Numbers decline rapidly in the autumn, as breeding ceases and mites take up their winter quarters. There is often considerable mortality of overwintering mites, as individuals are crushed between scales of the expanding buds prior to bud burst.

DAMAGE

Infestations tend to be most severe in sheltered situations and in hot, dry summers. **Raspberry:** mites feeding on the foliage cause distortion and a yellow blotching of the upper surface; leaf hair development beneath the yellow patches is arrested, the appearance of these areas changing from greyish to light green. Apical buds of young canes are sometimes killed, leading to the development of weak lateral shoots. Heavy attacks on the fruits cause irregular drupelet development, uneven ripening and malformation; fruit damage, however, is reduced by wet, cool weather during the green-fruit stage. Cultivars vary in their susceptibility – cv. Malling Jewel, for example, is particularly liable to be attacked. **Blackberry:** mites cause malformation of hairs on the underside of leaves; by late summer or autumn, heavy infestations may cause mildew-like blotches to develop on the upper surface of leaves.

Phyllocoptes malinus (Nalepa) (1098)

Apple leaf erineum mite

This species is sometimes regarded as a form of the hawthorn leaf erineum mite (*Phyllocoptes goniothorax* (Nalepa)). The mites induce the development of white to rusty-brown erinea on the underside of apple leaves, within which they feed and breed. Protogynes are 0.17 mm long, whitish and vermiform, with a pair of inwardly directed setae arising from tubercles located in front of the hind margin of the prodorsal shield; the hysterosoma bears microtubercles and the feather-claws are 4-rayed. Although galling on infested leaves is often

1097 Raspberry leaf & bud mite (*Phyllocoptes gracilis*) – damaged foliage.

1098 Apple leaf erineum mite (*Phyllocoptes malinus*) – galled leaf, viewed from below.

extensive, the mite is not considered of economic importance.

Phyllocoptruta oleivora (Ashmead)
Citrus rust mite
This virtually cosmopolitan, locally common species occurs on citrus, and can be an important pest, especially in warm, humid sites. In southern Europe, the pest is recorded from Italy and from the former Yugoslavia.

DESCRIPTION
Adult female: 0.15–0.16 mm long; body dirty yellowish, fusiform and dorsoventrally flattened; prodorsal shield with a moderately produced anterior lobe and with a pair of upwardly directed setae arising from tubercles located well in front of the hind margin; hysterosoma with moderately broad tergites and sternites, and the dorsal surface with a broad, longitudinal trough (cf. *Aceria sheldoni*); feather-claws 5-rayed.

LIFE HISTORY
This species reproduces parthenogenetically and breeding is continuous throughout the year. The opaque eggs are laid on the leaves, typically in depressions alongside the midrib; they are also laid on immature (green) fruits. Eggs hatch within a week. The nymphs then develop over a period of about a week before eventually moulting into adults. The adults occur on both sides of mature leaves, but tend to congregate during sunny days on the upper surface. They may also move onto the shoots. Adults survive for about 2 weeks, each depositing up to 20 eggs.

DAMAGE
Infested leaves, young shoots and fruits are discoloured – skins of lemon fruits become silvery, whereas those of oranges become brownish. In addition, such fruits may be smaller than normal and their skins thickened. Large populations can lead to significant bronzing of leaves and shoots and may also weaken host plants.

Family **TARSONEMIDAE**
(tarsonemid mites)

Small, elliptical, light brown to whitish (often more or less transparent) mites, with a distinct head-like gnathosoma, short, needle-like chelicerae and pronounced sexual dimorphism. The hind legs of females are reduced and clawless; those of males are robust, often clasper-like and each terminates in a stout claw and often bears a distinctive inwardly directed flange.

Phytonemus pallidus fragariae (Zimmermann) (**1099–1100**)
Strawberry mite
A widely distributed and potentially serious, but sporadic, pest of strawberry. Virtually cosmopolitan.

1099 Strawberry mite (*Phytonemus pallidus fragariae*) – infested plant (cv. Cambridge Favourite).

1100 Strawberry mite (*Phytonemus pallidus fragariae*) – damage in strawberry field (cv. Cambridge Favourite).

DESCRIPTION

Adult female: 0.25 mm long; body light brown, elongate-oval and somewhat barrel-shaped; hind legs very thin, each terminating in a long, whip-like seta. **Adult male:** 0.2 mm long; body light brown and oval; hind legs broad, each with a rounded inner flange and terminating in a strong claw. **Egg:** 0.125 × 0.075 mm; elliptical, somewhat translucent and whitish. **Larva:** whitish, with hind part of the body triangular; 6-legged.

LIFE HISTORY

Small numbers of adult females overwinter deep in the crowns of host plants. They begin to emerge in the early spring to feed amongst the hairs at the bases of furled and unfurling leaves. On young foliage, they are most numerous on the upper surface. Eggs, which are laid amongst the leaf hairs, soon hatch into larvae. These feed for a short period before entering a quiescent stage where, within the now bloated larval skin, the change to the adult form occurs. There are several overlapping generations annually, development in summer from egg to adult taking about 2–3 weeks. The mites are mainly parthenogenetic, males usually forming no more than 5% of summer populations, with few or none overwintering. The light-shy mites are most abundant on the young succulent growth and are usually absent from the oldest foliage, seeking alternative feeding and breeding sites as the leaves open out and harden. Adults will crawl from plant to plant where foliage is touching, but they do not travel across the soil; small numbers are also dispersed by passing insects, farm machinery, wind currents and so on. However, more significantly, the mites are often spread locally on the tiny runners developing at the tips of elongating stolons as these grow away from the crowns of infested mother plants. Populations tend to build up slowly in the spring, but they develop rapidly in summer as temperatures increase, to reach a peak in August or September. Numbers then decline as breeding ceases and the young females enter hibernation.

DAMAGE

Mites inject toxic saliva into the plant cells and this causes leaves to become roughened, wrinkled, discoloured and brittle; on some cultivars (e.g. cv. Cambridge Favourite), the lateral edges of the leaves roll tightly downwards. Severely infested plants become stunted and may even die. Damage is particularly evident from July onwards, and particularly severe on susceptible cultivars growing in hot, sunny situations. Mite damage is sometimes confused with that caused by leaf nematodes (*Aphelenchoides* spp.), but nematode damage, unlike that caused by tarsonemid mites, is most noticeable in the spring. In general, mite attacks are most severe in older, perennial fruiting beds, infested areas becoming larger and more obvious each year. Following heavy summer infestations, formation of fruit buds in autumn is badly affected and the next year's crop greatly reduced; fruits developing on such plants are undersized, dull coloured and leathery. Spring and early-summer populations of the mite are usually too small to affect fruiting in the current year; however, on double-cropping cultivars, autumn fruits on heavily infested plants may turn brown and wither, without reaching maturity.

Family **TETRANYCHIDAE**
(spider mites)

Spider-like, often reddish or greenish mites, with long, needle-like chelicerae and robust, 5-segmented palps (pedipalps). Each pedipalp bears a thumb-claw and a silk-producing spinneret. Development from egg to adult includes larval, protonymphal and deutonymphal stages.

Amphitetranychus viennensis (Zacher)
Hawthorn spider mite
A usually minor pest on unsprayed apple, pear and plum trees. Eurasiatic. Widely distributed in Europe.

DESCRIPTION
Adult female: 0.5–0.6 mm long; plum red (overwintering form bright red), with pale dorsal setae; striae on hysterosoma transverse (cf. *Tetranychus urticae*). **Adult male:** 0.37 mm long; light green. **Egg:** 0.17 mm in diameter; more or less spherical, but somewhat flattened; light greenish, becoming amber before hatching. **Immature stages:** pale yellowish to light green; larva 6-legged.

LIFE HISTORY
Adult females overwinter beneath loose bark or in other shelter on the trees, sometimes congregating in considerable numbers. Eggs are laid on the underside of leaves from May onwards. Mites live and breed in small, discrete colonies beneath slight canopies of webbing, with about five generations annually.

DAMAGE
The mites produce pale mottled areas on the leaves, visible from above. The discoloration may extend over a considerable part of the leaf lamina, but effects on tree growth are usually of little or no significance.

Bryobia cristata (Dugès)
Grass/pear bryobia mite
This mite is an uncommon fruit pest, usually breeding throughout the year on grasses and certain herbaceous plants. Adults may be distinguished from those of apple & pear bryobia mite (*Bryobia rubrioculus*) by their narrower dorsal setae and by other microscopical features. In May, females sometimes migrate to trees and shrubs, including apple and pear. Here, two summer generations may occur before individuals disperse to their more normal hosts. The pest is recorded from Europe, Australia, Japan, New Zealand and North Africa.

Bryobia ribis Thomas
Gooseberry bryobia mite
Formerly an important pest of gooseberry, particularly in gardens; nowadays, however, relatively uncommon. Widely distributed in Europe.

DESCRIPTION
Adult female: 0.7 mm long; dark reddish brown to red; body oval and rather flat, with spatulate dorsal setae; first pair of legs very long. **Egg:** 0.2 mm across; more or less spherical, dark red and without a dorsal spine. **Larva:** bright reddish orange; 6-legged. **Nymph:** dark red, brown or dark green; 8-legged.

LIFE HISTORY
Overwintering eggs occur under loose bark. They hatch from late February or early March to about mid-April. Mites then feed and develop on the underside of the leaves and are particularly active on warm, sunny days. They do not form webs and will migrate from the foliage to the bark in cool or wet weather. In spring, the mites tend to infest the lowermost branches, but they soon spread throughout the bush. Once mature, the adults deposit winter eggs, there being just one generation annually; most activity is completed by June. Males are unknown and reproduction is entirely parthenogenetic. Unlike colonies of many kinds of spider mite (e.g. two-spotted spider mite, *Tetranychus urticae*) those of bryobia mites are never protected by silken webs.

DAMAGE
Heavy infestations cause young foliage to turn pale within a matter of days. These leaves eventually become brown; they then wither and may also fall off so that bushes become defoliated. Fruits produced on severely affected bushes are small and of poor quality.

Bryobia rubrioculus (Scheuten) (**1101**)
Apple & pear bryobia mite
A minor pest of apple and plum, especially on older trees; pear and cherry are also attacked. Widely distributed in Europe; also present in South Africa, North America, South America, Asia and Australia.

1101 Apple & pear bryobia mite (*Bryobia rubrioculus*) – damage to cherry leaf.

DESCRIPTION

Structurally similar to *Bryobia ribis*, but biologically distinct.

LIFE HISTORY

This species overwinters as eggs laid in clusters on the spurs, branches and trunks of host trees. The eggs begin to hatch in early April, several weeks earlier than those of the more widely known fruit tree red spider mite (*Panonychus ulmi*). The larvae then attack the buds and unfurling leaves, feeding on the undersurface. Development from egg to adult takes about 6 weeks. The mites often wander over both leaf surfaces during sunny weather; however, much of their time is spent on the bark, where they also take shelter in cool, wet conditions. Aggregations of mite on the bark are most evident in late May and June, and again in August and September. The mites also cluster beneath the shoots and main branches in order to moult from one stage to the next; masses of greyish-white cast skins soon accumulate and these are a characteristic sign of an infestation. There are usually three overlapping generations in a season, summer eggs being laid on the shoots and, less frequently, on the leaves and petioles. Males are unknown, reproduction being entirely parthenogenetic. Summer eggs take about 3 weeks to hatch; however, the winter eggs, which are deposited on the bark during the summer and autumn months, do not hatch until the following spring.

DAMAGE

Leaves developing from damaged buds are often smaller than normal, misshapen and marked with pale speckles at the base of the veins. Severely damaged foliage is at first pale and silvery, but later turns brown, and becomes brittle and distorted. Attacked young pear leaves may exude a sticky, gum-like substance.

Bryobia rubrioculus redikorzevi Reck
Brown fruit mite

A polyphagous pest of fruit trees, particularly apple, cherry, peach, pear and plum; almond and *Prunus spinosa* are also attacked. Recorded from the drier parts of eastern Europe (e.g. Bulgaria, Hungary and the former Yugoslavia); also present in Asia.

DESCRIPTION

Structurally indistinguishable from *Bryobia rubrioculus*, but biologically distinct.

LIFE HISTORY

Eggs overwinter on the bark of host trees and hatch, often over a protracted period, in the early spring. During the day, the mites tend to inhabit the branches and twigs, but they do invade the foliage at night. Eggs are deposited on the underside of leaves and on the bark. They hatch about 1–4 weeks later, depending on temperature. There are several generations annually, with mites moulting either on the bark or on the underside of leaves. Females appearing in late summer or early autumn eventually deposit the winter eggs.

DAMAGE

Similar to that caused by apple & pear bryobia mite (*B. rubrioculus*), but tending to be concentrated in the uppermost parts of host trees.

Eotetranychus carpini (Oudemans) (**1102–1103**)
Hornbeam spider mite

A minor pest of apple, and most numerous on cultivars with hairy leaves; in some areas, infestations also occur on grapevine. Other recorded hosts include *Alnus*, *Carpinus betulus*, *Corylus avellana*, *Quercus*, *Rubus fruticosus* and *Salix*. Recorded most frequently from England and Germany, but present elsewhere in Europe. Also found in Mexico and the USA.

1102 Hornbeam spider mite (*Eotetranychus carpini*) – damage to apple leaf, viewed from below.

1103 Hornbeam spider mite (*Eotetranychus carpini*) – small colony.

DESCRIPTION
Adult female: 0.4 mm long; light green or greenish yellow (never red or orange). **Egg:** 0.1 mm in diameter and globular; light green.

LIFE HISTORY
Adult females hibernate in crevices or beneath loose bark, reappearing on leaves in the spring. The mites live in small, compact colonies beneath webs spun on the underside of the leaves, often (but not invariably) close to the midrib. Breeding continues from April to October, six generations being recorded in a season.

DAMAGE
Pale patches or bronzing develop on infested leaves, and severely affected leaves become brittle. However, colonies are often depleted by predatory mites that frequently inhabit the webbing, and this can limit the extent of damage.

Eotetranychus pruni (Oudemans) (**1104–1105**)
Polyphagous on various trees and shrubs, including neglected apple, cherry and plum trees; also occurs on grapevine. Widely distributed in Europe; also present in the USA.

DESCRIPTION
Adult: similar in general appearance to *Eotetranychus carpini*; structural differences between the two species are slight, requiring detailed microscopic examination of, for example, the male genitalia.

LIFE HISTORY
Adult females overwinter on the bark of host trees. They become active in April or early May, eggs then being laid on the underside of expanded leaves. Development from egg to adult takes about a month, depending on temperature. There are typically three or four generations annually, early-season survival and breeding potential being favoured by warm, dry conditions. Colonies are usually relatively small and occur between the major veins on the underside of leaves, sheltered beneath silk webbing.

DAMAGE
Similar to that caused by hornbeam spider mite (*E. carpini*).

Panonychus citri (McGregor) (**1106–1107**)
Citrus red spider mite
Although found mainly on citrus, this pest may also occur on peach and various other deciduous fruits. Widely distributed in tropical and subtropical regions, including parts of southern Europe; sometimes also reported in greenhouses in more northerly regions.

1104 *Eotetranychus pruni* – damage to leaf of plum, viewed from above.

1105 *Eotetranychus pruni* – damage to leaf of plum, viewed from below.

1106 Citrus red spider mite (*Panonychus citri*) – leaf damage on mandarin, viewed from above.

1107 Citrus red spider mite (*Panonychus citri*) – leaf damage on mandarin, viewed from below.

1108 European fruit tree red spider mite (*Panonychus ulmi*) – damage to apple leaf.

DESCRIPTION
Adult female: 0.4–0.7 mm long; body dark purplish red to purplish red, and distinctly rounded, with strong setae arising from distinct tubercles; tubercles pale, but not white-tipped (cf. those of *Panonychus ulmi*). **Adult male:** similar to female, but smaller, narrower-bodied and with tubercles less noticeable. **Egg:** bright red and more or less spherical. **Larva:** dark red; 6-legged. **Nymph:** brick-red; 8-legged.

LIFE HISTORY
Eggs are laid on the leaves of host plants, each held to the surface by several radiating strands of silk. They hatch within 1–3 weeks, depending upon temperature. The immature stages feed for up to 2 weeks before moulting into adults; however, again, development times vary considerably depending on temperature. Breeding is continuous throughout the year and reaches an optimum at temperatures of about 25°C; development from egg to adult is then completed in about 2 weeks.

DAMAGE
The mites cause a noticeable silvering, yellowing or speckling of infested leaves; especially under drought conditions, they may also cause premature leaf fall and fruit drop. Heavy attacks weaken host plants and, in severe cases, young shoots may be killed.

Panonychus ulmi (Koch) (**1108**)
European fruit tree red spider mite
A notorious pest of apple and other fruit trees, including cherry, damson, peach, pear and plum. Also occurs on currant, gooseberry, grapevine, loganberry, raspberry, walnut and many wild hosts. Present throughout Europe; also now established in many other parts of the world, including North Africa, South Africa, North America, South America, Australia, New Zealand and parts of Asia, including Japan.

DESCRIPTION
Adult female: 0.4 mm long; body oval, strongly convex and dark red, with long setae arising from light-tipped tubercles (cf. those of *Panonychus citri*); legs short and pale. **Adult male:** similar to female, but smaller; yellowish green to bright red and more or less pear-shaped, tapering posteriorly. **Egg:** 0.17 mm in diameter; dark red to red; spherical, with the top drawn into a thin spine (i.e. onion-shaped). **Immature stages:** pale yellowish green to bright red; the larva is 6-legged.

LIFE HISTORY
Eggs overwinter on the bark, and are most numerous on the smaller branches and spurs, where they often occur in considerable numbers. Sometimes, eggs are also present on the shoots, buds and trunk. Hatching begins in late April or May and is completed by about mid-June; however, because there are early- and late-hatching 'strains' of the mite, the date of peak egg hatch varies from orchard to orchard. On emerging from the egg, each mite moves to the underside of a leaf and begins feeding. Development from egg to adult usually includes larval, protonymphal and deutonymphal stages. Mites remain on the leaves whilst moulting from one stage to the next and cast skins often accumulate among the developing colonies. Summer eggs, which are paler than winter ones, are laid on the underside of leaves, development from egg to adult taking about a month and often omitting a deutonymphal stage. There are from five to eight overlapping generations annually. As with other spider mites, mated females produce male and female brood (from unfertilized and fertilized eggs, respectively), but unmated ones lay only male eggs. Adult females often spread infestations throughout the

tree by moving from leaf to leaf. Also, although mites do not form communal webs, an individual may produce a fine silken thread and then be carried by the wind to an adjacent tree. In September, with the arrival of short days and lower temperatures, mites begin to deposit winter eggs and breeding eventually ceases. Under poor breeding conditions, such as severe foliage bronzing resulting from heavy infestations, winter eggs are laid from August onwards.

DAMAGE

At first, immature and adult mites cause a light speckling of the foliage. Later, as mite populations and damage increases, the leaves become dull green, brownish and, finally, silvery bronze. Such foliage is brittle and may fall off prematurely. Leaf symptoms are usually most evident from July to September and attacks are particularly severe in hot, dry summers; however, bronzing can occur as early as June following high rates of overwintering egg survival and spring weather conditions favourable for mite development. Heavy infestations have an adverse effect on fruit yields; they also affect fruit bud formation, so the following year's crop may also be reduced.

Tetranychus mcdanieli McGregor
McDaniel's spider mite

This American species, first noted in Europe in 1981, is now established in southern Europe (e.g. southern France) on apple, plum and grapevine. The mites also occur on a wide range of herbaceous weeds and ornamental plants. The females survive the winter amongst debris in the soil, or beneath bark at the base or just beneath the canopy of host plants. In spring, the overwintered females attack the developing buds and eventually deposit eggs; on apple, they also invade young watershoots arising from the centre of host trees. Colonies occur throughout the summer, amongst an abundance of webbing. Development is rapid, and there are many generations annually, with mite numbers reaching their peak in July and August. Greatest damage to fruit trees typically occurs at the height of colony development, when large numbers of mites migrate outwards from the centre of infested trees, the foliage turning brownish grey. Severely damaged leaves on host plants may also turn red and fall off prematurely. Adult females are similar in appearance to those of two-spotted spider mite (*Tetranychus urticae*), but typically bright orange.

Tetranychus turkestani (Ugarov & Nikolski)
Strawberry spider mite

This species is very similar to the more widely known two-spotted spider mite (*Tetranychus urticae*), but

occurs only in warmer regions (including parts of southern Europe); it is particularly important as an agricultural pest in the USA. Infestations are reported on many herbaceous crops (including strawberry), but apparently not in greenhouses; insignificant attacks also occur on apple, pear, peach, plum and walnut trees, and on hop. Adult females are distinguished from those of *T. urticae* by the presence of four (rather than two) dark spots on the body – these comprise a pair of very large patches on either side of the hysterosoma, extending back from close behind the eyes, and a pair of small lateral spots near the hind end of the body. Hibernating adult females are initially green (cf. those of *T. urticae*), but eventually become bright orange.

Tetranychus urticae Koch (**1109–1111**)
Two-spotted spider mite

A generally abundant, polyphagous species and a notorious pest of greenhouse crops. It also attacks hop, walnut and many fruit crops, including black currant,

1109 Two-spotted spider mite (*Tetranychus urticae*) – damage to peach leaf.

1110 Two-spotted spider mite (*Tetranychus urticae*) – damage to currant leaf.

1111 Two-spotted spider mite (*Tetranychus urticae*) – damage to strawberry leaf.

gooseberry, blackberry, raspberry, strawberry and grapevine; sometimes troublesome on fruit trees growing against walls, particularly apricot, nectarine and peach. Less frequently, it occurs in apple, pear, cherry and plum orchards, but usually not on sprayed trees. Present throughout the temperate parts of the world.

DESCRIPTION
Adult female: 0.5–0.6 mm long; pale yellowish or greenish (the overwintering female orange to brick-red), with two dark patches on the body; body oval, with moderately long dorsal setae; striae on hysterosoma forming a diamond-shaped pattern (cf. *Amphitetranychus viennensis*). **Adult male:** similar to female, but body smaller, narrower and more pointed. **Egg:** 0.13 mm in diameter; globular and translucent, becoming pale reddish before hatching. **Immature stages:** light green, with darker markings; larva 6-legged.

LIFE HISTORY
Female mites overwinter in straw mulch and dead leaves, in dry crevices in the soil, in cracks on poles and stakes, and in many other similar situations. They become active in March or April and then invade the foliage, where they generally live on the underside of the leaves and eventually lay eggs. These hatch in about 2 weeks, but development is greatly retarded at temperatures below about 12°C. There are several (often up to seven) overlapping generations annually, mites passing though egg, larval, protonymphal and deutonymphal stages before maturing. Males, which form about 20% of summer populations, often bypass the deutonymphal stage in their development. Development from egg to adult takes about 3–4 weeks in summer, and is particularly rapid in hot, dry conditions. Colonies of the mites occur mainly on the underside of the leaves, sheltered by strands of fine webbing. In the case of heavy infestations, this webbing may cover complete leaves and also extend over the petioles and stems. Adults survive for about a month in summer, a female laying about 100 eggs in her lifetime. Those deposited by mated females give rise to males (developing from unfertilized eggs) or females (developing from fertilized eggs); however, as with other spider mites, those of unmated females will become males. Winter-female forms appear from September onwards, in response to short days. These become orange and then red and, after mating, they seek out suitable shelter in which to overwinter. Meanwhile, the males and summer females die and breeding ceases.

DAMAGE
General: mites pierce plant cells and suck out the sap, causing pale spots (visible from above) to develop on the leaves. More severe infestations cause distinct speckling, silvering or bronzing of the foliage and sometimes withering and premature leaf fall. Foliage of heavily infested host plants may also be contaminated by silken webbing. **Black currant:** infestations often build up rapidly in the spring, and this can eventually lead to bronzing of the foliage and premature leaf fall; reductions in both fruit yields and vigour of the bushes may then result. **Strawberry:** serious infestations are usually evident only after harvest, but in warm spring and on protected crops (e.g. those under cloches or growing in plastic tunnels) noticeable damage may occur well before fruiting. Severe attacks before harvest will reduce fruit size and quality; plant vigour is most likely to be affected during periods of hot, dry weather. Infestations are usually greatest on older fruiting beds, particularly where straw or other trash (affording ideal overwintering quarters) is allowed to remain on the beds from one year to the next. Some cultivars (e.g. cv. Cambridge Favourite) are very susceptible to attack. **Hop:** severe infestations cause silvering and bronzing of the foliage and browning of the burrs (strobili), reducing crop yields and adversely affecting quality. Newly planted sets are also liable to be damaged. The pest can be a particular problem on 'hedgerow' hops (dwarf hops).

Family **TENUIPALPIDAE**
(false spider mites)

Distinguished from members of the family Tetranychidae by the absence of thumb-claws and by the absence of spinnerets.

Cenopalpus pulcher (Canestrini & Fanzago) (**1112**)
Flat scarlet mite
A minor pest on old, unsprayed apple, pear and plum trees; also reported on walnut. Eurasiatic. Widely distributed in Europe.

1112 Flat scarlet mite (*Cenopalpus pulcher*) – typical site for a colony, sheltered by leaf hairs along the midrib.

DESCRIPTION
Adult female: up to 0.32 mm long; bright red and flattened; legs very short. **Adult male:** similar to female, but smaller and paler in colour. **Egg:** 0.11 × 0.07 mm; bright red and oval.

LIFE HISTORY
Females, which overwinter on the bark, become active in early spring and then begin to appear on the leaves. The first eggs are often laid on the bark, usually in late April; later, they are deposited along the veins on the underside of the leaves. Migration to the foliage is usually completed by May, the overwintered females continuing to lay eggs up to mid-July. Eggs hatch from late June onwards and the first young adults appear about a month later. Individuals occur in small groups on the underside of leaves, sheltering beneath the leaf hairs; however, unlike true spider mites, they do not produce webbing. Mating takes place in August and September, males and females occurring in about equal numbers. Afterwards, the males die and the impregnated females eventually hibernate. There is just one generation annually.

DAMAGE
Although sometimes present in large numbers on neglected trees, the mites are not harmful.

Order **CRYPTOSTIGMATA** (beetle mites)

1113   Cherry beetle mites (*Humerobates rostrolamellatus*).

Family **MYCOBATIDAE**

Beetle mites with relatively short chelicerae, a smooth idiosoma and hinged pteromorphs. Characteristically, during ecdysis the cuticle splits along a line of weakness around the sides of the hysterosoma.

Humerobates rostrolamellatus Grandjean (**1113**)
Cherry beetle mite

This widespread species is often abundant on unsprayed fruit trees, particularly apple, cherry, damson and plum. The adults are dark, shiny reddish to blackish and about 1 mm long; when clustered on the bark during the summer or as overwintering colonies on the trunks, branches and spurs, they are often mistaken for mite or insect eggs. Although occasionally reported feeding in masses on, for example, splitting or overripe cherries and plums, the mites do not attack sound fruits. They normally feed on mosses, algae or lichens growing on the bark of trees, and are harmless.

Wild or ornamental host plants cited in the text

Specific name	Vernacular name(s)	Family
Acacia	acacia	Mimosaceae
Acacia fanesiana	sweet acacia	Mimosaceae
Acer	e.g. maple	Aceraceae
Acer campestre	field maple	Aceraceae
Acer pseudoplatanus	sycamore	Aceraceae
Achillea millefolium	yarrow	Asteraceae
Aesculus hippocastanea	horse-chestnut	Hippocastanaceae
Agrimonia eupatoria	argimony	Rosaceae
Agropyron repens	common couch	Poaceae
Alisma plantago-aquatica	common water plantain	Alismataceae
Alnus	alder	Betulaceae
Anthemis	chamomile	Asteraceae
Anthriscus sylvestris	cow parsley	Umbelliferae
Arbutus unedo	strawberry-tree	Ericaceae
Arctium	burdock	Asteraceae
Artemisia gallica	–	Asteraceae
Artemisia vulgaris	mugwort	Asteraceae
Aster	aster	Asteraceae
Aster tripolium	sea aster	Asteraceae
Atriplex patula	common orache	Chenopodiaceae
Berberis vulgaris	barberry	Berberidaceae
Betula	birch	Betulaceae
Brachypodium sylvaticum	slender false-brome	Poaceae
Buxus	box	Buxaceae
Calluna vulgaris	ling	Ericaceae
Calycotoma spinosa	–	Fabaceae
Camellia	camellia	Theaceae
Camellia japonica	camellia	Theaceae
Carduus	thistle	Asteraceae
Carex	sedge	Cyperaceae
Carpinus betulus	hornbeam	Fagaceae
Castanea sativa	sweet chestnut	Fagaceae
Ceanothus	ceanothus	Rhamnaceae
Centaurea nigra	common knapweed	Asteraceae
Centaurea scabiosa	greater knapweed	Asteraceae
Ceratonia siliqua	carob	Fabaceae
Chaenomeles	–	Rosaceae
Chaerophyllum	chervil	Umbelliferae
Chaerophyllum bulbosum	–	Umbelliferae
Chaerophyllum hirsutum	–	Umbelliferae
Chamaenerion angustifolium	rose-bay	Onagraceae
Chenopodium album	fat hen	Chenopodiaceae
Chrysanthemum	e.g. chrysanthemum	Asteraceae
Chrysanthemum leucanthemum	ox-eye daisy	Asteraceae
Cirsium	thistle	Asteraceae
Cirsium arvense	creeping thistle	Asteraceae
Clematis	clematis	Ranunculaceae
Clematis vitalba	traveller's joy	Ranunculaceae
Convolvulus	bindweed	Convolvulaceae
Convolvulus arvensis	field bindweed	Convolvulaceae
Corylus avellana	wild hazel	Corylaceae
Cotoneaster	cotoneaster	Rosaceae

Specific name	Vernacular name(s)	Family
Crataegus	hawthorn	Rosaceae
Crataegus monogyna	hawthorn	Rosaceae
Crepis	hawksbeard	Asteraceae
Cynoglossum	houndstongue	Boraginaceae
Deschampsia cespitosa	tufted hair-grass	Poaceae
Dianthus	pink	Caryophyllaceae
Elaeagnus	–	Elaeagnaceae
Epilobium	willow-herb	Onagraceae
Epilobium montanum	hoary willow-herb	Onagraceae
Epilobium obscurum	short-fruited willow-herb	Onagraceae
Erigeron canadensis	Canadian fleabane	Asteraceae
Euonymus	spindle-tree	Celastraceae
Euonymus europaeus	spindle-tree	Celastraceae
Eupatorium cannabinum	hemp agrimony	Asteraceae
Euphrasia officinalis	eyebright	Scrophulariaceae
Fagus sylvatica	beech	Fagaceae
Ficus	fig-tree	Moraceae
Filipendula	e.g. meadowsweet	Rosaceae
Filipendula ulmaria	meadowsweet	Rosaceae
Filipendula vulgaris	dropwort	Rosaceae
Foeniculum vulgare	fennel	Umbelliferae
Forsythia	forsythia	Oleaceae
Fragaria vesca	wild strawberry	Rosaceae
Frangula alnus	alder buckthorn	Rhamnaceae
Fraxinus excelsior	ash	Oleaceae
Fuchsia	fuchsia	Onagraceae
Galeopsis tetrahit	common hemp-nettle	Labiatae
Galium	bedstraw	Rubiaceae
Galium aparine	goosegrass	Rubiaceae
Galium mollugo	hedge bedstraw	Rubiaceae
Genista	e.g. greenweed	Fabaceae
Geum	–	Rosaceae
Geum rivale	water avens	Rosaceae
Glechoma hederacea	ground ivy	Labiatae
Hedera helix	ivy	Araliaceae
Heracleum sphondylium	hogweed	Umbelliferae
Hieracium	hawkweed	Asteraceae
Humulus lupulus	wild hop	Cannabinaceae
Hydrangea	hydrangea	Saxifragaceae
Hypochoeris radicata	common catsear	Asteraceae
Ilex aquifolium	holly	Aquifoliaceae
Ilex crenata	Japanese holly	Aquifoliaceae
Impatiens	e.g. balsam	Balsaminaceae
Inula	–	Asteraceae
Ipomoea	e.g. morning glory	Convolvulaceae
Jasminum	jasmine	Oleaceae
Juglans regia	common walnut	Juglandaceae
Juncus	rush	Juncaceae
Juniperus	juniper	Cupressaceae
Laburnum	laburnum	Fabaceae
Lamium	dead-nettle	Labiatae
Lamium purpureum	red dead-nettle	Labiatae
Larix	larch	Pinaceae
Lathyrus pratensis	meadow pea	Fabaceae
Laurus nobilis	sweet bay	Lauraceae
Ligustrum vulgare	privet	Oleaceae
Lolium perenne	perennial rye-grass	Poaceae
Lonicera	honeysuckle	Caprifoliaceae
Lonicera periclymenum	honeysuckle	Caprifoliaceae
Lychnis flos-cuculi	ragged robin	Caryophyllaceae
Lythrum salicaria	purple loosestrife	Lythraceae
Malus	crab apple	Rosaceae

Specific name	Vernacular name(s)	Family
Malus sylvestris	crab apple	Rosaceae
Matricaria	mayweed	Asteraceae
Matricaria maritima	scentless mayweed	Asteraceae
Medicago	medick	Fabaceae
Medicago lupulina	black medick	Fabaceae
Medicago sativa	lucerne	Fabaceae
Mercurialis perennis	dog's mercury	Euphorbiaceae
Mimosa	mimosa	Mimosaceae
Myosotis	forget-me-not	Boraginaceae
Myrica gale	bog myrtle	Myricaceae
Myrtus communis	common myrtle	Myrtaceae
Nerium oleander	oleander	Apocynaceae
Nuphar lutea	yellow water-lily	Nymphaeaceae
Nymphae alba	white water-lily	Nymphaeaceae
Ostrya carpinifolia	European hop-hornbeam	Betulaceae
Paeonia	peony	Ranunculaceae
Pastinaca sativa	wild parsnip	Umbelliferae
Phillyrea angustifolia	–	Oleaceae
Phillyrea latifolia	mock privet	Oleaceae
Phillyrea media	–	Oleaceae
Phragmites communis	reed	Poaceae
Pinus	pine	Pinaceae
Pittosporum	Australian laurel	Pittosporaceae
Plantago	plantain	Plantaginaceae
Plantago lanceolata	ribwort plantain	Plantaginaceae
Platanus	plane tree	Platanaceae
Poa	meadow-grass	Poaceae
Poa annua	annual meadow-grass	Poaceae
Polygonum aviculare	knotgrass	Polygoneaceae
Populus	poplar	Salicaceae
Populus tremula	aspen	Salicaceae
Potentilla	cinquefoil	Rosaceae
Potentilla anserina	silverweed	Rosaceae
Potentilla palustris	marsh cinquefoil	Rosaceae
Potentilla recta	sulphur cinquefoil	Rosaceae
Poterium sanguisorba	salad burnet	Rosaceae
Prunus	e.g. cherry, plum	Rosaceae
Prunus avium	wild cherry	Rosaceae
Prunus cerasus	dwarf cherry	Rosaceae
Prunus domestica	wild plum	Rosaceae
Prunus insititia	wild bullace	Rosaceae
Prunus laurocerasus	cherry laurel	Rosaceae
Prunus padus	bird-cherry	Rosaceae
Prunus spinosa	blackthorn	Rosaceae
Pteridium aquilinum	bracken	Polypodiaceae
Pyracantha coccinea	firethorn	Rosaceae
Pyrus	pear	Rosaceae
Pyrus salicifolia	willow-leaved pear	Rosaceae
Quercus	oak	Fagaceae
Quercus ilex	holm oak	Fagaceae
Quercus robur	English oak	Fagaceae
Ranunculus	buttercup	Ranunculaceae
Ranunculus bulbosus	bulbous buttercup	Ranunculaceae
Rhamnus cathartica	common buckthorn	Rhamnaceae
Rhinanthus crista-galli	yellow-rattle	Scrophulariaceae
Rhinanthus major	greater yellow-rattle	Scrophulariaceae
Rhus	sumach	Anacardiaceae
Ribes	e.g. currant, gooseberry	Grossulariaceae
Ribes sanguineum	flowering currant	Grossulariaceae
Ricinus communis	castor-oil plant	Euphorbiaceae
Robinia pseudoacacia	false acacia	Fabaceae
Rosa	rose	Rosaceae

Specific name	Vernacular name(s)	Family
Rubus	e.g. blackberry, raspberry	Rosaceae
Rubus caesius	dewberry	Rosaceae
Rubus fruticosus	bramble	Rosaceae
Rubus idaeus	wild raspberry	Rosaceae
Rumex	dock	Polygonaceae
Sagittaria sagittifolia	arrow-head	Alismataceae
Salix	sallow, willow	Salicaceae
Salix atrocinerea	common sallow	Salicaceae
Salvia	sage	Labiatae
Sambucus nigra	elder	Caprifoliaceae
Scabiosa	scabious	Dipsaceae
Schinus molle	pepper tree	Anacardiaceae
Scirpus lacustris	bulrush	Cyperaceae
Senecio	e.g. groundsel, ragwort	Asteraceae
Solanum	nightshade	Solanaceae
Solanum nigrum	black nightshade	Solanaceae
Solidago virgaurea	golden-rod	Asteraceae
Sonchus	sowthistle	Asteraceae
Sonchus arvensis	corn sowthistle	Asteraceae
Sorbus	e.g. whitebeam	Rosaceae
Sorbus aria	whitebeam	Rosaceae
Sorbus aucuparia	rowan	Rosaceae
Sorbus torminalis	wild service tree	Rosaceae
Spiraea	e.g. spiraea	Rosaceae
Spiraea salicifolia	bridewort	Rosaceae
Stachys	woundwort	Labiatae
Stachys sylvatica	hedge woundwort	Labiatae
Stellaria	stitchwort	Caryophyllaceae
Symphoricarpos rivularis	snowberry	Caprifoliaceae
Symphytum	comfrey	Boraginaceae
Syringa vulgaris	lilac	Oleaceae
Taraxacum officinale	dandelion	Asteraceae
Thymus	thyme	Labiatae
Tilia	lime	Tiliaceae
Trifolium	clover	Fabaceae
Trifolium repens	white clover	Fabaceae
Triglochin palustris	marsh arrow-grass	Juncaginaceae
Tussilago farfara	coltsfoot	Asteraceae
Ulex europaeus	gorse	Fabaceae
Ulmus	elm	Ulmaceae
Urtica	nettle	Urticaceae
Urtica dioica	stinging nettle	Urticaceae
Vaccinium	e.g. bilberry, cranberry	Ericaceae
Vaccinium myrtillus	bilberry	Ericaceae
Vaccinium vitis-idaea	cowberry	Ericaceae
Verbascum	mullein	Scrophulariaceae
Veronica	speedwell	Scrophulariaceae
Viburnum	e.g. guelder rose, viburnum	Caprifoliaceae
Vinca	periwinkle	Apocynaceae
Wisteria	wisteria	Fabaceae

Selected bibliography

Alford, D. V. (1984). *A Colour Atlas of Fruit Pests. Their identification, biology and control.* Wolfe: London.

Alford, D. V. (1999). *A Textbook of Agricultural Entomology.* Blackwell: Oxford.

Balachowsky, A. S. (ed.) (1962–63). *Entomologie appliquée à l'agriculture. I. Coleoptera (2 vols).* Masson: Paris.

Balachowsky, A. S. (ed.) (1966–72). *Entomologie appliquée à l'agriculture. II. Lepidoptera (2 vols).* Masson: Paris.

Balachowsky, A. S. & Mesnil, L. (1935). *Les insectes nuisibles aux plantes cultivées (2 vols).* Busson: Paris.

Barnes, H. F. (1948). *Gall midges of economic importance. Vol. III: Gall midges of fruit.* Crosby Lockwood: London.

Barnes, H. F. (1949). *Gall midges of economic importance. Vol. VI: Gall midges of miscellaneous crops.* Crosby Lockwood: London.

Bierne, B. P. (1954). *British pyralid and plume moths.* Warne: London.

Blackman, R. L. & Eastop, V. F. (2000). *Aphids on the World's Crops. An Identification and Information Guide. 2nd edition.* Wiley: Chichester.

Bradley, J. D., Tremewan, W. G. & Smith, A. (1975–79). *British tortricoid moths. Cochylidae and Tortricidae (2 vols).* Ray Society: London.

Butler, E. A. (1923). *A biology of the British Hemiptera-Heteroptera.* Witherby: London.

Cameron, P. (1882–89). *A monograph of the British phytophagous Hymenoptera. Vols I–III. (Tenthredo, Sirex and Cynips, Linné).* Ray Society: London.

Carter, D. J. & Hargreaves, B. (1986). *A Field Guide to Caterpillars of Butterflies & Moths in Britain and Europe.* Collins: London.

Chinery, M. (1993). *A Field Guide to the Insects of Britain and Northern Europe. 3rd edition.* HarperCollins: London.

Edwards, J. (1896). *The Hemiptera-Homoptera of the British Islands.* Reeve: London.

Emmet, A. M. (1996). *The moths and butterflies of Great Britain and Ireland. Vol. 3 Yponomneutidae to Elachistidae.* Harley Books: Colchester.

Emmet, A. M. & Heath, J. (eds) (1990). *The moths and butterflies of Great Britain and Ireland. Vol. 7 (Part 1) Hesperiidae to Nymphalidae, The butterflies.* Harley Books: Colchester.

Emmet, A. M. & Heath, J. (eds) (1991). *The moths and butterflies of Great Britain and Ireland. Vol. 7 (Part 2) Lasiocampidae to Thyatiridae.* Harley Books: Colchester.

Emmet, A. M. & Langmaid, J. R. (eds) (2002). *The moths and butterflies of Great Britain and Ireland. Vol. 4 Oecophoridae to Scythrrididae.* Harley Books: Colchester.

Forster, W. & Wohlfahrt, Th. A. (1954–1981). *Die Schmetterlinge Mitteleuropas.* Franckh'sche Verlagshandlung: Stuttgart.

Frankenhuyzen, A. van (1992). *Schadelijke en nuttige insekten en mijten in fruitgewassen.* Nederlandse Fruittelers Organisatie: Wageningen.

Frankenhuyzen, A. van & Stigter, H. (2002). *Schädliche und nützliche Insekten und Milben an Kern- und Steinobst in Mitteleurope.* Ulmer: Stuttgart (Hohenheim).

Heath, J. (ed.) (1976). *The moths and butterflies of Great Britain and Ireland. Vol. 1 Micropterigidae to Heliozelidae.* Curwen Press: London.

Heath, J. & Emmet, A. M. (eds) (1980). *The moths and butterflies of Great Britain and Ireland. Vol. 9 Sphingidae to Noctuidae (Part I).* Curwen Press: London.

Heath, J. & Emmet, A. M. (eds) (1983). *The moths and butterflies of Great Britain and Ireland. Vol. 10 Noctuidae (Part II) and Agaristidae.* Harley Books: Colchester.

Heath, J. & Emmet, A. M. (eds) (1985). *The moths and butterflies of Great Britain and Ireland. Vol. 2 Cossidae to Heliodinidae.* Harley Books: Colchester.

Jeppson, L. R., Keiffer, H. H. & Baker, E. W. (1975). *Mites Injurious to Economic Plants.* University of California Press: Berkeley.

Joy, N. H. (1932). *A practical handbook of British beetles (2 vols).* Witherby: London.

Kosztarab, M. & Kozár, F. (1988). *Scale Insects of Central Europe.* Dr W. Junk: Dordrecht.

Lorenz, H. & Kraus, M. (1957). *Die Larvalsystematik der Blattwespen.* Akademie-Verlag: Berlin.

Massee, A. M. (1954). *The pests of fruit and hops.* Crosby Lockwood: London.

Newstead, R. (1901–3). *Monograph of the Coccidae of the British Isles (2 vols).* Ray Society: London.

Pittaway, A. R. (1993). The *Hawkmoths of the Western Palaearctic.* Harley Books: Colchester.

Reitter, E. (1908–16). *Fauna Germanica. Die Käfer des Deutschen Reiches (5 vols).* Lutz: Stuttgart.

Schwenke, W. (ed.) (1972–82). *Die Forstschädlinge Europas (4 vols).* Parey: Hamburg.

Sterk, G. (1991). *De geïntegreerde bestrijding in de fruitteelt.* Opzoekingsstation van Gorsem: St.-Truiden.

Theobald, F. V. (1909). *The insect and other allied pests of orchard, bush and hothouse fruits and their prevention and treatment*. Wye.

Tuovinen, T. (1997). *Hedelmä- ja marjakasvien tuhoeläimet*. Kasvinsuojeluseuran julkaisu no: 89.

Wheeler Jr, A. G. (2001). *Biology of the Plant Bugs (Hemiptera: Miridae)*. Cornell University Press: New York.

Host plant index

Abies 264, 349
acacia: see *Acacia*
–, false: see *Robinia pseudoacacia*
–, sweet: see *Acacia fanesiana*
Acacia 122
– *fanesiana* 296
Acer 117, 171, 264, 354
– *campestre* 246
– *pseudoplatanus* 370
Achillea millefolium 42, 271
Actinidia chinensis 21
ACTINIDIACEAE 21
Aesculus hippocastanea 19, 354
Agrimonia eupatoria 132, 198
agrimony: see *Agrimonia eupatoria*
–, hemp: see *Eupatorium cannabinum*
Agropyron repens 50
Aira cespitosa: see *Deschampsia cespitosa*
alder: see *Alnus*
Alectorolophus major: see *Rhinanthus major*
– *minor*: see *Rhinanthus crista-galli*
alfalfa: see lucerne
algae 428
Alisma plantago-aquatica 90
almond(s) 19, 29, 30, 35, 82, 86, 96, 99, 106, 119, 132, 147, 148, 172, 174, 206, 207, 215, 222, 236, 247, 249, 250, 266, 274, 279, 283, 296, 301, 303, 344, 357, 379, 381, 382, 385, 401, 406, 407, 410, 422
–, bitter 19
–, sweet 19
Alnus 42, 48, 53, 140, 166, 241, 281, 283 293, 312, 313, 317, 325, 341, 422
ANACARDIACEAE 20
Anthemis 39
Anthriscus sylvestris 78
APIACEAE 128
APOCYNACEAE 71
apple(s) 17, 23, 24, 26, 28, 29, 30, 32, 33, 34, 36, 37, 38, 39, 40, 41, 42, 43, 44, 46, 48, 49, 50, 51, 52, 53, 54, 56, 57, 58, 61, 68, 71, 72, 73, 79, 80, 81, 90, 94, 95, 101, 102, 103, 104, 105, 107, 108, 109, 112, 113, 119, 128, 129, 130, 131, 133, 134, 136, 139, 140, 142, 147, 148, 149, 150, 151, 152, 153, 154, 155, 157, 158, 159, 162, 165, 166, 167, 168, 172, 173, 174, 180, 185, 188, 190, 192, 193, 198, 199, 204, 205, 206, 208, 209, 210, 211, 212, 214,

216, 217, 218, 222, 225, 227, 228, 232, 233, 234, 236, 239, 240, 241, 242, 243, 244, 245, 247, 248, 249, 250, 251, 254, 255, 256, 257, 258, 259, 260, 261, 262, 263, 264, 265, 266, 267, 268, 269, 270, 271, 272, 274, 276, 277, 278, 279, 280, 281, 282, 283, 284, 285, 287, 288, 289, 290, 291, 292, 293, 294, 295, 297, 298, 303, 309, 310, 311, 312, 313, 316, 317, 318, 319, 320, 321, 322, 324, 325, 329, 330, 331, 333, 334, 340, 341, 343, 344, 345, 347, 349, 352, 353, 354, 355, 356, 357, 360, 363, 367, 368, 372, 373, 374, 375, 376, 380, 384, 390, 391, 399, 402, 403, 404, 411, 416, 418, 421, 422, 423, 424, 425, 426, 427, 428
–, crab: see *Malus sylvestris*
apricot(s) 17, 30, 46, 53, 74, 75, 82, 90, 96, 106, 112, 117, 118, 119, 132, 148, 150, 151, 158, 166, 172, 174, 178, 183, 188, 206, 236, 247, 248, 249, 258, 262, 263, 272, 278, 279, 283, 287, 291, 303, 311, 322, 344, 357, 379, 381, 389, 401, 410, 426
ARALIACEAE 114
Arbutus unedo 298, 307, 340
Arctium 298
arrow-grass, marsh: see *Triglochin palustris*
arrow-head: see *Sagittaria sagittifolia*
Artemisia gallica 159
– *vulgaris* 47, 164, 297
artichoke 298
ASCLEPIADACEAE 71
ash: see *Fraxinus excelsior*
aspen: see *Populus tremula*
aster: see *Aster*
–, sea: see *Aster tripolium*
Aster 75
– *tripolium* 271
ASTERACEAE 39, 42, 74
Atriplex patula 366
avens, water: see *Geum rivale*
avocado 21, 69, 109, 116

balsam: see *Impatiens*
barberry: see *Berberis vulgaris*
bay, sweet: see *Laurus nobilis*
bean(s), field 68, 151, 169
bedstraw: see *Galium*
–, hedge: see *Galium mollugo*
beech: see *Fagus sylvatica*

beet 365
–, red 298, 367
–, sugar 68, 85, 86, 87, 298
Berberis vulgaris 285
Betula 27, 107, 140, 146, 165, 166, 204, 211, 241, 243, 245, 256, 260, 264, 268, 294, 312, 313, 316, 317, 318, 320, 321, 322, 325, 330, 333, 340, 341, 347, 354, 355, 362, 364, 370, 386
bilberry: see *Vaccinium myrtillus*
bindweed: see *Convolvulus*
–, field: see *Convolvulus arvensis*
birch: see *Betula*
bird-cherry: see *Prunus padus*
blackberry 18, 29, 39, 40, 51, 52, 54, 63, 64, 67, 72, 84, 85, 91, 101, 115, 136, 137, 142, 149, 154, 155, 176, 179, 181, 184, 191, 194, 197, 201, 244, 255, 257, 262, 268, 269, 280, 289, 293, 300, 323, 353, 356, 373, 391, 392, 393, 398, 400, 406, 418, 426
–, wild: see *Rubus fruticosus*
blackthorn: see *Prunus spinosa*
blueberry 191
–, highbush 19, 163, 285, 295, 307, 330, 355, 377
BORAGINACEAE 74
box: see *Buxus*
boysenberry 18, 280, 418
Brachypodium sylvaticum 85
bracken: see *Pteridium aquilinum*
bramble: see *Rubus fruticosus*
bridewort: see *Spiraea salicifolia*
buckthorn, alder: see *Frangula alnus*
–, common: *Rhamnus cathartica*
bullace 18, 183, 328
–, wild: see *Prunus insititia*
bulrush: see *Scirpus lacustris*
burdock: see *Arctium*
burnet, salad: see *Poterium sanguisorba*
bush fruit(s) 18–19, 39, 158, 164
buttercup: see *Ranunculus*
–, bulbous: see *Ranunculus bulbosus*
Buxus 328

cabbage 367
CACTACEAE 22
Calluna vulgaris 314
Calycotoma spinosa 217
camellia: see *Camellia* and *Camellia japonica*

Camellia 106
– japonica 111
cane fruit(s) 18, 39, 42, 50, 175, 176, 246, 374
CANNABACEAE 21
CAPRIFOLIACEAE 114
Carduus 39, 74, 164
– arvensis: see Cirsium arvense
Carex 365
carob: see Ceratonia siliqua
Carpinus betulus 64, 140, 145, 164, 171, 241, 243, 316, 318, 330, 333, 422
carrot 39, 357
CARYOPHYLLACEAE 29
Castanea sativa 19, 200, 218, 243, 277, 288
castor-oil plant: see Ricinus communis
catsear, common: see Hypochoeris radicata
cauliflower 367
ceanothus: see Ceanothus
Ceanothus 121
Centaurea nigra 243
– obscura: see Centaurea nigra
– scabiosa 260
Ceratonia siliqua 296
cereals 25, 33
Chaenomeles 94, 217
Chaerophyllum 79
– bulbosum 79
– hirsutum 79
– sylvestre: see Anthriscus sylvestris
Chamaenerion angustifolium 64
chamomile: see Anthemis
Chenopodium album 125, 366, 384
cherry(ies) 17, 30, 34, 36, 39, 40, 43, 44, 48, 49, 50, 51, 52, 53, 56, 87, 88, 104, 107, 112, 117, 119, 139, 147, 148, 150, 151, 156, 157, 158, 165, 166, 167, 168, 172, 173, 174, 183, 188, 189, 191, 204, 205, 206, 209, 210, 215, 216, 217, 218, 222, 225, 228, 229, 233, 235, 236, 239, 242, 243, 244, 247, 257, 258, 261, 262, 263, 264, 266, 270, 274, 279, 283, 285, 288, 289, 291, 293, 300, 301, 311, 316, 317, 318, 319, 320, 322, 324, 328, 329, 333, 340, 343, 344, 345, 347, 349, 352, 356, 357, 360, 368, 376, 379, 384, 401, 410, 421, 422, 423, 424, 426, 428
–, dwarf: see Prunus cerasus
–, Morello 17, 190
–, ornamental: see Prunus
–, sour 17, 87, 88, 190
–, sweet 17, 87, 88, 189, 190
–, wild: see Prunus avium
cherry-plum: see myrobalan
chervil: see Chaerophyllum
chestnut 19, 155, 200, 204, 276, 277, 288, 343
–, Spanish 19
–, sweet 19
chrysanthemum 28

Chrysanthemum 74, 75, 366, 377
– leucanthemum 270, 271
cinquefoil: see Potentilla
–, marsh: see Potentilla palustris
–, sulphur: see Potentilla recta
Cirsium 39, 74, 164
– arvense 244
citrus 20, 28, 31, 33, 46, 63, 65, 69, 71, 73, 91, 99, 100, 102, 103, 105, 106, 108, 109, 111, 116, 118, 120, 121, 122, 123, 159, 188, 204, 205, 219, 230, 266, 267, 295, 296, 328, 409, 410, 419, 423
Citrus aurantium 20
– bergamia 20
– grandis 20
– limonum 20
– paradisi 20
– reticulana 20
– sinensis 20
clematis: see Clematis
Clematis 88, 328
– vitalba 285, 323
clementines 20
clover 49, 151
–, white: see Trifolium repens
–, wild: see Trifolium
cobnut 20
coltsfoot: see Tussilago farfara
comfrey: see Symphytum
Convolvulus 40
– arvensis 47, 336
corn, sweet 297
CORYLACEAE 19
Corylus 283
– avellana 19, 27, 34, 48, 64, 77, 93, 138, 145, 146, 156, 168, 170, 171, 213, 216, 241, 243, 281, 282, 285, 290, 293, 315, 318, 354–355, 372, 386, 405, 413, 422
– colurna 20
– maxima 20
cotoneaster: see Cotoneaster
Cotoneaster 90, 94, 217, 232, 259, 379
cotton 358
couch, common: see Agropyron repens
cowberry: see Vaccinium vitis-idaea
cranberry 292
Crataegus 32, 58, 90, 136, 217, 232, 274, 281, 282, 312, 313, 322, 325, 328, 374, 395, 402
– monogyna 32, 36, 37, 50, 107, 131, 147, 157, 158, 208, 209, 216, 235, 236, 239, 242, 243, 248, 255, 256, 259, 260, 274, 282, 283, 287, 288, 293, 309, 310, 318, 329, 332, 334, 347, 354, 360, 364, 380, 398
Crepis 89
cucumber 30
currant(s) 38, 39, 40, 41, 42, 50, 70, 73, 74, 78, 83, 84, 90, 95, 101, 102, 104, 106, 107, 115, 117, 121, 132, 158, 159, 160, 161, 162, 163, 202, 229, 247, 258, 262, 270, 289, 291, 301, 308, 315, 318, 320, 322, 327,

328, 329, 334, 342, 364, 367, 368, 370, 378, 424
–, black 18–19, 19, 26, 50, 73, 74, 77, 78, 83, 89, 118, 162, 163, 164, 182, 183, 186, 205, 223, 263, 269, 271, 272, 300, 304, 305, 320, 327, 372, 395, 396, 397, 399, 412, 413, 425, 426
–, flowering: see Ribes sanguineum
–, red 18, 40, 50, 73, 74, 77, 78, 83, 84, 89, 107, 121, 133, 186, 202, 223, 282, 327, 394, 395, 396, 399, 412, 413
–, white 18, 50, 78, 83, 86, 202, 394, 396
Cydonia oblonga 17
Cynoglossum 74

daisy, ox-eye: see Chrysanthemum leucanthemum
damson(s) 18, 32, 50, 51, 56, 74, 82, 83, 89, 90, 166, 167, 168, 199, 218, 227, 229, 232, 263, 272, 274, 285, 300, 305, 309, 312, 315, 316, 317, 328, 329, 334, 388, 389, 395, 410, 415, 417, 424, 428
dandelion: see Taraxacum officinale
Daphne gnidium 295
– mezereum 295
date plum: see Chinese persimmon
deadnettle: see Lamium
–, red: see Lamium purpureum
Deschampsia cespitosa 170
dewberry: see Rubus caesius
Dianthus 366
Diospyros kaki 21
dock: see Rumex
dog's mercury: see Mercurialis perennis
dropwort: see Filipendula vulgaris

EBENACEAE 21
Elaeagnus 171
elder: see Sambucus nigra
elm: see Ulmus
Epilobium 70, 338, 339, 378
– angustifolium: see Chamaenerion angustifolium
– montanum 70
– obscurum 70
ERICACEAE 19
Erigeron canadensis 50
Euonymus 106, 281, 342
– europaeus 68, 315
Eupatorium cannabinum 298, 323
Euphrasia officinalis 84, 88
eyebright: see Euphrasia officinalis

FABACEAE 46, 68, 169
FAGACEAE 19
Fagus 158, 312, 321
– sylvatica 132, 145, 164, 166, 171, 244, 245, 290, 316, 330, 355
false-brome, slender: see Brachypodium sylvaticum

fat hen: see *Chenopodium album*
fennel: see *Foeniculum vulgare*
ferns 361
Ficus 62, 111
– *carica* 21
fig 21, 31, 45, 46, 51, 62, 63, 65, 102,
    104, 109, 110, 118, 135, 139, 159,
    171, 188, 224, 225, 295, 408
–, barbary 22
fig-trees 110
filbert 20, 156
Filipendula 393
– *ulmaria* 37, 144, 260, 385
– *vulgaris* 132
firethorn: see *Pyracantha coccinea*
flax 142
fleabane, Canadian: see *Erigeron
    canadensis*
Foeniculum vulgare 122
forget-me-not: see *Myosotis*
forsythia: see *Forsythia*
Forsythia 65, 121
Fragaria × *ananassa* 19
– *vesca* 19, 198, 253, 259
Frangula alnus 252, 290
Fraxinus 264, 321
– *excelsior* 65, 107, 170, 171, 172,
    204, 261, 283, 297, 298, 340, 342
fruit bushes 25, 151, 263, 282, 359,
    364, 402
– trees 23, 25, 28, 32, 37, 41, 46, 49,
    50, 51, 56, 59, 68, 71, 73, 90, 99,
    105, 112, 117, 130, 132, 134, 138,
    139, 140, 148, 151, 157, 158, 164,
    165, 166, 167, 168, 169, 171, 172,
    173, 174, 185, 206, 218, 219, 243,
    244, 246, 249, 250, 254, 255, 258,
    259, 261, 263, 264, 269, 272, 282,
    290, 302, 305, 309, 310, 311, 312,
    313, 317, 318, 319, 320, 321, 324,
    325, 326, 328, 329, 330, 331, 332,
    333, 334, 335, 340, 341, 345, 348,
    349, 352, 354, 355, 356, 357, 359,
    360, 361, 362, 363, 364, 368, 372,
    376, 380, 384, 388, 402, 422, 424,
    425, 426, 428
fruits, citrus 20, 63, 65, 296
–, pome 17
–, stone 17–19
fuchsia: see *Fuchsia*
Fuchsia 338, 339

Galeopsis tetrahit 77
Galium 81, 88, 338, 339
– *aparine* 81, 358
– *mollugo* 81
Genista 363
Geum 387, 393
– *rivale* 144, 253
Glechoma hederacea 299
golden-rod: see *Solidago virgaurea*
gooseberry(ies) 19, 40, 41, 51, 52, 70,
    77, 84, 89, 94, 104, 107, 115, 117,
    132, 158, 159, 160, 161, 162, 163,
    164, 179, 196, 202, 223, 229, 258,

262, 263, 282, 298, 300, 301, 308,
    315, 316, 326, 327, 329, 334, 364,
    368, 378, 394, 395, 396, 399, 400,
    412, 421, 424, 426
–, Chinese 21
goosegrass: see *Galium aparine*
gorse: see *Ulex europaeus*
grapefruit 20, 409
grapes 19, 27, 53, 120, 143, 179, 190,
    191, 266, 286, 293, 295, 297, 403,
    404
–, table 19, 27, 143, 266, 286, 293
grapevine(s) 19, 24, 27, 28, 30, 38, 39,
    40, 46, 47, 48, 50, 52, 53, 55, 56,
    68, 97, 98, 99, 105, 115, 116, 117,
    120, 131, 132, 135, 139, 141, 143,
    146, 147, 158, 159, 162, 163, 164,
    179, 183, 184, 188, 203, 205, 207,
    208, 221, 236, 252, 253, 265, 266,
    268, 269, 270, 278, 285, 286, 293,
    295, 296, 332, 336, 337, 338, 339,
    340, 351, 352, 358, 359, 364, 366,
    369, 370, 371, 376, 378, 384, 412,
    413, 422, 423, 424, 425, 426
grasses 25, 83, 85, 90, 91, 128, 130,
    164, 165, 195, 196, 208, 260, 365,
    371, 421
greengage 18
greenweed: see *Genista*
GROSSULARIACEAE 18, 19
groundsel: see *Senecio*

hair-grass, tufted: see *Deschampsia
    cespitosa*
hawksbeard: see *Crepis*
hawkweed: see *Hieracium*
hawthorn: see *Crataegus monogyna*
hazel, Turkish 20
–, wild: see *Corylus avellana*
hazelnut 19–20, 33, 34, 51, 52, 56, 77,
    93, 104, 115, 138, 140, 145, 156,
    165, 166, 167, 168, 170, 213, 214,
    216, 241, 245, 271, 281, 282, 283,
    289, 290, 316, 329, 343, 348, 386,
    405, 413
Hedera helix 52, 114, 118, 122, 252,
    285, 286, 304, 332
hemp-nettle, common: see *Galeopsis
    tetrahit*
Heracleum sphondylium 68
Hieracium 89
hogweed: see *Heracleum sphondylium*
holly: see *Ilex aquifolium*
–, Japanese: see *Ilex crenata*
honeysuckle: see *Lonicera
    periclymenum*
hop 21, 37, 38, 43, 51, 52, 53, 54, 56,
    84, 89, 90, 133, 144, 162, 167, 168,
    175, 176, 177, 191, 195, 196, 197,
    262, 263, 268, 269, 271, 297, 308,
    320, 327, 345, 352, 365, 366, 374,
    375, 376, 425, 426
–, hedgerow 426
–, dwarf 426
–, wild: see *Humulus lupulus*

hop-hornbeam, European: see *Ostrya
    carpinifolia*
hornbeam: see *Carpinus betulus*
horse-chestnut: see *Aesculus
    hippocastanea*
houndstongue: see *Cynoglossum*
Humulus lupulus 21, 191, 308, 365
hydrangea: see *Hydrangea*
Hydrangea 117
Hypochoeris radicata 271

Ilex aquifoliium 292, 304, 328
– *crenata* 111
Impatiens 338, 339
Inula 164
Ipomoea 336
ivy: see *Hedera helix*
–, ground: see *Glechoma hederacea*

jasmine: see *Jasminum*
Jasminum 28, 230, 298
jostaberry 19, 78, 396
JUGLANDACEAE 20
Juglans regia 20, 171
Juncus 50, 91
juniper: see *Juniperus*
Juniperus 121, 164

kaki: see Chinese persimmon
kiwi fruit 21, 46, 69, 106, 161, 188,
    360
knapweed, common: see *Centaurea
    nigra*
–, greater: see *Centaurea scabiosa*
knotgrass: see *Polygonum aviculare*

laburnum: see *Laburnum*
Laburnum 121
Lamium 40, 165, 378
– *purpureum* 77
larch: see *Larix*
Larix 290, 294
Lathyrus pratensis 270
LAURACEAE 21
laurel, Australian: see *Pittosporum*
–, cherry: see *Prunus laurocerasus*
Laurus nobilis 111
leek 30
LEGUMINOSAE: see FABACEAE
lemon(s) 20, 28, 99, 159, 230, 409,
    419
lettuce 84, 89, 358, 367
lichens 428
Ligustrum vulgare 65, 171, 230, 252,
    282, 285, 298, 342
lilac: see *Syringa vulgaris*
lime: see *Tilia*
ling: see *Calluna vulgaris*
linseed 39, 142
loganberry 18, 29, 32, 40, 54, 70, 72,
    84, 91, 136, 137, 142, 143, 144,
    149, 160, 168, 176, 181, 184, 186,
    194, 198, 202, 255, 270, 280, 347,
    351, 356, 368, 392, 393, 398, 418,
    424

Lolium perenne 125
Lonicera 189, 252, 268, 285, 342
– *periclymenum* 64
loosestrife, purple: see *Lythrum salicaria*
lucerne 39, 151
–, wild: see *Medicago sativa*
Lychnis flos-cuculi 271
Lythrum salicaria 323, 339

maize 297
Malus 53
– *pumila* 17
– *sylvestris* 17, 153, 209, 236, 248, 256
MALVACEAE 34
mandarin 20, 63, 99, 106, 111, 112, 410, 423, 424
mangold 358
maple: see *Acer*
–, field: see *Acer campestre*
Matricaria 39, 74
– *maritima* 42
mayweed: see *Matricaria*
–, scentless: see *Matricaria maritima*
meadow-grass: see *Poa*
–, annual: see *Poa annua*
meadowsweet: see *Filipendula ulmaria*
Medicago 122
– *lupulina* 155
– *sativa* 46
medick: see *Medicago*
–, black: see *Medicago lupulina*
medlar 17, 90, 147, 148, 153, 303, 379
Mercurialis perennis 299
Mespilus germanica 17
mimosa: see *Mimosa*
Mimosa 122
mirabelle 17
MORACEAE 21
morning glory: see *Ipomoea*
Morus 21
– *alba* 21
– *nigra* 21
mosses 428
mugwort: see *Artemisia vulgaris*
mulberry 21, 49, 100, 102, 106, 135, 247, 352
–, black 21
–, white 21
mullein: see *Verbascum*
mustard 136
Myosotis 74
Myrica gale 256
myrobalan 17, 18, 74, 183, 227, 229, 389, 410
myrtle, bog: see *Myrica gale*
–, common: see *Myrtus communis*
Myrtus communis 110

nectarine 18, 23, 28, 30, 75, 86, 116, 132, 410, 426
Nepeta hederacea: see *Glechoma hederacea*

Nerium oleander 71, 100, 111, 118, 172, 337
nettle: see *Urtica*
–, stinging: see *Urtica dioica*
nightshade: see *Solanum*
–, black: see *Solanum nigrum*
Nuphar lutea 90
nuts 19, 296
Nymphae alba 90

oak: see *Quercus*
–, English: see *Quercus robur*
–, holm: see *Quercus ilex*
Olea europaea var. *europaea* 21
OLEACEAE 21, 114, 149, 159, 170, 186, 230, 297, 335
oleander: see *Nerium oleander*
olive(s) 21, 27, 28, 31, 34, 46, 61, 63, 65, 99, 100, 102, 104, 105, 109, 113, 114, 118, 122, 135, 139, 149, 159, 170 171, 172, 180, 184, 185, 186, 187, 188, 204, 205, 213, 230, 236, 266, 267, 297, 298, 328, 335, 342, 408
–, table 21, 188
onion 30, 365, 367
Opuntia ficus-indica 22
orache, common: see *Atriplex patula*
orange(s) 20, 99, 111, 296, 409, 410, 419
–, bergamot 20
–, navel 295
–, sour 20
–, sweet 20, 230
orchids 111
Ostrya carpinifolia 145

Paeonia 160
palms 102
PAPILIONACEAE: see FABACEAE
parsley, cow: see *Anthriscus sylvestris*
–, wild: see *Pastinaca sativa*
Pastinaca sativa 68
pea 169
–, meadow: see *Lathyrus pratensis*
peach(es) 18, 19, 23, 26, 28, 30, 38, 39, 41, 46, 49, 50, 51, 52, 56, 68, 69, 74, 75, 76, 82, 86, 87, 88, 89, 90, 96, 102, 103, 105, 106, 107, 108, 112, 115, 116, 117, 119, 132, 134, 135, 147, 151, 158, 170, 172, 174, 188, 190, 191, 206, 209, 210, 215, 218, 222, 236, 247, 249, 250, 258, 272, 274, 276, 279, 290, 295, 297, 301, 303, 311, 345, 379, 381, 403, 410, 422, 423, 424, 425, 426
pear(s) 17, 28, 30, 32, 33, 34, 35, 36, 37, 38, 39, 40, 41, 42, 43, 46, 48, 49, 50, 56, 58, 59, 60, 67, 68, 69, 71, 81, 84, 85, 90, 94, 95, 97, 99, 101, 102, 103, 104, 105, 106, 107, 108, 109, 112, 113, 119, 130, 131, 132, 134, 135, 136, 139, 146, 147, 148, 150, 151, 152, 153, 154, 157, 158, 162, 165, 166, 167, 168, 172,

173, 174, 178, 179, 181, 182, 185, 188, 189, 193, 200, 204, 205, 206, 208, 209, 210, 216, 222, 225, 232, 236, 243, 244, 248, 249, 250, 256, 257, 258, 260, 261, 262, 263, 264, 267, 268, 269, 270, 271, 272, 274, 275, 276, 277, 278, 279, 283, 284, 285, 287, 288, 289, 290, 291, 292, 293, 294, 297, 300, 303, 309, 311, 318, 319, 322, 325, 328, 329, 333, 340, 343, 344, 345, 347, 348, 349, 352, 354, 355, 356, 357, 363, 372, 374, 375, 376, 379, 380, 384, 388, 393, 398, 399, 403, 411, 414, 415, 416, 421, 422, 424, 425, 426, 427
–, prickly 22, 188
–, wild: see *Pyrus*
–, willow-leaved: see *Pyrus salicifolia*
peony: see *Paeonia*
pepper tree: see *Schinus molle*
periwinkle: see *Vinca*
Persea americana 21
persimmon, Chinese 21, 46
–, Japanese: see Chinese persimmon
Phillyrea 230
– *angustifolia* 213
– *latifolia* 63, 170
– *media* 213
Phragmites communis 50, 83
Picea 294, 349
pine: see *Pinus*
pink: see *Dianthus*
Pinus 122, 290, 349
pistachio 20, 109
Pistacia vera 20
Pittosporum 118
plane tree: see *Platanus*
Plantago 81, 158, 260, 270, 365, 378
– *lanceolata* 81
plantain: see *Plantago*
–, common water: see *Alisma plantago-aquatica*
–, ribwort: see *Plantago lanceolata*
plum(s) 17, 18, 26, 28, 30, 36, 39, 40, 46, 47, 48, 49, 50, 51, 52, 54, 56, 65, 74, 75, 76, 82, 83, 86, 89, 90, 96, 99, 100, 102, 103, 104, 106, 107, 109, 113, 115, 117, 119, 139, 147, 148, 149, 150, 151, 156, 158, 165, 166, 167, 168, 171, 172, 173, 174, 178, 183, 185, 188, 190, 199, 200, 204, 205, 206, 208, 209, 210, 214, 215, 216, 218, 222, 227, 228, 229, 235, 236, 239, 242, 243, 244, 247, 249, 254, 256, 258, 259, 260, 261, 262, 263, 264, 267, 268, 269, 272, 274, 278, 279, 282, 283, 285, 287, 288, 289, 291, 292, 293, 295, 298, 300, 303, 305, 306, 309, 310, 311, 312, 313, 315, 316, 317, 318, 319, 320, 324, 325, 328, 329, 331, 332, 333, 334, 343, 344, 345, 347, 348, 349, 352, 354, 355, 356, 357, 360, 368, 372, 376, 379, 384, 388, 389, 395, 398, 399, 401, 403, 406,

407, 410, 411, 414, 415, 417, 421,
422, 423, 424, 425, 426, 427, 428
–, wild: see *Prunus domestica*
Poa 165
– *annua* 85, 90
Polygonum 340
– *aviculare* 358, 384
pomegranate 22, 65, 69, 72, 100, 105,
118, 120, 122, 139, 295, 363
pomelo 20, 230
poplar: see *Populus*
Populus 35, 261, 264, 282, 312, 313,
320, 333, 341, 362, 374
– *tremula* 35, 260
potato 39, 84, 85, 86, 87, 335, 358,
365
Potentilla 192, 259
– *anserina* 144
– *palustris* 253, 385
– *recta* 132
Poterium sanguisorba 132, 259
privet: see *Ligustrum vulgare*
–, mock: see *Phillyrea latifolia*
Prunus 17, 89, 90, 91, 107, 112, 150,
178, 183, 190, 199, 200, 229, 279,
301, 306, 313, 325, 328, 332, 333,
362, 395, 401, 410, 415
– *armeniaca* 17
– *avium* 17, 88, 156, 190, 215, 285
– *cerasifera* 17, 18
– – *insititia* 18
– – *italica* 18
– *cerasus* 17, 88, 215, 233
– *communis* 19
– *domestica* 17, 233
– – *domestica* 18
– *insititia* 185, 227
– *laurocerasus* 161, 341
– *padus* 150, 156, 190, 233, 301, 417
– *persica* 18
– *spinosa* 18, 32, 51, 82, 89, 90, 96,
119, 150, 156, 158, 183, 185, 199,
209, 214, 218, 227, 229, 235, 239,
242, 243, 246, 255, 259, 260, 272,
282, 283, 285, 287, 293, 305, 306,
309, 310, 312, 314, 315, 318, 328,
331, 332, 334, 347, 354, 355, 360,
364, 379, 389, 395, 398, 415, 417,
421
Pteridium aquilinum 361
Punica granatum 22
PUNICACEAE 22
Pyracantha coccinea 217, 218
Pyrus 178, 402
– *communis* 17
– *malus*: see *Malus sylvestris*
– *salicifolia* 178

Quercus 30, 53, 56, 122, 134, 155,
168, 200, 218, 244, 245, 246, 261,
264, 268, 277, 282, 283, 287, 288,
293, 294, 312, 316, 317, 318, 319,
320, 321, 325, 326, 330, 333, 347,
349, 355, 360, 362, 372, 374, 400,
422

– *ilex* 218
– *robur* 27
quince 17, 41, 68, 71, 90, 94, 104,
109, 119, 134, 147, 148, 153, 167,
188, 205, 206, 209, 211, 274, 275,
276, 293, 303, 356

ragged robin: see *Lychnis flos-cuculi*
ragwort: see *Senecio*
Ranunculus 270
– *bulbosus* 47
rape, oilseed 136
raspberry(ies) 18, 32, 33, 34, 38, 40,
41, 42, 44, 51, 66, 70, 84, 91, 115,
128, 130, 131, 132, 133, 134, 135,
136, 137, 142, 143, 144, 149, 154,
155, 158, 159, 160, 161, 162, 164,
167, 168, 175, 176, 181, 184, 186,
187, 191, 192, 194, 195, 196, 198,
201, 202, 222, 238, 244, 255, 257,
262, 263, 266, 268, 270, 271, 278,
286, 289, 291, 293, 311, 314, 315,
322, 343, 347, 349, 351, 353, 355,
356, 361, 364, 365, 366, 368, 372,
373, 374, 378, 383, 384, 385, 387,
391, 392, 393, 394, 398, 400, 418,
424, 426
–, American black 18
–, black-red 18
–, red 18
–, white 18
–, wild: see *Rubus idaeus*
reed: see *Phragmites communis*
Rhamnus 267
– *cathartica* 305
– *frangula*: see *Frangula alnus*
Rhinanthus crista-galli 84
– *major* 84
– *minor*: see *Rhinanthus crista-galli*
rhubarb 365
Rhus 111
Ribes 77, 90, 94, 252, 286, 396
– *divaricatum* 19
– *grossulariae*: see *Ribes uva-crispa*
– *nigrum* 18, 19
– *sanguineum* 94
– *sativum* 18
– *uva-crispa* 19
Ricinus communis 363
Robinia pseudoacacia 68, 122, 171,
296
root crops 357, 358, 359
Rosa 51, 101, 130, 185, 243, 246, 255,
283, 284, 333, 383, 385, 386, 387
ROSACEAE 17, 18, 19, 29, 30, 107,
157, 192
rose: see *Rosa*
rose-bay: see *Chamaenerion
angustifolium*
rowan: see *Sorbus aucuparia*
Rubus 54, 130, 142, 149, 184, 201,
222, 240, 315, 392, 393, 398, 400,
406
– *caesius* 67, 400
– *dalmaticus* 363

– *fruticosus* 18, 37, 64, 84, 181, 184,
197, 223, 238, 246, 255, 294, 314,
315, 355, 373, 384, 400, 406, 422
– *idaeus* 18, 198, 238, 387, 400
– *loganobaccus* 18
– *occidentalis* 18
Rumex 40, 41, 43, 165, 170, 270, 340,
353, 364, 365, 370, 384
rush: see *Juncus*
RUTACEAE 20
rye-grass, perennial: see *Lolium
perenne*

sage: see *Salvia*
Sagittaria sagittifolia 90
Salix 42, 101, 204, 245, 255, 261, 264,
268, 282, 293, 310, 318, 322, 328,
333, 341, 362, 370, 372, 374, 422
– *atrocinerea* 314, 364, 373
sallow: see *Salix*
–, common: see *Salix atrocinerea*
Salvia 122
Sambucus 342
– *nigra* 332
scabious: see *Scabiosa*
Scabiosa 260
Schinus molle 111
Scirpus lacustris 50
sedge: see *Carex*
Senecio 40, 41, 42, 74, 378
service tree, wild: see *Sorbus
torminalis*
shallot 86
silverweed: see *Potentilla anserina*
snowberry: see *Symphoricarpos
rivularis*
SOLANACEAE 335
Solanum 40, 111
– *nigrum* 47
Solidago virgaurea 323
Sonchus 83
– *arvensis* 84
Sorbus 90, 94, 107, 131, 147, 157,
217, 241, 248, 261, 264, 279, 283,
293, 313, 385, 398, 402
– *aria* 216, 239, 242, 255
– *aucuparia* 50, 158, 208, 209, 216,
227, 233, 256, 398
– *torminalis* 208
sowthistle: see *Sonchus*
–, corn: *Sonchus arvensis*
speedwell: see *Veronica*
spindle-tree: see *Euonymus europaeus*
spiraea: see *Spiraea*
Spiraea 73
– *filipendula*: see *Filipendula vulgaris*
– *salicifolia* 158
– *ulmaria*: see *Filipendula ulmaria*
sprout, Brussels 367
Stachys 298, 299
– *sylvatica* 78
Stellaria 378
stitchwort: see *Stellaria*
strawberry(ies) 19, 25, 27, 28, 29, 30,
32, 38, 39, 40, 41, 42, 44, 45, 49,

51, 52, 56, 63, 64, 66, 69, 72, 76,
84, 85, 86, 91, 124, 125, 126, 127,
128, 129, 132, 133, 134, 135, 136,
137, 142, 143, 144, 149, 154,
154–155, 155, 158, 159, 160, 161,
163, 164, 166, 167, 168, 169, 170,
175, 176, 177, 188, 191, 192, 195,
196, 197, 198, 208, 240, 241, 253,
254, 255, 259, 260, 261, 266, 267,
268, 269, 270, 271, 272, 282, 286,
290, 293, 297, 299, 300, 322, 346,
351, 353, 355, 356, 357, 358, 359,
361, 365, 366, 367, 368, 371, 372,
373, 374, 376, 377, 378, 383, 385,
386, 394, 419, 425, 426
–, alpine 19
–, wild: see *Fragaria vesca*
strawberry-tree: see *Arbutus unedo*
sumach: see *Rhus*
swede 357, 358
sycamore: see *Acer pseudoplatanus*
Symphoricarpus racemosus: see
  Symphoricarpos rivularis
– *rivularis* 64, 189, 252, 342
Symphytum 74
Syringa 268
– *vulgaris* 65, 170, 171, 172, 342

Tamarix 295
tangerine(s) 20
Taraxacum officinale 40, 166, 353,
  358, 366, 369
tayberry 18, 280, 356
thistle: see *Carduus*; see *Cirsium*

–, creeping, see *Cirsium arvense*
thyme: see *Thymus*
Thymus 259
Tilia 34, 48, 140, 243, 261, 264, 294,
  312, 320, 321, 328, 333, 340, 341,
  355, 374
tobacco 358
tomato 30, 366, 367
traveller's joy: see *Clematis vitalba*
Trifolium 39, 49, 52, 75
– *repens* 155
Triglochin palustris 74
turnip 357
Tussilago farfara 67
Typha 90

Ulex europaeus 305
Ulmus 94, 95, 204, 241, 261, 264, 313,
  318, 320, 321, 325, 328, 333, 340,
  355, 362, 374
Urtica 37, 39, 40, 41, 42, 166, 167,
  299
– *dioica* 47, 166, 308

Vaccinium 255
– *corymbosum* 19
– *myrtillus* 256, 355
– *vitis-idaea* 256
vegetable brassicas 64, 87, 136
– crops 30, 33, 139, 160, 367
– root crops 359
Verbascum 122, 378
Veronica 88, 89
viburnum: see *Viburnum*

Viburnum 252, 285
– *tinus* 114
Vinca 337
vine(s), European 19, 98
VITACEAE 19
Vitis vinifera 19, 98

walnut(s) 20, 92, 102, 104, 105, 106,
  115, 138, 150, 164, 174, 190, 204,
  205, 212, 276,
277, 348, 352, 407, 409, 424, 425, 427
–, common: see *Juglans regia*
water-lily, white: see *Nymphae alba*
–, yellow: see *Nuphar lutea*
wheat 358
whitebeam: see *Sorbus aria*
willow: see *Salix*
willow-herb: see *Epilobium*
–, hoary: see *Epilobium montanum*
–, short-fruited: see *Epilibium
  obscurum*
wisteria: see *Wisteria*
Wisteria 112
Worcesterberry 19, 396
woundwort: see *Stachys*
–, hedge: see *Stachys sylvatica*

yang tao 21
yarrow: see *Achillea millefolium*
yellow-rattle: see *Rhinanthus crista-
  galli*
–, greater: see *Rhinanthus major*

General index

Abax ater: see *Abax parallelepipedus*
– *parallelepipedus* 124
abbreviata, Nematus: see
   Micronematus abbreviatus
–, *Pristiphora*: see *Micronematus*
   abbreviatus
abbreviatus, Micronematus 393
Abraxas grossulariata 315–316
Acalitus essigi 406
– *phloeocoptes* 227, 406–407
Acalla: see *Acleris*
– *contaminana*: see *Acleris rhombana*
Acanthosoma haemorrhoidale 32
ACANTHOSOMATIDAE 12, 32
ACARI 16
Aceria 409–410
– *brachytarsus* 409
– *erineus* 407
– *essigi*: see *Acalitus essigi*
– *ficus* 408
– *gracilis*: see *Phyllocoptes gracilis*
– *microcarpae* 409
– *oleae* 408
– *sheldoni* 409
– *tristriatus* 409
aceris, Acronicta 354–355
achatana, Ancylis 259
Acherontia atropos 335–336
Acleris comariana 253–254
– *cristana* 254–255
– *laterana* 255
– *latifasciana*: see *Acleris laterana*
– *lipsiana* 255–256
– *rhombana* 256–257
– *variegana* 257
Acocephalus nervosus: see *Aphrodes*
   bicinctus
acorn moth 277–278
Acroclita naevana: see *Rhopobota*
   naevana
ACRIDIDAE 12, 25
Acronicta aceris 354–355
– *alni* 355
– *auricoma* 355
– *psi* 355–356
– *rumicis* 356
– *tridens* 357
ACRONICTINAE 354
Acronycta: see *Acronicta*
ACTINOTRICHIDA 16
ACULEATA 15
Aculops pelekassi 410
Aculus cornutus 410
– *fockeui* 410–411
– *schlechtendali* 411–412

Acyrthosiphon malvae rogersii: see
   Acyrthosiphon rogersii
– *rogersii* 66, 84
adamsoni, Thrips: see *Thrips*
   minutissimus
adonidum, Pseudococcus: see
   Pseudococcus longispinus
Adoxophyes orana 258
– *reticulana*: see *Adoxophyes orana*
Adoxus obscurus: see *Bromius*
   obscurus
Aegeria myopaeformis: see
   Synanthedon myopaeformis
AEGERIIDAE: see SESIIDAE
aeneus, Harpalus: see *Harpalus affinis*
–, *Meligethes* 136
aequatus, Neocoenorrhinus 148–149
aerata, Batophila 142
aescularia, Alsophila 318–319
–, *Anisopteryx*: see *Alsophila*
   aescularia
aesculi, Pseudococcus: see
   Heliococcus bohemicus
–, *Zeuzera*: see *Zeuzera pyrina*
aethiops, Caliroa: see *Endelomyia*
   aethiops
–, *Endelomyia* 387–388
affinis, Harpalus 125
–, *Pseudococcus*: see *Pseudococcus*
   viburni
Agelastica alni 140–141
Aglaope infausta 206–207
Agrilus aurichalceus 130–131
– *derasofasciatus* 131
– *desarofasciatus*: see *Agrilus*
   derasofasciatus
– *sinuatus* 131–132
– *viridis* 132
Agriopis aurantiaria 316–317
– *leucophaearia* 317
– *marginaria* 317–318
Agriotes lineatus 133
– *obscurus* 133
– *sputator* 134
Agrius convolvuli 336–337
Agromyza flaviceps 191
– *igniceps* 191
– *potentillae* 192
– *spiraeae*: see *Agromyza potentillae*
AGROMYZIDAE 14, 191–193
Agrotis exclamationis 357–358
– *ipsilon* 358
– *puta* 358–359
– *segetum* 359–360
– *suffusa*: see *Agrotis ipsilon*

Aguriahana stellulata 48
albipes, Fenusa: see *Metallus albipes*
–, *Metallus* 391
albistria, Argyresthia 227
albithoracellus, Euhyponomeutoides
   229–230
Alcis repandata 318
alder flies 12
– leaf beetle 140–141
– moth 355
Aleurodes: see *Aleyrodes* [in part]
– *rubicola*: see *Asterobemisia carpini*
ALEURODIDAE: see
   ALEYRODIDAE
Aleurolobus olivinus 63
Aleurothrixus floccosus 63–64
Aleyrodes brassicae: see *Aleyrodes*
   proletella
– *fragariae*: see *Aleyrodes lonicerae*
– *lonicerae* 64
– *proletella* 64
ALEYRODIDAE 13, 63–65
ALEYRODOIDEA 13
algira, Dysgonia 363
Allantus cinctus 383–384
allied shade moth: see light grey
   tortrix moth
Allophyes oxyacanthae 360
almond & plum bud gall mite: see
   plum spur mite
– lace bug 35
– leaf skeletonizer moth 206–207
– sawfly 381–382
– seed wasp 401
alneti, Alnetoidia 48
–, *Phyllobius*: see *Phyllobius*
   pomaceus
Alnetoidia alneti 48
alni, Acronicta 355
–, *Agelastica* 140–141
–, *Bythoscopus*: see *Oncopsis alni*
–, *Oncopsis* 53–54
Alsophila aescularia 318–319
Altica ampelophaga 141
Amathes c-nigrum: see *Xestia c-*
   nigrum
ambiguella, Clysia: see *Eupoecilia*
   ambiguella
–, *Cochylis*: see *Eupoecilia ambiguella*
–, *Eupoecilia* 252–253
ambiguus, Psallus 42
ambrosia beetles 171
– fungi 171, 174
American blight aphid: see woolly
   aphid

– eastern cherry fruit fly 190
– grapevine leaf miner moth 221
– white moth 351–353
Ametastegia glabrata 384
ampelophaga, Altica 141
–, *Haltica*: see *Altica ampelophaga*
–, *Theresia*: see *Theresimima ampelophaga*
–, *Theresimima* 207
Amphimallon solstitialis 127–128
Amphimallus solstitialis: see *Amphimallon solstitialis*
Amphipyra pyramidea 360
Amphitetranychus viennensis 421
Amphorophora idaei 66–67
– *rubi* 67
amygdali, Brachycaudus: see *Brachycaudus schwartzi*
–, *Eurytoma* 401
–, *Hyalopterus* 82
–, *Scolytus* 172
ANACTINOTRICHIDA 16
Anarsia lineatella 247–248
anatipennella, Coleophora 239
Ancylis achatana 259
– *comptana* 259
– – *fragariae* 259
– *curvana*: see *Ancylis selenana*
– *selenana* 260
– *tineana* 260
angle-shades moth 376–377
angular leaf spot 43
angustiorana, Batodes: see *Ditula angustiorana*
–, *Capua*: see *Ditula angustiorana*
–, *Ditula* 278–279
angustipennella, Chrysocorys: see *Schreckensteinia festaliella*
Anisandrus dispar: see *Xyleborus dispar*
– *saxesenii*: see *Xyleborinus saxesenii*
Anisopteryx aescularia: see *Alsophila aescularia*
anomala, Pennisetia: see *Pennisetia hyaeiformis*
ANOPLURA 12
Anthocoptes ribis 412
ANTHOMYIIDAE 14, 194
Anthonomus cinctus: see *Anthonomus piri*
– *druporum*: see *Furcipes rectirostris*
– *piri* 152
– *pomorum* 153–154
– *pyri*: see *Anthonomus piri*
– *rectirostris*: see *Furcipes rectirostris*
– *rubi* 154–155
Anthophila: see *Choreutis*
anthrisci, Dysaphis 78, 79
antiqua, Orgyia 349–350
Antispila rivillei 203–204
antler sawfly 386
ants 12, 15, 60, 66, 68, 69, 70, 71, 76, 85, 88, 91, 92, 96, 110, 112, 113, 119, 120, 123, 205, 304, 402
Anuraphis farfarae 68

– *helichrysi*: see *Brachycaudus helichrysi*
– *masseei*: see *Brachycaudus persicae*
– *pyri*: see *Dysaphis pyri*
– *roseus*: see *Dysaphis plantaginea*
– *subterranea* 68
Aonidiella aurantii 99
– *citrina* 99–100
aonidum, Chrysomphalus 102
Apatele: see *Acronicta*
Aphanostigma piri 97
– *pyri*: see *Aphanostigma piri*
Aphelenchoides 420
Aphelia paleana 260–261
APHIDIDAE 12, 66–96
APHIDINAE 66–91
aphidivorus, Diastrophus: see *Diastrophus rubi*
APHIDOIDEA 13
aphids 10, 11, 66–96
Aphidula grossulariae: see *Aphis grossulariae*
Aphis citricola: see *Aphis spiraecola*
– *craccivora* 68
– *fabae* 68, 69
– *forbesi* 69
– *gossypii* 69–70
– *grossulariae* 70
– *idaei* 70
– *nerii* 71
– *pomi* 71–72
– *punicae* 72
– *ruborum* 72
– *schneideri* 73
– *spiraecola* 73–74
– *triglochinis* 74
Aphrodes bicinctus 49
Aphthona euphorbiae 142
Apion: see *Protapion*
– *dichroum*: see *Protapion fulvipes*
APIONIDAE 14, 151
Apocheima hispidaria 319
– *pilosaria*: see *Phigalia pilosaria*
APOCRITA 15
Apoderus coryli 145
Apolygus spinolae 37
Aporia crataegi 302–304
Appelia: see *Brachycaudus*
appendiculata, Nephrotoma 175
–, *Pristiphora*: see *Pristiphora rufipes*
apple & pear bryobia mite 421–422
– – plum casebearer moth 242–243
– blossom weevil 153–154
– borer: see apple clearwing moth
– bud moth: see bud moth
– – weevil 152
– canker 95, 251
– capsid 40–41
– clearwing moth 222–223, 279
– ermine moth: see apple small ermine moth
– fruit moth 227–228
– – rhynchites 148–149
– leaf & bud mite: see apple rust mite
– – blister moth 214–215

– – erineum mite 418–419
– – midge 180
– – miner 209–210
– – sawfly 399
– – skeletonizer 225–226
– – skeletonizer moth 225–226
apple leaf-curling midge: see apple leaf midge
– maggot 276
– proliferation (AP) phytoplasma 53, 59, 61
– psyllid: see apple sucker
– pygmy moth 198–199
– rust mite 411–412
– sawfly 390–391
– seed wasp 402
– small ermine moth 234–235
– sucker 57–58
– twig cutter 147
apple/anthriscus aphid 78, 79
apple/chervil aphid 78–79
apple/grass aphid 90
apple-wood longhorn beetle 138
apricans, Protapion 151
APTERYGOTA 11
ARACHNIDA 9, 16
araneiformis, Barypeithes 155
arbustorum, Plagiognathus 42
arcella, Argyresthia: see *Argyresthia curvella*
Archips crataegana 261
– *crataeganus*: see *Archips crataegana*
– *podana* 262–263
– *podanus*: see *Archips podana*
– *roborana*: see *Archips crataegana*
– *rosana* 263–264
– *rosanus*: see *Archips rosana*
– *xylosteana* 264–265
– *xylosteanus*: see *Archips xylosteana*
Arctia caja 351
ARCTIIDAE 15, 351–353
ARCTIINAE 351
argentatus, Phyllobius 165
argiolus, Celastrina 304–305
argyrana, Pammene 287
Argyresthia albistria 227
– *arcella*: see *Argyresthia curvella*
– *conjugella* 227–228
– *cornella*: see *Argyresthia curvella*
– *curvella* 228
– *ephippella*: see *Argyresthia pruniella*
– *pruniella* 228–229
– *spinosella* 229
Argyroploce: see *Hedya* [in part]
– *lacunana*: see *Celypha lacunana*
– *nubiferana*: see *Hedya dimidioalba*
– *pruniana*: see *Hedya pruniana*
– *variegana*: see *Hedya dimidioalba*
Argyrotaenia ljungiana: see *Argyrotaenia pulchellana*
– *pulchellana* 265–266
armoured scales 99–108
arrow-head scale: see Japanese citrus fruit scale
ARTHROPODA 9

arundinis, Cicadella: see *Cicadella viridis*
ascalonicus, Myzus 85–86
ash bark beetle 171–172
asperatus, Sciaphilus 168–169
Aspidiotus nerii 100
– *ostreaeformis*: see *Quadraspidiotus ostreaeformis*
– *perniciosus*: see *Quadraspidiotus pernicosus*
– *rosae*: see *Aulacaspis rosae*
asseclana, Cnephasia 270
assimilata, Eupithecia 327
Asterobemisia avellanae: see *Asterobemisia carpini*
– *carpini* 64–65
Asterochiton carpini: see *Asterobemisia carpini*
ASTEROLECANIIDAE 13, 109
ASTIGMATA 16
astigmatid mites 16
ater, Abax: see *Abax parallelepipedus*
Athysanus obsoletus: see *Euscelis obsoletus*
atra, Blastodacna: see *Spuleria atra*
–, *Chrysoclista*: see *Spuleria atra*
–, *Spuleria* 251
Atractotomus mali 37
atratus, Thrips 29
atropos, Acherontia 335–336
ATTELABIDAE 14, 145
attenuata, Psylliodes 144
AUCHENORRHYNCHA 12
augur, Graphiphora 364–365
Aulacaspis pentagona: see *Pseudaulacaspis pentagona*
– *rosae* 101
Aulacorthum solani 74
aurantiaria, Agriopis 316–317
–, *Erannis*: see *Agriopis aurantiaria*
aurantiaria, Hybernia: see *Agriopis aurantiaria*
aurantii, Aonidiella 99
–, *Toxoptera* 91
aurata, Cetonia 128
auratus, Rhynchites 150–151
aurella, Stigmella 197–198
aurichalceus, Agrilus 130–131
auricoma, Acronicta 355
auricularia, Forficula 26
Australian mealybug: see fluted scale
autumn apple tortrix moth 294–295
– leaf roller 294
avellanae, Asterobemisia: see *Asterobemisia carpini*
–, *Corylobium* 77
–, *Eriophyes*: see *Phytoptus avellanae*
–, *Phytocoptella*: see *Phytoptus avellanae*
–, *Phytoptus* 405
avellanella, Ornix: see *Parornix devoniella*

baccarum, Dolycoris 32
baccus, Rhynchites 151

bacteria 50
bacterial canker 171
Bactrocera oleae 187–188
Balaninus: see *Curculio*
banded rose sawfly 383–384
Barathra brassicae: see *Mamestra brassicae*
barbicornis, Magdalis 157
bark beetles 171–174
barred fruit tree tortrix moth 289–290
Barypeithes araneiformis 155
– *pellucidus* 155
Batia unitella 244
batis, Thyatira 315
Batodes angustiorana: see *Ditula angustiorana*
– *aerata* 142
– *rubi* 142–143
– *rubivora*: see *Batophila aerata*
beckii, Lepidosaphes 99, 103
beech jewel beetle 132
– leaf mining weevil 158
bees 9, 12, 15, 60
beetle mites 428
beetles 9, 10, 12, 124 et seq.
Bembecia hylaeiformis: see *Pennisetia hylaeiformis*
berlesiana, Lasioptera: see *Prolasioptera berlesiana*
–, *Prolasioptera* 184–185
betulae, Byctiscus 146–147
–, *Pulvinaria* 117
–, *Thecla* 306
–, *Zephrus*: see *Thecla betulae*
betularia, Biston 320
betuleti, Byctiscus: see *Byctiscus betulae*
–, *Rhynchites*: see *Byctiscus betulae*
Bibio 177
BIBIONIDAE 14, 177
bicinctus, Aphrodes 49
big bud mite: see black currant gall mite
big-beaked plum mite 414
bilunaria, Selenia: see *Selenia dentaria*
Biorhiza pallida 287
bipunctatus, Calocoris: see *Closterotomus norvegicus*
birds 124, 126, 202
bishop bug: see European tarnished plant bug
bisonia, Stictocephala 46
Biston betularia 320
– *hirtaria*: see *Lycia hirtaria*
– *strataria* 320–321
bistortata, Boarmia: see *Ectropis bistortata*
–, *Ectropis* 325
biting lice 12
black apple capsid 37
– arches moth 349
– bean aphid 68, 69
– citrus aphid 91
– – scale 106

– currant aphid: see European black currant aphid
– – flower midge 182
– – fruit sawfly 397
– – gall mite 412–413
– – leaf midge 182–183
– – leaf-curling midge: see black currant midge
– – sawfly 395–396
– – stem midge 186
– currant–sowthistle aphid: see currant/sowthistle aphid
– hairstreak butterfly 305
– legume aphid: see cowpea aphid
– peach aphid 75–76
– raspberry necrosis virus 67
– scale: see Mediterranean black scale
– vine weevil: see vine weevil
blackberry aphid: see blackberry/cereal aphid
– flower midge 179
– leaf midge 181
– leaf mite: see raspberry leaf & bud mite
– mite 406
– pygmy moth 197–198
– stem gall midge 184
blackberry/cereal aphid 91
blackflies 66
black-veined white butterfly 302–304
blancardella, Phyllonorycter 214–215
BLASTOBASIDAE 15, 250
Blastobasis decolorella 250
Blastodacna atra: see *Spuleria atra*
Blennocampa geniculata: see *Monophadnoides geniculatus*
blossom beetle: see pollen beetle
blue bug: see rosy apple aphid
Boarmia bistortata: see *Ectropis bistortata*
– *gemmaria*: see *Peribatodes rhomboidaria*
– *rhomboidaria*: see *Peribatodes rhomboidaria*
Bohemican mealybug 120
bohemicus, Heliococcus 120
BOMBYCOIDEA 15
booklice: see psocids
BOSTRICHOIDEA 13
BOSTRYCHIDAE 13, 135
botrana, Lobesia 285–286
–, *Polychrosis*: see *Lobesia botrana*
Brachycaudus amygdali: see *Brachycaudus schwartzi*
– *cardui* 74
– *helichrysi* 74–75
– *persicae* 75–76
– *persicaecola*: see *Brachycaudus persicae*
– *prunicola*: see *Brachycaudus schwartzi*
– *schwartzi* 76
Brachyrhinus: see *Otiorhynchus*
brachytarsus, Aceria 409
bractella, Oecophora 247

bramble shoot moth 280–281
– – webber 280
brassicae, Aleyrodes: see *Aleyrodes proletella*
–, *Barathra*: see *Mamestra brassicae*
–, *Mamestra* 367
–, *Melanchra*: see *Mamestra brassicae*
brevicollis, Nebria 126
brevicornis, Claremontia: see *Claremontia confusa*
Brevipalpus pulcher: see *Cenopalpus pulcher*
brevis, Hoplocampa 388
brimstone moth 331
brindled beauty moth 328–329
bristle-tails 11
broad-barred button moth 255
broad-bordered yellow underwing moth 370–371
broadleaved pinhole borer 174
Bromius obscurus 143
broom moth 361
brown apple budworm 293
– chafer 130
– citrus aphid 91
– fruit mite 422
– hairstreak butterfly 306
– leaf weevil 166
– mite: see apple & pear bryobia mite
– oak tortrix moth 264–265
– peach aphid 96
– rot 151, 386
– scale: see European brown scale
– soft scale 111–112
brown-tail moth 345–346
brullei, Priophorus: see *Priophorus morio*
brumata, Operophtera 319, 329–330, 363
brunnea, Serica 130
Bryobia cristata 421
– *ribis* 421
– *rubrioculus* 421–422
– – *redikorzevi* 422
bubalus, Ceresa: see *Stictocephala bisonia*
Bucculatrix crataegi 208–209
bucephala, Phalera 343
bud moth 293–294
buff ermine moth 353
buff-tip moth 343
buffalo treehopper 46
bulbous grapevine weevil 164
BUPRESTIDAE 13, 130–132
BUPRESTOIDEA 13
burning grape leafhopper 53
bush crickets 23–24
butterflies 10, 12, 15, 301–308
buttoned snout moth 365
Byctiscus betulae 146–147
– *betuleti*: see *Byctiscus betulae*
Bythoscopus alni: see *Oncopsis alni*
BYTURIDAE 13, 136–137
Byturus tomentosus 136–137
– *urbanus*: see *Byturus tomentosus*

cabbage moth 367
– whitefly 64
Cacoecia: see *Archips* [in part]
– *costana*: see *Clepsis spectrana*
– *hebenstreitella*: see *Choristoneura hebenstreitella*
– *lecheana*: see *Ptycholoma lecheana*
– *oporana*: see *Archips podana*
– *pronubana*: see *Cacoecimorpha pronubana*
Cacoecimorpha pronubana 266–267
Cacopsylla mali 57–58
– *melanoneura* 58–59
– *piri*: see *Cacopsylla pyri*
– *pirisuga*: see *Cacopsylla pyrisuga*
– *pyri* 59
– *pyricola* 59–60
– – *simulans* 59
– – *typica* 59
– *pyrisuga* 60
– *simulans*: see *Cacopsylla pyricola*
caddis flies 12
Caenorrhinus: see *Neocoenorrhinus*
caeruleocephala, Diloba 344
–, *Episema*: see *Diloba caeruleocephala*
caeruleus, Involvulus 147
caja, Arctia 351
Calathus cisteloides: see *Calathus fuscipes*
– *fuscipes* 124–125
c-album, Polygonia 308
calcaratus, Phyllobius: see *Phyllobius glaucus*
calceolariae, Pseudococcus 121
Calepitrimerus vitis 412
California red scale 99
– scale: see San José scale
californica, Pristiphora: see *Micronematus abbreviatus*
caliginosus, Epipolaeus: see *Plinthus caliginosus*
–, *Plintha* 167–168
–, *Plinthus*: see *Plintha caliginosus*
Caliroa aethiops: see *Endelomyia aethiops*
– *cerasi* 384–385
– *limacina*: see *Caliroa cerasi*
CALLAPHIDIDAE: see CALLAPHIDINAE
CALLAPHIDINAE 92–93
Callaphis juglandis 92
Callipterus juglandis: see *Callaphis juglandis*
Callisto denticulella 211–212
Calliteara pudibunda 345
Calocoris bipunctatus: see *Closterotomus norvegicus*
– *fulvomaculatus*: see *Closterotomus fulvomaculatus*
– *norvegicus*: see *Closterotomus norvegicus*
Caloptilia roscipennella 212–213
Calymnia: see *Cosmia*
Campaea margaritata 321

Campylomma verbasci 37
CANTHARIDAE 13, 134
Cantharis fusca 134
– *livida* 134
– *obscura* 134
– *pellucida* 134
CANTHAROIDEA 13
capitata, Ceratitis 188–189
capitella, Incurvaria: see *Lampronia capitella*
–, *Lampronia* 202
Capitophorus: see *Cryptomyzus* [in part]
– *fragariae*: see *Chaetosiphon fragaefolii*
capittella, Lampronia: see *Lampronia capitella*
Capnodis tenebrionis 132
capsids 37–42
Capsodes sulcatus 38
Capua angustiorana: see *Batodes angustiorana*
– *reticulana*: see *Adoxophyes orana*
CARABIDAE 13, 124–127
CARABOIDEA 13
Carcina quercana 244–245
cardui, Brachycaudus 74
carnation thrips 29
– tortrix moth 266–267
carpini, Asterobemisia 64–65
–, *Asterochiton*: see *Asterobemisia carpini*
–, *Eotetranychus* 422–423
Carpocapsa: see *Cydia*
CARSIDARIDAE 13, 62
casebearer moths 239–243
Cecidomyia pruni: see *Putoniella pruni*
CECIDOMYIIDAE 14, 177–187
Cecidophyopsis grossulariae 412
– *ribis* 412–413
– *selachodon* 413
– *vermiformis* 413
Celastrina argiolus 304–305
celerio, Hippotion 339
Celypha lacunana 268
Cemiostoma scitella: see *Leucoptera malifoliella*
Cenopalpus pulcher 427
CEPHIDAE 15, 380–381
CEPHOIDEA 15
Cephus compressus: see *Janus compressus*
CERAMBYCIDAE 13, 138–139
Ceramica pisi 361
cerasana, Pandemis 289–290
cerasi, Caliroa 384–385
–, *Magdalis* 157
–, *Myzus* 87–88
–, *Orthosia*: see *Orthosia incerta*
–, *Rhagoletis* 189–190
cerasicolella, Phyllonorycter 215–216
cerasiella, Swammerdamia: see *Swammerdamia pyrella*
Ceratitis capitata 188–189

ceratoniae, Ectomyelois 296
–, *Myelois*: see *Ectomyelois ceratoniae*
–, *Spectrobates*: see *Ectomyelois ceratoniae*
CERCOPIDAE 12, 43–44
CERCOPOIDEA 12
Cercopis sanguinea: see *Cercopis vulnerata*
– *vulnerata* 43
Ceresa bubalus: see *Stictocephala bisonia*
ceriferus, Ceroplastes 109
Ceroplastes ceriferus 109
– *nerii*: see *Ceroplastes rusci*
– *rusci* 109–110
– *sinensis* 110–111
Cerosipha forbesi: see *Aphis forbesi*
– *nerii*: see *Aphis nerii*
Cerostoma: see *Ypsolopha*
Cetonia aurata 128
chaerophylli, Dysaphis 78–79
Chaetocnema concinna 142
Chaetosiphon fragaefolii 76
chafers 127–130
chaff scale 106
chaffinches 202
CHALCIDOIDEA 15
Charaxes jasius 307
Cheimatobia: see *Operophtera*
cherry bark tortrix moth 279–280
– beetle mite 428
– fruit fly: see European cherry fruit fly
– – moth 228–229
– – weevil 156–157
– leaf blister moth 215–216
– leafhopper 48
– pistol casebearer moth 239
– small ermine moth 233
– stink bug: see forest bug
chestnut moth 361–362
Chimabacche: see *Diurnea*
Chinese scale: see San José scale
Chionaspis salicis 101–102
Chloroclysta siterata 321–322
– *truncata* 322
Chloroclystis coronata: see *Chloroclystis v-ata*
– *rectangulata* 322–323
– *v-ata* 323
Chloropulvinaria floccifera 111
chlorotic leaf roll disease 53
CHOREUTIDAE 14, 224–226
Choreutis nemorana 224–225
– *pariana* 225–226
Choristoneura diversana 267–268
– *hebenstreitella* 268
– *sorbiana*: see *Choristoneura hebenstreilella*
Chromaphis juglandicola 92
chrysanthemi, Plagiognathus 42
Chrysoclista atra: see *Spuleria atra*
Chrysocorys angustipennella: see *Schreckensteinia festaliella*
CHRYSOESTHIIDAE: see

SCHRECKENSTEINIIDAE
CHRYSOMELIDAE 13, 140–144
CHRYSOMELOIDEA 13
Chrysomphalus aonidum 102
– *dictyospermi* 102
– *ficus*: see *Chrysomphalus aonidum*
chrysorrhoea, Euproctis 345–346
Cicadella arundinis: see *Cicadella viridis*
– *stellulata*: see *Aguriahana stellulata*
– *viridis* 49–50
CICADELLIDAE 12, 47–57
CICADELLOIDEA 12
Cimbex quadrimaculata: see *Palaeocimbex quadrimaculata*
CIMBICIDAE 15, 381–382
cinctus, Allantus 383–384
–, *Anthonomus*: see *Anthonomus piri*
–, *Emphytus*: see *Allantus cinctus*
cingulata, Rhagoletis 190
Cionus fraxini: see *Stereonychus fraxini*
cisteloides, Calathus: see *Calathus fuscipes*
citrella, Phyllocnistis 219–220
citri, Dialeurodes 65
–, *Metatetranychus*: see *Panonychus citri*
–, *Panonychus* 423–424
–, *Planococcus* 120
–, *Prays* 230
–, *Pseudococcus*: see *Planococcus citri*
citricida, Toxoptera 91
citricola, Aphis: see *Aphis spiraecola*
–, *Lepidosaphes*: see *Lepidosaphes beckii*
–, *Lithocolletis*: see *Phyllocnistis citrella*
citrina, Aonidiella 99–100
citrophilus mealybug 121
citrus bud mite 409
– flower moth 230
– fruit mite: see citrus red spider mite
– leaf miner moth 219–220
– mealybug 120
– mussel scale 99, 103
– planthopper 45–46
– red mite: see citrus red spider mite
– – scale: see California red scale
– – spider mite 423–424
– rust mite 419
– scale 103
– tristeza closterovirus 70, 71, 74, 91
– wax scale 110–111
– whitefly 65
– yellow scale: see yellow scale
– young-fruit borers 230
CIXIIDAE 12, 47
Cladius difformis 385–386
– *pectinicornis* 386
Claremontia brevicornis: see *Claremontia confusa*
– *confusa* 386
clavipes, Otiorhynchus 158–159

clay-coloured weevil 162
clearwing moths 222–224
Cleora repandata: see *Alcis repandata*
Clepsis costana: see *Clepsis spectrana*
– *spectrana* 269
clerkella, Lyonetia 209–210
click beetles 133–134
climbing cutworms 371, 377
Clinodiplosis oculiperda: see *Resseliella oculiperda*
– *oleisuga*: see *Resseliella oleisuga*
Closterotomus fulvomaculatus 38
– *norvegicus* 38–39
clouded drab moth 374–375
clover seed weevil 151
Clysia ambiguella: see *Eupoecilia ambiguella*
Cneorrhinus plagiatus: see *Philopedon plagiatus*
Cnephasia asseclana 270
– *incertana* 270–271
– *interjectana*: see *Cnephasia asseclana*
– *longana* 271–272
– *virgaureana*: see *Cnephasia asseclana*
Cnephasiella incertana: see *Cnephasia incertana*
c-nigrum, Amathes: see *Xestia c-nigrum*
–, *Euxoa*: see *Xestia c-nigrum*
–, *Xestia* 378
COCCIDAE 13, 109–119
COCCOIDEA 13
Coccus hesperidum 111–112
COCHYLIDAE 15, 252–253
Cochylis ambiguella: see *Eupoecilia ambiguella*
cockchafers 128–129
cockroaches 12
codling moth 276–277
Coenorrhinus: see *Neocoenorrhinus*
coeruleus, Haplorhynchites: see *Involvulus caeruleus*
–, *Rhynchites*: see *Rhynchites caeruleus*
Coleophora anatipennella 239
– *coracipennella* 242
– *hemerobiella* 239–240
– *nigricella*: see *Coleophora spinella*
– *paripennella* 243
– *potentillae* 240–241
– *prunifoliae* 242
– *serratella* 241–242
– *spinella* 242–243
– *violacea* 243
COLEOPHORIDAE 15, 239–243
COLEOPTERA 9, 12, 13–14
COLLEMBOLA 11
Collotois pennaria: see *Colotois pennaria*
Colomerus vitis 413–414
Colotois pennaria 324
comariana, Acleris 253–254
comes, Noctua 369–370

comma butterfly 308
common black ant 402
– black ground beetle 126
– click beetles 133–134
– crane flies 175–176
– earwig 26
– froghopper 44
– gooseberry sawfly 396–397
– green capsid 39–40
– ground beetles 124–127
– leaf weevil 167
– marbled carpet moth 322
– quaker moth 376
– small ermine moth 235–236
– wasp 404
complanella, Tischeria: see Tischeria
    ekebladella
completa, Rhagoletis 190
compressus, Cephus: see Janus
    compressus
–, Janus 380–381
comptana, Ancylis 259
– fragariae, Ancylis 259
Comstockaspis perniciosa: see
    Quadraspidiotus pernicosus
conchyformis, Lepidosaphes: see
    Lepidosaphes minima
concinna, Chaetocnema 142
concomitella, Lithocolletis: see
    Phyllonorycter blancardella
confusa, Claremontia 386
–, Monophadnoides: see Claremontia
    confusa
congelatella, Exapate 282
conjugella, Argyresthia 227–228
Conistra vaccinii 361–362
Conopia myopaeformis: see
    Synanthedon myopaeformis
Conosanus obsoletus 52
consobrinus, Nematus: see Nematus
    leucotrochus
contaminana, Acalla: see Acleris
    rhombana
Contarinia humuli 177–178
– pruniflorum 178
– pyrivora 178–179
– ribis 179
– rubicola 179
– viticola 179
convolutella, Zophodia 300–301
convolvuli, Agrius 336–337
–, Herse: see Agrius convolvuli
–, Proctoparce: see Herse convolvuli
convolvulus hawk moth 336–337
copper underwing moth 360
coracipennella, Coleophora 242
cornella, Argyresthia: see Argyresthia
    curvella
corni, Eulecanium: see
    Parthenolecanium corni
–, Parthenolecanium 115–116
cornutus, Aculus 410
Coroebus elatus 132
coronata, Chloroclystis: see
    Chloroclystis v-ata

corylana, Pandemis 290
coryli, Apoderus 145
–, Eotetranychus: see Eotetranychus
    pruni
–, Eulecanium: see Eulecanium tiliae
–, Lecanium: see Eulecanium tiliae
–, Myzocallis 93
–, Phyllonorycter 216
–, Strophosomus: see Strophosoma
    melanogrammum
corylifoliella, Phyllonorycter 216–217
Corylobium avellanae 77
Cosmia pyralina 362
– trapezina 362–363
COSMOPTERYGIDAE: see
    MOMPHIDAE
COSSIDAE 14, 204–206
COSSOIDEA 14
Cossus cossus 204–205
– ligniperda: see Cossus cossus
cossus, Cossus 204–205
costalis, Psylla 61
costana, Cacoecia: see Clepsis
    spectrana
–, Clepsis: see Clepsis spectrana
cotton aphid: see melon & cotton
    aphid
cottony birch scale: see woolly vine
    scale
– currant scale: see woolly vine scale
– cushion scale: see fluted scale
– grape scale: see woolly vine scale
– hydrangea scale: see hydrangea scale
– maple scale 115
– vine scale: see woolly vine scale
cowpea aphid 68
crabro, Vespa 403
craccivora, Aphis 68
crane flies 175–176
crataegana, Archips 261
crataeganus, Archips: see Archips
    crataegana
crataegella, Recurvella: see
    Recurvella nanella
–, Scythropia 232
crataegi, Aporia 302–304
–, Bucculatrix 208–209
–, Edwardsiana 50
–, Trichiura 313
–, Typhlocyba: see Edwardsiana
    crataegi
cribricollis, Otiorhynchus 159
cribripennis, Neocoenorrhinus 149
crickets 10, 12, 23 et seq.
cristana, Acleris 254–255
cristata, Bryobia 421
Crocallis elinguaria 324–325
Croesia holmiana 272
Croesus septentrionalis 386–387
cruda, Orthosia 372
crudum, Eulecanium: see
    Parthenolecanium corni
–, Lecanium: see Parthenolecanium
    corni
Cryptoblabes gnidiella 295–296

Cryptomyzus galeopsidis 77
– ribis 78
CRYPTOSTIGMATA 16, 428
Ctenicera cuprea 134
Cuban laurel thrips 31
cuckoo-spit 43, 44
– bug 44
CUCUJOIDEA 13
cucullatella, Nola 354
cunea, Hyphantria 351–353
cuprea, Ctenicera 134
–, Feronia: see Poecilus cupreus
cupreus, Poecilus 126
–, Pterostichus: see Poecilus cupreus
Curculio elephas 155
– nucum 156
CURCULIONIDAE 14, 152–170
CURCULIONOIDEA 14
currant borer 223
– bud mite: see black currant gall mite
– – moth 229–230
– clearwing moth 223–224
– mealybug: see citrophilus mealybug
– moth: see magpie moth
– pug moth 327
– root aphid 95
– rust mite 412
– shoot borer 202
– stem aphid 90
currant/lettuce aphid 89
currant/sowthistle aphid 83
currant/yellow-rattle aphid 84
curvana, Ancylis: see Ancylis selenana
cyclamen tortrix moth: see straw-
    coloured tortrix moth
curvella, Argyresthia 228
cushion scale 111
cutworm(s) 358, 359, 364
–, climbing 371, 377
CYCLORRAPHA 14
Cydia funebrana 272–273
– janthinana 274
– lobarzewskii 274
– molesta 274–276
– pomonella 276–277
– prunivorana: see Cydia lobarzewskii
– pyrivora 277
– splendana 277–278
CYMATOPHORIDAE: see
    THYATIRIDAE
CYNIPIDAE 15, 400–401
CYNIPOIDEA 15

Dacus oleae: see Bactrocera oleae
daddy-longlegs 175
Daktulosphaira vitifoliae: see Viteus
    vitifoliae
damson/hop aphid 89–90
Daphnis nerii 337–338
dark dagger moth 357
– fruit-tree tortrix moth 291
– green apple capsid 41–42
– strawberry tortrix moth 269
– sword grass moth 358
Dasineura mali 180

– *oleae* 180
– *plicatrix* 181
– *pyri* 181–182
– *ribis* 182
– *tetensi* 182–183
– *tortrix* 183
Dasychira pudibunda: see *Calliteara pudibunda*
Dasyneura: see *Dasineura*
dealbana, *Gypsonoma* 282–283
death's head hawk moth 335–336
debilis, *Ribautiana* 54
–, *Typhlocyba*: see *Ribautiana debilis*
December moth 312–313
decipiens, *Empoasca* 52
decolorella, *Blastobasis* 250
defoliaria, *Erannis* 326–327
–, *Hibernia*: see *Erannis defoliaria*
Deilephila elpenor 338–339
– *nerii*: see *Daphnis nerii*
– *porcellus* 339
dentaria, *Selenia* 334
denticulella, *Callisto* 211–212
dentiens, *Pegomya*: see *Pegomya rubivora*
derasofasciatus, *Agrilus* 131
DERMAPTERA 12, 26
desarofasciatus, *Agrilus*: see *Agrilus derasofasciatus*
devecta, *Dysaphis* 79–80
devoniella, *Parornix* 213–214
Dialeurodes citri 65
DIASPIDIDAE 13, 99–108
Diaspis pentagona: see *Pseudaulacaspis pentagona*
– *rosae*: see *Aulacaspis rosae*
Diastrophus aphidivorus: see *Diastrophus rubi*
– *rubi* 400–401
Dichelomyia oenophila: see *Janetiella oenophila*
dichroum, *Apion*: see *Protapion fulvipes*
dictyospermi, *Chrysomphalus* 102
Didymella applanata 187
difformis, *Cladius* 385–386
Diloba caeruleocephala 344
DILOBIDAE 15, 344
Dilophus febrilis 177
dilutata, *Epirrita* 325–326
–, *Oporinia*: see *Epirrita dilutata*
dimidioalba, *Hedya* 283–284
Diplosis humuli: see *Contarinia humuli*
DIPLURA 11
diplurans 11
Diptacus gigantorhynchus 414
DIPTERA 12, 14
DISCYOPTERA 12
dispar, *Anisandrus*: see *Xyleborus dispar*
–, *Lymantria* 348–349
–, *Porthetria*: see *Lymantria dispar*
–, *Xyleborus* 174
DITRYSIA 14–15

Ditula angustiorana 278–279
Diurnea fagella 245
– *lipsiella* 246
– *phryganella*: see *Diurnea lipsiella*
diversana, *Choristoneura* 267–268
dock sawfly 384
Dolycoris baccarum 32
dot moth 368
dotted border moth 317–318
double dart moth 364–365
double-striped pug moth: see olive pug moth
dragonflies 11
Drepanothrips reuteri 27
Drosophila 190–191
– *suzukii* 191
DROSOPHILIDAE 14, 190–191
druporum, *Anthonomus*: see *Furcipes rectirostris*
dryberry mite: see raspberry leaf & bud mite
dry-wood termites 27
dumetata, *Pandemis* 290
dun-bar moth 362–363
Dysaphis anthrisci 78, 79
– *chaerophylli* 78–79
– *devecta* 79–80
– *plantaginea* 80–81
– *pyri* 81
Dysgonia algira 363

early moth 334–335
– thorn moth 334
earwigs 9, 10, 12, 26
Ectomyelois ceratoniae 296
Ectropis bistortata 325
Edwardsiana crataegi 50
– *prunicola* 50–51
– *rosae* 51
ekabladella, *Tischeria* 200–201
ELATERIDAE 13, 133–134
ELATEROIDEA 13
elatus, *Coroebus* 132
elephant bug: see strawberry blossom weevil
– hawk moth 338–339
elephas, *Curculio* 155
elinguaria, *Crocallis* 324–325
elm balloon-gall aphid 95
elm leaf aphid: see currant root aphid
elm–currant aphid: see currant root aphid
elpenor, *Deilephila* 338–339
EMBIOPTERA 12
emperor moth 314
Emphytus cinctus: see *Allantus cinctus*
Empoasca decipiens 52
– *flavescens*: see *Empoasca vitis*
– *lybica*: see *Jacobiasca lybica*
– *vitis* 52
Empria tridens 387
Enarmonia formosana 279–280
– *woeberiana*: see *Enarmonia formosana*
Endelomyia aethiops 387–388

ENDOPTERYGOTA 11, 12
engrailed moth 325
Eotetranychus carpini 422–423
– *coryli*: see *Eotetranychus pruni*
– *pomi*: see *Eotetranychus pruni*
– *pruni* 423
– *viticola*: see *Eotetranychus pruni*
EPHEMEROPTERA 11
Ephestia gnidiella: see *Cryptoblabes gnidiella*
– *parasitella* 296–297
– – *parasitella* 296
– – *unicolorella* 296
– *pinguis*: see *Euzerophera pinguis*
– *woodiella* 296
ephippella, *Argyresthia*: see *Argyresthia pruniella*
Epiblema penkleriana: see *Epinotia tenerana*
– *uddmanniana* 280–281
Epidiaspis leperii 102–103
Epinotia tenerana 281
Epiphyas postvittana 281–282
Epipolaeus caliginosus: see *Plinthus caliginosus*
Epirrita dilutata 325–326
Episema caeruleocephala: see *Diloba caeruleocephala*
Epitrimerus gigantorhynchus: see *Diptacus gigantorhynchus*
– *piri* 414–415
– *pyri*: see *Epitrimerus piri*
– *vitis*: see *Calepitrimerus vitis*
Erannis aurantiaria: see *Agriopis aurantiaria*
– *defoliaria* 326–327
– *leucophaearia*: see *Agriopis leucophaearia*
– *progemmaria*: see *Agriopis marginaria*
erineus, *Aceria* 407
–, *Eriophyes*: see *Aceria erineus*
Eriocampoides limacina: see *Caliroa cerasi*
Eriogaster lanestris 309–310
– *populi*: see *Poecilocampa populi*
Eriophyes avellanae: see *Phytoptus avellanae*
– *erineus*: see *Aceria erineus*
– *ficus*: see *Aceria ficus*
– *gracilis*: see *Phyllocoptes gracilis*
– *oleae*: see *Aceria oleae*
– *padi* 415–416
– *phloeocoptes*: see *Acalitus phloeocoptes*
– *pyri* 416–417
– *ribis*: see *Cecidophyopsis ribis*
– *sheldoni*: see *Aceria sheldoni*
– *similis* 417
– *tristriatus*: see *Aceria tristriatus*
– *vermiformis*: see *Cecidophyopsis vermiformis*
– *vitis*: see *Colomerus vitis*
ERIOPHYIDAE 16, 406–419
ERIOPHYOIDEA 16

Eriosoma grossulariae 94
– *lanigerum* 94–95
– *lanuginosum* 95
– *ulmi* 95
ERIOSOMATINAE 94–95
ermine moths 351, 353
Erythroneura flammigera: see *Zygina flammigera*
Esperia sulphurella 246–247
essigi, Acalitus 406
–, *Aceria*: see *Acalitus essigi*
–, *Phytoptus*: see *Acalitus essigi*
Euacanthus interruptus: see *Evacanthus interruptus*
Eucosma penkleriana: see *Epinotia tenerana*
Eudia pavonia: see *Saturnia pavonia*
Euhyponomeutoides albithoracellus 229–230
– *rufellus*: see *Euhyponomeutoides albithoracellus*
Eulecanium corni: see *Parthenolecanium corni*
– *coryli*: see *Eulecanium tiliae*
– *crudum*: see *Parthenolecanium corni*
– *excrescens* 112–113
– *mali*: see *Eulecanium tiliae*
– *persicae*: see *Parthenolecanium persicae*
– *prunastri*: see *Sphaerolecanium prunastri*
– *tiliae* 113
Eulia politana: see *Argyrotaenia pulchellana*
Eulithis prunata 327
euphorbiae, Aphthona 142
–, *Macrosiphum* 66, 84
Euphyllura olivina 61
Eupista: see *Coleophora*
EUPISTIDAE: see COLEOPHORIDAE
Eupithecia assimilata 327
– *exiguata* 328
Eupoecilia ambiguella 252–253
Euproctis chrysorrhoea 345–346
– *phaeoerrhoea*: see *Euproctis chrysorrhoea*
– *similis* 347
Eupsilia transversa 363–364
Eupteroidea stellulata: see *Aguriahana stellulata*
Eupteryx stellulata: see *Aguriahana stellulata*
Eupulvinaria hydrangeae: see *Pulvinaria hydrangeae*
European black currant aphid 77
– brown scale 115–116
– cherry fruit fly 189–190
– corn borer 297
– corn borer moth 297–298
– dry-wood termite 27
– fruit scale: see yellow plum scale
– – tree red spider mite 424–425
– gooseberry aphid 70
– mole cricket: see mole cricket

– peach scale 116
– pear sucker: see pear sucker
– permanent currant aphid 73
– red mite: see fruit tree red spider mite
– shot-hole borer: see broadleaved pinhole borer
– tarnished plant bug 41
– vine moth 285–286
Eurytoma amygdali 401
EURYTOMIDAE 15, 401
Euscelis incisus 52
Eutromula pariana: see *Choreutis pariana*
Euxoa c-nigrum: see *Xestia c-nigrum*
– *exclamationis*: see *Agrotis exclamationis*
– *tritici* 364
Euzetes lapidarius: see *Humerobates rostrolamellatus*
Euzophera fischeri: see *Euzerophera pinguis*
– *pinguis* 297
Evacanthus interruptus 52–53
evonymella, Yponomeuta 233
Exapate congelatella 282
exclamationis, Agrotis 357–358
–, *Euxoa*: see *Agrotis exclamationis*
excrescens, Eulecanium 112–113
exiguata, Eupithecia 328
EXOPTERYGOTA 11
eyed hawk moth 341
eye-spotted bud moth: see bud moth

fabae, Aphis 68, 69
fagata, Operophtera 330–331
fagella, Diurnea 245
fagi, Orchestes 158
–, *Rhynchaenus*: see *Orchestes fagi*
fall webworm 351
false red spider mites: see false spider mites
– spider mites 427
farfarae, Anuraphis 68
fasciana, Pammene 288
feathered thorn moth 324
febrilis, Dilophus 177
Fenusa albipes: see *Metallus albipes*
– *pumilio*: see *Metallus pumilus*
Feronia: see *Pterostichus* [in part]
– *cuprea*: see *Poecilus cupreus*
ferrugalis, Udea 298–299
ferruginea, Hoplocampa: see *Hoplocampa flava*
festaliella, Schreckensteinia 238
fever fly 177
ficorum, Gynaikothrips 31
ficus, Aceria 408
–, *Chrysomphalus*: see *Chrysomphalus aonidum*
–, *Eriophyes*: see *Aceria ficus*
–, *Homotoma* 62
–, *Hypoborus* 139, 171
–, *Lepidosaphes*: see *Lepidosaphes minima*

Fieberiella florii 53
fig bud mite 408
– mite: see fig bud mite
– mosaic virus 408
– mussel scale 104
– sucker 62
– wax scale 109–110
fig-tree scale: see fig wax scale
– skeletonizer moth 224–225
figure of eight moth 344
filbert aphid: see hazel aphid
– big-bud mite: see filbert bud mite
– bud mite 405
Filippia follicularis 113–114
fimbriata, Noctua 370–371
finitimella, Parornix 214
firethorn leaf blister moth 217–218
fischeri, Euzophera: see *Euzerophera pinguis*
flammigera, Erythroneura: see *Zygina flammigera*
–, *Zygina* 56–57
flat scarlet mite 427
flat-headed woodborer 132
FLATIDAE 12, 45–46
flava, Hoplocampa 388–389
flavescens, Empoasca: see *Empoasca vitis*
flaviceps, Agromyza 191
flavicollis, Kalotermes 27
flaviventris, Neurotoma: see *Neurotoma saltuum*
flavus, Thrips 29
flax flea beetle 142
– tortrix moth 270
flea beetles 140, 141, 142, 144
fleas 12
flies 61, 118
–, true 10, 12, 175 et seq.
floccifera, Chloropulvinaria 111
–, *Pulvinaria*: see *Chloropulvinaria floccifera*
floccosus, Aleurothrixus 63–64
Florida red scale 102
florii, Fieberiella 53
flower thrips 27
fluted scale 122–123
fockeui, Aculus 410–411
–, *Phyllocoptes*: see *Aculus fockeui*
–, *Phytoptus*: see *Aculus fockeui*
–, *Vasates*: see *Aculus fockeui*
follicularis, Filippia 113–114
–, *Philippia*: see *Filippia follicularis*
forbesi, Aphis 69
–, *Cerosipha*: see *Aphis forbesi*
forest bug 34
Forficula auricularia 26
FORFICULIDAE 12, 26
FORMICIDAE 15, 402
FORMICOIDEA 15
formosana, Enarmonia 279–280
formosus, Polydrusus 168
forsterana, Lozotaenia 286–287
four-spot bush cricket 24
fragaefolii, Chaetosiphon 76

–, *Pentatrichus*: see *Chaetosiphon
  fragaefolii*
fragariae, Aleyrodes: see *Aleyrodes
  lonicerae*
–, *Capitophorus*: see *Chaetosiphon
  fragaefolii*
–, *Pentatrichus*: see *Chaetosiphon
  fragaefolii*
–, *Sitobion* 91
–, *Steneotarsonemus*: see *Phytonemus
  pallidus fragariae*
–, *Tarsonemus*: see *Phytonemus
  pallidus fragariae*
fragariella, Stigmella 198
fragilis, Pseudococcus: see
  Pseudococcus calceolariae
Frankliniella intonsa 27
– *occidentalis* 28
fraxini, Cionus: see *Stereonychus
  fraxini*
–, *Stereonychus* 170
froggatti, Typhlocybai: see
  Edwardsiana crataegi
froghoppers 43–44
fruit bark beetle 173–174
– tree casebearer moth 239–240
– – leafhoppers 48, 50, 54, 56–57
– – red spider mite: see European fruit
  tree red spider mite
– – tortrix moth 262–263
fruitlet-mining tortrix moth 288–289
fruit-tree wood ambrosia beetle 174
FULGOROIDEA 12
fulvicornis, Hoplocampa: see
  Hoplocampa minuta
fulvipes, Protapion 151
fulvomaculatus, Calocoris: see
  Closterotomus fulvomaculatus
–, *Closterotomus* 38
funebrana, Cydia 272–273
funestum, Macrosiphum 84–85
fungi, ambrosia 171, 174
–, pathogenic 35, 47, 95, 151, 156,
  187, 189, 191, 223, 224, 251, 263,
  266, 286, 384
–, vectors of 151, 191
Furcipes rectirostris 156–157
fusca, Cantharis 134
fuscipennis, Thrips 29
fuscipes, Calathus 124–125

gahani, Pseudococcus: see
  Pseudococcus calceolariae
galeopsidis, Cryptomyzus 77
Galerucella sagittariae 143
– *tenella* 144
gall midges 177–187
– mites 405, 406–419
– wasps 15, 400–401
garden chafer 129–130
– swift moth 196–197
– tiger moth 351
Gastropacha quercifolia 310–311
Gelechia rhombella 248
GELECHIIDAE 15, 247–249

GELECHIOIDEA 14–15
gemmaria, Boarmia: see *Peribatodes
  rhomboidaria*
geniculata, Blennocampa: see
  Monophadnoides geniculatus
geniculatus, Monophadnoides
  393–394
Geoktapia pyraria: see *Melanaphis
  pyraria*
geometer moths 315–335
GEOMETRIDAE 15, 315–335
GEOMETROIDEA 15
German wasp 403–404
germanica, Vespula 403–404
germanicus, Neocoenorrhinus 149
geum sawfly 393–394
ghost swift moth 195–196
giant scales 122–123
– silk moths 314
gigantorhynchus, Diptacus 414
–, *Epitrimerus*: see *Diptacus
  gigantorhynchus*
glabrata, Ametastegia 384
glabratus, Taxonus: see *Ametastegia
  glabrata*
glasshouse & potato aphid 74
– mealybug 121
– orthezia: see fluted scale
– red spider mite: see two-spotted
  spider mite
glaucus, Phyllobius 165
Glischrochilus hortensis 135
– *quadripunctatus*: see *Glischrochilus
  hortensis*
– *quadrisignatus* 135
Gloeosporium album 95
– *perennans* 95
gloeosporium rots 95
gloverii, Lepidosaphes 103
GLYPHIPTERYGIDAE: see
  CHOREUTIDAE
Glyphodes unionalis: see *Palpita
  unionalis*
gnidiella, Cryptoblabes 295–296
goat moth 204–205
gold-tail moth: see yellow-tail moth
goniothorax, Phyllocoptes 418
gooseberry aphid: see European
  gooseberry aphid
– bryobia mite 421
– bud mite 412
– flower midge 179
– mite: see gooseberry bryobia mite
– moth 300–301
– red spider mite: see gooseberry
  bryobia mite
– root aphid 94
– sawfly: see common gooseberry
  sawfly
– vein-banding virus 89
gooseberry/sowthistle aphid 84
Gortyna micacea: see *Hydraecia
  micacea*
gossypii, Aphis 69–70
gothic moth 368–369

gothica, Orthosia 372–373
Gracilaria: see *Gracillaria*
GRACILARIIDAE: see
  GRACILLARIIDAE
gracilis, Aceria: see *Phyllocoptes
  gracilis*
–, *Eriophyes*: see *Phyllocoptes gracilis*
–, *Orthosia* 373–374
–, *Phyllocoptes* 418
–, *Phytoptus*: see *Phyllocoptes gracilis*
Gracillaria juglandella: see *Caloptilia
  roscipennella*
– *roscipennella*: see *Caloptilia
  roscipennella*
GRACILLARIIDAE 14, 211–219
graminella, Psyche: see *Pachythelia
  unicolor*
grape bud moth: see vine moth
– erineum mite: see vine leaf blister
  mite
– gall mite: see vine leaf blister mite
– phylloxera 97–98
– rust mite see vine rust mite
grapevine bois noir phytoplasma 47
– flavescence dorée phytoplasma 55
– vesperus beetle 139
– yellows 46, 54
Graphiphora augur 364–365
Grapholitha: see *Cydia* [in part]
– *ocellana: see Spilonota ocellana*
grasshoppers 9, 12, 23, 25
grass/pear bryobia mite 421
green apple aphid 71–72
– budworm: see spotted apple
  budworm
– citrus aphid 73–74
– leaf weevil 168
– leafhoppers 52
– petal virus 49, 52
– pug moth 322–323
– shield bug 33–34
– stink bug: see green vegetable bug
– vegetable bug 33
green-brindled crescent moth 360
greenflies 66
greenhouse mealybug: see glasshouse
  mealybug
– orthezia: see fluted scale
– red spider mite: see two-spotted
  spider mite
grey bud weevil 164
– dagger moth 355–356
– pear scale: see Italian pear scale
griseus, Peritelus: see *Peritelus
  sphaeroides*
–, *Trichoferus* 139
grossulariae, Aphidula: see *Aphis
  grossulariae*
–, *Aphis* 70
–, *Cecidophyopsis* 412
–, *Eriosoma* 94
–, *Medoralis*: see *Aphis grossulariae*
grossulariata, Abraxas 315–316
ground beetles 124–127
– bugs 34

GRYLLOBLATTODEA 11
grylloblattodeans 11
Gryllotalpa gryllotalpa 24–25
gryllotalpa, Gryllotalpa 24–25
GRYLLOTALPIDAE 12, 24–25
Gueriniella serratulae 122
gummosis 41
guttea, Ornix: see *Callisto denticulella*
Gymnoscelis pumilata: see
    Gymnoscelis rufifasciata
– *rufifasciata* 328
Gynaikothrips ficorum 31
Gypsonoma dealbana 282–283
gypsy moth 348–349

Hadena oleracea: see *Lacanobia
    oleracea*
haemorrhoidale, Acanthosoma 32
hairy broad-nosed weevil 155
Haltica ampelophaga: see *Altica
    ampelophaga*
Haplorhynchites coeruleus: see
    Involvulus caeruleus
hard scales: see armoured scales
Harpalus aeneus: see *Harpalus affinis*
– *affinis* 125
– *pubescens*: see *Harpalus rufipes*
– *rufipes* 125–126
hawk moths 335–342
hawthorn leaf erineum mite 418
– red spider mite: see hawthorn spider
    mite
– shield bug 32
– spider mite 421
– sucker 58–59
– webber moth 232
hazel aphid 93
– casebearer moth 241–242
– leaf-roller weevil 145
– longhorn beetle 138
– nut weevil: see nut weevil
– sawfly 386–387
heart & dart moth 357–358
hebenstreitella, Cacoecia: see
    Choristoneura hebenstreitella
–, *Choristoneura* 268
hebrew character moth 372–373
Hedya dimidioalba 283–284
– *nubiferana*: see *Hedya dimidioalba*
– *ochroleucana* 284
– *pruniana* 284–285
– *variegana*: see *Hedya dimidioalba*
helichrysi, Anuraphis: see
    Brachycaudus helichrysi
–, *Brachycaudus* 74–75
Heliococcus bohemicus 120
– *hystrix*: see *Heliococcus bohemicus*
HELIOZELIDAE 14, 203–204
hemerobiella, Coleophora 239–240
HEMIMETABOLA 11
HEMIPTERA 12
heparana, Pandemis 291
HEPIALIDAE 14, 195–197
HEPIALOIDEA 14
Hepialus humuli 195–196

– *lupulinus* 196–197
heringiana, Phytomyza 192–193
Herse convolvuli: see *Agrius
    convolvuli*
hesperidum, Coccus 111–112
–, *Lecanium*: see *Coccus hesperidum*
HETEROPTERA 12
Hibernia defoliaria: see *Erannis
    defoliaria*
hippocastani, Melolontha 128–129
Hippotion celerio 339
hirta, Lagria 137–138
hirtaria, Biston: see *Lycia hirtaria*
–, *Lycia* 328–329
hispidaria, Apocheima 319
hispidulus, Pogonocherus 138
holly blue butterfly 304–305
– leaf tier 292
– tortrix moth 292–293
holmiana, Croesia 272
Holocacista rivillei: see *Antispila
    rivillei*
HOLOMETABOLA 11, 12
HOLOTHYRIDA 16
holothyridid mites 16
Homotoma ficus 62
honeydew 46, 57, 58, 59, 60, 61, 63,
    64, 65, 66, 70, 71, 74, 76, 77, 78,
    81, 83, 86, 88, 90, 91, 92, 93, 100,
    102, 109, 110, 112, 113, 114, 116,
    117, 118, 119, 120, 121, 123, 189,
    295, 402, 404
honeysuckle thrips: see yellow flower
    thrips
– whitefly 64
hop asparagus 21
– capsid 38
– flea beetle 144
– froghopper: see hop leafhopper
– leaf miners 191
– leafhopper 52–53
– mosaic virus 84, 90
– red spider mite: see two-spotted
    spider mite
– root weevil 167–168
– strig maggot 177–178
– – midge 177–178
hop-cat 320
hop–damson aphid: see damson/hop
    aphid
hop-dogs 345
Hoplia philanthus 128
Hoplocampa brevis 388
– *ferruginea*: see *Hoplocampa flava*
– *flava* 388–389
– *fulvicornis*: see *Hoplocampa minuta*
– *minuta* 389–390
– *testudinea* 390–391
hornbeam spider mite 422–423
– whitefly 64–65
hornet 403
hortensis, Glischrochilus 135
horticola, Phyllopertha 129–130
Humerobates rostrolamellatus 428
humuli, Contarinia 177–178

–, *Diplosis*: see *Contarinia humuli*
–, *Hepialus* 195–196
–, *Phorodon* 89–90
Hyalesthes obsoletus 47
Hyalopterus amygdali 82
– *pruni* 82–83
Hybernia aurantiaria: see *Agriopis
    aurantiaria*
– *rupricapraria*: see *Theria primaria*
Hydraecia micacea 365
hydrangea scale 117
hydrangeae, Eupulvinaria: see
    Pulvinaria hydrangeae
–, *Pulvinaria* 117
hylaeiformis, Bembecia: see *Pennisetia
    hylaeiformis*
–, *Pennisetia* 222
Hyles livornica 340
Hylesinus oleiperda 171
HYMENOPTERA 12, 15
Hypena rostralis 365–366
Hyperomyzus lactucae 83
– *pallidus* 84
– *rhinanthi* 84
Hyphantria cunea 351–353
Hypoborus ficus 139, 171
Hyponomeuta: see *Yponomeuta*
HYPONOMEUTIDAE: see
    YPONOMEUTIDAE
hystrix, Heliococcus: see *Heliococcus
    bohemicus*
–, *Phenacoccus*: see *Heliococcus
    bohemicus*

Icerya purchasi 122–123
idaei, Amphorophora 66–67
–, *Aphis* 70
igniceps, Agromyza 191
incerta, Orthosia 374–375
incertana, Cnephasia 270–271
–, *Cnephasiella*: see *Cnephasia
    incertana*
inchworms 315
incisus, Euscelis 52
incognitella, Stigmella: see *Stigmella
    pomella*
inconsequens, Taeniothrips 28–29
Incurvaria capitella: see *Lampronia
    capitella*
– *rubiella*: see *Lampronia rubiella*
INCURVARIIDAE 14, 202–203
INCURVARIOIDEA 14
Indian wax scale: see Japanese wax
    scale
infausta, Aglaope 206–207
–, *Zygaena infausta*: see *Aglaope –
    infausta*
INSECTA 9, 11
insects 9–15, 23 et seq.
innumerabilis, Neopulvinaria 115
insertum, Rhopalosiphum 90
instabilis, Orthosia: see *Orthosia
    incerta*
interjectana, Cnephasia: see
    Cnephasia asseclana

interruptus, Euacanthus: see
    Evacanthus interruptus
–, *Evacanthus* 52–53
intonsa, Frankliniella 27
inustus, Polydrusus 168
Involvulus caeruleus 147
Iphiclides podalirius 301–302
ipsilon, Agrotis 358
ISOPTERA 12, 27
Italian pear scale 102–103
Itame wauaria: see *Semiothisa*
    wauaria
IXODIDA 16

Jacobiasca lybica 53
Janetiella oenophila 183–184
janthinana, Cydia 274
Janus compressus 380–381
Japanese citrus fruit scale 108
– wax scale 109
jasius, Charaxes 307
jasmine moth 298
jewel beetles 130–132
juglandella, Gracillaria: see *Caloptilia*
    roscipennella
juglandicola, Chromaphis 92
juglandis, Callaphis 92
–, *Callipterus*: see *Callaphis juglandis*
–, *Panaphis*: see *Callaphis juglandis*
juliana, Pammene: see *Pammene*
    fasciana
June bug: see garden chafer

Kalotermes flavicollis 27
KALOTERMITIDAE 12, 27
Kelly's thrips 28
kellyanus, Pezothrips 28
knotgrass moth 356
Korscheltellus lupulinus: see *Hepialus*
    lupulinus

Lacanobia oleracea 366–367
lace bugs 35–36
lacewings 12
LACHNIDAE: see LACHNINAE
LACHNINAE 96
lackey moth 311–312
lactucae, Hyperomyzus 83
–, *Myzus*: see *Nasonovia ribisnigri*
lacunana, Argyroploce: see *Celypha*
    lacunana
–, *Celypha* 268
–, *Olethreutes*: see *Celypha lacunana*
Lagria hirta 137–138
Lampra: see *Noctua*
Lampronia capitella 202
– *capittella*: see *Lampronia capitella*
– *rubiella* 202–203
lanestris, Eriogaster 309–310
lanigerum, Eriosoma 94–95
lanuginosum, Eriosoma 95
lapidarius, Euzetes: see *Humerobates*
    rostrolamellatus
lappet moth 310–311
large blackberry aphid 67

– European raspberry aphid 66–67
– flax flea beetle 142
– fruit bark beetle 172–173
– – flies 187–190
– hazel aphid 77
– pear sucker 60
– raspberry aphid: see large European
    raspberry aphid
– walnut aphid 92
– yellow underwing moth 371–372
LASIOCAMPIDAE 15, 309–313
Lasioptera berlesiana: see
    Prolasioptera berlesiana
– *rubi* 184
Lasius niger 402
Laspeyresia: see *Cydia* [in part]
– *woeberiana*: see *Enarmonia*
    formosana
laterana, Acleris 255
latifasciana, Acleris: see *Acleris*
    laterana
latifoliella, Metriochroa 213
–, *Parectopa*: see *Metriochroa*
    latifoliella
lavaterae, Oxycarenus 34
leaf beetles 140–144
– nematodes 420
leaf-curling plum aphid 74–75
leafhoppers 47–56
leaf-insects 12
leatherjackets 175, 176
LECANIDAE: see COCCIDAE
Lecanium: see *Parthenolecanium* [in
    part]
– *coryli*: see *Eulecanium tiliae*
– *crudum*: see *Parthenolecanium corni*
– *hesperidum*: see *Coccus hesperidum*
– *oleae*: see *Saissetia oleae*
Leche's twist moth 291–292
lecheana, Cacoecia: see *Ptycholoma*
    lecheana
–, *Ptycholoma* 291–292
lecheanum, Ptycholoma: see
    Ptycholoma lecheana
leopard moth 205–206
Leperesinus varius 171–172
leperii, Epidiaspis 102–103
LEPIDOPTERA 12, 14–15
Lepidosaphes beckii 99, 103
– *citricola*: see *Lepidosaphes beckii*
– *conchyformis*: see *Lepidosaphes*
    minima
– *ficus*: see *Lepidosaphes minima*
– *gloverii* 103
– *minima* 104
– *ulmi* 104–105
Leptophyes punctatissima 23
Leptosphaeria coniothyrium 187
lesser antler sawfly 385–386
– apple foliage weevil 158
– ash bark beetle 171
– bud moth 249
– strawberry weevils 160–161, 161
– yellow underwing moth 369–370
leucatella, Recurvaria 248–249

leucographella, Phyllonorycter
    217–218
leucophaearia, Agriopis 317
–, *Erannis*: see *Agriopis leucophaearia*
leucophthalmus, Philaenus: see
    Philaenus spumarius
Leucoptera malifoliella 209
– *scitella*: see *Leucoptera malifoliella*
leucotrochus, Nematus 394
Lichtensia viburni 114–115
light brown apple moth 281–282
– emerald moth 321
– grey tortrix moth 270–271
ligniperda, Cossus: see *Cossus cossus*
limacina, Caliroa: see *Caliroa cerasi*
–, *Eriocampoides*: see *Caliroa cerasi*
ligustri, Sphinx 342
lime hawk moth 340
linearis, Obera 138
lineatella, Anarsia 247–248
lineatus, Agriotes 133
–, *Sitona* 169
linnets 126
Liothrips oleae 31
lipsiana, Acleris 255–256
lipsiella, Diurnea 246
LITHOCOLLETIDAE: see
    GRACILLARIIDAE
Lithocolletis: see *Phyllonorycter* [in
    part]
– *citricola*: see *Phyllocnistis citrella*
– *concomitella*: see *Phyllonorycter*
    blancardella
little longhorn beetle 139
littoralis, Prodenia: see *Spodoptera*
    littoralis
–, *Scaphoideus*: see *Scaphoideus*
    titanus
–, *Spodoptera* 377–378
livida, Cantharis 134
livornica, Hylesa 340
ljungiana, Argyrotaenia: see
    Argyrotaenia pulchellana
lobarzewskii, Cydia 274
Lobesia botrana 285–286
locust bean moth 296
Locusta migratoria 25
locusts 9, 12, 23, 25
loganberry beetle: see raspberry beetle
– cane fly 194
– leafhopper 54
longana, Cnephasia 271–272
longhorn beetles 138–139
long-horned grasshoppers: see bush
    crickets
longispinus, Pseudococcus 121
Longitarsus parvulus 142
long-tailed mealybug 121
Longuinguis pyrarius: see *Melanaphis*
    pyraria
lonicerae, Aleyrodes 64
Loxostege martialis: see *Udea*
    ferrugalis
Lozotaenia forsterana 286–287
lubricipeda lutea, Phalaena: see

Spilosoma lutea
lucidus, Nematus 395
lugdunensis, Otiorhynchus: see
   Otiorhynchus clavipes
lunar-spotted pinion moth 362
lupulinus, Hepialus 196–197
–, *Korscheltellus*: see *Hepialus
   lupulinus*
lutea, Spilosoma 353
luteolata, Opisthograptis 331
luteum, Spilosoma: see *Spilosoma
   lutea*
lybica, Empoasca: see *Jacobiasca
   lybica*
–, *Jacobiasca* 53
LYCAENIDAE 15, 304–306
Lycia hirtaria 328–329
Lyda: see *Neurotoma*
LYGAEIDAE 12, 34
Lygocoris pabulinus 39–40
– *rugicollis* 40–41
– *spinolae*: see *Apolygus spinolae*
Lygris prunata: see *Eulithis prunata*
Lygus pabulinus: see *Lygocoris
   pabulinus*
– *pratensis*: see *Lygus rugulipennis*
– *rugulipennis* 41
– *spinolae*: see *Apolygus spinolae*
Lymantria dispar 348–349
– *monacha* 349
LYMANTRIIDAE 15, 345–350
Lyonetia clerkella 209–210
– *prunifoliella* 210–211
LYONETIIDAE 14, 208–201

Macrophoma dalmatica 185
Macrosiphum euphorbiae 66, 84
– *funestum* 84–85
– *rubiellum*: see *Sitobion fragariae*
– *rubifolium*: see *Macrosiphum
   funestrum*
maculata, Nephrotoma: see
   Nephrotoma appendiculata
–, *Pales*: see *Nephrotoma
   appendiculata*
madidus, Pterostichus 126
maesta, Pristiphora 399
Magdalis barbicornis 157
– *cerasi* 157
– *pruni*: see *Magdalis ruficornis*
– *ruficornis* 158
magpie moth 315–316
major, Thrips 30
Malacosoma neustria 311–312
malella, Stigmella 198–199
mali, Atractotomus 37
–, *Cacopsylla* 57–58
–, *Dasineura* 180
–, *Eulecanium*: see *Eulecanium tiliae*
–, *Psylla*: see *Cacopsylla mali*
–, *Sappaphis*: see *Dysaphis
   plantaginea*
–, *Scolytus* 172–173
malifoliella, Leucoptera 209
malinellus, Yponomeuta 234–235

malinus, Phyllocoptes 418–419
MALLOPHAGA 12
malvae rogersii, Acyrthosiphon: see
   Acyrthosiphon rogersii
Mamestra brassicae 367
– *oleracea*: see *Lacanobia oleracea*
– *persicariae*: see *Melanchra
   persicariae*
– *pisi*: see *Ceramica pisi*
mammals 124
mangold flea beetle 142
mantids 12
marani, Quadraspidiotus 107
marbled orchard tortrix moth 283–284
March moth 318–319
margaritata, Campaea 321
MARGARODIDAE 13, 122–123
Margaronia unionalis: see *Palpita
   unionalis*
marginalis, Orthotylus 41–42
marginaria, Agriopis 317–318
marginatus, Polydrusus 168
marginea, Tischeria 201
marsupialus, Putoniella: see
   Putoniella pruni
martialis, Loxostege: see *Udea
   ferrugalis*
–, *Mesographe*: see *Udea ferrugalis*
–, *Pyrausta*: see *Udea ferrugalis*
masseei, Anuraphis: see *Brachycaudus
   persicae*
mayflies 10, 11
McDaniel's spider mite 425
mcdanieli, Tetranychus 425
mealy peach aphid 82
– plum aphid 82–83
mealybugs 120–121
Mecinus pyraster 158
MECOPTERA 12
Med fly: see Mediterranean fruit fly
Mediterranean black scale 118–119
– brocade moth 377–378
– carnation tortrix moth: see carnation
   tortrix moth
– climbing cutworm 377
– fruit fly 188–189
Medoralis grossulariae: see *Aphis
   grossulariae*
MEGALODONTOIDEA 15
Meganephria oxyacanthae: see
   Allophyes oxyacanthae
Melanaphis pyraria 85
melanarius, Pterostichus 127
Melanchra brassicae: see *Mamestra
   brassicae*
– *persicariae* 368
melanogrammum, Strophosoma 170
melanogrammus, Strophosomus: see
   Strophosoma melanogrammum
melanoneura, Cacopsylla 58–59
–, *Psylla*: see *Cacopsylla melanoneura*
Meligethes aeneus 136
Melolontha hippocastani 128–129
– *melolontha* 128–129
– *vulgaris*: see *Melolontha melolontha*

melolontha, Melolontha 128–129
melon & cotton aphid 69–70
– aphid: see melon & cotton aphid
MEMBRACIDAE 12, 46
MEMBRACOIDEA 12
meridiana, Mesembrina 193
meridionalis, Otiorhynchus 159
–, *Thrips* 30
Mesembrina meridiana 193
Mesographe martialis: see *Udea
   ferrugalis*
MESOSTIGMATA 16
mesostigmatid mites 16
messaniella, Phyllonorycter 218
Metallus albipes 391
– *pumilus* 392–393
Metatetranychus citri: see *Panonychus
   citri*
– *pilosus*: see *Panonychus ulmi*
– *ulmi*: see *Panonychus ulmi*
Metcalfa pruinosa 45–46
 meticulosa, Phlogophora 376–377
Metriochroa latifoliella 213
micacea, Gortyna: see *Hydraecia
   micacea*
–, *Hydraecia* 365
microcarpae, Aceria 409
Micronematus abbreviatus 393
midge blight 187
migratoria, Locusta 25
migratory locust 25
Mimas tiliae 340
minima, Lepidosaphes 104
–, *Mytilaspis*: see *Lepidosaphes
   minima*
minuta, Hoplocampa 389–390
minutissimus, Thrips 30
MIRIDAE 12, 37–42
mirids 37–42
mites 16, 405 et seq.
MLOs 48
mole cricket(s) 24–25
molesta, Cydia 274–276
mollis, Polydrusus 168
MOMPHIDAE 15, 251
monacha, Lymantria 349
Monima: see *Orthosia*
Monophadnoides confusa: see
   Claremontia confusa
– *geniculatus* 393–394
Monostira unicostata 35
MONOTRYSIA 14
morio, Priophorus 398
moths 10, 12, 14–15, 195 et seq.
mottled beauty moth 318
– pug moth 328
– umber moth 326–327
mulberry moth: see American white
   moth
mullein capsid 37
munda, Orthosia 375–376
muricatus, Sciaphilus: see *Sciaphilus
   asperatus*
MUSCIDAE 14, 193
musculana, Syndemis 294–295

mussel scale 104–105
MYCOBATIDAE 16, 428
mycoplasma-like organisms 48
Myelois ceratoniae: see *Ectomyelois*
    ceratoniae
– *pinguis*: see *Euzerophera pinguis*
myopaeformis, Aegeria: see
    Synanthedon myopaeformis
–, *Conopia*: see *Synanthedon*
    myopaeformis
–, *Synanthedon* 222–223, 279
myopiformis, Synanthedon: see
    Synanthedon myopaeformis
Mytilaspis minima: see *Lepidosaphes*
    minima
– *pomorum*: see *Lepidosaphes ulmi*
Mytilococcus: see *Lepidosaphes*
Myzocallis coryli 93
Myzus ascalonicus 85–86
– *cerasi* 87–88
– *lactucae*: see *Nasonovia ribisnigri*
– *ornatus* 86
– *persicae* 86–87
– *pruniavium* 87–88
– *varians* 88

Naenia typica 368–369
naevana, Acroclita: see *Rhopobota*
    naevana
–, *Rhopobota* 292–293
NALEPELLIDAE: see
    PHYTOPTIDAE
nana, Phaneroptera 24
nanella, Recurvaria 249
Nasonovia ribisnigri 89
Nebria brevicollis 126
Nectria galligena 95, 251
needle-bug: see strawberry blossom
    weevil
needle-nosed hop bug: see hop capsid
neglectus, Oecophyllembius: see
    Metriochroa latifoliella
NEMATOCERA 14
Nematus abbreviata: see
    Micronematus abbreviatus
– *consobrinus*: see *Nematus*
    leucotrochus
– *leucotrochus* 394
– *lucidus* 395
– *olfaciens* 395–396
– *ribesii* 396–397
– *ventricosus*: see *Nematus ribesii*
nemoralis, Neurotoma 379
nemorana, Choreutis 224–225
–, *Simaethis*: see *Choreutis nemorana*
Neocoenorrhinus aequatus 148–149
– *cribripennis* 149
– *germanicus* 149
– *pauxillus* 150
Neopulvinaria innumerabilis 115
Neosphaleroptera nubilana 287
Nephrotoma appendiculata 175
– *maculata*: see *Nephrotoma*
    appendiculata
Nepticula: see *Stigmella*

NEPTICULIDAE 14, 197–200
NEPTICULOIDEA 14
nerii, Aphis 71
–, *Aspidiotus* 100
–, *Ceroplastes*: see *Ceroplastes rusci*
–, *Cerosipha*: see *Aphis nerii*
–, *Daphnis* 337–338
–, *Deilephila*: see *Daphnis nerii*
nervosus, Acocephalus: see *Aphrodes*
    bicinctus
nettle leaf weevil 166–167
NEUROPTERA 12
Neurotoma flaviventris: see *Neurotoma*
    saltuum
– *nemoralis* 379
– *saltuum* 379–380
neustria, Malacosoma 311–312
Nezara viridula 33
niger, Lasius 402
nigricella, Coleophora: see
    Coleophora spinella
nigricolle, Poecilosoma: see
    Endelomyia aethiops
nigritarse, Protapion 151
NITIDULIDAE 13, 135–136
Noctua comes 369–370
– *fimbriata* 370–371
– *pronuba* 371–372
NOCTUIDAE 15, 354–378
NOCTUOIDEA 15
Nola cucullatella 354
NOLIDAE 15, 354
North American cherry fruit fly: see
    American eastern cherry fruit fly
northern strawberry leaf beetle 143
– winter moth 330–331
norvegicus, Calocoris: see
    Closterotomus norvegicus
–, *Closterotomus* 38–39
Notocelia uddmanniana: see *Epiblema*
    uddmanniana
NOTODONTIDAE 15, 343
NOTODONTOIDEA 15
NOTOSTIGMATA 16
notostigmatid mites 16
November moth 325–326
noxius, Peritelus 164
nubiferana, Argyroploce: see *Hedya*
    dimidioalba
–, *Hedya*: see *Hedya dimidioalba*
nubilalis, Pyrausta: see *Ostrinia*
    nubilalis
–, *Ostrinia* 297–298
nubilana, Neosphaleroptera 287
nucum, Curculio 156
nut bud moth 281
– bud tortrix moth: see nut bud moth
– gall mite: see filbert bud mite
– leaf blister moth 216
– – weevil 170
– scale 113
– weevil 156
Nygmia phaeorrhoea: see *Euproctis*
    chrysorrhoea
nymphaeae, Rhopalosiphum 90–91

NYMPHALIDAE 15, 307

oak beauty moth 320–321
oak-apple gall wasp 287
oat–apple aphid: see apple/grass aphid
Obera linearis 138
oblongus, Phyllobius 166
obscura, Cantharis 134
obscurus, Adoxus: see *Bromius*
    obscurus
–, *Agriotes* 133
–, *Bromius* 143
–, *Pseudococcus*: see *Pseudococcus*
    viburni
obsoletus, Athysanus: see *Euscelis*
    obsoletus
–, *Conosanus* 52
–, *Hyalesthes* 47
occidentalis, Frankliniella 28
ocellana, Grapholitha: see *Spilonota*
    ocellana
–, *Spilonota* 293–294
ocellata, Smerinthus 341
ochroleucana, Hedya 284
oculiperda, Clinodiplosis: see
    Resseliella oculiperda
–, *Resseliella* 185–186
ODONATA 11
Odonestis pruni 312
Oecophora bractella 247
– *sulphurella*: see *Oecophora bractella*
OECOPHORIDAE 15, 244–247
Oecophyllembius neglectus: see
    Metriochroa latifoliella
oenophila, Dichelomyia: see *Janetiella*
    oenophila
–, *Janetiella* 183–184
oleae, Aceria 408
–, *Bactrocera* 187–188
–, *Dacus*: see *Bactrocera oleae*
–, *Dasineura* 180
–, *Eriophyes*: see *Aceria oleae*
–, *Lecanium*: see *Saissetia oleae*
–, *Liothrips* 31
–, *Parlatoria* 105
–, *Phloeotribus*: see *Phloeotribus*
    scarabaeoides
–, *Prays* 230–231
–, *Saissetia* 118–119
oleander aphid 71
– hawk moth 337–338
– scale 100
oleastrella, Zelleria 236–237
oleellus, Prays: see *Prays oleae*
oleiperda, Hylesinus 171
oleisuga, Clinodiplosis: see *Resseliella*
    oleisuga
–, *Resseliella* 186
oleivora, Phyllocoptruta 419
oleracea, Hadena: see *Lacanobia*
    oleracea
–, *Lacanobia* 366–367
–, *Mamestra*: see *Lacanobia oleracea*
–, *Polia*: see *Lacanobia oleacea*
–, *Tipula* 175–176

Olethreutes lacunana: see *Celypha lacunana*
– *pruniana*: see *Hedya pruniana*
– *splendana*: see *Cydia splendana*
OLETHREUTINAE 253
olfaciens, Nematus 395–396
Oligonochus ulmi: see *Panonychus ulmi*
olivalis, Udea 299
olive bark beetle 172
– bud mite 408
– cushion scale 113–114
– fruit fly 187–188
– – midge 184–185
– – rhynchites 149
– geometrid moth: see olive pug moth
– leaf gall midge 180
– – miner moth 213
– – weevil 170
– moth 230–231
– parlatoria scale 105
– pit scale 109
– pug moth 328
– pyralid moth 297
– scale: see Mediterranean black scale
– scolytid: see olive bark beetle
– stem midge 186
– sucker 61
– thrips 31
– weevil 159
– whitefly 63
olivina, Euphyllura 61
olivinus, Aleurolobus 63
omnivorous leaf tier 271
Oncopsis alni 53–54
onion thrips 30
Operophtera brumata 319, 329–330, 363
– *fagata* 330–331
Ophonus rufipes: see *Harpalus rufipes*
Opisthograptis luteolata 331
oporana, Cacoecia: see *Archips podana*
Oporinia dilutata: see *Epirrita dilutata*
orana, Adoxophyes 258
orchard ermine moth: see common small ermine moth
Orchestes fagi 158
Orgyia antiqua 349–350
oriental citrus scale: see Japanese citrus fruit scale
– fruit moth 274–276
– peach moth: see oriental fruit moth
ornatus, Myzus 86
Ornix avellanella: see *Parornix devoniella*
– *guttea*: see *Callisto denticulella*
ORTHOPTERA: see SALTATORIA
Orthosia cerasi: see *Orthosia incerta*
– *cruda* 372
– *gothica* 372–373
– *gracilis* 373–374
– *incerta* 374–375
– *instabilis*: see *Orthosia incerta*
– *munda* 375–376

– *stabilis* 376
Orthotylus marginalis 41–42
ostreaeformis, Aspidiotus: see *Quadraspidiotus ostreaeformis*
–, *Quadraspidiotus* 107
Ostrinia nubilalis 297–298
Otiorhynchus 152
– *clavipes* 158–159
– *cribricollis* 159
– *lugdunensis*: see *Otiorhynchus clavipes*
– *meridionalis* 159
– *ovatus* 160
– *picipes*: see *Otiorhynchus singularis*
– *raucus* 160
– *rugifrons* 160–161
– *rugosostriatus* 161
– *salicicola* 161
– *singularis* 162
– *sulcatus* 162–163
Otiorrhynchus: see *Otiorhynchus*
Ourapteryx sambucaria 332
ovatus, Otiorhynchus 160
oxyacanthae, Allophyes 360
–, *Meganephria*: see *Allophyes oxyacanthae*
Oxycarenus lavaterae 34
oyster scale: see yellow plum scale

pabulinus, Lygocoris 39–40
–, *Lygus*: see *Lygocoris pabulinus*
Pachynematus pumilio 397
Pachythelia unicolor 208
padella, Yponomeuta 235–236
padi, Eriophyes 415–416
–, *Phytoptus*: see *Eriophyes padi*
Palaeocimbex quadrimaculata 381–382
pale brindled beauty moth 333
– eggar moth 313
– oak eggar moth: see pale eggar moth
– tussock moth 345
paleana, Aphelia 260–261
Pales maculata: see *Nephrotoma appendiculata*
pale-spotted gooseberry sawfly 394
pallida, Biorhiza 287
pallidus fragariae, Phytonemus 419–420
– *fragariae, Tarsonemus*: see *Phytonemus pallidus fragariae*
–, *Hyperomyzus* 84
–, *Phytonemus*: see *Phytonemus pallidus fragariae*
–, *Tarsonemus*: see *Phytonemus pallidus fragariae*
pallipes, Priophorus 398–399
–, *Pristiphora*: see *Pristiphora rufipes*
palm scale 102
Palomena prasina 33–34
Palpita unionalis 298
paludosa, Tipula 176
Pammene argyrana 287
– *fasciana* 288
– *juliana*: see *Pammene fasciana*

– *rhediella* 288–289
PAMPHILIIDAE 15, 379–380
Panaphis juglandis: see *Callaphis juglandis*
Pandemis cerasana 289–290
– *corylana* 290
– *dumetata* 290
– *heparana* 291
– *ribeana*: see *Pandemis cerasana*
Panonychus citri 423–424
– *ulmi* 424–425
Papilio podalirius: see *Iphiclides podalirius*
PAPILIONIDAE 15, 301–302
PAPILIONINAE 301
PAPILIONOIDEA 15
parallelepipedus, Abax 124
parallel-sided ground beetle 124
parasitella, Ephestia 296–297
–, *Ephestia parasitella* 296
–, *Ephestia unicolorella* 296
PARASITICA 15
Paratetranychus pilosus: see *Panonychus ulmi*
Paravespula: see *Vespula*
Parectopa latifoliella: see *Metriochroa latifoliella*
pariana, Choreutis 225–226
–, *Eutromula*: see *Choreutis pariana*
–, *Simaethis*: see *Choreutis pariana*
paripennella, Coleophora 243
Parlatoria oleae 105
– *pergandii* 106
– *ziziphi* 106
– *zizyphi*: see *Parlatoria ziziphi*
Parornix devoniella 213–214
– *finitimella* 214
– *torquillella* 214
Parthenolecanium corni 115–116
– *persicae* 116
parvulus, Longitarsus 142
Pasiphila rectangulata: see *Chloroclystis rectangulata*
passenger moth 363
pathogens 135, 149, 172, 267, 296, 321, 365
–, fungal 35, 46, 95, 151, 156, 187, 189, 191, 223, 224, 251, 263, 286, 384
pauxillus, Neocoenorrhinus 150
pavonia, Eudia: see *Saturnia pavonia*
–, *Saturnia* 314
pea & bean weevil 169
peach aphid 76
– leaf-roll aphid 88
– scale: see European peach scale
– silver mite 410
– thrips 30
– twig borer moth 247–248
peach-blossom moth 315
peach/potato aphid 86–87
pear & cherry sawfly: see pear slug sawfly
– – – slugworm 384
– bark aphid: see pear phylloxera

– bryobia mite: see grass/pear bryobia mite
– decline 59
– jewel beetle 131–132
– lace bug 36
– leaf blister mite 416–417
– – – moth 209
– – midge 181–182
– – sawfly 393
– leaf-curling midge: see pear leaf midge
– leaf-roller weevil 146–147
– midge 178–179
– phylloxera 97
– psylla: see pear sucker
– psyllid: see pear sucker
– pygmy moth 200
– rust mite 414–415
– sawfly 388
– scab 415
– scale: see yellow pear scale
– shoot sawfly 380–381
– slug sawfly 384–385
– sucker 59–60
– thrips 28–29
– weevil 157
pear/bedstraw aphid 81
pear/coltsfoot aphid 68
pear/grass aphid 85
pear–hogweed aphid: see pear/parsnip aphid
pear/parsnip aphid 68
pectinicornis, Cladius 386
pedaria, Phigalia: see *Phigalia pilosaria*
Pegomya dentiens: see *Pegomya rubivora*
– *rubivora* 194
Pegomyia: see *Pegomya*
pelekassi, Aculops 410
pellucida, Cantharis 134
pellucidus, Barypeithes 155
PEMPHIGIDAE: see ERIOSOMATINAE
penkleriana, Epiblema: see *Epinotia tenerana*
–, *Eucosma:* see *Epinotia tenerana*
pennaria, Collotois: see *Colotois pennaria*
–, *Colotois* 324
Pennisetia anomala: see *Pennisetia hyaeiformis*
– *hylaeiformis* 222
pentagona, Aulacaspis: see *Pseudaulacaspis pentagona*
–, *Diaspis:* see *Pseudaulacaspis pentagona*
–, *Pseudaulacaspis* 106–107
Pentatoma rufipes 34
PENTATOMIDAE 12, 32–34
Pentatrichus fragaefolii: see *Chaetosiphon fragaefolii*
– *fragariae:* see *Chaetosiphon fragaefolii*
peppered moth 320

pergandii, Parlatoria 106
Peribatodes rhomboidaria 332–333
Peritelus griseus: see *Peritelus sphaeroides*
– *noxius* 164
– *sphaeroides* 164
permanent apple aphid: see green apple aphid
– blackberry aphid 72
– currant aphid: see European permanent currant aphid
perniciosa, Comstockaspis: see *Quadraspidiotus pernicosus*
perniciosus, Aspidiotus: see *Quadraspidiotus pernicosus*
–, *Quadraspidiotus* 107–108
pernicious scale: see San José scale
Peronea: see *Acleris*
Perrisia: see *Dasineura*
Persian walnut blister mite: see walnut leaf erineum mite
persicae, Brachycaudus 75–76
–, *Eulecanium:* see *Parthenolecanium persicae*
–, *Myzus* 86–87
–, *Parthenolecanium* 116
–, *Pterochloroides* 96
persicaecola, Brachycaudus: see *Brachycaudus persicae*
persicariae, Mamestra: see *Melanchra persicariae*
–, *Melanchra* 368
–, *Polia:* see *Melanchra persicariae*
persicella, Ypsolopha 236
Pezothrips kellyanus 28
phaeoerrhoea, Euproctis: see *Euproctis chrysorrhoea*
phaeorrhoea, Nygmia: see *Euproctis chrysorrhoea*
Phalaena lubricipeda lutea: see *Spolinota lutea*
Phalera bucephala 343
PHALONIIDAE: see COCHYLIDAE
Phaneroptera nana 24
– *quadripunctata:* see *Phaneroptera nana*
PHASMIDA 12
Phenacoccus hystrix: see *Heliococcus bohemicus*
Phigalia pedaria: see *Phigalia pilosaria*
– *pilosaria* 333
Philaenus leucophthalmus: see *Philaenus spumarius*
– *spumarius* 44
philanthus, Hoplia 128
Philippia follicularis: see *Filippia follicularis*
Philopedon plagiatum 164
– *plagiatus:* see *Philopedon plagiatum*
Philudoria pruni: see *Odonestis pruni*
PHLAEOTHRIPIDAE 13, 31
phloeocoptes, Acalitus 227, 406–407
–, *Eriophyes:* see *Acalitus phloeocoptes*

Phloeotribus oleae: see *Phloeotribus scarabaeoides*
– *scarabaeoides* 172
Phlogophora meticulosa 376–377
phoenix moth 327
Phomopsis 224
Phorodon humuli 89–90
– *pruni:* see *Phorodon humuli*
phryganella, Diurnea: see *Diurnea lipsiella*
Phyllobius 152
– *alneti:* see *Phyllobius pomaceus*
– *argentatus* 165
– *calcaratus:* see *Phyllobius glaucus*
– *glaucus* 165
– *oblongus* 166
– *piri:* see *Phyllobius pyri*
– *pomaceus* 166–167
– *pyri* 167
– *urticae:* see *Phyllobius pomaceus*
PHYLLOCNISTIDAE 14, 219–221
Phyllocnistis citrella 219–220
– *vitegenella* 221
Phyllocoptes fockeui: see *Aculus fockeui*
– *goniothorax* 418
– *gracilis* 418
– *malinus* 418–419
– *schlechtendali:* see *Aculus schlechtendali*
– *vitis* 412
Phyllocoptruta oleivora 419
Phyllonorycter blancardella 214–215
– *cerasicolella* 215–216
– *coryli* 216
– *corylifoliella* 216–217
– *leucographella* 217–218
– *messaniella* 218
– *pomonella:* see *Phyllonorycter spinicolella*
– *spinicolella* 218–219
Phyllopertha horticola 129–130
Phylloxera vastatrix: see *Viteus vitifoliae*
– *vitifolii:* see *Viteus vitifoliae*
phylloxeras 97–98
PHYLLOXERIDAE 13, 97–98
Phytocoptella avellanae: see *Phytoptus avellanae*
Phytomyza heringiana 192–193
Phytonemus pallidus: see *Phytonemus pallidus fragariae*
– *pallidus fragariae* 419–420
phytoplasmas 46, 47, 48, 53, 54, 55, 59, 61
–, vectors of 46, 47, 48, 53, 54, 55, 59, 61
PHYTOPTIDAE 16, 405
Phytoptus avellanae 405
– *essigi:* see *Acalitus essigi*
– *fockeui:* see *Aculus fockeui*
– *gracilis:* see *Phyllocoptes gracilis*
– *padi:* see *Eriophyes padi*
– *piri:* see *Eriophyes pyri*
– *pyri:* see *Eriophyes pyri*

– *ribis*: see *Cecidophyopsis ribis*
– *similis*: see *Eriophyes similis*
– *tristriatus*: see *Aceria tristriatus*
– *vitis*: see *Colomerus vitis*
picipes, Otiorhynchus: see
   Otiorhynchus singularis
Pierce's disease 50
PIERIDAE 15, 302–304
pilleriana, Sparganothis 293
pilosaria, Apocheima: see *Phigalia*
   pilosaria
–, *Phigalia* 333
pilosus, Metatetranychus: see
   Panonychus ulmi
–, *Paratetranychus*: see *Panonychus*
   ulmi
pinguis, Ephestia: see *Euzerophera*
   pinguis
–, *Euzophera* 297
–, *Myelois*: see *Euzerophera pinguis*
pink citrus rust mite 410
Pionea: see *Udea*
piri, Anthonomus 152
–, *Aphanostigma* 97
–, *Epitrimerus* 414–415
–, *Phyllobius*: see *Phyllobius pyri*
–, *Phytoptus*: see *Eriophyes pyri*
–, *Psylla*: see *Cacopsylla pyri*
pirisuga, Psylla: see *Cacopsylla*
   pyrisuga
pisi, Ceramica 361
–, *Mamestra*: see *Ceramica pisi*
–, *Polia*: see *Ceramica pisi*
pit scales 109
pith moth 251
plagiatum, Philopedon 164
plagiatus, Cneorrhinus: see
   Philopedon plagiatus
–, *Philopedon*: see *Philopedon*
   plagiatum
plagicolella, Stigmella 199
Plagiognathus arbustorum 42
– *chrysanthemi* 42
PLANIPENNIA: see NEUROPTERA
Planococcus citri 120
plantaginea, Dysaphis 80–81
planthoppers 12, 45–46, 47
plebeja, Thamnotettix: see *Euscelis*
   incisus
PLECOPTERA 11
Plesiocoris rugicollis: see *Lygocoris*
   rugicollis
plicatrix, Dasineura 181
Plintha caliginosus 167–168
Plinthus caliginosus: see *Plintha*
   caliginosus
plum fruit moth 272–273
– fruit-bud midge 178
– gall mite: see big-beaked plum mite
– lappet moth 312
– leaf gall midge 185
– – gall mite 415–416
– – sawfly 398–399
– leaf-curling midge 183
– lecanium scale 119

– pouch-gall mite 417
– pox 74, 75, 83, 86, 88, 90
– rust mite 410–411
– sawfly 388–389
– spur mite 227, 406–407
– tortrix moth 284–285
– weevil 158
plum-tree leaf midge: see plum fruit-
   bud midge
PLUSIINAE 354
PLUTELLINAE 227
podalirius, Iphiclides 301–302
–, *Papilio*: see *Iphiclides podalirius*
podana, Archips 262–263
podanus, Archips: see *Archips podana*
Poecilocampa populi 312–313
Poecilosoma nigricolle: see
   Endelomyia aethiops
Poecilus cupreus 126
Pogonochaerus: see *Pogonocherus*
Pogonocherus hispidulus 138
Polia oleracea: see *Lacanobia oleacea*
– *persicariae*: see *Melanchra*
   persicariae
– *pisi*: see *Ceramica pisi*
politana, Eulia: see *Argyrotaenia*
   pulchellana
pollen beetle 136
pollini, Pollinia 109
Pollinia pollini 109
Polychrosis botrana: see *Lobesia*
   botrana
Polydrosus: see *Polydrusus*
Polydrusus 152
– *formosus* 168
– *inustus* 168
– *marginatus* 168
– *mollis* 168
– *pterygomalis* 168
– *sericeus*: see *Polydrusus formosus*
Polygonia c-album 308
pomaceus, Phyllobius 166–167
pomegranate aphid 72
pomella, Stigmella 199–200
pomi, Aphis 71–72
–, *Eotetranychus*: see *Eotetranychus*
   pruni
pomonella, Cydia 276–277
–, *Phyllonorycter*: see *Phyllonorycter*
   spinicolella
pomorum, Anthonomus 153–154
–, *Mytilaspis*: see *Lepidosaphes ulmi*
populi, Eriogaster: see *Poecilocampa*
   populi
–, *Poecilocampa* 312–313
porcellus, Deilephila 339
Porthesia: see *Euproctis* [in part]
– *dispar*: see *Lymantria dispar*
postvittana, Epiphyas 281–282
potato aphid 66, 84
– capsid 38–39
– stem borer 365
potentillae, Agromyza 192
–, *Coleophora* 240–241
powdered quaker moth 373–374

praeusta, Tetrops 139
prasina, Palomena 33–34
pratensis, Lygus: see *Lygus*
   rugulipennis
Prays citri 230
– *oleae* 230–231
– *oleellus*: see *Prays oleae*
primaria, Theria 334–335
Priophorus brullei: see *Priophorus*
   morio
– *morio* 398
– *pallipes* 398–399
– *tener*: see *Priophorus morio*
– *varipes*: see *Priophorus pallipes*
Pristiphora abbreviata: see
   Micronematus abbreviatus
– *appendiculata*: see *Pristiphora*
   rufipes
– *californica*: see *Micronematus*
   abbreviatus
– *maesta* 399
– *pallipes*: see *Pristiphora rufipes*
– *rufipes* 399–400
Pristophora: see *Pristiphora*
privet hawk moth 342
Proctoparce convolvuli: see *Herse*
   convolvuli
Prodenia littoralis: see *Spodoptera*
   littoralis
progemmaria, Erannis: see *Agriopis*
   marginaria
Prolasioptera berlesiana 184–185
proletella, Aleyrodes 64
pronuba, Noctua 371–372
pronubana, Cacoecia: see
   Cacoecimorpha pronubana
–, *Cacoecimorpha* 266–267
PROSTIGMATA 16, 395–419
prostigmatid mites 16
Protapion apricans 151
– *fulvipes* 151
– *nigritarse* 151
Protopulvinaria pyriformis 116–117
PROTURA 11
proturans 11
pruinosa, Metcalfa 45–46
prunalis, Udea 300
prunastri, Eulecanium: see
   Sphaerolecanium prunastri
–, *Sphaerolecanium* 119
prunata, Eulithis 327
–, *Lygris*: see *Eulithis prunata*
pruni, Cecidomyia: see *Putoniella*
   pruni
–, *Eotetranychus* 423
–, *Hyalopterus* 82–83
–, *Magdalis*: see *Magdalis ruficornis*
–, *Odonestis* 312
–, *Philudoria*: see *Odonestis pruni*
–, *Phorodon*: see *Phorodon humuli*
–, *Putoniella* 185
–, *Satyrium* 305
–, *Scolytus*: see *Scolytus mali*
–, *Strymon*: see *Satyrium pruni*
–, *Thecla*: see *Satyrium pruni*

pruniana, Argyroploce: see *Hedya pruniana*
–, *Hedya* 284–285
–, *Olethreutes*: see *Hedya pruniana*
pruniavium, Myzus 87–88
prunicola, Brachycaudus: see *Brachycaudus schwartzi*
–, *Edwardsiana* 50–51
–, *Typhlocyba*: see *Edwardsiana prunicola*
pruniella, Argyresthia 228–229
pruniflorum, Contarinia 178
prunifoliae, Coleophora 242
prunifoliella, Lyonetia 210–211
prunivorana, Cydia: see *Cydia lobarzewskii*
Psallus ambiguus 42
Pseudaulacaspis pentagona 106–107
PSEUDOCOCCIDAE 13, 120–121
Pseudococcus adonidum: see *Pseudococcus longispinus*
– *aesculi*: see *Heliococcus bohemicus*
– *affinis*: see *Pseudococcus viburni*
– *calceolariae* 121
– *citri*: see *Planococcus citri*
– *fragilis*: see *Pseudococcus calceolariae*
– *gahani*: see *Pseudococcus calceolariae*
– *longispinus* 121
– *obscurus*: see *Pseudococcus viburni*
– *viburni* 121
Pseudomonas syringiae pv. *mors-prunorum* 171
Pseudoophonus rufipes: see *Harpalus rufipes*
Pseudophonus rufipes: see *Harpalus rufipes*
psi, Acronicta 355–356
psocids 12
PSOCOPTERA 12
Psyche graminella: see *Pachythelia unicolor*
PSYCHIDAE 14, 208
Psylla [in part]: see *Cacopsylla*
– *costalis* 61
PSYLLIDAE 13, 57–61
Psylliodes attenuata 144
PSYLLOIDEA 13
Pterochloroides persicae 96
Pteronidea: see *Nematus*
Pterostichus 124
– *cupreus*: see *Poecilus cupreus*
– *madidus* 126
– *melanarius* 127
– *vulgaris*: see *Pterostichus melanarius*
pterygomalis, Polydrusus 168
PTERYGOTA 11, 11–12
Ptycholoma lecheana 291–292
– *lecheanum*: see *Ptycholoma lecheana*
pubescens, Harpalus: see *Harpalus rufipes*
pudibunda, Calliteara 345

–, *Dasychira*: see *Calliteara pudibunda*
pulchellana, Argyrotaenia 265–266
pulcher, Brevipalpus: see *Cenopalpus pulcher*
–, *Cenopalpus* 427
Pulvinaria betulae 117
– *floccifera*: see *Chloropulvinaria floccifera*
– *hydrangeae* 117
– *ribesiae* 117
– sp. 118
– *vitis* 117–118
pumilata, Gymnoscelis: see *Gymnoscelis rufifasciata*
pumilio, Fenusa: see *Metallus pumilus*
–, *Pachynematus* 397
pumilus, Metallus 392–393
punctatissima, Leptophyes 23
punicae, Aphis 72
purchasi, Icerya 122–123
purple scale: see Florida red scale
puta, Agrotis 358–359
Putoniella marsupialus: see *Putoniella pruni*
– *pruni* 185
PYRALIDAE 15, 295–301
pyralina, Cosmia 362
PYRALOIDEA 15
pyramidea, Amphipyra 360
pyraria, Geoktapia: see *Melanaphis pyraria*
–, *Melanaphis* 85
pyrarius, Longuinguis: see *Melanaphis pyraria*
pyraster, Mecinus 158
Pyrausta: see *Udea* [in part]
– *martialis*: see *Udea ferrugalis*
– *nubilalis*: see *Ostrinia nubilalis*
pyrella, Swammerdamia 232–233
pyri, Anthonomus: see *Anthonomus piri*
–, *Anuraphis*: see *Dysaphis pyri*
–, *Aphanostigma*: see *Aphanostigma piri*
–, *Cacopsylla* 59
–, *Dasineura* 181–182
–, *Dysaphis* 81
–, *Epitrimerus*: see *Epitrimerus piri*
–, *Eriophyes* 416–417
–, *Phyllobius* 167
–, *Phytoptus*: see *Eriophyes pyri*
–, *Psylla*: see *Cacopsylla pyri*
–, *Quadraspidiotus* 108
–, *Stephanitis* 36
–, *Stigmella* 200
–, *Tygnis*: see *Stephanitis pyri*
pyricola, Cacopsylla 59–60
pyriform scale 116–117
pyriformis, Protopulvinaria 116–117
pyrina, Zeuzera 205–206
pyrisuga, Cacopsylla 60
–, *Psylla*: see *Cacopsylla pyrisuga*
pyrivora, Contarinia 178–179
–, *Cydia* 277

–, *Laspeyresia*: see *Cydia pyrivora*

Quadraspidiotus marani 107
– *ostreaeformis* 107
– *pernicosus* 107–108
– *pyri* 108
– *schneideri*: see *Quadraspidiotus marani*
quadrimaculata, Cimbex: see *Palaeocimbex quadrimaculata*
–, *Palaeocimbex* 381–382
quadripunctata, Phaneroptera: see *Phaneroptera nana*
quadripunctatus, Glischrochilus: see *Glischrochilus hortensis*
quadrisignatus, Glischrochilus 135
quercana, Carcina 244–245
quercifolia, Gastropacha 310–311
quercus, Typhlocyba 56

raspberry aphid: see small European raspberry aphid
– beetle 136–137
– cane blight 187
– – midge 186–187
– clearwing moth 222
– crown borer 222
– flea beetles 142–143
– jewel beetle 130–131
– leaf & bud mite 418
– – mottle virus 67
– – spot virus 67
– leaf-mining sawflies 391, 392–393
– moth 202–203
– sawfly 387
– shoot borer 202–203
– spur blight 187
– stem gall midge: see blackberrry sten gall midge
– vein chlorosis virus 70
raucus, Otiorhynchus 160
rectangulata, Chloroclystis 322–323
–, *Pasiphila*: see *Chloroclystis rectangulata*
rectirostris, Anthonomus: see *Furcipes rectirostris*
–, *Furcipes* 156–157
Recurvaria crataegella: see *Recurvella nanella*
– *leucatella* 248–249
– *nanella* 249
red & black froghopper 43
– apple capsid 42
– bud borer 185–186
– currant blister aphid 78
– – gall mite 413
– currant/arrow-grass aphid 74
– plum maggot 272, 273
– spider mites: see spider mites
red-belted clearwing moth: see apple clearwing moth
red-berry disease 406
– mite: see blackberry mite
red-green carpet moth 321–322
red-legged weevil 158–159

repandata, Alcis 318
–, *Cleora*: see *Alcis repandata*
Resseliella oculiperda 185–186
– *oleisuga* 186
– *ribis* 186
– *theobaldi* 186–187
reticulana, Adoxophyes: see
    Adoxophyes orana
–, *Capua*: see *Adoxophyes orana*
reuteri, Drepanothrips 27
reversion disease 413
Rhagoletis cerasi 189–190
– *cingulata* 190
– *completa* 190
rhamni, Zygina 57
rhediella, Pammene 288–289
rhinanthi, Hyperomyzus 84
rhombana, Acleris 256–257
rhombella, Gelechia 248
rhomboidaria, Boarmia: see
    Peribatodes rhomboidaria
–, *Peribatodes* 332–333
Rhopalosiphoninus ribesinus 90
Rhopalosiphum insertum 90
– *nymphaeae* 90–91
Rhopobota naevana 292–293
– *unipunctana*: see *Rhopobota*
    naevana
Rhynchaenus fagi: see *Orchestes fagi*
Rhynchites auratus 150–151
– *baccus* 151
– *betuleti*: see *Byctiscus betulae*
– *coeruleus*: see *Rhynchites caeruleus*
RHYNCHITIDAE 14, 146–151
Ribautiana debilis 54
– *tenerrima* 54
ribeana, Pandemis: see *Pandemis*
    cerasana
ribesiae, Pulvinaria 117
ribesii, Nematus 396–397
ribesinus, Rhopalosiphoninus 90
ribis, Anthocoptes 412
–, *Bryobia* 421
–, *Cecidophyopsis* 412–413
–, *Contarinia* 179
–, *Cryptomyzus* 78
–, *Dasineura* 182
–, *Eriophyes*: see *Cecidophyopsis ribis*
–, *Phytoptus*: see *Cecidophyopsis ribis*
–, *Resseliella* 186
ribisnigri, Nasonovia 89
rivillei, Antispila 203–204
–, *Holocacista*: see *Antispila rivillei*
roborana, Archips: see *Archips*
    crataegana
rogersii, Acyrthosiphon 66, 84
rosae Diaspis: see *Aulacaspis rosae*
–, *Aspidiotus*: see *Aulacaspis rosae*
–, *Aulacaspis* 100
–, *Edwardsiana* 51
–, *Typhlocyba*: see *Edwardsiana rosae*
rosana, Archips 263–264
rosanus, Archips: see *Archips rosana*
roscipennella, Caloptilia 212–213
–, *Gracillaria*: see *Caloptilia*

roscipennella
rose chafer 128
– leafhopper 51
– scale 100
– slug sawfly 387–388
– thrips 29
– tortrix moth 263–264
roseus, Anuraphis: see *Dysaphis*
    plantaginea
rostralis, Hypena 365–366
rostrolamellatus, Humerobates 428
rosy apple aphid 80–81
– leaf-curling aphid 79–80
– rustic moth 365
rubi, Amphorophora 67
–, *Anthonomus* 154–155
–, *Batophila* 142–143
–, *Diastrophus* 400–401
–, *Lasioptera* 184
rubicola, Aleurodes: see *Asterobemisia*
    carpini
–, *Contarinia* 179
rubiella, Incurvaria: see *Lampronia*
    rubiella
–, *Lampronia* 202–203
rubiellum, Macrosiphum: see *Sitobion*
    fragariae
rubifolium, Macrosiphum: see
    Macrosiphum funestrum
rubivora, Batophila: see *Batophila*
    aerata
–, *Pegomya* 194
ruborum, Aphis 72
rubrioculus redikorzevi, Bryobia 422
–, *Bryobia* 421–422
rubus aphid: see large blackberry
    aphid
– gall wasp 400–401
– thrips 30
– yellow net virus 67
rufellus, Euhyponomeutoides: see
    Euhyponomeutoides albithoracellus
ruficornis, Magdalis 158
rufifasciata, Gymnoscelis 328
rufipes, Harpalus 125–126
–, *Ophonus*: see *Harpalus rufipes*
–, *Pentatoma* 34
–, *Pristiphora* 399–400
–, *Pseudoophonus*: see *Harpalus*
    rufipes
–, *Pseudophonus*: see *Harpalus rufipes*
rugicollis, Lygocoris 40–41
–, *Plesiocoris*: see *Lygocoris rugicollis*
rugifrons, Otiorhynchus 160–161
rugosostriatus, Otiorhynchus 161
rugulipennis, Lygus 41
Ruguloscolytus: see *Scolytus*
rugulosus, Scolytus 173–174
rumicis, Acronicta 356
rupricapraria, Hybernia: see *Theria*
    primaria
–, *Theria*: see *Theria primaria*
rusci, Ceroplastes 109–110
rust mites 406–419
rusty dot moth 298–299

sagittariae, Galerucella 143
St. Mark's flies 177
Saissetia oleae 118–119
salicicola, Otiorhynchus 161
salicis, Chionaspis 101–102
salmachus, Synanthedon: see
    Synanthedon tipuliformis
SALTATORIA 12, 23 et seq.
saltuum, Neurotoma 379–380
sambucaria, Ourapteryx 332
San José scale 107–108
sand weevil 164
sanguinea, Cercopis: see *Cercopis*
    vulnerata
Sappaphis mali: see *Dysaphis*
    plantaginea
–: see *Dysaphis*
satellite moth 363–364
Saturnia pavonia 314
SATURNIIDAE 15, 314
Satyrium pruni 305
sawflies 11, 12, 15, 379 et seq.
saxeseni, Anisandrus: see *Xyleborinus*
    saxeseni
–, *Xyleborinus* 174
–, *Xyleborus*: see *Xyleborinus saxeseni*
scabrella, Ypsolopha 236
scalloped oak moth 324–325
scaly strawberry weevil 168–169
Scaphoideus littoralis: see
    Scaphoideus titanus
– *titanus* 55
SCARABAEIDAE 13, 127–130
SCARABAEOIDEA 13
scarabaeoides, Phloeotribus 172
scarce blackberry aphid 84–85
– dagger moth 355
– umber moth 316–317
Schizoneura: see *Eriosoma*
schlechtendali, Aculus 411–412
–, *Phyllocoptes*: see *Aculus*
    schlechtendali
schneideri, Aphis 73
–, *Quadraspidiotus*: see
    Quadraspidiotus marani
Schreckensteinia festaliella 238
SCHRECKENSTEINIIDAE 14, 238
schwartzi, Brachycaudus 76
Sciaphilus asperatus 168–169
– *muricatus*: see *Sciaphilus asperatus*
scitella, Cemiostoma: see *Leucoptera*
    malifoliella
–, *Leucoptera*: see *Leucoptera*
    malifoliella
Sclerotinia fructigena 151, 386
SCOLYTIDAE: see SCOLYTINAE
SCOLYTINAE 171–174
Scolytus amygdali 172
– *mali* 172–173
– *pruni*: see *Scolytus mali*
– *rugulosus* 173–174
scorpion flies 12
Scotia: see *Agrotis*
scurfy scale: see rose scale
Scythropia crataegella 232

seed wasps 15, 401, 402
segetum, Agrotis 359–360
selachodon, Cecidophyopsis 413
selenana, Ancylis 260
Selenia bilunaria: see *Selenia dentaria*
– *dentaria* 334
Semiothisa wauaria 334
septentrionalis, Croesus 386–387
Serica brunnea 130
sericeus, Polydrusus: see *Polydrusus formosus*
serratella, Coleophora 241–242
serratulae, Gueriniella 122
SESIIDAE 14, 222–224
setaceous hebrew character moth 378
sexdentatum, Sinoxylon 135
shallot aphid 85–86
Sharka 74, 75, 83, 86, 88, 90
sheldoni, Aceria 408
–, *Eriophyes*: see *Aceria sheldoni*
shield bugs 32–34
short-cloaked moth 354
shot-hole borers 171
shuttle-shaped dart moth 358–359
silverfish 11
silver-green leaf weevil 165
silver-leaf disease 52
silver-striped hawk moth 339
Simaethis: see *Choreutis*
similis, Eriophyes 417
–, *Euproctis* 347
–, *Phytoptus*: see *Eriophyes similis*
simulans, Psylla: see *Cacopsylla pyricola*
sinensis, Ceroplastes 110–111
singularis, Otiorhynchus 162
Sinoxylon sexdentatum 135
sinuatus, Agrilus 131–132
SIPHONAPTERA 12
siterata, Chloroclysta 321–322
Sitobion fragariae 91
Sitona lineatus 169
skipjacks: see click beetles
sloe bug 32
small almond-tree borer 172
– brindled beauty moth 319
– eggar moth 309
– elephant hawk moth 339
– European raspberry aphid 70
– fruit flies 190–191
– gooseberry sawfly 399–400
– olive leaf miner moth: see olive leaf miner moth
– quaker moth 372
– raspberry aphid: see small European raspberry aphid
– – sawfly 398
– walnut aphid 92
Smerinthus ocellata 341
smooth broad-nosed weevil 155
social peach sawfly 379
– pear sawfly 379–380
– wasps 15, 403–404
soft scales 109–119
solani, Aulacorthum 74

solstitialis, Amphimallon 127–128
–, *Amphimallus*: see *Amphimallon solstitialis*
sooty moulds 36, 46, 56, 60, 61, 63, 64, 65, 70, 71, 74, 78, 83, 86, 88, 90, 91, 93, 100, 102, 109, 110, 112, 113, 114, 116, 117, 118, 119, 120, 121, 123
sorbiana, Choristoneura: see *Choristoneura hebenstreilella*
sour-cherry aphid 88
southern strawberry tortrix moth 259
Sparganothis pilleriana 293
sparrows 202
speckled bush cricket 23
spectrana, Clepsis 269
Spectrobates ceratoniae: see *Ectomyelois ceratoniae*
sphaeroides, Peritelus 164
Sphaerolecanium prunastri 119
Sphaeropsis dalmatica 185
SPHINGIDAE 15, 335–342
SPHINGOIDEA 15
Sphinx ligustri 342
spider mites 421–426
Spilonota ocellana 293–294
– *uddmanniana*: see *Epiblema uddmanniana*
Spilosoma lutea 353
–, *luteum* see *Spilosoma lutea*
spinella, Coleophora 242–243
spinicolella, Phyllonorycter 218–219
spinolae, Apolygus 37
–, *Lygocoris*: see *Apolygus spinolae*
–, *Lygus*: see *Apolygus spinolae*
spinosella, Argyresthia 229
spiraeae, Agromyza: see *Agromyza potentillae*
spiraecola, Aphis 73–74
splendana, Cydia 277–278
–, *Olethreutes*: see *Cydia splendana*
Spodoptera littoralis 377–378
spotted apple budworm 283
– crane fly 175
spring usher moth 317
springtails 11
Spuleria atra 251
spumarius, Philaenus 44
sputator, Agriotes 134
stabilis, Orthosia 376
stellulata, Aguriahana 48
–, *Cicadella*: see *Aguriahana stellulata*
–, *Eupteroidea*: see *Aguriahana stellulata*
–, *Eupteryx*: see *Aguriahana stellulata*
stem sawflies 380–381
Steneotarsonemus fragariae: see *Phytonemus pallidus fragariae*
Stephanitis pyri 36
Stereonychus fraxini 170
STERNORRHYNCHA 12–13
stick-insects 12
Stictocephala bisonia 46
Stigmella aurella 197–198
– *fragariella* 198

– *incognitella*: see *Stigmella pomella*
– *malella* 198–199
– *plagicolella* 199
– *pomella* 199–200
– *pyri* 200
STIGMELLIDAE: see NEPTICULIDAE
STIGMELLOIDEA: see NEPTICULOIDEA
stink bugs 32
stolbur phytoplasma: see grapevine bois noir phytoplasma.
stoneflies 10, 11
stony-pit symptom 38, 41, 42
strataria, Biston 320–321
strawberry aphid 76
– blossom weevil 154–155
– casebearer moth 240–242
– crinkle virus 76, 86
– fruit weevil: see smooth broad-nosed weevil
– ground beetles 126, 126–127
– jewel beetle 132
– leaf beetle 144
– – miner 192
– – rollers 259
– – sawfly 386
– leafhopper 49
– mite 419–420
– mottle virus 66
– rhynchites 149
– root aphid 69
– – weevils: see lesser strawberry weevils, scaly strawberry weevil and strawberry weevil
– seed beetle 125–126
– spider mite 425
– tortrix moth 253–254
– weevil 160
– whitefly: see honeysuckle whitefly
– yellow-edge virus 76
straw-coloured apple moth 250
– tortrix moth 269
STREPSIPTERA 12
striped hawk moth 340
Strophosoma melanogrammum 170
Strophosomus coryli: see *Strophosoma melanogrammum*
– *melanogrammus*: see *Strophosoma melanogrammum*
Strymon pruni: see *Satyrium pruni*
stylopids 12
subterranea, Anuraphis 68
sucking lice 12
suffusa, Agrotis: see *Agrotis ipsilon*
sulcatus, Capsodes 38
–, *Otiorhynchus* 163–164
sulphurella, Esperia 246–247
–, *Oecophora*: see *Oecophora bractella*
summer apple sucker 61
– chafer 127–128
– fruit tortrix moth 258
suzukii, Drosophila 191
swallow-tailed moth 332

Swammerdamia cerasiella: see
  Swammerdamia pyrella
– *pyrella* 232–233
sweet-cherry aphid 87–88
swift moths 195–197
sycamore moth 354–355
SYMPHYTA 15
Synanthedon myopaeformis 222–223,
  279
– *myopiformis*: see *Synanthedon
  myopaeformis*
– *salmachus*: see *Synanthedon
  tipuliformis*
– *tipuliformis* 223–224
Syndemis musculana 294–295

tabaci, Thrips 30
Taeniocampa: see *Orthosia*
Taeniothrips inconsequens 28–29
tarnished plant bug: see European
  tarnished plant bug
tarsonemid mites 419–420
TARSONEMIDAE 16, 419–420
TARSONEMOIDEA 16
Tarsonemus fragariae: see
  Phytonemus pallidus fragariae
– *pallidus fragariae*: see *Phytonemus
  pallidus fragariae*
Taxonus glabratus: see *Ametastegia
  glabrata*
TENEBRIONIDAE 13, 137–138
tenebrionis, Capnodis 132
tenella, Galerucella 144
tener, Priophorus: see *Priophorus
  morio*
tenerana, Epinotia 281
tenerrima, Ribautiana 54
–, *Typhlocyba*: see *Ribautiana
  tenerrima*
TENTHREDINIDAE 15, 383–400
TENTHREDINOIDEA 15
TENUIPALPIDAE 16, 427
TEPHRITIDAE 14, 187–190
TEREBRANTIA 13
termites 12, 27
testudinea, Hoplocampa 390–391
tetensi, Dasineura 182–183
TETRANYCHIDAE 16, 421–426
TETRANYCHOIDEA 16
Tetranychus mcdanieli 425
– *turkestani* 425
– *urticae* 425–426
– *viennensis*: see *Amphitetranychus
  viennensis*
Tetrops praeusta 139
Tettigonia viridis: see *Cicadella viridis*
TETTIGONIIDAE 12, 23–24
Thamnonoma wauaria: see *Semiothisa
  wauaria*
Thamnotettix plebeja: see *Euscelis
  incisus*
Thecla betulae 306
– *pruni*: see *Satyrium pruni*
theobaldi, Resseliella 186–187
Theresia ampelophaga: see

Theresimima ampelophaga
Theresimima ampelophaga 207
Theria primaria 334–335
– *rupricapraria*: see *Theria primaria*
thistle aphid 74
Thomasiniana: see *Resseliella*
THRIPIDAE 13, 27–30
thrips 10, 12, 27 et seq.
Thrips adamsoni: see *Thrips
  minutissimus*
– *atratus* 29
– *flavus* 29
– *fuscipennis* 29
– *major* 30
– *meridionalis* 30
– *minutissimus* 30
– *tabaci* 30
thunderflies: see thrips
Thyatira batis 315
THYATIRIDAE 15, 315
THYSANOPTERA 12, 13, 27 et seq.
THYSANURA 11
ticks 16
tiger moths 351
tiliae, Eulecanium 113
–, *Mimas* 340
tineana, Ancylis 260
TINEOIDEA 14
TINGIDAE 12, 35–36
Tipula oleracea 175–176
– *paludosa* 176
TIPULIDAE 14, 175–176
tipuliformis, Synanthedon 223–224
Tischeria complanella: see *Tischeria
  ekebladella*
– *ekabladella* 200–201
– *marginea* 201
TISCHERIIDAE 14, 200–201
titanus, Scaphoideus 55
tomato moth 366–367
tomentosus, Byturus 136–137
torquillella, Parornix 214
TORTRICIDAE 15, 253–295
TORTRICINAE 253
TORTRICOIDEA 15
tortrix, Dasineura 183
tortrix moths 253–295
TORYMIDAE 15, 402
Torymus varians 402
Toxoptera aurantii 91
– *citricida* 91
transversa, Eupsilia 363–364
trapezina, Cosmia 362–363
treehoppers 12, 46
Trichiura crataegi 313
Trichoferus griseus 139
TRICHOPTERA 12
tridens, Acronicta 357
–, *Empria* 387
triglochinis, Aphis 74
Triphaena: see *Noctua*
tristriatus, Aceria 409
–, *Eriophyes*: see *Aceria tristriatus*
–, *Phytoptus*: see *Aceria tristriatus*
tritici, Euxoa 364

tropical citrus aphid: see brown citrus
  aphid
true bugs 12, 32 et seq.
– flies 12, 175 et seq.
– wasps 403–404
– weevils 152–170
truncata, Chloroclysta 322
TUBULIFERA 13
turkestani, Tetranychus 425
turnip moth 359–360
twin-spotted quaker moth 375–376
two-spotted spider mite 425–426
two-tailed pasha butterfly 307
Tygnis pyri: see *Stephanitis pyri*
Typhlocyba crataegi: see *Edwardsiana
  crataegi*
– *debilis*: see *Ribautiana debilis*
– *froggatti*: see *Edwardsiana crataegi*
– *prunicola*: see *Edwardsiana
  prunicola*
– *quercus* 56
– *rosae*: see *Edwardsiana rosae*
– *tenerrima*: see *Ribautiana tenerrima*
typica, Naenia 368–369

uddmanniana, Epiblema 280–281
–, *Notocelia*: see *Epiblema
  uddmanniana*
–, *Spilonota*: see *Epiblema
  uddmanniana*
Udea ferrugalis 298–299
– *olivalis* 299
– *prunalis* 300
ulmi, Eriosoma 95
–, *Lepidosaphes* 104–105
–, *Metatetranychus*: see *Panonychus
  ulmi*
–, *Oligonochus*: see *Panonychus ulmi*
–, *Panonychus* 424–425
Unaspis yanonensis 108
unicolor, Pachythelia 208
unicolorella, Ephestia parasitella
  296–297
unicostata, Monostira 35
unionalis, Glyphodes: see *Palpita
  unionalis*
–, *Margaronia*: see *Palpita unionalis*
–, *Palpita* 298
unipunctana, Rhopobota: see
  Rhopobota naevana
unitella, Batia 244
upland click beetle 134
– wireworm 134
urbanus, Byturus: see *Byturus
  tomentosus*
urticae, Phyllobius: see *Phyllobius
  pomaceus*
–, *Tetranychus* 425–426
urticaria 346

vaccinii, Conistra 361–362
vapourer moth 349–350
varians, Myzus 88
–, *Torymus* 402
variegana, Acleris 257

–, *Argyroploce*: see *Hedya dimidioalba*
–, *Hedya*: see *Hedya dimidioalba*
varipes, *Priophorus*: see *Priophorus pallipes*
varius, *Leperesinus* 171–172
Vasates fockeui: see *Aculus fockeui*
vastatrix, *Phylloxera*: see *Viteus vitifoliae*
v-ata, *Chloroclystis* 323
vectors of plant pathogens 46, 47, 48, 49, 50, 51, 52, 53, 54, 55, 59, 61, 66, 67, 70, 71, 74, 75, 76, 83, 86, 87, 88, 89, 90, 91, 100, 115, 151, 191, 408, 413
ventricosus, *Nematus*: see *Nematus ribesii*
Venturia pirina 415
verbasci, *Campylomma* 37
vermiformis, *Eriophyes*: see *Cecidophyopsis vermiformis*
–, *Cecidophyopsis* 413
Vespa crabro 403
– *vulgaris*: see *Vespula vulgaris*
Vesperus xatarti 139
VESPIDAE 15, 403–404
VESPINAE 403
VESPOIDEA 15
Vespula 403, 404
– *germanica* 403–404
– *vulgaris* 404
viburni, *Lichtensia* 114–115
–, *Pseudococcus* 121
viburnum cushion scale 114–115
viennensis, *Amphitetranychus* 421
–, *Tetranychus*: see *Amphitetranychus viennensis*
vine erineum mite: see vine leaf blister mite
– flower midge 179
– jewel beetle 131
– leaf beetle 143
– – blister mite 413–414
– – flea beetle 141
– – gall midge 183–184
– – skeletonizer moth 207
– leafhopper 55
– louse: see grape phylloxera
– moth (see also European vine moth) 252–253
– rust mite 412
– thrips 27
– tortrix moth 278–279
– weevil 163–164
vinegar flies 190
violacea, *Coleophora* 243
violet aphid 86
virgaureana, *Cnephasia*: see *Cnephasia asseclana*
viridis, *Agrilus* 132
–, *Cicadella* 49–50
–, *Tettigonia*: see *Cicadella viridis*
viridula, *Nezara* 33
virus diseases 10, 49, 52, 53, 66, 67, 70, 71, 74, 76, 84, 86, 89, 90, 91, 408, 413

– vectors 10, 49, 52, 66, 67, 70, 71, 74, 75, 76, 83, 84, 86, 87, 88, 89, 90, 91, 100, 115, 408, 413
vitegenella, *Phyllocnistis* 221
Viteus vitifoliae 97–98
– *vitifolii*: see *Viteus vitifoliae*
viticola, *Contarinia* 179
–, *Eotetranychus*: see *Eotetranychus pruni*
vitifoliae, *Daktulosphaira*: see *Viteus vitifoliae*
–, *Viteus* 97–98
vitifolii, *Phylloxera*: see *Viteus vitifoliae*
–, *Viteus*: see *Viteus vitifoliae*
vitis, *Calepitrimerus* 412
–, *Colomerus* 413–414
–, *Empoasca* 52
–, *Epitrimerus*: see *Calepitrimerus vitis*
–, *Eriophyes*: see *Colomerus vitis*
–, *Phyllocoptes* 412
–, *Phytoptus*: see *Colomerus vitis*
–, *Pulvinaria* 117–118
v-moth 334
v-pug moth 323
vulgaris, *Melolontha*: see *Melolontha melolontha*
–, *Pterostichus*: see *Pterostichus melanarius*
–, *Vespa*: see *Vespula vulgaris*
–, *Vespula* 404
vulnerata, *Cercopis* 43

walnut husk fly 190
– leaf erineum mite 407
– – gall mite 409
– – miner 212–213
wasps 9, 12, 60, 118
–, gall 15
–, seed 15, 401, 402
–, social 15, 403–404
–, true 403–404
water-lily aphid 90–91
wauaria, *Itame*: see *Semiothisa wauaria*
–, *Semiothisa* 334
–, *Thamnonoma*: see *Semiothisa wauaria*
wax 46, 57, 58, 59, 61, 63, 64, 65, 66, 68, 70, 71, 73, 79, 81, 82, 90, 94, 95, 99, 106, 107, 109, 111, 112, 114, 118, 120, 121, 122, 123
– glands 61
– plates 94, 95
– scales 109–119
web-spinners 12
weevils 9, 146–151, 151, 152–170
Welsh chafer 128
West Indian peach scale: see white peach scale
western flower thrips 28
white ants 12
– clover seed weevil 151
– peach scale 106–107

whiteflies 63–65
white-line dart moth 364
willow beauty moth 332–333
– scale 101–102
– scurfy scale: see willow scale
winter moth 319, 329–330, 363
wisteria scale 112–113
witches' broom 53, 412
woeberiana, *Enarmonia*: see *Enarmonia fiormosana*
–, *Laspeyresia*: see *Enarmonia formosana*
woodiella, *Ephestia* 296
woolly aphid 94–95
– apple aphid: see woolly aphid
– birch scale: see woolly vine scale
– currant scale: see woolly vine scale
– bears 351
– vine scale 117–118
– whitefly 63–64

xatarti, *Vesperus* 139
Xestia c-nigrum 378
Xyleborinus saxeseni 174
Xyleborus dispar 174
– *saxesenii*: see *Xyleborinus saxesenii*
Xylella fastidiosa 50
xylosteana, *Archips* 264–265
xylosteanus, *Archips*: see *Archips xylosteana*

yanonensis, *Unaspis* 108
yellow flower thrips 29
– pear scale 108
– plum scale 107
– scale 99–100
yellow-tail moth 347
Yponomeuta 227
– *evonymella* 233
– *malinellus* 234–235
– *padella* 235–236
YPONOMEUTIDAE 14, 227–237
YPONOMEUTINAE 227
YPONOMEUTOIDEA 14
Ypsolopha 227
– *persicella* 236
– *scabrella* 236

Zahradnik's pear scale 107
Zelleria oleastrella 236–237
Zephrus betulae: see *Thecla betulae*
– *aesculi*: see *Zeuzera pyrina*
– *pyrina* 205–206
ziziphi, *Parlatoria* 106
–, *Parlatoria*: see *Parlatoria ziziphi*
Zophodia convolutella 300–301
ZORAPTERA 12
zorapterans 12
Zygaena infausta: see *Aglaope infausta*
ZYGAENIDAE 14, 206–207
ZYGAENOIDEA 14
Zygina flammigera 56–57
– *rhamni* 57

T - #0439 - 101024 - C462 - 261/194/25 - PB - 9781138034228 - Gloss Lamination